MATH

BASIC ALGEBRA

SOLUTION KEY

Richard G. Brown
Geraldine D. Smith
Mary P. Dolciani

BEN FRANKLIN HIGH SCHOOL

Houghton Mifflin Company · Boston
Atlanta   Dallas   Geneva, Illinois
Palo Alto   Princeton   Toronto

ISBN: 0-395-63748-1

123456789—CS—96 95 94 93 92

CONTENTS

Pages 6–7 • WRITTEN EXERCISES

A

1. $5 - 4 = 1$
2. $7 - 4 = 3$
3. $14 - 4 = 10$
4. $27 - 4 = 23$
5. $6 \div 2 = 3$
6. $16 \div 2 = 8$
7. $20 \div 2 = 10$
8. $30 \div 2 = 15$
9. $4 \cdot 4 = 16$
10. $4 + 4 = 8$
11. $4 - 4 = 0$
12. $4 \div 4 = 1$
13. $12 - 3 = 9$
14. $12 + 7 = 19$
15. $2 \cdot 12 = 24$
16. $12 \div 4 = 3$
17. $3 \cdot 8 = 24$
18. $4 \cdot 8 = 32$
19. $8 \div 2 = 4$
20. $8 \div 4 = 2$
21. $15 + 2 = 17$
22. $15 \div 3 = 5$
23. $15 - 5 = 10$
24. $2 \cdot 15 = 30$
25. $20 - 10 = 10$
26. $5 \cdot 20 = 100$
27. $20 \div 2 = 10$
28. $30 - 20 = 10$
29. $12 + 8 + 6 = 26$
30. $12 - 8 + 6 = 10$
31. $12 \cdot 8 = 96$
32. $12 + 4 - 8 = 8$
33. $x + 7$
34. $x - 4$
35. $2n$
36. $n \div 2$
37. ba
38. $a - 7$
39. Subtract 4.
40. Multiply by 3.
41. Divide by 8.
42. x
43. t

Page 9 • WRITTEN EXERCISES

A

1. $(16 - 4) \cdot 2 = 12 \cdot 2 = 24$
2. $16 - (4 \cdot 2) = 16 - 8 = 8$
3. $3 \cdot (5 + 7) = 3 \cdot 12 = 36$
4. $(16 - 4) \div (3 + 1) = 12 \div 4 = 3$
5. $(8 + 4) \div (4 + 2) = 12 \div 6 = 2$
6. $9 \div 1 + 3 = 9 + 3 = 12$
7. $3 \cdot 4 + 5 = 12 + 5 = 17$
8. $6 + 8 \cdot 4 = 6 + 32 = 38$
9. $(8 + 6) \div (7 - 5) = 14 \div 2 = 7$
10. $2 \cdot 2 - 1 = 4 - 1 = 3$
11. $3 + 7 \cdot 4 = 3 + 28 = 31$
12. $8 \cdot 2 + 5 \cdot 4 = 16 + 20 = 36$
13. $8 + (2 \cdot 3) = 8 + 6 = 14$
14. $(2 + 3) \cdot (4 + 1) = 5 \cdot 5 = 25$
15. $4 \cdot 4 + 4 \div 2 = 16 + 2 = 18$
16. $4 + 8 \div 2 = 4 + 4 = 8$
17. $6 + 2 \div 2 + 2 = 6 + 1 + 2 = 9$
18. $(3 - 2) \cdot (9 - 3) = 1 \cdot 6 = 6$
19. $2(4 + 6) = 2 \cdot 10 = 20; \quad 2 \cdot 4 + 6 = 8 + 6 = 14$
20. $5(7 + 1) = 5 \cdot 8 = 40; \quad 5 \cdot 7 + 1 = 35 + 1 = 36$
21. $8(9 - 4) = 8 \cdot 5 = 40; \quad 8 \cdot 9 - 4 = 72 - 4 = 68$
22. $3(8 + 2) = 3 \cdot 10 = 30; \quad 3 \cdot 8 + 2 = 24 + 2 = 26$
23. $4(5 - 2) = 4 \cdot 3 = 12; \quad 4 \cdot 5 - 2 = 20 - 2 = 18$
24. $6(3 + 4) = 6 \cdot 7 = 42; \quad 6 \cdot 3 + 4 = 18 + 4 = 22$
25. $2 \cdot 3 = 6; \quad 6 + 7 = 13$
26. $2 \cdot 4 = 8; \quad 8 + 7 = 15$
27. $2x; \quad 2x + 7$
28. $3 + 7 = 10; \quad 2 \cdot 10 = 20$
29. $4 + 7 = 11; \quad 2 \cdot 11 = 22$
30. $x + 7; \quad 2(x + 7)$

Page 11 · WRITTEN EXERCISES

A
1. $2x + 4x = 6x$ 2. $3x + 5x = 8x$ 3. $3a + 7a = 10a$
4. $5b - 2b = 3b$ 5. $9y - 6y = 3y$ 6. $6xy - 2xy = 4xy$
7. $2a + 5a + 6 = 7a + 6$ 8. $5y + 3y + 4y = 12y$
9. $3a + 5a + 4a = 12a$ 10. $6b + 3b - b = 8b$
11. $4 + 5y + 6y = 4 + 11y$ 12. $2x + 4x + 7 = 6x + 7$
13. $8s - 2s - 2 = 6s - 2$ 14. $3a + a - 2 = 4a - 2$
15. $3 + 4y + 5y = 3 + 9y$ 16. $3a + 3a + 3 = 6a + 3$
17. $3a + a - 4 = 4a - 4$ 18. $2a + 5a + 3a = 10a$
19. $6 - 2 + 4x = 4 + 4x$ 20. $y + 2y - 7 = 3y - 7$
21. $6xy - xy + 7 = 5xy + 7$ 22. $9c - 2c + 4c = 11c$
23. $5d + 6d - 3d = 8d$ 24. $8y + 3y - 4y = 7y$
25. $7y - 2y + 9x = 5y + 9x$ 26. $2x + x + 4y = 3x + 4y$
27. $7ab - 2ab + 5b = 5ab + 5b$ 28. $cd + 4cd - 2a = 5cd - 2a$
29. $3xy - xy + 2x = 2xy + 2x$ 30. $2m + 5n + 6 - 4 = 2m + 5n + 2$
31. not possible   32. $4s - 4s = 0$ 33. not possible
34. $2a + 5a = 7a$ 35. $ar + 4ar = 5ar$ 36. not possible
37. not possible   38. not possible

Page 13 · WRITTEN EXERCISES

A
1. $26 + 5 + 15 = 26 + (5 + 15) = 26 + 20 = 46$
2. $7 + 63 + 3 = (7 + 3) + 63 = 10 + 63 = 73$
3. $49 + 93 + 1 = (49 + 1) + 93 = 50 + 93 = 143$
4. $7 + 25 + 5 + 3 = (7 + 3) + (25 + 5) = 10 + 30 = 40$
5. $998 + 571 + 2 = (998 + 2) + 571 = 1000 + 571 = 1571$
6. $187 + 38 - 87 = (187 - 87) + 38 = 100 + 38 = 138$
7. $3n + 6 + 5n = (3n + 5n) + 6 = 8n + 6$
8. $2n + 3 + 4n = (2n + 4n) + 3 = 6n + 3$
9. $5c + 7 - 3 + 2c = (5c + 2c) + (7 - 3) = 7c + 4$
10. $4x + 3 - 2 + 4x = (4x + 4x) + (3 - 2) = 8x + 1$
11. $30 + 5a + a + 1 = (30 + 1) + (5a + a) = 31 + 6a$
12. $9 + 2ac + 3ac + 7 = (9 + 7) + (2ac + 3ac) = 16 + 5ac$
13. $8m + 5 + 6 + 2m = (8m + 2m) + (5 + 6) = 10m + 11$
14. $4xy + 5 - xy + 2 = (4xy - xy) + (5 + 2) = 3xy + 7$
15. $7ab + 7 - 6ab + 3 = (7ab - 6ab) + (7 + 3) = ab + 10$
16. $12m + 3 - 5m + 8m = (12m - 5m + 8m) + 3 = 15m + 3$
17. $6s + 5 + 2s - 4s = (6s + 2s - 4s) + 5 = 4s + 5$
18. $12a - 3a + 5 - 2a = (12a - 3a - 2a) + 5 = 7a + 5$

19. $6k + 9n - 5n - 2k = (6k - 2k) + (9n - 5n) = 4k + 4n$

20. $4p - 2p + 9a + 3p = (4p - 2p + 3p) + 9a = 5p + 9a$

21. $4n - 9b + 3n - 2n = (4n + 3n - 2n) - 9b = 5n - 9b$

22. $3c + 7d + 5c + 5d = (3c + 5c) + (7d + 5d) = 8c + 12d$

23. $8a + 9b - 3b - a = (8a - a) + (9b - 3b) = 7a + 6b$

24. $3p + 4q + 7p - 2q = (3p + 7p) + (4q - 2q) = 10p + 2q$

25. $8a + 2b + 4 + 3b = 8a + (2b + 3b) + 4 = 8a + 5b + 4$

26. $7x + 3y - 2y + 5 = 7x + (3y - 2y) + 5 = 7x + y + 5$

27. $4d - 7a - 2d + 5d = (4d - 2d + 5d) - 7a = 7d - 7a$

28. $5m + 3m - 3 + 4n = (5m + 3m) - 3 + 4n = 8m - 3 + 4n$

29. $6a - 2a + 4b - a = (6a - 2a - a) + 4b = 3a + 4b$

30. $cd + 4a + 5cd - 2 = (cd + 5cd) + 4a - 2 = 6cd + 4a - 2$

Page 15 · WRITTEN EXERCISES

A

1. $3^2 = 3 \cdot 3 = 9$ **2.** $3^3 = 3 \cdot 3 \cdot 3 = 27$ **3.** $4^1 = 4$

4. $5^2 = 5 \cdot 5 = 25$ **5.** $10^2 = 10 \cdot 10 = 100$ **6.** $x \cdot x \cdot x = x^3$

7. $y \cdot y \cdot y \cdot y = y^4$ **8.** $a \cdot a \cdot a \cdot a = a^4$ **9.** $x \cdot x \cdot y \cdot y \cdot y = x^2 y^3$

10. $p \cdot q \cdot q = pq^2$ **11.** $r \cdot s \cdot s \cdot s = rs^3$ **12.** $c \cdot c \cdot c \cdot c \cdot c = c^5$

13. $t \cdot t \cdot m \cdot m = t^2 m^2$ **14.** $n \cdot r \cdot r \cdot r = nr^3$

15. $f \cdot f \cdot f \cdot g \cdot g \cdot h = f^3 g^2 h$ **16.** $a \cdot a \cdot b \cdot b \cdot b \cdot c = a^2 b^3 c$

17. $d \cdot e \cdot e \cdot f \cdot f = de^2 f^2$ **18.** $5 \cdot n \cdot n = 5n^2$

19. $7 \cdot m \cdot m \cdot m = 7m^3$ **20.** $6 \cdot a \cdot b \cdot b = 6ab^2$

21. $2 \cdot a \cdot a \cdot m \cdot m = 2a^2 m^2$ **22.** $c \cdot c \cdot c \cdot m \cdot m = c^3 m^2$

23. $5 \cdot n \cdot n \cdot n \cdot p = 5n^3 p$ **24.** $7 \cdot s \cdot s \cdot t \cdot t \cdot t = 7s^2 t^3$

25. $3 \cdot 2 \cdot a \cdot a \cdot b = 6a^2 b$ **26.** $9 \cdot r \cdot r \cdot s \cdot s = 9r^2 s^2$

Page 15 · PUZZLE PROBLEMS

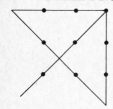

Page 17 · WRITTEN EXERCISES

A

1. $4 \cdot 29 \cdot 25 = (4 \cdot 25) \cdot 29 = 100 \cdot 29 = 2900$

2. $92 \cdot 5 \cdot 2 = 92 \cdot (5 \cdot 2) = 92 \cdot 10 = 920$

3. $(33 \cdot 25) \cdot 4 = 33 \cdot (25 \cdot 4) = 33 \cdot 100 = 3300$

4. $(50 \cdot 83) \cdot 2 = (50 \cdot 2) \cdot 83 = 100 \cdot 83 = 8300$

5. $2 \cdot (73 \cdot 5) = (2 \cdot 5) \cdot 73 = 10 \cdot 73 = 730$

6. $10 \cdot 3 \cdot 10 \cdot 2 \cdot 10 = (10 \cdot 10 \cdot 10) \cdot (3 \cdot 2) = 1000 \cdot 6 = 6000$

7. $2 \cdot (3x) = (2 \cdot 3) \cdot x = 6x$ $\qquad$ 8. $3 \cdot (4x) = (3 \cdot 4) \cdot x = 12x$

9. $5 \cdot (3a) = (5 \cdot 3) \cdot a = 15a$ $\qquad$ 10. $4 \cdot (4a) = (4 \cdot 4) \cdot a = 16a$

11. $b \cdot 2b = 2 \cdot (b \cdot b) = 2b^2$ $\qquad$ 12. $c \cdot 3c = 3 \cdot (c \cdot c) = 3c^2$

13. $(3y) \cdot 6 = (3 \cdot 6) \cdot y = 18y$ $\qquad$ 14. $(5k) \cdot 2 = (5 \cdot 2) \cdot k = 10k$

15. $(4n) \cdot 5 = (4 \cdot 5) \cdot n = 20n$ $\qquad$ 16. $2s \cdot s = 2 \cdot (s \cdot s) = 2s^2$

17. $8 \cdot 9f = (8 \cdot 9) \cdot f = 72f$ $\qquad$ 18. $m \cdot 6n = 6mn$

19. $(2a)(2a) = (2 \cdot 2)(a \cdot a) = 4a^2$ $\qquad$ 20. $(3x)(5y) = (3 \cdot 5) \cdot x \cdot y = 15xy$

21. $(2a)(4b) = (2 \cdot 4) \cdot a \cdot b = 8ab$ $\qquad$ 22. $(6s)(3s) = (6 \cdot 3)(s \cdot s) = 18s^2$

23. $8 \cdot x \cdot x \cdot 2 = (8 \cdot 2)(x \cdot x) = 16x^2$ $\qquad$ 24. $3 \cdot b \cdot b \cdot 2 = (3 \cdot 2)(b \cdot b) = 6b^2$

25. $5 \cdot a \cdot 7 \cdot a \cdot a = (5 \cdot 7)(a \cdot a \cdot a) = 35a^3$

26. $4 \cdot y \cdot y \cdot y \cdot 2 = (4 \cdot 2)(y \cdot y \cdot y) = 8y^3$

27. $2 \cdot b \cdot 5 \cdot b \cdot b = (2 \cdot 5)(b \cdot b \cdot b) = 10b^3$

28. $3 \cdot a \cdot a \cdot 4 \cdot a = (3 \cdot 4)(a \cdot a \cdot a) = 12a^3$

29. $2 \cdot 3x = 6x$ $\qquad$ 30. $3 \cdot 5a = 15a$ $\qquad$ 31. $x \cdot 2x = 2x^2$

32. $3c$ $\qquad$ 33. Multiply by 4. $\qquad$ 34. Multiply by 6.

Page 19 · WRITTEN EXERCISES

A

1. $3(4 + 5) = 3 \cdot 9 = 27$; $(3 \cdot 4) + (3 \cdot 5) = 12 + 15 = 27$; yes

2. $4(8 + 3) = 4 \cdot 11 = 44$; $(4 \cdot 8) + (4 \cdot 3) = 32 + 12 = 44$; yes

3. $5(7 - 3) = 5 \cdot 4 = 20$; $(5 \cdot 7) - (5 \cdot 3) = 35 - 15 = 20$; yes

4. $6(21 - 11) = 6 \cdot 10 = 60$; $(6 \cdot 21) - (6 \cdot 11) = 126 - 66 = 60$; yes

5. $2(x + 3) = 2x + 2 \cdot 3 = 2x + 6$ $\qquad$ 6. $3(x + 5) = 3x + 15$

7. $4(x + 2) = 4x + 8$ $\qquad$ 8. $5(a + 4) = 5a + 20$

9. $7(a + 3) = 7a + 21$ $\qquad$ 10. $8(b + 1) = 8b + 8$

11. $9(x + 4) = 9x + 36$ $\qquad$ 12. $6(y + 2) = 6y + 12$

13. $3(x - 2) = 3x - 6$ $\qquad$ 14. $4(y - 7) = 4y - 28$

15. $c(c - 3) = c^2 - 3c$ $\qquad$ 16. $d(d - 4) = d^2 - 4d$

17. $7(2x + 1) = 14x + 7$ $\qquad$ 18. $4(3x + 2) = 12x + 8$

19. $5(3x - 2) = 15x - 10$ $\qquad$ 20. $3(7x - 3) = 21x - 9$

21. $2(3x + 4y) = 6x + 8y$ $\qquad$ 22. $b(4a + 2b) = 4ab + 2b^2$

23. $x(3x - y) = 3x^2 - xy$ $\qquad$ 24. $a(2a - 3b) = 2a^2 - 3ab$

25. $2(x + 5) + 3 = 2x + 10 + 3 = 2x + 13$

26. $3(x - 1) + 5x = 3x - 3 + 5x = (3x + 5x) - 3 = 8x - 3$

27. $5(x + 3) + 2x = 5x + 15 + 2x = 7x + 15$

28. $3(2x - 1) + 4x = 6x - 3 + 4x = 10x - 3$

29. $4(9 + 2y) - 3y = 36 + 8y - 3y = 36 + 5y$

30. $7(1 + 2c) - 8c = 7 + 14c - 8c = 7 + 6c$

B   31. $5(3x + 5) - 7x - 2 = 15x + 25 - 7x - 2 = (15x - 7x) + (25 - 2) = 8x + 23$

32. $2(1 + 3x) + 7x - 1 = 2 + 6x + 7x - 1 = 1 + 13x$

33. $5(2 + 7x) - 3x - 2 = 10 + 35x - 3x - 2 = 8 + 32x$

34. $4(a + 3) + 3(a - 2) = 4a + 12 + 3a - 6 = 7a + 6$

35. $8(2b + 3) + 4(b - 5) = 16b + 24 + 4b - 20 = 20b + 4$

36. $6(c + 3) + 4(1 - c) = 6c + 18 + 4 - 4c = 2c + 22$

37. $5(m + 5) + 2(m - 8) = 5m + 25 + 2m - 16 = 7m + 9$

38. $3(4p + 3) + 4(p - 1) = 12p + 9 + 4p - 4 = 16p + 5$

39. $x(x + 2y) - xy + 2 = x^2 + 2xy - xy + 2 = x^2 + xy + 2$

Page 21 · WRITTEN EXERCISES

A   1. 0   2. 0   3. $18 - 18 + 35 = 0 + 35 = 35$ 4. 16   5. 0

6. impossible   7. $\frac{5}{5 \cdot 0} = \frac{5}{0}$; impossible   8. 0

9. $5(1 - 1) = 5 \cdot 0 = 0$ 10. $(2 - 2) \div 4 = 0 \div 4 = 0$

11. $(2 \cdot 3 - 6) \div 4 = (6 - 6) \div 4 = 0 \div 4 = 0$

12. $0 \div 3 = 0$

13. $3x + 6 + 5x - 6 = 3x + 5x + 6 - 6 = 8x + 0 = 8x$

14. $5a + 3 + 8a - 3 = 5a + 8a + 3 - 3 = 13a + 0 = 13a$

15. $6 + 4c + 2c - 6 = 6 - 6 + 4c + 2c = 0 + 6c = 6c$

16. $4(x + 2) - 8 = 4x + 8 - 8 = 4x + 0 = 4x$

17. $3(2 + 5a) - 6 = 6 + 15a - 6 = 6 - 6 + 15a = 0 + 15a = 15a$

18. $2(c + 5) - 10 = 2c + 10 - 10 = 2c + 0 = 2c$

19. x 20. x 21. x 22. x 23. x 24. x 25. Add 7.

26. Add 13.   27. Subtract 9.   28. Subtract 4.   29. Add 16.

30. Subtract 14.   31. Add 11.

Page 23 · WRITTEN EXERCISES

A   1. $\frac{4}{4} = 1$ 2. $4 \cdot 4 = 16$ 3. $(4 \div 4) \cdot 1 = 1 \cdot 1 = 1$

4. $\frac{4 + 2}{6} = \frac{6}{6} = 1$ 5. $b \div b = 1$ 6. $5x \div 5 = \frac{5x}{5} = \frac{5}{5}x = x$

7. $3x \div x = \frac{3x}{x} = 3 \cdot \frac{x}{x} = 3$ 8. $10x \div 10 = \frac{10x}{10} = \frac{10}{10}x = x$

9. $\frac{5c}{5} = \frac{5}{5}c = c$ 10. $\frac{3}{3}x = x$ 11. $\frac{4n}{n} = 4 \cdot \frac{n}{n} = 4$

12. $\frac{6}{6}y = y$ 13. $5 \cdot \frac{x}{5} = \frac{5}{5}x = x$ 14. $\frac{x}{7} \cdot 7 = x \cdot \frac{7}{7} = x$

15. $\frac{2a}{2} = \frac{2}{2}a = a$ 16. $\frac{c}{9} \cdot 9 = c \cdot \frac{9}{9} = c$ 17. $\frac{4x}{4} = \frac{4}{4}x = x$

18. $\frac{5x}{5} = \frac{5}{5}x = x$ 19. $\frac{3x}{3} = \frac{3}{3}x = x$ 20. Divide by 7.

21. Divide by 6.   **22.** $8 \cdot \dfrac{x}{8} = \dfrac{8}{8}x = x$ **23.** $4 \cdot \dfrac{x}{4} = \dfrac{4}{4}x = x$

24. $9 \cdot \dfrac{x}{9} = \dfrac{9}{9}x = x$ **25.** Multiply by 7.   **26.** Multiply by 10.

27. $4(3 - 2) = 4 \cdot 1 = 4$

28. $(2 \cdot 5 - 10) \div 1 = (10 - 10) \div 1 = 0 \div 1 = 0$

Page 24 · CALCULATOR ACTIVITIES

1. 2   **2.** 5   **3.** 32   **4.** 3   **5.** 21   **6.** 56

Page 26 · WRITTEN EXERCISES

A   **1.** If $x = 3$: $3 + 9 = 12$; if $x = 21$: $21 + 9 = 30$; 3

2. If $y = 8$: $8 - 7 = 1$; if $y = 22$: $22 - 7 = 15$; 22

3. If $n = 12$: $3 \cdot 12 = 36$; if $n = 33$: $3 \cdot 33 = 99$; 12

4. If $x = 6$: $4 \cdot 6 = 24$; if $x = 20$: $4 \cdot 20 = 80$; 6

5. If $a = 2$: $3 \cdot 2 + 7 = 6 + 7 = 13$; if $a = 4$: $3 \cdot 4 + 7 = 12 + 7 = 19$; 2

6. If $a = 4$: $2 \cdot 4 - 6 = 8 - 6 = 2$; if $a = 5$: $2 \cdot 5 - 6 = 10 - 6 = 4$; 5

7. If $r = 3$: $5(3 + 2) = 5 \cdot 5 = 25$; if $r = 4$: $5(4 + 2) = 5 \cdot 6 = 30$; 4

8. If $s = 5$: $7(5 - 3) = 7 \cdot 2 = 14$; if $s = 6$: $7(6 - 3) = 7 \cdot 3 = 21$; 6

9. If $x = 2$: $2 + 9 = 4 \cdot 2$? $11 = 8$? No.

　　If $x = 3$: $3 + 9 = 4 \cdot 3$? $12 = 12$? Yes.

　　If $x = 4$: $4 + 9 = 4 \cdot 4$? $13 = 16$? No.   3

10. If $b = 1$: $2 \cdot 1 + 5 = 1 + 7$? $2 + 5 = 8$? $7 = 8$? No.

　　If $b = 2$: $2 \cdot 2 + 5 = 2 + 7$? $4 + 5 = 9$? $9 = 9$? Yes.

　　If $b = 3$: $2 \cdot 3 + 5 = 3 + 7$? $6 + 5 = 10$? $11 = 10$? No.   2

11. If $a = 5$: $5 \cdot 5 - 1 = 3 \cdot 5 + 9$? $25 - 1 = 15 + 9$? $24 = 24$? Yes.

　　If $a = 6$: $5 \cdot 6 - 1 = 3 \cdot 6 + 9$? $30 - 1 = 18 + 9$? $29 = 27$? No.

　　If $a = 7$: $5 \cdot 7 - 1 = 3 \cdot 7 + 9$? $35 - 1 = 21 + 9$? $34 = 30$? No.   5

12. If $a = 6$: $4(6 - 5) = 6 + 1$? $4 \cdot 1 = 7$? $4 = 7$? No.

　　If $a = 7$: $4(7 - 5) = 7 + 1$? $4 \cdot 2 = 8$? $8 = 8$? Yes.

　　If $a = 8$: $4(8 - 5) = 8 + 1$? $4 \cdot 3 = 9$? $12 = 9$? No.   7

13. $0^2 = 0$, $5^2 = 25$, $10^2 = 100$; 5   **14.** $0^3 = 0$, $1^3 = 1$, $2^3 = 8$; 1

B   Answers to Exs. 15–18 may vary.   **15.** $x + 2 = 11$ **16.** $x - 5 = 6$

17. $2x + 3 = 3$ **18.** $x + 1 = x$ **19.** 8   **20.** 37   **21.** 6   **22.** 6

23. 8   **24.** 25   **25.** any number   **26.** 6   **27.** 40   **28.** 20

29. any number   **30.** 36

C   **31.** 2   **32.** 6   **33.** 0   **34.** 0 or 1

Page 26 · PUZZLE PROBLEMS

The lion tamer must be bald; since Geraldine and Amos have hair, Christopher is the lion tamer. Then, by clue 3, the ringmaster is Geraldine. Thus, Amos is the elephant trainer.

Page 28 · WRITTEN EXERCISES

A   1. $6 < 7$ 2. $0 < 4$ 3. $3 > 1$

     4. $4 < 9$ 5. $7 > 4$ 6. $5 < 11$

     7. $6 > 3$ 8. $5 > 0$ 9. 4

   10. $3, 5, 7$ 11. $1, 3, 7$ 12. $10, 14$

   13. 20 14. $6, 2, 4$ 15. $6, 2$

   16. $12, 35$ 17. $4, 6$ 18. $6, 7, 11$

   19. $8 + 7 = 15 < 16;$ $9 + 7 = 16;$ $10 + 7 = 17 > 16;$ $11 + 7 = 18 > 16;$ $10, 11$

   20. $3 \cdot 4 - 1 = 12 - 1 = 11 > 9;$ $3 \cdot 3 - 1 = 9 - 1 = 8 < 9;$

      $3 \cdot 6 - 1 = 18 - 1 = 17 > 9;$ $3 \cdot 2 - 1 = 6 - 1 = 5 < 9;$ $3, 2$

   21. $3 \cdot 2 + 3 = 6 + 3 = 9 < 24;$ $3 \cdot 5 + 3 = 15 + 3 = 18 < 24;$

      $3 \cdot 7 + 3 = 21 + 3 = 24;$ $3 \cdot 9 + 3 = 27 + 3 = 30 > 24;$ $2, 5$

   22. $4 \cdot 2 - 7 = 8 - 7 = 1 < 13;$ $4 \cdot 4 - 7 = 16 - 7 = 9 < 13;$

      $4 \cdot 5 - 7 = 20 - 7 = 13;$ $4 \cdot 7 - 7 = 28 - 7 = 21 > 13;$ 7

   23. $2 \cdot 4 - 8 = 8 - 8 = 0 < 6;$ $2 \cdot 5 - 8 = 10 - 8 = 2 < 6;$

      $2 \cdot 6 - 8 = 12 - 8 = 4 < 6;$ $2 \cdot 7 - 8 = 14 - 8 = 6;$ $4, 5, 6$

   24. $4 \cdot 1 > 1 + 6?$ $4 > 7?$ No. $4 \cdot 0 > 0 + 6?$ $0 > 6?$ No.

      $4 \cdot 2 > 2 + 6?$ $8 > 8?$ No. $4 \cdot 5 > 5 + 6?$ $20 > 11?$ Yes. 5

   25. $3(8 + 4) = 3 \cdot 12 = 36;$ $7 \cdot 6 - 5 = 42 - 5 = 37;$ $36 < 37;$ $<$

   26. $2 \cdot 3 + 7 = 6 + 7 = 13;$ $10 + 6 \div 2 = 10 + 3 = 13;$ $13 = 13;$ $=$

   27. $83 \cdot 5 \cdot 0 \div 9 = 0 \div 9 = 0;$ $0 = 0;$ $=$

   28. $791(17 - 17) = 791 \cdot 0 = 0;$ $0 < 2;$ $<$

   29. $(24 \div 8) \cdot 8 = 3 \cdot 8 = 24;$ $3^3 = 3 \cdot 3 \cdot 3 = 27;$ $24 < 27;$ $<$

   30. $(5 \cdot 4 - 2) \div 6 = (20 - 2) \div 6 = 18 \div 6 = 3;$ $4^2 = 16;$ $3 < 16;$ $<$

Page 29 · READING ALGEBRA

Answers to Exs. 1–4 may vary.

1. fifteen minus eight;   subtract eight from fifteen;   the difference of fifteen and eight;   eight less than fifteen

2. five times twelve;   multiply twelve by five;   the product of five and twelve

3. seventeen plus four;   add four to seventeen;   the sum of seventeen and four;   four more than seventeen

4. eighteen divided by nine;   divide eighteen by nine;   the quotient of eighteen and nine

5. $30 - 5;$ 25 6. $9 \times 0;$ 0 7. $3^2;$ 9

8. $\dfrac{32}{4};$ 8 9. $80 + 8;$ 88 10. $100 \div 10;$ 10

Page 30 · SKILLS REVIEW

   1. 56 2. 399 3. 993 4. 880 5. 229 6. 395 7. 712

   8. 980 9. 822 10. 1556 11. 34 12. 32 13. 77 14. 78

15. 535   **16.** 337   **17.** 354   **18.** 784   **19.** 217   **20.** 219

21. 288   **22.** 236

23.
```
   73
 ×25
  365
 146
 1825
```

24.
```
   87
 ×42
  174
 348
 3654
```

25.
```
  120
 × 17
  840
  120
 2040
```

26.
```
   311
 × 54
  1244
 1555
 16794
```

27.
```
   672
 × 65
  3360
 4032
 43680
```

28.
```
   247
 ×132
   494
  741
  247
 32604
```

29.
```
   721
 ×105
  3605
 7210
 75705
```

30.
```
   630
 ×116
  3780
  630
  630
 73080
```

31.
```
      17
 6) 102
      6
     42
     42
      0
```

32.
```
     158
 4) 632
     4
     23
     20
      32
      32
       0
```

33. 68; 68 R 2
```
 7) 478
    42
    58
    56
     2
```

34. 40; 40 R 7
```
 9) 367
    360
      7
```

35. 74; 74 R 1
```
 2) 149
    140
      9
      8
      1
```

36. 70; 70 R 2
```
 7) 492
    490
      2
```

37. 237; 237 R 5
```
 6) 1427
    12
    22
    18
     47
     42
      5
```

38. 2140; 2140 R 2
```
 4) 8562
    8
    5
    4
    16
    16
     2
```

39. 104; 104 R 13
```
 57) 5941
     57
     241
     228
      13
```

40. 28; 28 R 3
```
 24) 675
     48
     195
     192
       3
```

41. $\overline{221}$; 221 R 14   **42.** $\overline{78}$; 78 R 13
38$\overline{)\,8412}$ 72$\overline{)\,5629}$
   76   504
    81   589
    76   576
    52   13
    38
    14

Pages 31–32 · CHAPTER REVIEW EXERCISES

1. $8 + 3 = 11$　　　　　2. $2 \cdot 8 = 16$　　　　　3. $8 \div 4 = 2$

4. $2 \cdot 8 + 5 = 16 + 5 = 21$　　　　5. $2(8 + 5) = 2 \cdot 13 = 26$

6. $3 \cdot 8 - 3 = 24 - 3 = 21$　　　　7. $3(8 - 3) = 3 \cdot 5 = 15$

8. $5 \cdot 8 - 2 = 40 - 2 = 38$　　　　9. $3 \cdot 4 + 2 = 12 + 2 = 14$

10. $5 + 8 \cdot 2 = 5 + 16 = 21$　　　　11. $(3 + 4)(5 - 2) = 7 \cdot 3 = 21$

12. $(8 + 4) \div 4 - 2 = 12 \div 4 - 2 = 3 - 2 = 1$

13. $8 + 4 \div 4 - 2 = 8 + 1 - 2 = 9 - 2 = 7$

14. $8 + 4 \div (4 - 2) = 8 + 4 \div 2 = 8 + 2 = 10$

15. $2x + 5x = 7x$　　　　　16. $8a - 6a = 2a$

17. $3b + 4b + 2 = 7b + 2$　　　　18. $7c - c + 3c = 9c$

19. $7c - c + 3 = 6c + 3$　　　　20. $8y - 2y + 7 + 5 = 6y + 12$

21. $5x + 7 + 3x - 4 = (5x + 3x) + (7 - 4) = 8x + 3$

22. $4a - 4 + 6a - a = (4a + 6a - a) - 4 = 9a - 4$

23. $3a + 2b + 7a - b = (3a + 7a) + (2b - b) = 10a + b$

24. $2^4 = 2 \cdot 2 \cdot 2 \cdot 2 = 16$　　　　25. $3^5 = 3 \cdot 3 \cdot 3 \cdot 3 \cdot 3 = 9 \cdot 27 = 243$

26. $5^3 = 5 \cdot 5 \cdot 5 = 125$　　　　27. $3^3 = 3 \cdot 3 \cdot 3 = 27$

28. $5^2 = 5 \cdot 5 = 25$　　　　29. $4^3 = 4 \cdot 4 \cdot 4 = 64$

30. $7^2 = 7 \cdot 7 = 49$　　　　31. $1^1 = 1$

32. $2 \cdot (5x) = (2 \cdot 5) \cdot x = 10x$　　　　33. $3 \cdot 4a = (3 \cdot 4) \cdot a = 12a$

34. $8 \cdot 2b = (8 \cdot 2) \cdot b = 16b$　　　　35. $4 \cdot x \cdot x \cdot x = 4x^3$

36. $3 \cdot b \cdot b \cdot b \cdot 2 = (3 \cdot 2)(b \cdot b \cdot b) = 6b^3$

37. $a \cdot b \cdot b \cdot a \cdot a = (a \cdot a \cdot a)(b \cdot b) = a^3 b^2$

38. $(2x)(3x) = (2 \cdot 3)(x \cdot x) = 6x^2$

39. $(4y)(3y) = (4 \cdot 3)(y \cdot y) = 12y^2$

40. $n(2n)(3n) = (2 \cdot 3)(n \cdot n \cdot n) = 6n^3$

41. $2(x + 5) = 2x + 2 \cdot 5 = 2x + 10$　　　　42. $3(2a - 7) = 6a - 21$

43. $5(4x - 2y) = 20x - 10y$　　　　44. $5(3y - 4) = 15y - 20$

45. $4(2c - 3d) = 8c - 12d$　　　　46. $2(9x + 7y) = 18x + 14y$

47. $3(a - 6b) = 3a - 18b$　　　　48. $6(4x + 3) = 24x + 18$

49. $6(4 - 4) = 6 \cdot 0 = 0$　　　　50. $\dfrac{4 - 4}{4 + 4} = \dfrac{0}{8} = 0$

51. $\dfrac{4-4}{4-4} = \dfrac{0}{0}$; impossible

52. $\dfrac{4-4}{4} = \dfrac{0}{4} = 0$

53. $\dfrac{4}{4-4} = \dfrac{4}{0}$; impossible

54. $\dfrac{4}{4} = 1$

55. $0 \div 4 = 0$

56. $4 \div 0$; impossible

57. $3x + 7 + 5x - 7 = 3x + 5x + 7 - 7 = 8x + 0 = 8x$

58. $9 + 4a - 4a - 3 = 9 - 3 + 4a - 4a = 6 + 0 = 6$

59. $4a + 7b - 3a - 7b = 4a - 3a + 7b - 7b = a + 0 = a$

60. $2b + 5c - 5c - 2b = 2b - 2b + 5c - 5c = 0 + 0 = 0$

61. $3(2x + 3) - 9 = 6x + 9 - 9 = 6x + 0 = 6x$

62. $3(x + 4) + 2(x - 6) = 3x + 12 + 2x - 12 = 3x + 2x + 12 - 12 = 5x + 0 = 5x$

63. $4s + 9 + 6s - 9 = 4s + 6s + 9 - 9 = 10s + 0 = 10s$

64. $8t + 6 - 8t - 5 = 8t - 8t + 6 - 5 = 0 + 1 = 1$

65. $5(m + n) - 5m = 5m + 5n - 5m = 5m - 5m + 5n = 0 + 5n = 5n$

66. If $x = 3$: $5 \cdot 3 - 7 = 15 - 7 = 8$; if $x = 4$: $5 \cdot 4 - 7 = 20 - 7 = 13$;
if $x = 5$: $5 \cdot 5 - 7 = 25 - 7 = 18$; 4

67. If $x = 5$: $3 \cdot 5 + 9 = 15 + 9 = 24$; if $x = 7$: $3 \cdot 7 + 9 = 21 + 9 = 30$;
if $x = 9$: $3 \cdot 9 + 9 = 27 + 9 = 36$; 9

68. If $b = 3$: $2(3 + 6) = 2 \cdot 9 = 18$; if $b = 6$: $2(6 + 6) = 2 \cdot 12 = 24$;
if $b = 9$: $2(9 + 6) = 2 \cdot 15 = 30$; 3

69. If $x = 0$: $0 + 3 = 4 \cdot 0$? $3 = 0$? No.
If $x = 1$: $1 + 3 = 4 \cdot 1$? $4 = 4$? Yes.
If $x = 3$: $3 + 3 = 4 \cdot 3$? $6 = 12$? No. 1

70. $6 + 8 = 14 < 15$; $7 + 8 = 15$; $8 + 8 = 16 > 15$; 8

71. $3 \cdot 3 - 4 = 9 - 4 = 5 < 6$; $3 \cdot 4 - 4 = 12 - 4 = 8 > 6$;
$3 \cdot 5 - 4 = 15 - 4 = 11 > 6$; 3

Page 33 · CHAPTER TEST

1. $9 - 2 = 7$

2. $6 \div 2 = 3$

3. $9 \cdot 6 = 54$

4. $9 + 6 - 8 = 7$

5. $7 \cdot 6 + 2 = 42 + 2 = 44$

6. $7(6 + 2) = 7 \cdot 8 = 56$

7. $15 - 9 \div 3 = 15 - 3 = 12$

8. $9 \cdot (8 + 2) = 9 \cdot 10 = 90$

9. $7k - k - 3k = 3k$

10. not possible

11. $9 + 3x - 3x = 9 + 0 = 9$

12. $6yz - 2yz + 4y = 4yz + 4y$

13. $8r + 12 + r = (8r + r) + 12 = 9r + 12$

14. $5h + 7t - 6t + 6h = (5h + 6h) + (7t - 6t) = 11h + t$

15. $3b + 4c - 2d - c = 3b + (4c - c) - 2d = 3b + 3c - 2d$

16. $v + 4v + 7y - 3v = (v + 4v - 3v) + 7y = 2v + 7y$

17. $4 \cdot a \cdot a \cdot a \cdot b = 4a^3 b$

18. $d \cdot e \cdot e \cdot e \cdot e = de^4$

19. $(5z) \cdot 7 = (5 \cdot 7) \cdot z = 35z$

20. $9 \cdot b \cdot 2 \cdot b = (9 \cdot 2)(b \cdot b) = 18b^2$

21. $(4x)(3y) = (4 \cdot 3) \cdot x \cdot y = 12xy$

22. $50 \cdot (7 \cdot 2) = (50 \cdot 2) \cdot 7 = 100 \cdot 7 = 700$

23. $6(r + 3) = 6r + 18$

24. $2(9x - 8) = 18x - 16$ **25.** $t(7t + 2q) = 7t^2 + 2tq$

26. $4(2y + 1) + 6 - 7y = 8y + 4 + 6 - 7y = 8y - 7y + 4 + 6 = y + 10$

27. $\dfrac{4 + 4}{4 - 4} = \dfrac{8}{0};$ impossible   **28.** $\dfrac{0 \cdot 1}{0 + 1} = \dfrac{0}{1} = 0$

29. $\dfrac{2 \cdot 4 - 8}{5} = \dfrac{8 - 8}{5} = \dfrac{0}{5} = 0$ **30.** $5x \div x = \dfrac{5x}{x} = 5 \cdot \dfrac{x}{x} = 5$

31. $4t \div 1 = \dfrac{4t}{1} = 4t$ **32.** $\dfrac{r}{8} \cdot 8 = r \cdot \dfrac{8}{8} = r$

33. If $j = 3$: $5 \cdot 3 - 2 = 2 \cdot 3 + 7?$ $15 - 2 = 6 + 7?$ $13 = 13?$ Yes.

     If $j = 4$: $5 \cdot 4 - 2 = 2 \cdot 4 + 7?$ $20 - 2 = 8 + 7?$ $18 = 15?$ No. 3

34. $1^3 = 1, 2^3 = 8, 3^3 = 27;$ 2   **35.** 10

36. $2 \cdot 0 = 0 < 18;$ $2 \cdot 5 = 10 < 18;$ $2 \cdot 8 = 16 < 18;$ $2 \cdot 10 = 20 > 18;$ 0, 5, 8

37. $0 + 1 < 2 \cdot 0?$ $1 < 0?$ No. $5 + 1 < 2 \cdot 5?$ $6 < 10?$ Yes.

     $8 + 1 < 2 \cdot 8?$ $9 < 16?$ Yes. $10 + 1 < 2 \cdot 10?$ $11 < 20?$ Yes. 5, 8, 10

38. $2 + 5 \cdot 3 = 2 + 15 = 17;$ $21 > 17;$ $>$

Pages 34–35 · MIXED REVIEW

Arithmetic Review   **1.** 43   **2.** 18,756   **3.** 1140   **4.** 39   **5.** 94

   6. 37   **7.** 333   **8.** 56   **9.** 490   **10.** 508

11. 7489   **12.** 783   **13.** 75.9   **14.** $\dfrac{2}{9} + \dfrac{4}{9} = \dfrac{6}{9} = \dfrac{2}{3};$ f

15. $\dfrac{7}{6} - \dfrac{5}{6} = \dfrac{2}{6} = \dfrac{1}{3};$ a   **16.** $\dfrac{7}{8} \cdot \dfrac{4}{3} = \dfrac{28}{24} = \dfrac{7}{6};$ d

17. $1\dfrac{7}{8} \div 1\dfrac{1}{4} = \dfrac{15}{8} \div \dfrac{5}{4} = \dfrac{15}{8} \cdot \dfrac{4}{5} = \dfrac{60}{40} = \dfrac{3}{2};$ g

18. $6 \div \dfrac{2}{3} = \dfrac{6}{1} \cdot \dfrac{3}{2} = \dfrac{18}{2} = 9;$ e

19. $4\dfrac{1}{6} - 2\dfrac{4}{5} = \dfrac{25}{6} - \dfrac{14}{5} = \dfrac{125}{30} - \dfrac{84}{30} = \dfrac{41}{30};$ b

20. $\dfrac{7}{12} + \dfrac{3}{4} = \dfrac{7}{12} + \dfrac{9}{12} = \dfrac{16}{12} = \dfrac{4}{3};$ c

21. $12 = 2^2 \cdot 3, 9 = 3^2, 18 = 2 \cdot 3^2;$ LCD $= 2^2 \cdot 3^2 = 36$

22. $42 = 2 \cdot 3 \cdot 7, 60 = 2^2 \cdot 3 \cdot 5;$ GCF $= 2 \cdot 3 = 6$

23. $\dfrac{129}{10,000} = 0.0129$ **24.** $3.175 = 3\dfrac{175}{1000} = 3\dfrac{7}{40}$ **25.** 8.009

26. 17.837   **27.** 0.27408   **28.** $(4)(31.71) = 126.84$

29. 427.5   **30.** $\dfrac{2.08}{6.5} = 0.32$

Chapter 1 Review   **1.** $7(3 + 5x) - 21 = 21 + 35x - 21 = (21 - 21) + 35x = 35x$

2. $5ab - 2a - 4ab = (5ab - 4ab) - 2a = ab - 2a$

3. $2 \cdot 5 \cdot x \cdot x \cdot x \cdot y = 10x^3 y$ **4.** $\dfrac{c}{6} \cdot 6 = c \cdot \dfrac{6}{6} = c$

5. $y \cdot 3 \cdot y \cdot 8 \cdot x \cdot y = 3 \cdot 8 \cdot x \cdot y \cdot y \cdot y = 24xy^3$

6. $9k + 8 - 3k + 2kt = (9k - 3k) + 8 + 2kt = 6k + 8 + 2kt$

7. x **8.** $n \div 5$ **9.** Subtract 1.   **10.** Multiply by b.

11. x **12.** Divide by 5.   **13.** $7(6 - 4) = 7 \cdot 2 = 14$

14. $7 \cdot 6 - 4 = 42 - 4 = 38$ **15.** $18 \div 6 = 3$ **16.** $6^1 = 6$

17. $\dfrac{6 - 6}{6 + 6} = \dfrac{0}{12} = 0$ **18.** $\dfrac{6}{6} \cdot 3 = 3$ **19.** $9 \cdot 6 = 54$

20. $6 \cdot 6 = 36$ **21.** $9 + 6 \div 3 = 9 + 2 = 11$

22. $8 - 4 \div 2 + 1 = 8 - 2 + 1 = 7$

23. $496 + 13 + 4 = (496 + 4) + 13 = 500 + 13 = 513$

24. $4^3 = 4 \cdot 4 \cdot 4 = 64$ **25.** $\dfrac{3 \cdot 3}{3 - 3} = \dfrac{9}{0}$;   impossible

26. $(20 \cdot 18) \cdot 5 = (20 \cdot 5) \cdot 18 = 100 \cdot 18 = 1800$

27. $(7 \cdot 4 + 2) \div 5 = (28 + 2) \div 5 = 30 \div 5 = 6$; $2^3 = 8$; $6 < 8$; $<$

28. $7 + 4 - 10 = 1$ **29.** no   **30.** Answers may vary.   18

31. $6(2x + 1) + 5(2 - x) = 12x + 6 + 10 - 5x = 12x - 5x + 6 + 10 = 7x + 16$

32. Answers may vary. $5x = 5$

33. If $x = 1$: $4 \cdot 1 + 7 = 2 \cdot 1 + 11$? $4 + 7 = 2 + 11$? $11 = 13$?   No.

    If $x = 2$: $4 \cdot 2 + 7 = 2 \cdot 2 + 11$? $8 + 7 = 4 + 11$? $15 = 15$?   Yes.

    If $x = 3$: $4 \cdot 3 + 7 = 2 \cdot 3 + 11$? $12 + 7 = 6 + 11$? $19 = 17$?   No.   2

34. $0 + 4 = 4 > 2$; $1 + 4 = 5 > 2$; $2 + 4 = 6 > 2$; $3 + 4 = 7 > 2$;   0, 1, 2, 3

35. If $a = 0$: $5 \cdot 0 < 0 + 9$? $0 < 9$?   Yes.

    If $a = 2$: $5 \cdot 2 < 2 + 9$? $10 < 11$?   Yes.

    If $a = 4$: $5 \cdot 4 < 4 + 9$? $20 < 13$?   No.

    If $a = 6$: $5 \cdot 6 < 6 + 9$? $30 < 15$?   No.   0, 2

36. If $b = 8$: $5(8 - 8) = 8 - 4$? $5 \cdot 0 = 4$? $0 = 4$?   No.

    If $b = 9$: $5(9 - 8) = 9 - 4$? $5 \cdot 1 = 5$? $5 = 5$?   Yes.

    If $b = 11$: $5(11 - 8) = 11 - 4$? $5 \cdot 3 = 7$? $15 = 7$?   No.   9

Page 39 · WRITTEN EXERCISES

A **1.** $x - 5 = 11$; $x - 5 + 5 = 11 + 5$; $x = 16$ **2.** $x - 7 = 5$; $x - 7 + 7 = 5 + 7$; $x = 12$ **3.** $x - 8 = 2$; $x - 8 + 8 = 2 + 8$; $x = 10$ **4.** $y - 3 = 17$; $y - 3 + 3 = 17 + 3$; $y = 20$ **5.** $x - 3 = 9$; $x - 3 + 3 = 9 + 3$; $x = 12$ **6.** $y - 4 = 7$; $y - 4 + 4 = 7 + 4$; $y = 11$ **7.** $a - 7 = 6$; $a - 7 + 7 = 6 + 7$; $a = 13$ **8.** $y - 6 = 2$; $y - 6 + 6 = 2 + 6$; $y = 8$ **9.** $b - 9 = 15$; $b - 9 + 9 = 15 + 9$; $b = 24$ **10.** $a - 12 = 12$; $a - 12 + 12 = 12 + 12$; $a = 24$ **11.** $x - 9 = 10$; $x - 9 + 9 = 10 + 9$; $x = 19$ **12.** $x - 3 = 12$; $x - 3 + 3 = 12 + 3$; $x = 15$ **13.** $y - 2 = 13$; $y - 2 + 2 = 13 + 2$; $y = 15$ **14.** $x - 5 = 2$; $x - 5 + 5 = 2 + 5$; $x = 7$ **15.** $y - 7 = 2$; $y - 7 + 7 = 2 + 7$; $y = 9$ **16.** $a - 9 = 4$; $a - 9 + 9 = 4 + 9$; $a = 13$ **17.** $n - 1 = 8$; $n - 1 + 1 = 8 + 1$; $n = 9$ **18.** $x - 8 = 20$; $x - 8 + 8 = 20 + 8$; $x = 28$ **19.** $x - 9 = 21$; $x - 9 + 9 = 21 + 9$; $x = 30$ **20.** $x - 7 = 19$; $x - 7 + 7 = 19 + 7$; $x = 26$ **21.** $x - 15 = 10$; $x - 15 + 15 = 10 + 15$; $x = 25$ **22.** $a - 9 = 25$; $a - 9 + 9 = 25 + 9$; $a = 34$ **23.** $6 = x - 2$; $6 + 2 = x - 2 + 2$; $8 = x$ **24.** $9 = y - 4$; $9 + 4 = y - 4 + 4$; $13 = y$ **25.** $10 = x - 3$; $10 + 3 = x - 3 + 3$; $13 = x$ **26.** $6 = x - 5$; $6 + 5 = x - 5 + 5$; $11 = x$ **27.** $0 = y - 5$; $0 + 5 = y - 5 + 5$; $5 = y$ **28.** $15 = x - 5$; $15 + 5 = x - 5 + 5$; $20 = x$ **29.** $21 = b - 14$; $21 + 14 = b - 14 + 14$; $35 = b$ **30.** $43 = x - 21$; $43 + 21 = x - 21 + 21$; $64 = x$ **31.** $36 = y - 15$; $36 + 15 = y - 15 + 15$; $51 = y$ **32.** $x - 52 = 100$; $x - 52 + 52 = 100 + 52$; $x = 152$

B **33.** $x - 3 = 7$; $x = 10$ **34.** $x - 4 = 13$; $x = 17$

35. $x - 2 = 4$; $x = 6$ **36.** $x - 5 = 3$; $x = 8$

37. $y - 2 = 9$; $y = 11$ **38.** $n - 6 = 4$; $n = 10$

39. $n - 10 = 11$; $n = 21$ **40.** $x - 3 = 0$; $x = 3$

41. $a - 5 = 10$; $a = 15$ **42.** $x - 7 = 20$; $x = 27$

43. $7 = y - 2$; $9 = y$ **44.** $10 = x - 7$; $17 = x$

45. $12 = x - 3$; $15 = x$ **46.** $20 = n - 5$; $25 = n$

47. $22 = x - 3$; $25 = x$ **48.** $14 = y - 9$; $23 = y$

Page 41 · WRITTEN EXERCISES

A **1.** $x + 5 = 15$; $x + 5 - 5 = 15 - 5$; $x = 10$ **2.** $x + 8 = 20$; $x + 8 - 8 = 20 - 8$; $x = 12$ **3.** $x + 2 = 10$; $x + 2 - 2 = 10 - 2$; $x = 8$ **4.** $a + 7 = 25$; $a + 7 - 7 = 25 - 7$; $a = 18$ **5.** $y + 7 = 11$; $y + 7 - 7 = 11 - 7$; $y = 4$ **6.** $n + 6 = 14$; $n + 6 - 6 = 14 - 6$; $n = 8$ **7.** $n + 4 = 16$; $n + 4 - 4 = 16 - 4$; $n = 12$ **8.** $y + 6 = 21$; $y + 6 - 6 = 21 - 6$; $y = 15$

9. $x + 20 = 34$; $x + 20 - 20 = 34 - 20$; $x = 14$ **10.** $8 + y = 15$; $8 + y - 8 = 15 - 8$; $y = 7$ **11.** $5 + x = 20$; $5 + x - 5 = 20 - 5$; $x = 15$ **12.** $28 + y = 42$; $28 + y - 28 = 42 - 28$; $y = 14$ **13.** $24 = x + 7$; $24 - 7 = x + 7 - 7$; $17 = x$ **14.** $30 = y + 8$; $30 - 8 = y + 8 - 8$; $22 = y$ **15.** $60 = b + 48$; $60 - 48 = b + 48 - 48$; $12 = b$ **16.** $88 = x + 7$; $88 - 7 = x + 7 - 7$; $81 = x$

B   **17.** $x + 4 = 9$; $x = 5$ **18.** $x + 7 = 14$; $x = 7$

19. $y + 3 = 20$; $y = 17$ **20.** $n + 8 = 12$; $n = 4$

21. $x + 3 = 15$; $x = 12$ **22.** $y + 6 = 10$; $y = 4$

23. $n + 4 = 17$; $n = 13$ **24.** $y + 9 = 25$; $y = 16$

25. $6 + x = 16$; $x = 10$ **26.** $9 + y = 10$; $y = 1$

27. $8 + x = 21$; $x = 13$ **28.** $4 + x = 27$; $x = 23$

29. $18 = x + 3$; $15 = x$ **30.** $20 = x + 4$; $16 = x$

31. $14 = x + 6$; $8 = x$ **32.** $24 = x + 6$; $18 = x$

33. $15 = 8 + x$; $7 = x$ **34.** $x + 12 = 16$; $x = 4$

35. $20 + x = 24$; $x = 4$ **36.** $50 = x + 10$; $40 = x$

37. $19 + y = 19$; $y = 0$ **38.** $x + 14 = 37$; $x = 23$

39. $z + 52 = 76$; $z = 24$ **40.** $34 = m + 9$; $25 = m$

Page 41 · MIXED PRACTICE EXERCISES

1. $x - 2 = 7$; $x - 2 + 2 = 7 + 2$; $x = 9$ **2.** $x - 5 = 4$; $x - 5 + 5 = 4 + 5$; $x = 9$ **3.** $x - 9 = 9$; $x - 9 + 9 = 9 + 9$; $x = 18$ **4.** $y - 7 = 3$; $y - 7 + 7 = 3 + 7$; $y = 10$ **5.** $y - 6 = 5$; $y = 11$ **6.** $x - 4 = 6$; $x = 10$ **7.** $y - 7 = 11$; $y = 18$ **8.** $x - 2 = 14$; $x = 16$ **9.** $x + 5 = 11$; $x + 5 - 5 = 11 - 5$; $x = 6$ **10.** $x + 6 = 13$; $x + 6 - 6 = 13 - 6$; $x = 7$ **11.** $y + 7 = 12$; $y + 7 - 7 = 12 - 7$; $y = 5$ **12.** $n + 2 = 17$; $n + 2 - 2 = 17 - 2$; $n = 15$ **13.** $n + 8 = 14$; $n = 6$ **14.** $x + 3 = 22$; $x = 19$

15. $n + 9 = 26$; $n = 17$ **16.** $x + 12 = 19$; $x = 7$

17. $y - 2 = 17$; $y = 19$ **18.** $x - 12 = 42$; $x = 54$

19. $y + 6 = 35$; $y = 29$ **20.** $n + 17 = 31$; $n = 14$

21. $x - 6 = 27$; $x = 33$ **22.** $y + 7 = 40$; $y = 33$

23. $x - 18 = 12$; $x = 30$ **24.** $y + 9 = 37$; $y = 28$

25. $20 = 16 + x$; $20 - 16 = 16 + x - 16$; $4 = x$

26. $12 = x - 7$; $12 + 7 = x - 7 + 7$; $19 = x$

27. $15 = x - 2$; $15 + 2 = x - 2 + 2$; $17 = x$

28. $25 = y + 6$; $25 - 6 = y + 6 - 6$; $19 = y$

29. $36 = y + 9$; $27 = y$ **30.** $40 = x - 16$; $56 = x$

31. $29 = x - 17$; $46 = x$ **32.** $48 = n + 26$; $22 = n$

33. $45 = z - 15$; $60 = z$ **34.** $28 = x + 12$; $16 = x$

35. $x + 18 = 36$; $x = 18$ **36.** $x - 16 = 54$; $x = 70$

Page 43 • WRITTEN EXERCISES

A   1. $3x = 12$; $\dfrac{3x}{3} = \dfrac{12}{3}$; $x = 4$ 2. $5x = 15$; $\dfrac{5x}{5} = \dfrac{15}{5}$; $x = 3$

    3. $6x = 42$; $x = 7$ 4. $4a = 32$; $a = 8$

    5. $6b = 72$; $b = 12$ 6. $4y = 16$; $y = 4$

    7. $7x = 35$; $x = 5$ 8. $5x = 25$; $x = 5$

    9. $2x = 20$; $x = 10$ 10. $3x = 15$; $x = 5$

   11. $6y = 18$; $y = 3$ 12. $7y = 28$; $y = 4$

   13. $8m = 24$; $m = 3$ 14. $6x = 48$; $x = 8$

   15. $7n = 49$; $n = 7$ 16. $4a = 36$; $a = 9$

   17. $9a = 81$; $a = 9$ 18. $8x = 72$; $x = 9$

   19. $9x = 108$; $x = 12$ 20. $10a = 100$; $a = 10$

   21. $7x = 91$; $x = 13$ 22. $12a = 84$; $a = 7$

   23. $8x = 112$; $x = 14$ 24. $9b = 513$; $b = 57$

   25. $3b = 540$; $b = 180$ 26. $11n = 231$; $n = 21$

   27. $7y = 175$; $y = 25$ 28. $4m = 244$; $m = 61$

   29. $\dfrac{x}{2} = 9$; $2 \cdot \dfrac{x}{2} = 2 \cdot 9$; $x = 18$ 30. $\dfrac{x}{3} = 8$; $3 \cdot \dfrac{x}{3} = 3 \cdot 8$; $x = 24$

   31. $\dfrac{x}{4} = 11$; $x = 44$ 32. $\dfrac{x}{5} = 7$; $x = 35$

   33. $\dfrac{a}{6} = 9$; $a = 54$ 34. $\dfrac{a}{3} = 7$; $a = 21$

   35. $\dfrac{n}{2} = 6$; $n = 12$ 36. $\dfrac{x}{6} = 4$; $x = 24$

   37. $\dfrac{n}{5} = 3$; $n = 15$ 38. $\dfrac{x}{2} = 10$; $x = 20$

   39. $\dfrac{n}{7} = 8$; $n = 56$ 40. $\dfrac{x}{3} = 11$; $x = 33$

   41. $\dfrac{x}{6} = 8$; $x = 48$ 42. $\dfrac{a}{9} = 9$; $a = 81$

   43. $\dfrac{a}{9} = 8$; $a = 72$ 44. $\dfrac{x}{10} = 13$; $x = 130$

Pages 44–45 • MIXED PRACTICE EXERCISES

   1. $x = 11$ 2. $x = 2$ 3. $x = 5$

   4. $x = 13$ 5. $x = 4$ 6. $y = 12$

   7. $x = 7$ 8. $y = 4$ 9. $x = 20$

  10. $y = 14$ 11. $x = 6$ 12. $x = 12$

  13. $x = 23$ 14. $n = 15$

  15. $x - 4 = 10$; $x - 4 + 4 = 10 + 4$; $x = 14$

  16. $y + 2 = 7$; $y + 2 - 2 = 7 - 2$; $y = 5$

17. $x + 4 = 10$; $x = 6$

18. $y - 6 = 3$; $y = 9$

19. $n - 7 = 4$; $n = 11$

20. $n + 8 = 17$; $n = 9$

21. $x - 7 = 14$; $x = 21$

22. $y + 9 = 11$; $y = 2$

23. $x - 9 = 12$; $x = 21$

24. $4x = 20$; $\dfrac{4x}{4} = \dfrac{20}{4}$; $x = 5$

25. $8y = 16$; $y = 2$

26. $n + 6 = 32$; $n = 26$

27. $\dfrac{a}{7} = 3$; $7 \cdot \dfrac{a}{7} = 7 \cdot 3$; $a = 21$

28. $\dfrac{x}{4} = 9$; $x = 36$

29. $\dfrac{n}{5} = 2$; $n = 10$

30. $\dfrac{n}{8} = 9$; $n = 72$

31. $8 + y = 13$; $y = 5$

32. $a - 6 = 18$; $a = 24$

33. $x + 12 = 26$; $x = 14$

34. $y + 17 = 30$; $y = 13$

35. $a - 15 = 12$; $a = 27$

36. $12x = 48$; $x = 4$

37. $m - 14 = 37$; $m = 51$

38. $n + 36 = 52$; $n = 16$

39. $\dfrac{x}{6} = 3$; $x = 18$

40. $\dfrac{n}{8} = 1$; $n = 8$

41. $\dfrac{x}{5} = 6$; $x = 30$

42. $\dfrac{a}{7} = 9$; $a = 63$

43. $y + 10 = 15$; $y = 5$

44. $x - 3 = 20$; $x = 23$

45. $9x = 27$; $x = 3$

46. $7n = 63$; $n = 9$

47. $3y = 63$; $y = 21$

48. $7x = 42$; $x = 6$

49. $5a = 100$; $a = 20$

50. $9x = 90$; $x = 10$

51. $7 + x = 20$; $x = 13$

52. $a - 12 = 14$; $a = 26$

53. $x + 9 = 28$; $x = 19$

54. $x - 15 = 22$; $x = 37$

55. $x - 3 = 18$; $x = 21$

56. $y + 7 = 26$; $y = 19$

57. $3x = 57$; $x = 19$

58. $2n = 104$; $n = 52$

59. $35 = x + 7$; $28 = x$

60. $29 = y - 6$; $35 = y$

61. $54 = 9x$; $6 = x$

62. $7 = 7x$; $1 = x$

Page 45 • PUZZLE PROBLEMS

If Melba and Fran are both wearing black hats, then Brent would know he must be wearing red (since there are only 2 black hats). Since Brent does not know, either Melba or Fran (or both) must be wearing red. If Melba is wearing a black hat, Fran would thus know she must be wearing red. Since Fran does not know, Melba must be wearing a red hat.

Pages 47–49 • WRITTEN EXERCISES

A   1. $5x = 10$; $x = 2$ 2. $2a = 2$; $a = 1$ 3. $3b = 3$; $b = 1$

4. $4y = 8$; $y = 2$ 5. $2a = 18$; $a = 9$ 6. $5b = 40$; $b = 8$

7. $4c = 12$; $c = 3$ 8. $8a = 48$; $a = 6$ 9. $6n = 18$; $n = 3$

10. $2x - 4 = 0$; $2x - 4 + 4 = 0 + 4$; $2x = 4$; $x = 2$

11. $5x - 8 = 7$; $5x - 8 + 8 = 7 + 8$; $5x = 15$; $x = 3$

12. $2x + 1 = 5$; $2x + 1 - 1 = 5 - 1$; $2x = 4$; $x = 2$

13. $5x - 1 = 14$; $5x - 1 + 1 = 14 + 1$; $5x = 15$; $x = 3$

14. $3x - 2 = 10$; $3x - 2 + 2 = 10 + 2$; $3x = 12$; $x = 4$

15. $2x - 1 = 7$; $2x - 1 + 1 = 7 + 1$; $2x = 8$; $x = 4$

16. $3x - 5 = 4$; $3x - 5 + 5 = 4 + 5$; $3x = 9$; $x = 3$

17. $4x - 7 = 1$; $4x - 7 + 7 = 1 + 7$; $4x = 8$; $x = 2$

18. $2x - 4 = 8$; $2x - 4 + 4 = 8 + 4$; $2x = 12$; $x = 6$

19. $4x - 8 = 12$; $4x - 8 + 8 = 12 + 8$; $4x = 20$; $x = 5$

20. $2x - 5 = 1$; $2x - 5 + 5 = 1 + 5$; $2x = 6$; $x = 3$

21. $4x - 5 = 15$; $4x - 5 + 5 = 15 + 5$; $4x = 20$; $x = 5$

22. $2x + 5 = 11$; $2x + 5 - 5 = 11 - 5$; $2x = 6$; $x = 3$

23. $3x + 4 = 22$; $3x + 4 - 4 = 22 - 4$; $3x = 18$; $x = 6$

24. $4x + 1 = 1$; $4x + 1 - 1 = 1 - 1$; $4x = 0$; $x = 0$

25. $4y + 7 = 27$; $4y + 7 - 7 = 27 - 7$; $4y = 20$; $y = 5$

26. $3y + 2 = 8$; $3y + 2 - 2 = 8 - 2$; $3y = 6$; $y = 2$

27. $2a + 5 = 25$; $2a + 5 - 5 = 25 - 5$; $2a = 20$; $a = 10$

28. $5n - 1 = 9$; $5n - 1 + 1 = 9 + 1$; $5n = 10$; $n = 2$

29. $4n - 7 = 13$; $4n - 7 + 7 = 13 + 7$; $4n = 20$; $n = 5$

30. $2y + 6 = 22$; $2y + 6 - 6 = 22 - 6$; $2y = 16$; $y = 8$

31. $4y + 7 = 15$; $4y + 7 - 7 = 15 - 7$; $4y = 8$; $y = 2$

32. $6y + 4 = 34$; $6y + 4 - 4 = 34 - 4$; $6y = 30$; $y = 5$

33. $8 + 3r = 23$; $8 + 3r - 8 = 23 - 8$; $3r = 15$; $r = 5$

34. $2x - 3 = 5$; $2x - 3 + 3 = 5 + 3$; $2x = 8$; $x = 4$

35. $8y + 9 = 33$; $8y + 9 - 9 = 33 - 9$; $8y = 24$; $y = 3$

36. $2a - 5 = 11$; $2a - 5 + 5 = 11 + 5$; $2a = 16$; $a = 8$

37. $6s - 1 = 17$; $6s - 1 + 1 = 17 + 1$; $6s = 18$; $s = 3$

38. $3c - 7 = 14$; $3c - 7 + 7 = 14 + 7$; $3c = 21$; $c = 7$

39. $3t - 8 = 13$; $3t - 8 + 8 = 13 + 8$; $3t = 21$; $t = 7$

40. $7y + 2 = 23$; $7y + 2 - 2 = 23 - 2$; $7y = 21$; $y = 3$

41. $5b - 9 = 21$; $5b - 9 + 9 = 21 + 9$; $5b = 30$; $b = 6$

42. $6r - 3 = 21$; $6r - 3 + 3 = 21 + 3$; $6r = 24$; $r = 4$

43. $4c + 1 = 13$; $4c + 1 - 1 = 13 - 1$; $4c = 12$; $c = 3$

44. $8a + 5 = 53$; $8a + 5 - 5 = 53 - 5$; $8a = 48$; $a = 6$

45. $8x - 5 = 35$; $8x - 5 + 5 = 35 + 5$; $8x = 40$; $x = 5$

46. $3y + 11 = 41$; $3y + 11 - 11 = 41 - 11$; $3y = 30$; $y = 10$

47. $4x - 5 = 23$; $4x - 5 + 5 = 23 + 5$; $4x = 28$; $x = 7$

48. $8a - 5 = 75$; $8a - 5 + 5 = 75 + 5$; $8a = 80$; $a = 10$

49. $4b + 6 = 30$; $4b + 6 - 6 = 30 - 6$; $4b = 24$; $b = 6$

50. $3x - 5 = 28;$ $3x - 5 + 5 = 28 + 5;$ $3x = 33;$ $x = 11$

51. $3b - 12 = 45;$ $3b - 12 + 12 = 45 + 12;$ $3b = 57;$ $b = 19$

52. $7s - 9 = 68;$ $7s - 9 + 9 = 68 + 9;$ $7s = 77;$ $s = 11$

53. $8y + 5 = 37;$ $8y + 5 - 5 = 37 - 5;$ $8y = 32;$ $y = 4$

54. $8t - 14 = 26;$ $8t - 14 + 14 = 26 + 14;$ $8t = 40;$ $t = 5$

B   **55.** $2n + 8 = 32;$ $2n = 24;$ $n = 12$ **56.** $3n + 7 = 25;$ $3n = 18;$ $n = 6$

57. $2n + 4 = 16;$ $2n = 12;$ $n = 6$ **58.** $5n - 17 = 53;$ $5n = 70;$ $n = 14$

59. $4n + 5 = 17;$ $4n = 12;$ $n = 3$ **60.** $3n - 6 = 21;$ $3n = 27;$ $n = 9$

61. $2n - 15 = 25;$ $2n = 40;$ $n = 20$ **62.** $5n + 11 = 66;$ $5n = 55;$ $n = 11$

C

[1] 1	[2] 5	▓	[3] 3	0	[4] 2
▓	[5] 4	[6] 3	[7] 2	9	[8] 3 / [9] 4
[10] 1	▓	[11] 1	7	2	[12] 6 / 1
[13] 8	6	5	▓	[14] 1	[15] 3 / ▓
▓	[16] 2	[17] 1	▓	[18] 2	0 / [19] 4
[20] 2	[21] 9	▓	[22] 1	[23] 1	0 / 6
[24] 1	2	▓	[25] 2	0	4 / [26] 2
[27] 5	0	0	▓	[28] 1	4

Page 51 • WRITTEN EXERCISES

A   **1.**

Arith.	Algebra
8	n
40	$5n$
44	$5n + 4$
880	$20(5n + 4)$, or $100n + 80$
800	$100n$

2.

Arith.	Algebra
6	n
15	$n + 9$
60	$4(n + 9)$, or $4n + 36$
30	$4n + 6$
750	$25(4n + 6)$, or $100n + 150$

3. Steps: $n;$ $n + 8;$ $2(n + 8)$, or $2n + 16;$ $2n + 10$, or $2(n + 5);$ $n + 5;$ $n.$ Thus, $n = 9.$

4. Steps: $n;$ $n + 5;$ $2(n + 5)$, or $2n + 10;$ $2n + 4;$ $5(2n + 4)$, or $10n + 20;$ $10n + 16.$ Thus, $10n + 16 = 56;$ $10n = 40;$ $n = 4.$

5. Steps: $n;$ $5n;$ $5n + 6;$ $4(5n + 6)$, or $20n + 24;$ $20n + 33;$ $5(20n + 33)$, or $100n + 165.$ Thus, $100n + 165 = 1365;$ $100n = 1200;$ $n = 12.$

6. Answers will vary. For example: think of a number;   multiply it by 3;   add 5; multiply by 2;   subtract 4;   divide by 6.

B   **7.** If n is *any* number chosen, then the steps are: $n;$ $n + 10;$ $2n + 20;$ $2n + 25;$ $8n + 100;$ $8n + 160;$ $n + 20;$ 20.

Pages 52–53 • WRITTEN EXERCISES

A   **1.** $4a + 9a - 6a = 42$; $7a = 42$; $a = 6$. Check: $4 \cdot 6 + 9 \cdot 6 - 6 \cdot 6 = 24 + 54 -$
$36 = 42$ **2.** $6y - 4y + y = 18$; $3y = 18$; $y = 6$. Check: $6 \cdot 6 - 4 \cdot 6 + 6 =$
$36 - 24 + 6 = 18$ **3.** $9c + 4c - c = 48$; $12c = 48$; $c = 4$. Check: $9 \cdot 4 +$
$4 \cdot 4 - 4 = 36 + 16 - 4 = 48$ **4.** $9 + 6a - 2a = 21$; $9 + 4a = 21$; $9 + 4a -$
$9 = 21 - 9$; $4a = 12$; $a = 3$. Check: $9 + 6 \cdot 3 - 2 \cdot 3 = 9 + 18 - 6 = 21$
5. $8a + 7 - 5a = 31$; $3a + 7 = 31$; $3a + 7 - 7 = 31 - 7$; $3a = 24$; $a = 8$.
Check: $8 \cdot 8 + 7 - 5 \cdot 8 = 64 + 7 - 40 = 31$ **6.** $9a - 4 + 3a = 32$; $12a - 4 =$
32; $12a - 4 + 4 = 32 + 4$; $12a = 36$; $a = 3$. Check: $9 \cdot 3 - 4 + 3 \cdot 3 = 27 -$
$4 + 9 = 32$ **7.** $13 + 5b - 2b = 37$; $13 + 3b = 37$; $3b = 24$; $b = 8$. Check:
$13 + 5 \cdot 8 - 2 \cdot 8 = 13 + 40 - 16 = 37$ **8.** $8b + 4b - b = 55$; $11b = 55$; $b = 5$.
Check: $8 \cdot 5 + 4 \cdot 5 - 5 = 40 + 20 - 5 = 55$ **9.** $12c - 8 - 5c = 27$; $7c - 8 = 27$;
$7c = 35$; $c = 5$. Check: $12 \cdot 5 - 8 - 5 \cdot 5 = 60 - 8 - 25 = 27$ **10.** $8x + 4 -$
$3x = 24$; $5x + 4 = 24$; $5x = 20$; $x = 4$. Check: $8 \cdot 4 + 4 - 3 \cdot 4 = 32 + 4 -$
$12 = 24$ **11.** $7n + 2 + 3n = 12$; $10n + 2 = 12$; $10n = 10$; $n = 1$. Check:
$7 \cdot 1 + 2 + 3 \cdot 1 = 7 + 2 + 3 = 12$ **12.** $13x + 5 - 4x = 23$; $9x + 5 = 23$; $9x =$
18; $x = 2$. Check: $13 \cdot 2 + 5 - 4 \cdot 2 = 26 + 5 - 8 = 23$ **13.** $4x + 5x - 3 =$
15; $9x - 3 = 15$; $9x = 18$; $x = 2$. Check: $4 \cdot 2 + 5 \cdot 2 - 3 = 8 + 10 - 3 =$
15 **14.** $4x - x + 1 = 22$; $3x + 1 = 22$; $3x = 21$; $x = 7$. Check: $4 \cdot 7 - 7 +$
$1 = 28 - 7 + 1 = 22$ **15.** $7x + 3x - 5 = 15$; $10x - 5 = 15$; $10x = 20$; $x = 2$.
Check: $7 \cdot 2 + 3 \cdot 2 - 5 = 14 + 6 - 5 = 15$ **16.** $12 = 3c - 12 + 5c$; $12 = 8c -$
12; $24 = 8c$; $3 = c$. Check: $3 \cdot 3 - 12 + 5 \cdot 3 = 9 - 12 + 15 = 12$ **17.** $9 =$
$13x - 7 - 5x$; $9 = 8x - 7$; $16 = 8x$; $2 = x$. Check: $13 \cdot 2 - 7 - 5 \cdot 2 =$
$26 - 7 - 10 = 9$ **18.** $35 = 19m - 4m + 5$; $35 = 15m + 5$; $30 = 15m$; $2 = m$.
Check: $19 \cdot 2 - 4 \cdot 2 + 5 = 38 - 8 + 5 = 35$ **19.** $18 = 8y - 7 - 3y$; $18 = 5y - 7$;
$25 = 5y$; $5 = y$. Check: $8 \cdot 5 - 7 - 3 \cdot 5 = 40 - 7 - 15 = 18$ **20.** $21 = 4a - 3a +$
$6a$; $21 = 7a$; $3 = a$. Check: $4 \cdot 3 - 3 \cdot 3 + 6 \cdot 3 = 12 - 9 + 18 = 21$ **21.** $16 =$
$3x + 2x - 9$; $16 = 5x - 9$; $25 = 5x$; $5 = x$. Check: $3 \cdot 5 + 2 \cdot 5 - 9 = 15 +$
$10 - 9 = 16$ **22.** $21 = 4n - 3 - n$; $21 = 3n - 3$; $24 = 3n$; $8 = n$. Check:
$4 \cdot 8 - 3 - 8$; $32 - 3 - 8 = 21$ **23.** $32 = 8x + 4 - 4x$; $32 = 4x + 4$; $28 = 4x$;
$7 = x$. Check: $8 \cdot 7 + 4 - 4 \cdot 7 = 56 + 4 - 28 = 32$ **24.** $7 = 3a + 5a - a + 7$;
$7 = 7a + 7$; $0 = 7a$; $0 = a$. Check: $3 \cdot 0 + 5 \cdot 0 - 0 + 7 = 0 + 0 - 0 + 7 = 7$
25. $4x + 3x = 63$; $7x = 63$; $x = 9$ **26.** $7y - 3y = 20$; $4y = 20$; $y = 5$
27. $2x + 5 = 73$; $2x = 68$; $x = 34$ **28.** $3x - 7 = 83$; $3x = 90$; $x = 30$
B   **29.** $2x + 9 = 25$; $2x = 16$; $x = 8$ **30.** $3x - 11 = 37$; $3x = 48$; $x = 16$

Pages 55–57 • WRITTEN EXERCISES

A   **1.** $n + 5$ **2.** $j - 5$ **3.** $j + 2$ **4.** $r - 2$
       5. $b + 11$ **6.** $g - 11$ **7.** $3x$ **8.** $x + 180$

9. $4x$ 10. $t + 12$ 11. $3x$ 12. $n + 6$

13. $x + 1$ 14. $x + 10$ 15. $r - 1$ 16. $t - 7$

17. $m - 1$ 18. $15 + x$

B   19. $x + 2$; $(x + 2) + 4$, or $x + 6$ 20. $y + 4$; $y - 3$

21. $k + 3$; $k - 1$; $(k + 3) - 1$, or $k + 2$ 22. $n + 2$; $n + 4$

23. $x + 2$; $x + 4$ 24. $k + 1$; $2(k + 1)$, or $2k + 2$

Pages 59–61 · WRITTEN EXERCISES

A

1. Let x = the number of black and white monitors. Then $x + 32$ = the number of color monitors. $x + x + 32 = 80$; $2x + 32 = 80$; $2x = 48$; $x = 24$.   24 computers have black and white monitors.

2. Let x = Linda's score. Then $x - 6$ = Carmen's score. $x + x - 6 = 156$; $2x - 6 = 156$; $2x = 162$; $x = 81$ and $x - 6 = 75$.   Linda: 81;   Carmen: 75

3. Let x = the number of students at South High. Then $x - 85$ = the number at North High. $x + x - 85 = 1495$; $2x - 85 = 1495$; $2x = 1580$; $x = 790$.   790 students at South High

4. Let x = the number of games won. Then $x - 22$ = the number lost. $x + x - 22 = 64$; $2x - 22 = 64$; $2x = 86$; $x = 43$ and $x - 22 = 21$.   43 games won, 21 games lost

5. Let x = the number of cars sold. Then $x - 5$ = the number of trucks sold. $x + x - 5 = 37$; $2x - 5 = 37$; $2x = 42$; $x = 21$.   21 cars sold

6. Let x = the amount Luis has. Then $x - 48$ = the amount Ed has. $x + x - 48 = 200$; $2x - 48 = 200$; $2x = 248$; $x = 124$ and $x - 48 = 76$.   Luis: \$124;   Ed: \$76

7. Let x = the sheep's weight. Then $2x$ = the hog's weight. $x + 2x = 285$; $3x = 285$; $x = 95$ and $2x = 190$.   sheep: 95 kg;   hog: 190 kg

8. Let x = the number of calories in a peach. Then $2x$ = the number in an apple. $x + 2x = 105$; $3x = 105$; $x = 35$ and $2x = 70$.   peach: 35 calories;   apple: 70 calories

9. Let x = the number of games won. Then $x - 17$ = the number lost. $x + x - 17 = 153$; $2x - 17 = 153$; $2x = 170$; $x = 85$.   85 games won

10. Let x = the number of points scored in the first game. Then $2x$ = the number scored in the second. $x + 2x = 114$; $3x = 114$; $x = 38$.   38 points in the first game

11. Let x and $x + 1$ = the consecutive whole numbers. $x + x + 1 = 75$; $2x + 1 = 75$; $2x = 74$; $x = 37$ and $x + 1 = 38$.   37 and 38

12. Let x, $x + 1$, and $x + 2$ = the consecutive whole numbers. $x + x + 1 + x + 2 = 75$; $3x + 3 = 75$; $3x = 72$; $x = 24$, $x + 1 = 25$, and $x + 2 = 26$.   24, 25, and 26

B

13. Let x = the length of Route 1. Then $2x$ = the length of Route 2 and $2x + 30$ = the length of Route 3. $x + 2x + 2x + 30 = 280$; $5x + 30 = 280$; $5x = 250$; $x = 50$, so $2x + 30 = 2 \cdot 50 + 30 = 100 + 30 = 130$.   Route 3 is 130 km long.

14. Let x = the number of women. Then $x + 4$ = the number of men and $(x + x + 4) + 10$, or $2x + 14$ = the number of children. $x + x + 4 + 2x + 14 = 82$; $4x + 18 = 82$; $4x = 64$; $x = 16$, so $2x + 14 = 2 \cdot 16 + 14 = 32 + 14 = 46$.   46 children

C   **15.** Let x = the amount Sue gets. Then $2x$ = the amount for Liz and $3x$ = the amount for Wendy. $x + 2x + 3x = 450$; $6x = 450$; $x = 75$, $2x = 150$, and $3x = 225$. Sue: \$75; Liz: \$150; Wendy: \$225

Page 63 · WRITTEN EXERCISES

A   **1.** $2x = 14 + x$; $2x - x = 14 + x - x$; $x = 14$

2. $7x = 2x + 5$; $7x - 2x = 2x + 5 - 2x$; $5x = 5$; $x = 1$

3. $9x = 5x + 16$; $9x - 5x = 5x + 16 - 5x$; $4x = 16$; $x = 4$

4. $6y = 18 - 3y$; $6y + 3y = 18 - 3y + 3y$; $9y = 18$; $y = 2$

5. $4n = 21 - 3n$; $4n + 3n = 21 - 3n + 3n$; $7n = 21$; $n = 3$

6. $6x = 20 - 4x$; $6x + 4x = 20 - 4x + 4x$; $10x = 20$; $x = 2$

7. $8x = 25 + 3x$; $8x - 3x = 25 + 3x - 3x$; $5x = 25$; $x = 5$

8. $9x = 26 - 4x$; $9x + 4x = 26 - 4x + 4x$; $13x = 26$; $x = 2$

9. $12x = 10 + 7x$; $12x - 7x = 10 + 7x - 7x$; $5x = 10$; $x = 2$

10. $6 - 2x = x$; $6 - 2x + 2x = x + 2x$; $6 = 3x$; $2 = x$

11. $16 - 3x = x$; $16 - 3x + 3x = x + 3x$; $16 = 4x$; $4 = x$

12. $20 - 6a = 4a$; $20 - 6a + 6a = 4a + 6a$; $20 = 10a$; $2 = a$

13. $7 - 3x = 4x$; $7 - 3x + 3x = 4x + 3x$; $7 = 7x$; $1 = x$

14. $10 + x = 6x$; $10 + x - x = 6x - x$; $10 = 5x$; $2 = x$

15. $12 + 3y = 9y$; $12 + 3y - 3y = 9y - 3y$; $12 = 6y$; $2 = y$

16. $16 + 5a = 13a$; $16 + 5a - 5a = 13a - 5a$; $16 = 8a$; $2 = a$

17. $14 - 6y = 8y$; $14 - 6y + 6y = 8y + 6y$; $14 = 14y$; $1 = y$

18. $25 - 2n = 3n$; $25 - 2n + 2n = 3n + 2n$; $25 = 5n$; $5 = n$

19. $6y - 3 = 3y$; $6y - 3 + 3 = 3y + 3$; $6y = 3y + 3$; $6y - 3y = 3y + 3 - 3y$; $3y = 3$; $y = 1$

20. $4x - 2 = 3x$; $4x - 2 + 2 = 3x + 2$; $4x = 3x + 2$; $4x - 3x = 3x + 2 - 3x$; $x = 2$

21. $8n - 6 = 5n$; $8n - 6 + 6 = 5n + 6$; $8n = 5n + 6$; $8n - 5n = 5n + 6 - 5n$; $3n = 6$; $n = 2$

22. $5x - 9 = 2x$; $5x - 9 + 9 = 2x + 9$; $5x = 2x + 9$; $5x - 2x = 2x + 9 - 2x$; $3x = 9$; $x = 3$

23. $12n - 20 = 2n$; $12n - 20 + 20 = 2n + 20$; $12n = 2n + 20$; $12n - 2n =$ $2n + 20 - 2n$; $10n = 20$; $n = 2$

24. $15x - 12 = 12x$; $15x - 12 + 12 = 12x + 12$; $15x = 12x + 12$; $15x - 12x =$ $12x + 12 - 12x$; $3x = 12$; $x = 4$

25. $6x - 5 = 4x + 9$; $6x - 5 + 5 = 4x + 9 + 5$; $6x = 4x + 14$; $6x - 4x =$ $4x + 14 - 4x$; $2x = 14$; $x = 7$

26. $13x - 7 = 5x + 1$; $13x - 7 + 7 = 5x + 1 + 7$; $13x = 5x + 8$; $13x - 5x =$ $5x + 8 - 5x$; $8x = 8$; $x = 1$

27. $8a + 6 = 3a + 21$; $8a + 6 - 6 = 3a + 21 - 6$; $8a = 3a + 15$; $8a - 3a = 3a + 15 - 3a$; $5a = 15$; $a = 3$

28. $4x + 4 = 2x + 6$; $4x + 4 - 4 = 2x + 6 - 4$; $4x = 2x + 2$; $4x - 2x = 2x + 2 - 2x$; $2x = 2$; $x = 1$

29. $7a - 3 = 3a + 5$; $7a - 3 + 3 = 3a + 5 + 3$; $7a = 3a + 8$; $7a - 3a = 3a + 8 - 3a$; $4a = 8$; $a = 2$

30. $10n - 8 = 3n - 1$; $10n - 8 + 8 = 3n - 1 + 8$; $10n = 3n + 7$; $10n - 3n = 3n + 7 - 3n$; $7n = 7$; $n = 1$

31. $5b + 8 = 3b + 26$; $5b + 8 - 8 = 3b + 26 - 8$; $5b = 3b + 18$; $5b - 3b = 3b + 18 - 3b$; $2b = 18$; $b = 9$

32. $18 + 3x = x + 30$; $18 + 3x - 18 = x + 30 - 18$; $3x = x + 12$; $3x - x = x + 12 - x$; $2x = 12$; $x = 6$

33. $7 - 4x = 5x - 2$; $7 - 4x + 2 = 5x - 2 + 2$; $9 - 4x = 5x$; $9 - 4x + 4x = 5x + 4x$; $9 = 9x$; $1 = x$

34. $5y + 9 = 8y - 15$; $5y + 9 + 15 = 8y - 15 + 15$; $5y + 24 = 8y$; $5y + 24 - 5y = 8y - 5y$; $24 = 3y$; $8 = y$

35. $7c + 9 = 11c - 31$; $7c + 9 + 31 = 11c - 31 + 31$; $7c + 40 = 11c$; $7c + 40 - 7c = 11c - 7c$; $40 = 4c$; $10 = c$

36. $8a - 5 = 3a + 25$; $8a - 5 + 5 = 3a + 25 + 5$; $8a = 3a + 30$; $8a - 3a = 3a + 30 - 3a$; $5a = 30$; $a = 6$

B   **37.** $7n = 3n + 12$; $7n - 3n = 3n + 12 - 3n$; $4n = 12$; $n = 3$

38. $4n = n + 18$; $4n - n = n + 18 - n$; $3n = 18$; $n = 6$

39. $2n + 24 = 8n$; $2n + 24 - 2n = 8n - 2n$; $24 = 6n$; $4 = n$

40. $30 - n = 4n$; $30 - n + n = 4n + n$; $30 = 5n$; $6 = n$

41. $n + 10 = 6n$; $n + 10 - n = 6n - n$; $10 = 5n$; $2 = n$

42. $3n + 2 = n + 8$; $3n + 2 - n = n + 8 - n$; $2n + 2 = 8$; $2n + 2 - 2 = 8 - 2$; $2n = 6$; $n = 3$

Page 65 • WRITTEN EXERCISES

A   **1.** $3(2x - 5) = 6x - 15$

2. $5(6 - 2a) = 30 - 10a$

3. $8(2x - 5) = 16x - 40$

4. $7(3 - 5a) = 21 - 35a$

5. $6(8y + 4) = 48y + 24$

6. $7(4x + 12) = 28x + 84$

7. $8(3 + 6b) = 24 + 48b$

8. $9(2 + 5y) = 18 + 45y$

9. $2(x + 3) = 10$; $2x + 6 = 10$; $2x + 6 - 6 = 10 - 6$; $2x = 4$; $x = 2$

10. $3(2x - 1) = 9$; $6x - 3 = 9$; $6x - 3 + 3 = 9 + 3$; $6x = 12$; $x = 2$

11. $2(6 + 2a) = 24$; $12 + 4a = 24$; $4a = 12$; $a = 3$

12. $5(x - 2) = 0$; $5x - 10 = 0$; $5x = 10$; $x = 2$

13. $8(y - 4) = 0$; $8y - 32 = 0$; $8y = 32$; $y = 4$

14. $6(x + 1) = 36$; $6x + 6 = 36$; $6x = 30$; $x = 5$

15. $2(x - 5) = x$; $2x - 10 = x$; $2x - 10 + 10 = x + 10$; $2x = x + 10$; $2x - x = x + 10 - x$; $x = 10$

16. $3(x - 2) = 2x$; $3x - 6 = 2x$; $3x - 6 + 6 = 2x + 6$; $3x = 2x + 6$; $3x - 2x = 2x + 6 - 2x$; $x = 6$

17. $4(x - 1) = 2x$; $4x - 4 = 2x$; $4x - 4 + 4 = 2x + 4$; $4x = 2x + 4$; $4x - 2x = 2x + 4 - 2x$; $2x = 4$; $x = 2$

18. $3(y + 4) = 5y$; $3y + 12 = 5y$; $3y + 12 - 3y = 5y - 3y$; $12 = 2y$; $6 = y$

19. $6(y - 2) = 3y$; $6y - 12 = 3y$; $6y = 3y + 12$; $3y = 12$; $y = 4$

20. $2(3 - y) = 4y$; $6 - 2y = 4y$; $6 = 6y$; $1 = y$

21. $4(x - 3) = 0$; $4x - 12 = 0$; $4x = 12$; $x = 3$

22. $2(2x + 2) = 6x$; $4x + 4 = 6x$; $4 = 2x$; $2 = x$

23. $7(2 + y) = 14$; $14 + 7y = 14$; $7y = 0$; $y = 0$

24. $5(x - 3) = 2x + 3$; $5x - 15 = 2x + 3$; $5x - 15 + 15 = 2x + 3 + 15$; $5x = 2x + 18$; $5x - 2x = 2x + 18 - 2x$; $3x = 18$; $x = 6$

25. $5(x - 1) = 7 + x$; $5x - 5 = 7 + x$; $5x - 5 + 5 = 7 + x + 5$; $5x = x + 12$; $5x - x = x + 12 - x$; $4x = 12$; $x = 3$

26. $7(x + 1) = 9 + 5x$; $7x + 7 = 9 + 5x$; $7x + 7 - 7 = 9 + 5x - 7$; $7x = 5x + 2$; $7x - 5x = 5x + 2 - 5x$; $2x = 2$; $x = 1$

27. $4(y - 1) = 2y + 6$; $4y - 4 = 2y + 6$; $4y - 4 + 4 = 2y + 6 + 4$; $4y = 2y + 10$; $2y = 10$; $y = 5$

28. $3(x - 3) = x + 1$; $3x - 9 = x + 1$; $3x - 9 + 9 = x + 1 + 9$; $3x = x + 10$; $2x = 10$; $x = 5$

29. $5(y + 4) = 2y + 20$; $5y + 20 = 2y + 20$; $5y + 20 - 20 = 2y + 20 - 20$; $5y = 2y$; $5y - 2y = 0$; $3y = 0$; $y = 0$

30. $2(z + 2) = 3z + 2$; $2z + 4 = 3z + 2$; $2z + 4 - 2 = 3z + 2 - 2$; $2z + 2 = 3z$; $2 = z$

31. $4(m + 3) = 6m + 8$; $4m + 12 = 6m + 8$; $4m + 12 - 8 = 6m + 8 - 8$; $4m + 4 = 6m$; $4 = 2m$; $2 = m$

32. $3(x - 4) = 2x + 5$; $3x - 12 = 2x + 5$; $3x - 12 + 12 = 2x + 5 + 12$; $3x = 2x + 17$; $x = 17$

B   **33.** $6(x - 1) = 3(x + 1)$; $6x - 6 = 3x + 3$; $6x - 6 + 6 = 3x + 3 + 6$; $6x = 3x + 9$; $3x = 9$; $x = 3$.   Check: $6(3 - 1) = 3(3 + 1)$? $6 \cdot 2 = 3 \cdot 4$? $12 = 12\checkmark$

34. $2(x + 3) = 3(x - 3)$; $2x + 6 = 3x - 9$; $2x + 6 + 9 = 3x - 9 + 9$; $2x + 15 = 3x$; $15 = x$.   Check: $2(15 + 3) = 3(15 - 3)$? $2 \cdot 18 = 3 \cdot 12$? $36 = 36\checkmark$

35. $8(x - 1) = 4(x + 4)$; $8x - 8 = 4x + 16$; $8x - 8 + 8 = 4x + 16 + 8$; $8x = 4x + 24$; $4x = 24$; $x = 6$.   Check: $8(6 - 1) = 4(6 + 4)$? $8 \cdot 5 = 4 \cdot 10$? $40 = 40\checkmark$

36. $7(2a - 4) = 2(a + 4)$; $14a - 28 = 2a + 8$; $14a - 28 + 28 = 2a + 8 + 28$; $14a = 2a + 36$; $12a = 36$; $a = 3$.   Check: $7(2 \cdot 3 - 4) = 2(3 + 4)$? $7(6 - 4) = 2 \cdot 7$? $7 \cdot 2 = 14$? $14 = 14\surd$

37. $3(a - 2) + a = 2(a + 1)$; $3a - 6 + a = 2a + 2$; $4a - 6 = 2a + 2$; $4a - 6 + 6 = 2a + 2 + 6$; $4a = 2a + 8$; $2a = 8$; $a = 4$.   Check: $3(4 - 2) + 4 = 2(4 + 1)$? $3 \cdot 2 + 4 = 2 \cdot 5$? $6 + 4 = 10$? $10 = 10\surd$

38. $5(x + 2) = x + 6(x - 3)$; $5x + 10 = x + 6x - 18$; $5x + 10 = 7x - 18$; $5x + 10 + 18 = 7x - 18 + 18$; $5x + 28 = 7x$; $28 = 2x$; $14 = x$.   Check: $5(14 + 2) = 14 + 6(14 - 3)$? $5 \cdot 16 = 14 + 6 \cdot 11$? $80 = 14 + 66$? $80 = 80\surd$

39. $4(m + 3) - 2m = 3(m - 3)$; $4m + 12 - 2m = 3m - 9$; $2m + 12 = 3m - 9$; $2m + 12 + 9 = 3m - 9 + 9$; $2m + 21 = 3m$; $21 = m$.   Check: $4(21 + 3) - 2 \cdot 21 = 3(21 - 3)$? $4 \cdot 24 - 42 = 3 \cdot 18$? $96 - 42 = 54$? $54 = 54\surd$

40. $2(a + 4) = 2(a - 4) + 4a$; $2a + 8 = 2a - 8 + 4a$; $2a + 8 = 6a - 8$; $2a + 8 + 8 = 6a - 8 + 8$; $2a + 16 = 6a$; $16 = 4a$; $4 = a$.   Check: $2(4 + 4) = 2(4 - 4) + 4 \cdot 4$? $2 \cdot 8 = 2 \cdot 0 + 16$? $16 = 0 + 16$? $16 = 16\surd$

41. $7(b + 2) - 4b = 2(b + 10)$; $7b + 14 - 4b = 2b + 20$; $3b + 14 = 2b + 20$; $3b + 14 - 14 = 2b + 20 - 14$; $3b = 2b + 6$; $b = 6$.   Check: $7(6 + 2) - 4 \cdot 6 = 2(6 + 10)$? $7 \cdot 8 - 24 = 2 \cdot 16$? $56 - 24 = 32$? $32 = 32\surd$

42. $3(x + 2) = 3(x - 2) + 3x$; $3x + 6 = 3x - 6 + 3x$; $3x + 6 = 6x - 6$; $3x + 6 + 6 = 6x - 6 + 6$; $3x + 12 = 6x$; $12 = 3x$; $4 = x$.   Check: $3(4 + 2) = 3(4 - 2) + 3 \cdot 4$? $3 \cdot 6 = 3 \cdot 2 + 12$? $18 = 6 + 12$? $18 = 18\surd$

C **43.** $2(x + 5) = 3x + 4 - x$; $2x + 10 = 2x + 4$; $2x + 10 - 2x = 2x + 4 - 2x$; $10 = 4$.   a

44. $6(2x + 3) = 2(9 + 6x)$; $12x + 18 = 18 + 12x$; $12x = 12x$; $x = x$.   c

45. $7(x - 5) = 2(x + 5)$; $7x - 35 = 2x + 10$; $7x - 35 + 35 = 2x + 10 + 35$; $7x = 2x + 45$; $5x = 45$; $x = 9$.   b

46. $4(x + 1) = 4x + 1$; $4x + 4 = 4x + 1$; $4 = 1$.   a

47. $6(x + 3) = 2(3x + 1)$; $6x + 18 = 6x + 2$; $18 = 2$.   a

48. $2(8x + 6) = 4(4x + 3)$; $16x + 12 = 16x + 12$; $16x = 16x$; $x = x$.   c

49. $8(x + 1) = 4(x + 5)$; $8x + 8 = 4x + 20$; $8x + 8 - 8 = 4x + 20 - 8$; $8x = 4x + 12$; $4x = 12$; $x = 3$.   b

50. $5(2x + 3) = 10(x + 1)$; $10x + 15 = 10x + 1$; $15 = 1$.   a

51. $2(3x + 9) = 3(2x + 6)$; $6x + 18 = 6x + 18$; $6x = 6x$; $x = x$.   c

52. $4(x + 2) = 5x + 3$; $4x + 8 = 5x + 3$; $4x + 8 - 3 = 5x + 3 - 3$; $4x + 5 = 5x$; $5 = x$.   b

Pages 66–67 · WRITTEN EXERCISES

A **1.** Let x = the number of calories in a pear. Then $x + 20$ = the number of calories in a melon. $6x = 5(x + 20)$; $6x = 5x + 100$; $6x - 5x = 5x + 100 - 5x$; $x = 100$ and $x + 20 = 120$.   pear: 100 calories;   melon: 120 calories

2. Let x = the number of hours Marco works in a week. Then $x + 7$ = the hours Sonya works. $4x = 3(x + 7)$; $4x = 3x + 21$; $4x - 3x = 3x + 21 - 3x$; $x = 21$ and $x + 7 = 28$. Marco: 21 hours; Sonya: 28 hours

3. Let x = the smaller number. Then $x + 7$ = the larger number. $2(x + 7) = 4x - 22$; $2x + 14 = 4x - 22$; $2x + 14 + 22 = 4x - 22 + 22$; $2x + 36 = 4x$; $36 = 2x$; $18 = x$ and $x + 7 = 25$. 18 and 25

4. Let x = the larger number. Then $x - 5$ = the smaller number. $5(x - 5) = 3x - 1$; $5x - 25 = 3x - 1$; $5x - 25 + 25 = 3x - 1 + 25$; $5x = 3x + 24$; $2x = 24$; $x = 12$ and $x - 5 = 7$. 12 and 7

5. Let x = the amount Mona has. Then $x + 8$ = the amount Paul has and $2(x + 8)$ = the amount Tran has. $x + x + 8 + 2(x + 8) = 236$; $2x + 8 + 2x + 16 = 236$; $4x + 24 = 236$; $4x = 212$; $x = 53$, $x + 8 = 61$, and $2(x + 8) = 122$. Mona: \$53; Paul: \$61; Tran: \$122

6. Let x = Carla's age. Then $x + 3$ = Rich's age and $2(x + 3)$ = Ruth's age. $x + x + 3 + 2(x + 3) = 33$; $2x + 3 + 2x + 6 = 33$; $4x + 9 = 33$; $4x = 24$; $x = 6$, $x + 3 = 9$, and $2(x + 3) = 18$. Carla: 6; Rich: 9; Ruth: 18

B   7. Let x = Kent's age now. Then $x - 14$ = Harlin's age now, and in one year their ages will be $x + 1$ and $(x - 14) + 1$, or $x - 13$. $x + 1 = 3(x - 13)$; $x + 1 = 3x - 39$; $x + 1 + 39 = 3x - 39 + 39$; $x + 40 = 3x$; $40 = 2x$; $20 = x$. Kent is 20.

8. Let x = Peggy's age now. Then $x + 14$ = Beth's age now, and last year their ages were $x - 1$ and $(x + 14) - 1$, or $x + 13$. $x + 13 = 3(x - 1)$; $x + 13 = 3x - 3$; $x + 13 + 3 = 3x - 3 + 3$; $x + 16 = 3x$; $16 = 2x$; $8 = x$. Peggy is 8.

C   9. Let d = the daughter's age now. Then $d + 3$ = the son's age now, and $6d - 1$ = the man's age now. $(6d - 1) + 10 = (d + 10) + (d + 13) + 14$; $6d + 9 = 2d + 37$; $6d = 2d + 28$; $4d = 28$; $d = 7$, so $6d - 1 = 6 \cdot 7 - 1 = 42 - 1 = 41$. The man is 41.

Page 69 • PROBLEM SOLVING STRATEGIES

1. **a.** Let m = the number of minutes for a one-way bike trip; $2m = 40$; $m = 20$. 20 min

   b. 20 min; $25 - 20 = 5$. 5 min

   c. 5 min; $2 \cdot 5 = 10$. 10 min

2. **a.** $2p$; $20p$; $12 \cdot 2p = 24p$

   b. $20p + 24p = 44$

   c. $20p + 24p = 44$; $44p = 44$; $p = 1$ and $2p = 2$. \$2

3. **a.** $2p$; $2p + 4$

   b. number of oranges = 20; $2p + 4 = 20$

   c. $2p + 4 = 20$; $2p = 16$; $p = 8$ and $2p = 16$. 8 pears and 16 apples

   d. $8 + 16 + 20 = 44$. 44 pieces of fruit in all

Page 70 • SKILLS REVIEW

1. $<$ 2. $>$ 3. $>$ 4. $<$ 5. $>$ 6. $>$

7. $>$ 8. $<$ 9. $<$ 10. $>$ 11. $<$ 12. $>$

13.–20.

21. $0, \frac{1}{2}, 2$ 22. $18, 23, 32$ 23. $\frac{1}{7}, 7, 17$ 24. $5, 15, 51$

25. $\frac{2}{5}, 1, \frac{6}{5}$ 26. $100, 107, 170$ 27. $\frac{1}{4}, \frac{1}{2}, \frac{3}{4}$

28. $44, 45, 54$ 29. $\frac{5}{6}, \frac{5}{3}, \frac{5}{1}$ 30. $\frac{9}{10}, 1, \frac{7}{5}$

31. $2\frac{1}{5}, 2\frac{1}{4}, 2\frac{1}{3}$ 32. $2^4, 5^2, 3^3$

Pages 71–72 • CHAPTER REVIEW EXERCISES

1. $a - 5 = 13$; $a - 5 + 5 = 13 + 5$; $a = 18$
2. $x + 9 = 24$; $x + 9 - 9 = 24 - 9$; $x = 15$
3. $3b = 24$; $\frac{3b}{3} = \frac{24}{3}$; $b = 8$ 4. $5m = 35$; $m = 7$
5. $2x - 7 = 9$; $2x - 7 + 7 = 9 + 7$; $2x = 16$; $x = 8$
6. $3y + 4 = 19$; $3y = 15$; $y = 5$ 7. $\frac{b}{2} = 7$; $2 \cdot \frac{b}{2} = 2 \cdot 7$; $b = 14$
8. $\frac{x}{3} = 13$; $x = 39$ 9. $\frac{m}{4} = 7$; $m = 28$
10. $5a + 9 = 29$; $5a = 20$; $a = 4$ 11. $3x - 11 = 52$; $3x = 63$; $x = 21$
12. $8y - 15 = 41$; $8y = 56$; $y = 7$ 13. $12a - 7 = 17$; $12a = 24$; $a = 2$
14. $19 = x + 7$; $19 - 7 = x + 7 - 7$; $12 = x$
15. $25 = b - 9$; $25 + 9 = b - 9 + 9$; $34 = b$
16. $3n + 4 = 40$; $3n = 36$; $n = 12$ 17. $6x - 9 = 21$; $6x = 30$; $x = 5$
18. $4s - 11 = 41$; $4s = 52$; $s = 13$ 19. $2n + 7 = 25$; $2n = 18$; $n = 9$
20. $9y - 3y = 42$; $6y = 42$; $y = 7$
21. $4x - 2x + 5x = 14$; $7x = 14$; $x = 2$
22. $15y - 12 + 3y = 24$; $18y - 12 = 24$; $18y - 12 + 12 = 24 + 12$; $18y = 36$;
    $y = 2$
23. $20 = 5z - 8 - z$; $20 = 4z - 8$; $20 + 8 = 4z - 8 + 8$; $28 = 4z$; $7 = z$
24. $x + 5$ 25. $a - 3$
26. Let x = the amount of money Phil has. Then $3x$ = the amount Marta has. $x + 3x = 44$; $4x = 44$; $x = 11$, and $3x = 33$.   Phil: \$11;   Marta: \$33
27. Let x = the number of games lost. Then $2x$ = the number won. $x + 2x = 48$; $3x = 48$; $x = 16$, so $2x = 32$.   32 games won
28. Let x = the number of girls. Then $x - 20$ = the number of boys. $x + x - 20 = 900$; $2x - 20 = 900$; $2x = 920$; $x = 460$.   460 girls
29. Let x = the length (in cm) of the shorter piece. Then $x + 66$ = the length of the longer piece. $x + x + 66 = 400$; $2x + 66 = 400$; $2x = 334$; $x = 167$.   167 cm
30. Let x = the amount Bert has. Then $2x$ = the amount Cindy has and $2x + 8$ = the amount Mona has. $x + 2x + 2x + 8 = 63$; $5x + 8 = 63$; $5x = 55$; $x = 11$, $2x = 22$, and $2x + 8 = 30$.   Bert: \$11;   Cindy: \$22;   Mona: \$30
31. Let n = the number. $3(n + 7) = 36$; $3n + 21 = 36$; $3n = 15$; $n = 5$.

32. Let n = the number. $6(n - 8) = 24$; $6n - 48 = 24$; $6n = 72$; $n = 12$.

33. $8x = 25 + 3x$; $8x - 3x = 25 + 3x - 3x$; $5x = 25$; $x = 5$.   Check: $8 \cdot 5 = 25 + 3 \cdot 5$? $40 = 25 + 15$? $40 = 40\checkmark$

34. $9t = 40 - t$; $9t + t = 40 - t + t$; $10t = 40$; $t = 4$.   Check: $9 \cdot 4 = 40 - 4$? $36 = 36\checkmark$

35. $7x - 10 = 2x$; $7x - 10 + 10 = 2x + 10$; $7x = 2x + 10$; $7x - 2x = 2x + 10 - 2x$; $5x = 10$; $x = 2$.   Check: $7 \cdot 2 - 10 = 2 \cdot 2$? $14 - 10 = 4$? $4 = 4\checkmark$

36. $3y + 12 = 7y$; $3y + 12 - 3y = 7y - 3y$; $12 = 4y$; $3 = y$.   Check: $3 \cdot 3 + 12 = 7 \cdot 3$? $9 + 12 = 21$? $21 = 21\checkmark$

37. $11a - 6 = 5a + 24$; $11a - 6 + 6 = 5a + 24 + 6$; $11a = 5a + 30$; $11a - 5a = 5a + 30 - 5a$; $6a = 30$; $a = 5$.   Check: $11 \cdot 5 - 6 = 5 \cdot 5 + 24$? $55 - 6 = 25 + 24$? $49 = 49\checkmark$

38. $3(a - 4) = a$; $3a - 12 = a$; $3a - 12 + 12 = a + 12$; $3a = a + 12$; $2a = 12$; $a = 6$.   Check: $3(6 - 4) = 6$? $3 \cdot 2 = 6$? $6 = 6\checkmark$

39. $4(m - 4) = 2m - 2$; $4m - 16 = 2m - 2$; $4m - 16 + 16 = 2m - 2 + 16$; $4m = 2m + 14$; $2m = 14$; $m = 7$.   Check: $4(7 - 4) = 2 \cdot 7 - 2$? $4 \cdot 3 = 14 - 2$? $12 = 12\checkmark$

40. $5m + 7 = 3(m + 5)$; $5m + 7 = 3m + 15$; $5m + 7 - 7 = 3m + 15 - 7$; $5m = 3m + 8$; $2m = 8$; $m = 4$.   Check: $5 \cdot 4 + 7 = 3(4 + 5)$? $20 + 7 = 3 \cdot 9$; $27 = 27\checkmark$

41. $x - 3 = 4(3 - x)$; $x - 3 = 12 - 4x$; $x - 3 + 3 = 12 - 4x + 3$; $x = 15 - 4x$; $5x = 15$; $x = 3$.   Check: $3 - 3 = 4(3 - 3)$? $0 = 4 \cdot 0$? $0 = 0\checkmark$

42. $6(y + 2) = 5y + 18$; $6y + 12 = 5y + 18$; $6y + 12 - 12 = 5y + 18 - 12$; $6y = 5y + 6$; $y = 6$.   Check: $6(6 + 2) = 5 \cdot 6 + 18$? $6 \cdot 8 = 30 + 18$? $48 = 48\checkmark$

Page 73 · CHAPTER TEST

1. $n - 5 = 5$; $n - 5 + 5 = 5 + 5$; $n = 10$ **2.** $0 = x - 12$; $0 + 12 = x - 12 + 12$; $12 = x$ **3.** $y - 7 = 2$; $y - 7 + 7 = 2 + 7$; $y = 9$ **4.** $n + 10 = 18$; $n + 10 - 10 = 18 - 10$; $n = 8$ **5.** $3 + a = 9$; $3 + a - 3 = 9 - 3$; $a = 6$ **6.** $20 = x + 8$; $20 - 8 = x + 8 - 8$; $12 = x$ **7.** $6m = 42$; $\dfrac{6m}{6} = \dfrac{42}{6}$; $m = 7$ **8.** $5x = 85$; $\dfrac{5x}{5} = \dfrac{85}{5}$; $x = 17$ **9.** $70b = 210$; $\dfrac{70b}{70} = \dfrac{210}{70}$; $b = 3$ **10.** $\dfrac{x}{2} = 8$; $2 \cdot \dfrac{x}{2} = 2 \cdot 8$; $x = 16$ **11.** $\dfrac{m}{12} = 3$; $12 \cdot \dfrac{m}{12} = 12 \cdot 3$; $m = 36$ **12.** $\dfrac{y}{5} = 5$; $5 \cdot \dfrac{y}{5} = 5 \cdot 5$; $y = 25$ **13.** $4x - 1 = 19$; $4x - 1 + 1 = 19 + 1$; $4x = 20$; $x = 5$ **14.** $7c + 8 = 36$; $7c + 8 - 8 = 36 - 8$; $7c = 28$; $c = 4$ **15.** $9t + 2 = 11$; $9t + 2 - 2 = 11 - 2$; $9t = 9$; $t = 1$

16.

Arith.	Algebra
10	n
20	$2n$
32	$2n + 12$, or $2(n + 6)$
16	$n + 6$
11	$n + 1$

17. $13 = 3a + 7 - a$; $13 = 2a + 7$; $13 - 7 = 2a + 7 - 7$; $6 = 2a$; $3 = a$.
Check: $13 = 3 \cdot 3 + 7 - 3$? $13 = 9 + 7 - 3$? $13 = 13\sqrt{}$

18. $50 = 5t - 2t + 7t$; $50 = 10t$; $5 = t$.   Check: $50 = 5 \cdot 5 - 2 \cdot 5 + 7 \cdot 5$? $50 = 25 - 10 + 35$? $50 = 50\sqrt{}$

19. $4p$ **20.** $m + 4$; $(m + 4) + 1$, or $m + 5$

21. Let $x =$ the number of calories in a stalk of celery. Then $x + 25 =$ the number of calories in a carrot. $x + x + 25 = 35$; $2x + 25 = 35$; $2x = 10$; $x = 5$ and $x + 25 = 30$.   celery: 5 calories;   carrot: 30 calories

22. $12 - x = 2x$; $12 - x + x = 2x + x$; $12 = 3x$; $4 = x$

23. $9n - 7 = 5n + 17$; $9n - 7 + 7 = 5n + 17 + 7$; $9n = 5n + 24$; $9n - 5n = 5n + 24 - 5n$; $4n = 24$; $n = 6$

24. $40 + a = 8a + 5$; $40 + a - 5 = 8a + 5 - 5$; $35 + a = 8a$; $35 + a - a = 8a - a$; $35 = 7a$; $5 = a$

25. $4(x - 3) = 3x$; $4x - 12 = 3x$; $4x - 12 + 12 = 3x + 12$; $4x = 3x + 12$; $4x - 3x = 3x + 12 - 3x$; $x = 12$

26. $6(y + 5) = 30$; $6y + 30 = 30$; $6y + 30 - 30 = 30 - 30$; $6y = 0$; $y = 0$

27. $8(x + 1) = x + 15$; $8x + 8 = x + 15$; $8x + 8 - 8 = x + 15 - 8$; $8x = x + 7$; $8x - x = x + 7 - x$; $7x = 7$; $x = 1$

28. Let $x =$ the amount Jim has. Then $3x =$ the amount Rita has and $3x + 4 =$ the amount Pedro has. $x + 3x + 3x + 4 = 46$; $7x + 4 = 46$; $7x + 4 - 4 = 46 - 4$; $7x = 42$; $x = 6$, $3x = 18$, and $3x + 4 = 22$.   Jim: \$6;   Rita: \$18;   Pedro: \$22

Page 74 • CUMULATIVE REVIEW

1. $5 \cdot 2 = 10$ **2.** $\dfrac{2 - 2}{7} = \dfrac{0}{7} = 0$ **3.** $6 \div 2 = 3$

4. $\dfrac{2 + 5}{7} = \dfrac{7}{7} = 1$ **5.** $2 + 8 \div 2 = 2 + 4 = 6$ **6.** $(6 + 4) \div 5 = 10 \div 5 = 2$

7. $5^3 = 5 \times 5 \times 5 = 125$ **8.** $(25 \cdot 73) \cdot 4 = (25 \cdot 4) \cdot 73 = 100 \cdot 73 = 7300$

9. $5t + 12t = 17t$ **10.** $7n + 2 - 4n = (7n - 4n) + 2 = 3n + 2$

11. $4x + 8y + 2x - 3y = (4x + 2x) + (8y - 3y) = 6x + 5y$

12. $6 \cdot a \cdot a \cdot b \cdot b \cdot b = 6a^2b^3$ **13.** $r \cdot 3 \cdot r \cdot 2 = (3 \cdot 2) \cdot (r \cdot r) = 6r^2$

14. $(3t)(8y) = (3 \cdot 8) \cdot t \cdot y = 24ty$ **15.** $y(3x - 4y) = 3xy - 4y^2$

16. $8(a - 2) + 4a = 8a - 16 + 4a = 12a - 16$

17. $5 + 3y - 5 - y = (5 - 5) + (3y - y) = 0 + 2y = 2y$

18. If $t = 6$: $6 - 5 = 11$? $1 - 11$?   No.

If $t = 16$: $16 - 5 = 11$? $11 = 11$?   Yes.   16

19. If $x = 5$: $3(5 - 4) = 5 + 2$? $3 \cdot 1 = 7$? $3 = 7$?   No.

If $x = 6$: $3(6 - 4) = 6 + 2$? $3 \cdot 2 = 8$? $6 = 8$?   No.

If $x = 7$: $3(7 - 4) = 7 + 2$? $3 \cdot 3 = 9$? $9 = 9$?   Yes.   7

20. If $y = 10$: $10 - 3 < 8$? $7 < 8$?   Yes.

If $y = 11$: $11 - 3 < 8$? $8 < 8$?   No.

If $y = 12$: $12 - 3 < 8$? $9 < 8$?   No.   10

21. If $n = 0$: $2 \cdot 0 + 5 > 8$? $0 + 5 > 8$? $5 > 8$?   No.

If $n = 1$: $2 \cdot 1 + 5 > 8$? $2 + 5 > 8$? $7 > 8$?   No.

If $n = 2$: $2 \cdot 2 + 5 > 8$? $4 + 5 > 8$? $9 > 8$?   Yes.   2

22. $y - 3 = 8$; $y - 3 + 3 = 8 + 3$; $y = 11$ **23.** $n + 5 = 24$; $n + 5 - 5 = 24 - 5$; $n = 19$ **24.** $3x = 81$; $\dfrac{3x}{3} = \dfrac{81}{3}$; $x = 27$ **25.** $\dfrac{b}{7} = 21$; $7 \cdot \dfrac{b}{7} = 7 \cdot 21$; $b = 147$ **26.** $5y + 3 = 18$; $5y + 3 - 3 = 18 - 3$; $5y = 15$; $y = 3$ **27.** $7t - 2 = 19$; $7t - 2 + 2 = 19 + 2$; $7t = 21$; $t = 3$ **28.** $5n + 8n - 2 = 50$; $13n - 2 = 50$; $13n - 2 + 2 = 50 + 2$; $13n = 52$; $n = 4$ **29.** $7m - 4 = 3m$; $7m - 4 + 4 = 3m + 4$; $7m = 3m + 4$; $7m - 3m = 3m + 4 - 3m$; $4m = 4$; $m = 1$ **30.** $3(z + 2) = 8z + 1$; $3z + 6 = 8z + 1$; $3z + 6 - 1 = 8z + 1 - 1$; $3z + 5 = 8z$; $3z + 5 - 3z = 8z - 3z$; $5 = 5z$; $1 = z$ **31.** $4n - 17 = 11$; $4n - 17 + 17 = 11 + 17$; $4n = 28$; $n = 7$ **32.** $8n + 5 = 101$; $8n + 5 - 5 = 101 - 5$; $8n = 96$; $n = 12$

33. Let x = the amount Jon has. Then $x + 18$ = the amount Miyoshi has.
$x + (x + 18) = 56$; $2x + 18 = 56$; $2x + 18 - 18 = 56 - 18$; $2x = 38$;
$x = 19$ and $x + 18 = 37$.   Jon: \$19;   Miyoshi: \$37

34. Let x = Lester's age now. Then $x + 8$ = Lea's age now, and in 4 years their ages
will be $x + 4$ and $x + 8 + 4$, or $x + 12$. $x + 12 = 2(x + 4)$; $x + 12 = 2x + 8$;
$x + 12 - 8 = 2x + 8 - 8$; $x + 4 = 2x$; $4 = x$ and $x + 8 = 12$.   Lester is 4 and
Lea is 12.

Page 79 · WRITTEN EXERCISES

A　1. +25　2. +20　3. −10　4. +5　5. −200　6. −7　7. <

　　8. <　9. >　10. >　11. >　12. <　13. <　14. >　15. <

　　16. <　17. >　18. <　19. −274

B　20. 30° + 30° = 60°　　　　　　　　21. Answers will vary.

Pages 81–82 · WRITTEN EXERCISES

A　1. $-2 < 3$ or $3 > -2$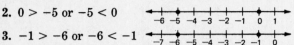

　　2. $0 > -5$ or $-5 < 0$

　　3. $-1 > -6$ or $-6 < -1$

　　4. $7 > -2$ or $-2 < 7$

　　5. $5 > -4$ or $-4 < 5$

　　6. $-7 < -1$ or $-1 > -7$

　　7. $-1 > -7$ or $-7 < -1$

　　8. $0 > -4$ or $-4 < 0$

　　9. $8 > -8$ or $-8 < 8$

　　10. $-5 < -2$ or $-2 > -5$

　　11. $x + 1 = 7$;　$x = 6$

　　12. $y - 2 = 5$;　$y = 7$

　　13. $x + 6 = 9$;　$x = 3$

　　14. $5 + x = 10$;　$x = 5$

　　15. $y - 6 = 2$;　$y = 8$

　　16. $a + 3 = 12$;　$a = 9$

　　17. $x - 6 = 4$;　$x = 10$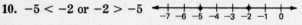

　　18. $y - 2 = 7$;　$y = 9$

19.

20.

21.

22.

23.

24.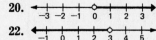

25.

26.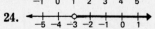

27.

28.

29.

30.

B   **31.** $-4 < 0 < 4$ **32.** $-6 < -1 < 2$ **33.** $-5 < -3 < 0$

34. $-6 < -5 < 3$ **35.** $-4 < -2 < 6$ **36.** $-6 < -2 < -1$

C   **37.**    **38.**

39.   **40.**

41.   **42.**

43.   **44.**

Page 82 • PUZZLE PROBLEMS

Answers may vary.

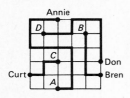

Page 84 • WRITTEN EXERCISES

A   **1.** $-2 + (-3) = -5$ **2.** $-7 + (-4) = -11$ **3.** $-4 + 3 = -1$

4. $2 + (-4) = -2$ **5.** $-6 + (-2) = -8$ **6.** $2 + (-2) = 0$

7. $-8 + 0 = -8$ **8.** $-1 + (-2) = -3$ **9.** $6 + (-3) = 3$

10. $-5 + (-4) = -9$ **11.** $-4 + 6 = 2$ **12.** $3 + (-5) = -2$

13. $-2 + (-4) = -6$ **14.** $0 + (-9) = -9$ **15.** $-7 + 4 = -3$

16. $-3 + 7 = 4$ **17.** $-5 + (-1) = -6$ **18.** $-6 + (-4) = -10$

19. $-8 + 8 = 0$ **20.** $-2 + 2 = 0$ **21.** $4 + (-4) = 0$

22. $6 + (-6) = 0$ **23.** $3 + (-9) = -6$ **24.** $-5 + (-8) = -13$

25. $-7 + (-2) = -9$ **26.** $-5 + (-12) = -17$ **27.** $6 + (-11) = -5$

28. $12 + (-6) = 6$ **29.** $5 + (-2) + (-6) = 3 + (-6) = -3$

30. $-10 + 2 + (-5) = -8 + (-5) = -13$ **31.** $1 + (-5) + 3 = -4 + 3 = -1$

32. $-6 + 6 + (-2) = 0 + (-2) = -2$ **33.** $9 + (-6) + 7 = 3 + 7 = 10$

34. $12 + (-12) + 0 = 0 + 0 = 0$

Page 87 • WRITTEN EXERCISES

A   **1.** $-10;\ 5$ **2.** $-15;\ -5$ **3.** $-7;\ -23$

4. $-7;\ -10$ **5.** $2;\ 6$ **6.** $1;\ 7$

7. $2 - 3 = 2 + (-3) = -1$ **8.** $1 - 5 = 1 + (-5) = -4$

9. $6 - 9 = 6 + (-9) = -3$ **10.** $8 - 10 = 8 + (-10) = -2$

11. $-3 - 4 = -3 + (-4) = -7$ **12.** $-5 - 2 = -5 + (-2) = -7$

13. $-7 - 6 = -7 + (-6) = -13$ **14.** $-3 - 7 = -3 + (-7) = -10$

15. $-3 - 9 = -3 + (-9) = -12$ **16.** $-1 - 6 = -1 + (-6) = -7$

000347

17. $6 - (-2) = 6 + (2) = 8$

18. $7 - (-1) = 7 + (1) = 8$

19. $3 - (-5) = 3 + (5) = 8$

20. $8 - (-3) = 8 + (3) = 11$

21. $5 - (-8) = 5 + (8) = 13$

22. $9 - (-4) = 9 + (4) = 13$

23. $-2 - (-1) = -2 + (1) = -1$

24. $-7 - (-3) = -7 + (3) = -4$

25. $-10 - (-4) = -10 + (4) = -6$

26. $-3 - (-2) = -3 + (2) = -1$

27. $-6 - (-5) = -6 + (5) = -1$

28. $-2 - (-5) = -2 + (5) = 3$

29. $-8 - (-6) = -8 + (6) = -2$

30. $-10 - (-1) = -10 + (1) = -9$

31. $5 - (-3) = 5 + (3) = 8$

32. $8 - (-4) = 8 + (4) = 12$

33. $7 - 10 = 7 + (-10) = -3$

34. $-4 - 6 = -4 + (-6) = -10$

35. $-2 - 5 = -2 + (-5) = -7$

36. $3 - 9 = 3 + (-9) = -6$

37. $-9 - (-4) = -9 + (4) = -5$

38. $6 - (-5) = 6 + (5) = 11$

39. $30 - (-4) = 30 + 4 = 34; \quad 34°$

40. $16 - (-34) = 16 + 34 = 50; \quad 50°$

41. $7 + 3 = 10; \quad 10 - 16 = 10 + (-16) = -6$

42. $-4 + 9 = 5; \quad 5 - 10 = 5 + (-10) = -5$

43. $-7 + (-17) = -24; \quad -24 - (-77) = -24 + (77) = 53$

44. $35 + (-50) = -15; \quad -15 - (-3) = -15 + 3 = -12$

45. $n = -18 - 12 = -18 + (-12) = -30$

46. $n = 10 - 100 = 10 + (-100) = -90$

Page 88 • MIXED PRACTICE EXERCISES

1. $2 + (-3) = -1$

2. $-5 + (-4) = -9$

3. $2 - 7 = 2 + (-7) = -5$

4. $-3 - (-2) = -3 + (2) = -1$

5. $8 - (-1) = 8 + (1) = 9$

6. $-7 + (-2) = -9$

7. $-3 - 4 = -3 + (-4) = -7$

8. $6 - (-4) = 6 + (4) = 10$

9. $8 + (-4) = 4$

10. $-2 + (-5) = -7$

11. $6 - (-2) = 6 + (2) = 8$

12. $-5 + (-5) = -10$

13. $-2 + 2 = 0$

14. $-8 - (-6) = -8 + (6) = -2$

15. $4 - 10 = 4 + (-10) = -6$

16. $6 + (-2) = 4$

17. $-5 + 4 = -1$

18. $-7 + 0 = -7$

19. $-1 + 1 = 0$

20. $2 - (-3) = 2 + (3) = 5$

21. $8 - (-7) = 8 + (7) = 15$

22. $-10 - (2) = -10 + (-2) = -12$

23. $-7 + (-2) = -9$

24. $-3 - (-4) = -3 + (4) = 1$

25. $5 - (-6) = 5 + (6) = 11$

26. $-12 - (-3) = -12 + (3) = -9$

27. $8 - 12 = 8 + (-12) = -4$

28. $7 - (-1) = 7 + (1) = 8$

29. $9 - (-2) = 9 + (2) = 11$

30. $8 + (-6) = 2$

31. $4 - (-6) = 4 + (6) = 10$

32. $-12 - (-9) = -12 + (9) = -3$

Page 90 • WRITTEN EXERCISES

A

1. $x + 3x = (1 + 3)x = 4x$

2. $6x + 7x = (6 + 7)x = 13x$

3. $3x + 7x = (3 + 7)x = 10x$

4. $x - 3x = (1 - 3)x = -2x$

5. $6y - 7y = (6 - 7)y = -y$ 6. $a - 4a = (1 - 4)a = -3a$

7. $-x^2 + 3x^2 = (-1 + 3)x^2 = 2x^2$ 8. $-b + 5b = (-1 + 5)b = 4b$

9. $-2c^2 + 8c^2 = (-2 + 8)c^2 = 6c^2$ 10. $-6y + 3y = (-6 + 3)y = -3y$

11. $a - 2a = (1 - 2)a = -a$ 12. $7c^2 - 9c^2 = (7 - 9)c^2 = -2c^2$

13. $-2x + 8x = (-2 + 8)x = 6x$ 14. $-5b - 6b = (-5 - 6)b = -11b$

15. $-8a - 2a = (-8 - 2)a = -10a$ 16. $-x + 5x = (-1 + 5)x = 4x$

17. $2n^2 - 5n^2 = (2 - 5)n^2 = -3n^2$ 18. $-10x + 4x = (-10 + 4)x = -6x$

19. $y - 10y + 10 = (1 - 10)y + 10 = -9y + 10$

20. $x + 3 - 3x = (1 - 3)x + 3 = -2x + 3$

21. $-3y + 3y + 1 = (-3 + 3)y + 1 = 0 + 1 = 1$

22. $-7x - 8x - 2 = (-7 - 8)x - 2 = -15x - 2$

23. $-6x - (-7x) = -6x + 7x = (-6 + 7)x = x$

24. $10x - (-7x) = 10x + 7x = 17x$

25. $-4x - (-5x) = -4x + 5x = (-4 + 5)x = x$

26. $2a - (-a) + 4 = 2a + a + 4 = 3a + 4$

27. $x^2 - 3x^2 - (-y) = (1 - 3)x^2 + y = -2x^2 + y$

28. $y - 2x^2 - (-3y) = y - 2x^2 + 3y = 4y - 2x^2$

29. $4x - (-2x) + x = 4x + 2x + x = 7x$

30. $y - 5y - (-x^2) = (1 - 5)y + x^2 = -4y + x^2$

B   31. $10 - r - 2r = 10 - 3r; \quad 10 - 3r - 4r = 10 - 7r$

32. $12 + 2x - 5x = 12 - 3x; \quad 12 - 3x - 3x = 12 - 6x$

33. $6n - 2 - n = 5n - 2; \quad (5n - 2)$ kilometers

34. $2y - 8 - y = y - 8; \quad (y - 8)$ dollars

Page 92 · WRITTEN EXERCISES

A   1. $4 \cdot 8 = 32$ 2. $-2(100) = -200$ 3. $-3(-40) = 120$

4. $6 \cdot 8 = 48$ 5. $-2 \cdot 7 = -14$ 6. $-16 \cdot 0 = 0$

7. $-4(-5) = 20$ 8. $5(-2) = -10$ 9. $-6(4) = -24$

10. $-9(-1) = 9$ 11. $6(-1) = -6$ 12. $-7(-3) = 21$

13. $5 \cdot 6 = 30$ 14. $-8(3) = -24$ 15. $-1(7) = -7$

16. $-2(4) = -8$ 17. $3(-4) = -12$ 18. $(0)(-2) = 0$

19. $4(-1) = -4$ 20. $8(-6) = -48$ 21. $-1(-4) = 4$

22. $2(-10) = -20$ 23. $7(-4) = -28$ 24. $-3 \cdot 9 = -27$

25. $8(-2) = -16$ 26. $-7(-7) = 49$ 27. $5(-5) = -25$

28. $9(-2) = -18$ 29. $(-5)^2 = (-5)(-5) = 25$

30. $6^2 = 6 \cdot 6 = 36$ 31. $(-6)^2 = (-6)(-6) = 36$

32. $4^2 = 4 \cdot 4 = 16$ 33. $(-4)^2 = (-4)(-4) = 16$

34. $-(-1)^2 = -(-1 \cdot -1) = -1$ 35. $-(10)^2 = -(10 \cdot 10) = -100$

36. $-(7)^2 = -(7 \cdot 7) = -49$

37. $-(-5)^2 = -(-5 \cdot -5) = -25$

38. $-(-3)^2 = -(-3 \cdot -3) = -9$

39. $(-10)^2 = (-10)(-10) = 100$

40. $(-8)^2 = (-8)(-8) = 64$

41. $-(6)^2 = -(6 \cdot 6) = -36$

42. $-(-8)^2 = -(-8 \cdot -8) = -64$

43. $-(-10)^2 = -(-10 \cdot -10) = -100$

B   **44.** $3(-10) + 3(6) = -30 + 18 = -12$

45. $-1(6) + (-1)(-6) = -6 + 6 = 0$

46. $-7(-1) + (-4)(-1) = 7 + 4 = 11$

47. $8(-7) - 4(-2) = -56 - (-8) = -56 + 8 = -48$

48. $3(-2) - 7(-4) = -6 - (-28) = -6 + 28 = 22$

49. $(4)(-2)(-3) + (-1)(-1) = (-8)(-3) + 1 = 24 + 1 = 25$

50. $(-3)^3 = [(-3)(-3)] \cdot (-3) = 9(-3) = -27$

51. $(-2)^4 = [(-2)(-2)] \cdot [(-2)(-2)] = 4 \cdot 4 = 16$

52. $(-1)^5 = [(-1)(-1)] \cdot [(-1)(-1)] \cdot (-1) = 1 \cdot 1 \cdot (-1) = -1$

Page 94 • WRITTEN EXERCISES

A   **1.** $-1(-2x) = 2x$ **2.** $-4(-3x) = 12x$ **3.** $2(-3x) = -6x$

4. $-1(7y) = -7y$ **5.** $-6a \cdot 0 = 0$ **6.** $-4a(2) = -8a$

7. $-4(3x) = -12x$ **8.** $-5(2x) = -10x$

9. $3a(-2b) = 3(-2)(a)(b) = -6ab$ **10.** $-a(4b) = (-1)(4)(a)(b) = -4ab$

11. $-3x(-2y) = (-3)(-2)(x)(y) = 6xy$ **12.** $-7y(-6z) = (-7)(-6)(y)(z) = 42yz$

13. $4x(-2y) = 4(-2)(xy) = -8xy$ **14.** $-5x(-3y) = (-5)(-3)(xy) = 15xy$

15. $7a(-2b) = 7(-2)(ab) = -14ab$ **16.** $-4r(-5s) = (-4)(-5)(rs) = 20rs$

17. $4(9 - x) = (4 \cdot 9) + (4 \cdot -x) = 36 - 4x$

18. $-4(9 - x) = (-4 \cdot 9) + (-4 \cdot -x) = -36 + 4x$

19. $2(a - b) = (2 \cdot a) + (2 \cdot -b) = 2a - 2b$

20. $-2(a - b) = (-2 \cdot a) + (-2 \cdot -b) = -2a + 2b$

21. $-x(6 - x) = (-x \cdot 6) + (-x \cdot -x) = -6x + x^2$

22. $-y(y - 8) = (-y \cdot y) + (-y \cdot -8) = -y^2 + 8y$

23. $(a + b)(-3) = (a \cdot -3) + (b \cdot -3) = -3a - 3b$

24. $(c + d)(-2) = (c \cdot -2) + (d \cdot -2) = -2c - 2d$

25. $-(a + 4) = -1(a + 4) = (-1 \cdot a) + (-1 \cdot 4) = -a - 4$

26. $-(x + 6) = -1(x + 6) = (-1 \cdot x) + (-1 \cdot 6) = -x - 6$

27. $-(y - 1) = -1(y - 1) = (-1 \cdot y) + (-1 \cdot -1) = -y + 1$

28. $-(x - 3) = -1(x - 3) = (-1 \cdot x) + (-1 \cdot -3) = -x + 3$

29. $-(-x + 5) = -1(-x + 5) = (-1 \cdot -x) + (-1 \cdot 5) = x - 5$

30. $-(4 - y) = -1(4 - y) = (-1 \cdot 4) + (-1 \cdot -y) = -4 + y$

31. $-(7 - a) = -1(7 - a) = (-1 \cdot 7) + (-1 \cdot -a) = -7 + a$

32. $-(x - y) = -1(x - y) = (-1 \cdot x) + (-1 \cdot -y) = -x + y$

33. $x - (x + 3) = x + (-1)(x + 3) = x + (-1 \cdot x) + (-1 \cdot 3) = x - x - 3 = -3$

34. $y - (y - 1) = y + (-1)(y - 1) = y + (-1 \cdot y) + (-1 \cdot -1) = y - y + 1 = 1$

35. $2x - (x + 4) = 2x + (-1)(x + 4) = 2x + (-1 \cdot x) + (-1 \cdot 4) = 2x - x - 4 = x - 4$

36. $3a - (a - 5) = 3a + (-1)(a - 5) = 3a + (-1 \cdot a) + (-1 \cdot -5) = 3a - a + 5 = 2a + 5$

37. $8a - (a - 4) = 8a + (-1)(a - 4) = 8a + (-1 \cdot a) + (-1 \cdot -4) = 8a - a + 4 = 7a + 4$

38. $2a - (a + 4) = 2a + (-1)(a + 4) = 2a + (-1 \cdot a) + (-1 \cdot 4) = 2a - a - 4 = a - 4$

39. $6x - (3 - x) = 6x + (-1)(3 - x) = 6x + (-1 \cdot 3) + (-1 \cdot -x) = 6x - 3 + x = 7x - 3$

40. $5x - (7 - 2x) = 5x + (-1)(7 - 2x) = 5x + (-1 \cdot 7) + (-1 \cdot -2x) = 5x - 7 + 2x = 7x - 7$

B **41.** $x + 11 = 2(x - 1);\quad x + 11 = (2 \cdot x) + (2 \cdot -1);\quad x + 11 = 2x - 2;\quad x + 11 + 2 = 2x - 2 + 2;\quad x + 13 = 2x;\quad 13 = x$

42. $3(n - 1) = 2n + 8;\quad (3 \cdot n) + (3 \cdot -1) = 2n + 8;\quad 3n - 3 = 2n + 8;\quad 3n - 3 + 3 = 2n + 8 + 3;\quad 3n = 2n + 11;\quad n = 11$

43. $-(n - 1) = -2(n + 3);\quad (-1 \cdot n) + (-1 \cdot -1) = (-2 \cdot n) + (-2 \cdot 3);\quad -n + 1 = -2n - 6;\quad -n + 1 - 1 = -2n - 6 - 1;\quad -n = -2n - 7;\quad -n + 2n = -2n - 7 + 2n;\quad n = -7$

44. $p - (p - 2) = p;\quad p + (-1 \cdot p) + (-1 \cdot -2) = p;\quad p - p + 2 = p;\quad 2 = p$

45. $-3(5 - y) = 4(1 - 4y);\quad (-3 \cdot 5) + (-3 \cdot -y) = (4 \cdot 1) + (4 \cdot -4y);\quad -15 + 3y = 4 - 16y;\quad -15 + 3y + 15 = 4 - 16y + 15;\quad 3y = 19 - 16y;\quad 3y + 16y = 19 - 16y + 16y;\quad 19y = 19;\quad y = 1$

46. $6a - (a - 2) = 3a - 8;\quad 6a + (-1)(a - 2) = 3a - 8;\quad 6a + (-1 \cdot a) + (-1 \cdot -2) = 3a - 8;\quad 6a - a + 2 = 3a - 8;\quad 5a + 2 = 3a - 8;\quad 5a + 2 - 3a = 3a - 8 - 3a;\quad 2a + 2 = -8;\quad 2a + 2 - 2 = -8 - 2;\quad 2a = -10;\quad a = -5$

C **47.** The correct solution is as follows: $4(7x - 1) - 3(8x - 2) = 4 - (-3x + 2);$
$(4 \cdot 7x) + (4 \cdot -1) + (-3 \cdot 8x) + (-3 \cdot -2) = 4 + (-1 \cdot -3x) + (-1 \cdot 2);\quad 28x - 4 - 24x + 6 = 4 + 3x - 2;\quad 4x + 2 = 3x + 2;\quad x = 0$

Page 96 · WRITTEN EXERCISES

A **1.** $3 \div -3 = -1$

2. $-3 \div -3 = 1$

3. $-56 \div 7 = -8$

4. $56 \div -7 = -8$

5. $4 \div -2 = -2$

6. $-6 \div 2 = -3$

7. $-10 \div -5 = 2$

8. $-15 \div -3 = 5$

9. $42 \div -6 = -7$

10. $-16 \div 4 = -4$

11. $-20 \div -4 = 5$

12. $25 \div -5 = -5$

13. $49 \div -7 = -7$

14. $-18 \div 6 = -3$

15. $-21 \div -7 = 3$

16. $24 \div -4 = -6$

17. $\dfrac{12}{-2} = -6$

18. $\dfrac{14}{7} = 2$

19. $\dfrac{-10}{5} = -2$

20. $\dfrac{-15}{-3} = 5$

21. $\dfrac{-21}{-3} = 7$

22. $\dfrac{18}{-2} = -9$ **23.** $\dfrac{-9}{-3} = 3$ **24.** $\dfrac{-20}{5} = -4$

25. $\dfrac{35}{7} = 5$ **26.** $\dfrac{56}{-8} = -7$

Page 96 · CALCULATOR ACTIVITIES

1. 3   **2.** −3   **3.** −12   **4.** 5   **5.** −136   **6.** 216   **7.** −4   **8.** 16

Page 98 · WRITTEN EXERCISES

A   **1.** $x + 2 = -4$; $x + 2 - 2 = -4 - 2$; $x = -6$. Check: $-6 + 2 = -4\sqrt{}$

2. $x + 6 = 3$; $x + 6 - 6 = 3 - 6$; $x = -3$. Check: $-3 + 6 = 3\sqrt{}$

3. $y + 12 = -12$; $y + 12 - 12 = -12 - 12$; $y = -24$. Check: $-24 + 12 = -12\sqrt{}$

4. $y - 12 = 12$; $y - 12 + 12 = 12 + 12$; $y = 24$. Check: $24 - 12 = 12\sqrt{}$

5. $a - 3 = -2$; $a - 3 + 3 = -2 + 3$; $a = 1$. Check: $1 - 3 = -2\sqrt{}$

6. $3x = -15$; $\dfrac{3x}{3} = \dfrac{-15}{3}$; $x = -5$. Check: $3(-5) = -15\sqrt{}$

7. $-4x = 12$; $\dfrac{-4x}{-4} = \dfrac{12}{-4}$; $x = -3$. Check: $-4(-3) = 12\sqrt{}$

8. $-6x = 18$; $\dfrac{-6x}{-6} = \dfrac{18}{-6}$; $x = -3$. Check: $-6(-3) = 18\sqrt{}$

9. $5y - 3y = -10$; $2y = -10$; $\dfrac{2y}{2} = \dfrac{-10}{2}$; $y = -5$. Check: $5(-5) - 3(-5) =$

$-25 + 15 = -10\sqrt{}$

10. $5x - 2x = -18$; $3x = -18$; $\dfrac{3x}{3} = \dfrac{-18}{3}$; $x = -6$. Check: $5(-6) - 2(-6) =$

$-30 + 12 = -18\sqrt{}$

11. $15x = -45 + 30$; $15x = -15$; $\dfrac{15x}{15} = \dfrac{-15}{15}$; $x = -1$. Check: $15(-1) =$

$-45 + 30?$ $-15 = -15\sqrt{}$

12. $3x - 2x = -2$; $x = -2$. Check: $3(-2) - 2(-2) = -6 + 4 = -2\sqrt{}$

13. $-x + 2x = 9$; $x = 9$ **14.** $-n + 4n = 15$; $3n = 15$; $n = 5$

15. $-2n + 5n = 12$; $3n = 12$; $n = 4$ **16.** $-x + 5x = 16$; $4x = 16$; $x = 4$

17. $-2a + 6a = 12$; $4a = 12$; $a = 3$ **18.** $-2n + 3n = -12$; $n = -12$

19. $4c - 3c = -6$; $c = -6$

20. $4x + 7 = 3x$; $4x + 7 - 4x = 3x - 4x$; $7 = -x$; $-1 \cdot 7 = (-1)(-x)$; $-7 = x$

21. $2n + 5 = n$; $2n + 5 - 2n = n - 2n$; $5 = -n$; $-1 \cdot 5 = (-1)(-n)$; $-5 = n$

22. $3y + 4 = 2y$; $3y + 4 - 3y = 2y - 3y$; $4 = -y$; $-1 \cdot 4 = (-1)(-y)$; $-4 = y$

23. $7y - 2 - 5y = 0$; $2y - 2 = 0$; $2y = 2$; $y = 1$

24. $-6m + 5 = -7m$; $-6m + 5 + 6m = -7m + 6m$; $5 = -m$; $-1 \cdot 5 =$

$(-1)(-m)$; $-5 = m$

25. $-3x - 8 = -5x$; $-3x - 8 + 3x = -5x + 3x$; $-8 = -2x$; $\dfrac{-8}{-2} = \dfrac{-2x}{-2}$; $4 = x$

26. $3x + 1 = 2x$; $3x + 1 - 3x = 2x - 3x$; $1 = -x$; $-1 = x$

27. $6y + 4 = 4y$; $6y + 4 - 6y = 4y - 6y$; $4 = -2y$; $\dfrac{4}{-2} = \dfrac{-2y}{-2}$; $-2 = y$

28. $-14 = -6x - 2$; $-14 + 2 = -6x - 2 + 2$; $-12 = -6x$; $\dfrac{-12}{-6} = \dfrac{-6x}{-6}$; $2 = x$

29. $2n + 16 = n + 2$; $2n + 16 - n = n + 2 - n$; $n + 16 = 2$; $n + 16 - 16 = 2 - 16$; $n = -14$

30. $-3x + 2 = -x + 4$; $-3x + 2 + x = -x + 4 + x$; $-2x + 2 = 4$; $-2x + 2 - 2 = 4 - 2$; $-2x = 2$; $\dfrac{-2x}{-2} = \dfrac{2}{-2}$; $x = -1$

31. $2x - (3x + 2) = -7$; $2x - 3x - 2 = -7$; $-x - 2 = -7$; $-x - 2 + 2 = -7 + 2$; $-x = -5$; $x = 5$

32. $5y - (2y - 1) = -2$; $5y - 2y + 1 = -2$; $3y + 1 = -2$; $3y + 1 - 1 = -2 - 1$; $3y = -3$; $y = -1$

33. $-x - (5x - 7) = -5$; $-x - 5x + 7 = -5$; $-6x + 7 = -5$; $-6x = -12$; $x = 2$

34. $7 - (2 - x) = -4$; $7 - 2 + x = -4$; $5 + x = -4$; $x = -9$

35. $6 - (4 - x) = 3x$; $6 - 4 + x = 3x$; $2 + x = 3x$; $2 = 2x$; $1 = x$

36. $2x - (3 - x) = x - 7$; $2x - 3 + x = x - 7$; $3x - 3 = x - 7$; $2x - 3 = -7$; $2x = -4$; $x = -2$

Page 100 · WRITTEN EXERCISES

A

1. $|-3| = 3$ and $|-2| = 2$; $3 > 2$, so $|-3| > |-2|$;   false

2. $|-6| = 6$ and $|2| = 2$; $6 > 2$, so $|-6| > |2|$;   true

3. $|-4| = 4$ and $|-3| = 3$; $4 > 3$, so $|-4| > |-3|$;   true

4. $|5| = 5$ and $|-1| = 1$; $5 > 1$, so $|5| > |-1|$;   true

5. $|0| = 0$ and $|6| = 6$; $0 < 6$, so $|0| < |6|$;   true

6. $|-3| = 3$ and $|0| = 0$; $3 > 0$, so $|-3| > |0|$;   false

7. $|1| = 1$ and $|-7| = 7$; $1 < 7$, so $|1| < |-7|$;   false

8. $|-1| = 1$ and $|-7| = 7$; $1 < 7$, so $|-1| < |-7|$;   false

9. $|2| = 2$ and $|-2| = 2$; $2 = 2$, so $|2| = |-2|$;   false

10. $|-6| = 6$ and $|4| = 4$; $6 > 4$, so $|-6| > |4|$;   false

11. $|-7| = 7$ and $|7| = 7$; $7 = 7$, so $|-7| = |7|$;   false

12. $|3| = 3$ and $|-2| = 2$; $3 > 2$, so $|3| > |-2|$;   true

13. $|5| < |6|$ **14.** $|2| < |-3|$ **15.** $|-7| > |-3|$

16. $|0| < |-4|$ **17.** $|9| < |-10|$ **18.** $|-6| > |-2|$

19. $|-11| > |7|$ **20.** $|-1| = |1|$ **21.** $|-5| > |2|$

22. $-|3| = -3$ **23.** $-|-2| = -2$

24. $-|3| + |-2| = -3 + 2 = -1$ **25.** $7 + |-4| = 7 + 4 = 11$

26. $-|5| + |-5| = -5 + 5 = 0$ **27.** $|-2| - |3| = 2 - 3 = -1$

B   **28.** $|x| + 4 = 9$; $|x| = 5$; $x = 5$ or -5 **29.** $|y| + 3 = 7$; $|y| = 4$; $y = 4$ or -4

30. $|x| - 1 = 9$; $|x| = 10$; $x = 10$ or -10 **31.** $|a| + 5 = 11$; $|a| = 6$; $a = 6$ or -6

32. $|n| - 2 = 5$; $|n| = 7$; $n = 7$ or -7

33. $|x| - 7 = 7$; $|x| = 14$; $x = 14$ or -14

34. $|n| - 3 = 12$; $|n| = 15$; $n = 15$ or -15

35. $|c| + 1 = 4$; $|c| = 3$; $c = 3$ or -3

36. $|r| - 9 = 18$; $|r| = 27$; $r = 27$ or -27 **37.** $|z| + 6 = 13$; $|z| = 7$; $z = 7$ or -7

38. $|a| - 2 = 22$; $|a| = 24$; $a = 24$ or -24

39. $|x| + 5 = 35$; $|x| = 30$; $x = 30$ or -30

Page 100 · PUZZLE PROBLEMS

Fill the large pitcher. Pour water from it to fill the small pitcher. Empty the small pitcher. Pour the 2 L from the large pitcher into it. Fill the large pitcher again. Pour water from it into the small pitcher until the small one is full. The large pitcher now has 4 L.

Page 101 · READING ALGEBRA

1. positive and negative numbers **2.** integers **3.** to the left of zero

4. the positive numbers, the negative numbers, and zero **5.** negative three

6. no; their purpose is to explain in words the inequalities written with symbols.

Page 103 · COMPUTER ACTIVITIES

1. Yes, the expressions are equivalent.

For Exs. 2–4, modifications to the program shown on page 102 of the student book are given.

2. a.
```
80 LET A = -1 * X
100 LET B = 0 - X
Yes
```

b.
```
80 LET A = -(X - 5)
100 LET B = -X - 5
No
```

c.
```
80 LET A = -(X - 5)
100 LET B = -X + 5
Yes
```

d.
```
80 LET A = 10 - X
100 LET B = X - 10
No
```

e.
```
80 LET A = (-X) * (-X)
100 LET B = -(X * X)
No
```

f.
```
80 LET A = X
100 LET B = -(-X)
Yes
```

g. 80 LET A = X - (-X)
100 LET B = 0
No

h. 80 LET A = (7 - X) - 4
100 LET B = (4 - X) - 7
No

i. 80 LET A = X - X
100 LET B = X + (-X)
Yes

j. 80 LET A = -1 * (6 - X)
100 LET B = X - 6
Yes

3. For Ex. 3, add the line 125 PRINT "EXPRESSION C:";C to the program.

a. 80 LET A = -X
100 LET B = 1 - X
105 LET C = X/(-1)
$1 - x$

b. 80 LET A = (-2) * X
100 LET B = 2 * (-X)
105 LET C = (-2) * (-X)
$(-2)(-x)$

c. 80 LET A = -8X - 5
100 LET B = -8 * (X - 5)
105 LET C = -8 * X + 40
$-8x - 5$

d. 80 LET A = 2 * X - (X - 3)
100 LET B = X + 3
105 LET C = X - 3
$x - 3$

4. For Ex. 4, delete lines 105 and 125 (added in Ex. 3) by typing the line number and pressing RETURN.

a. 80 LET A = X * X
100 LET B = X + X
0 and 2

b. 80 LET A = 6 * X
100 LET B = X + 10
2

c. 80 LET A = 2 - X
100 LET B = X - 2
2

d. 80 LET A = X/5
100 LET B = 5/X
5 and -5

e. 80 LET A = X * X
100 LET B = -4 * X
0 and −4

f. 80 LET A = ABS(X)
100 LET B = -X
$x \le 0$

Page 104 · SKILLS REVIEW

1. about 80   **2.** about 60   **3.** about 110   **4.** about 110   **5.** about 90   **6.** about 120

7. about 120   **8.** about 150   **9.** about 200   **10.** about 190   **11.** about 190

12. about 280   **13.** about 120   **14.** about 420   **15.** about 80   **16.** about 300

17. about \$1.60   **18.** about \$5.00   **19.** about \$1.40   **20.** about \$4.00   **21.** about \$3.60

22. about \$2.40   **23.** about \$2.70   **24.** about \$2.10   **25.** no

26. yes   **27.** no   **28.** yes   **29.** no   **30.** no   **31.** no   **32.** no

33. no   **34.** no   **35.** yes   **36.** no

37. $12 + 8 + 19 = 39; \quad 39 \div 3 = 13$ \qquad **38.** $3 + 14 + 16 = 33; \quad 33 \div 3 = 11$

39. $98 + 20 + 32 = 150; \quad 150 \div 3 = 50$ \qquad **40.** $10 + 12 + 20 = 42; \quad 42 \div 3 = 14$

41. $2 + 4 + 10 + 20 = 36; \quad 36 \div 4 = 9$ \qquad **42.** $16 + 5 + 7 + 12 = 40; \quad 40 \div 4 = 10$

43. $5 + 10 + 20 + 21 = 56; \quad 56 \div 4 = 14$

44. $10 + 12 + 16 + 20 + 27 = 85; \quad 85 \div 5 = 17$

Pages 105–106 · CHAPTER REVIEW EXERCISES

1. <   **2.** >   **3.** >   **4.** >   **5.** >   **6.** <   **7.** >   **8.** <

9. $x + 2 = 4; \quad x = 2$

10. $x + 6 = 7; \quad x = 1$

11. $x + 9 = 3; \quad x = -6$ **12.**

13.   **14.**

15. $(-6) + (-7) = -13$ **16.** $(-3) + (-8) = -11$ **17.** $(-4) + 1 = -3$

18. $6 + (-6) = 0$ **19.** $3 + (-2) = 1$ **20.** $(-8) + 5 = -3$

21. $(-5) + (-6) = -11$ **22.** $(-9) + 5 = -4$ **23.** $4 + (-10) = -6$

24. $(-5) + 9 = 4$ **25.** $(-7) + (-4) = -11$ **26.** $7 + (-11) = -4$

27. $8 - (-10) = 8 + (10) = 18$ **28.** $-6 - (-2) = -6 + (2) = -4$

29. $3 - (-1) = 3 + (1) = 4$ **30.** $6 - 7 = 6 + (-7) = -1$

31. $-8 - 5 = -8 + (-5) = -13$ **32.** $13 - (-10) = 13 + (10) = 23$

33. $-9 - 3 = -9 + (-3) = -12$ **34.** $8 - (-5) = 8 + (5) = 13$

35. $6 - 11 = 6 + (-11) = -5$ **36.** $-9 - (-5) = -9 + (5) = -4$

37. $-5 - (-6) = -5 + (6) = 1$ **38.** $3 - (-12) = 3 + (12) = 15$

39. $x - 4x = (1 - 4)x = -3x$

40. $x - 10 - 4x = (1 - 4)x - 10 = -3x - 10$

41. $5a - 15 - 5a = (5 - 5)a - 15 = 0 - 15 = -15$

42. $h - 2h + 3 = (1 - 2)h + 3 = -h + 3$

43. $-x - y + 3x = (-1 + 3)x - y = 2x - y$

44. $6b - 9b + 3 = (6 - 9)b + 3 = -3b + 3$

45. $-b^2 + 9b^2 = (-1 + 9)b^2 = 8b^2$

46. $-a - 10 + 8a = (-1 + 8)a - 10 = 7a - 10$

47. $-4x + 3y - 2x = (-4 - 2)x + 3y = -6x + 3y$

48. $-8(3) = -24$ **49.** $6(-9) = -54$ **50.** $(-3)(-8) = 24$

51. $(-9)(-3) = 27$ **52.** $-9(8) = -72$ **53.** $18(-2) = -36$

54. $(y - 5)y = (y \cdot y) + (-5 \cdot y) = y^2 - 5y$

55. $x(x - 4) = (x \cdot x) + (x \cdot -4) = x^2 - 4x$

56. $-2(x + 3) = (-2 \cdot x) + (-2 \cdot 3) = -2x - 6$

57. $(x - 4)(-5) = (x \cdot -5) + (-4 \cdot -5) = -5x + 20$

58. $-3(a - b) = (-3 \cdot a) + (-3 \cdot -b) = -3a + 3b$

59. $6(-m + n) = (6 \cdot -m) + (6 \cdot n) = -6m + 6n$

60. $-(x)^2 = -(x \cdot x) = -x^2$ **61.** $-(-b)^2 = -(-b \cdot -b) = -b^2$

62. $-(y - 7) = -1(y - 7) = (-1 \cdot y) + (-1 \cdot -7) = -y + 7$

63. $-a - (2a + b) = -a + (-1 \cdot 2a) + (-1 \cdot b) = -a - 2a - b = -3a - b$

64. $-(-x - 9) = (-1 \cdot -x) + (-1 \cdot -9) = x + 9$

65. $-x - (-3 + y) = -x + (-1 \cdot -3) + (-1 \cdot y) = -x + 3 - y$

66. $-81 \div 3 = -27$ **67.** $-72 \div 8 = -9$ **68.** $-12 \div -12 = 1$

69. $-16 \div 4 = -4$ **70.** $27 \div 9 = 3$ **71.** $30 \div -3 = -10$

72. $-16 \div -4 = 4$ **73.** $24 \div -6 = -4$ **74.** $\dfrac{48}{-6} = -8$

75. $\dfrac{-25}{5} = -5$ **76.** $\dfrac{-24}{-2} = 12$ **77.** $\dfrac{18}{-9} = -2$

78. $6x = -30;\quad \dfrac{6x}{6} = \dfrac{-30}{6};\quad x = -5$ **79.** $8x = -64;\quad \dfrac{8x}{8} = \dfrac{-64}{8};\quad x = -8$

80. $5x - 6x = -8;\quad -x = -8;\quad (-1)(-x) = (-1)(-8);\quad x = 8$

81. $5x + 9 = 6x;\quad 5x + 9 - 5x = 6x - 5x;\quad 9 = x$

82. $-9x = 36;\quad \dfrac{-9x}{-9} = \dfrac{36}{-9};\quad x = -4$

83. $-6 + x = -9;\quad -6 + x + 6 = -9 + 6;\quad x = -3$

84. $8x - 9x = -2;\quad -x = -2;\quad x = 2$

85. $4x + 10 = 6x;\quad 4x + 10 - 4x = 6x - 4x;\quad 10 = 2x;\quad 5 = x$

86. $9x - 4x = -15;\quad 5x = -15;\quad \dfrac{5x}{5} = \dfrac{-15}{5};\quad x = -3$

87. $-|-5| = -5$ **88.** $-|2| + |-2| = -2 + 2 = 0$

89. $|-4| - |-1| = 4 - 1 = 3$

Page 107 · CHAPTER TEST

1. -5 **2. a.** $>$ **b.** $<$ **3.** $x + 7 = 11;\quad x = 4$ 0 1 2 3 4 5 6

4.

5. $-5 < 0 < 2$ **6.** $-9 + 2 = -7$ **7.** $-5 + (-5) = -10$

8. $7 + (-5) + (-3) = 2 + (-3) = -1$ **9.** $1 - (-6) = 1 + 6 = 7$

10. $8 - 12 = 8 + (-12) = -4$ **11.** $-2 - 4 = -2 + (-4) = -6$

12. $n = -5 - 15 = -5 + (-15) = -20$ **13.** $3x - 5 - 7x = (3 - 7)x - 5 = -4x - 5$

14. $-7y - (-y) = -7y + y = (-7 + 1)y = -6y$ **15.** $-2a - 8b + 9a =$
$(-2 + 9)a - 8b = 7a - 8b$ **16.** $4(-2) = -8$ **17.** $(-8)(-3) = 24$

18. $-10(5) = -50$ **19.** $0(-40) = 0$ **20. a.** $-2(-3) + (-2)7 = 6 + (-14) = -8$

b. $(-2)^5 = [(-2)(-2)] \cdot [(-2)(-2)] \cdot (-2) = 4 \cdot 4 \cdot (-2) = -32$ **21. a.** $-3a(4b) =$
$(-3)(4)(ab) = -12ab$ **b.** $-x(x - 1) = (-x \cdot x) + (-x \cdot -1) = -x^2 + x$

22. $5y - (4 - y) = 5y + (-1)(4 - y) = 5y + (-1 \cdot 4) + (-1 \cdot -y) =$
$5y - 4 + y = 6y - 4$ **23.** $-(8 - x) = 2(x + 1);$ $(-1)(8) + (-1)(-x) =$
$2 \cdot x + 2 \cdot 1;$ $-8 + x = 2x + 2;$ $-10 + x = 2x;$ $-10 = x$

24. $9 \div -3 = -3$ **25.** $-2 \div -1 = 2$ **26.** $\frac{-40}{5} = -8$ **27.** $\frac{-36}{-9} = 4$

28. $2x - 5x = -18;$ $-3x = -18;$ $\frac{-3x}{-3} = \frac{-18}{-3};$ $x = 6.$

Check: $2(6) - 5(6) = -18?$ $12 - 30 = -18?$ $-18 = -18\checkmark$

29. $x + 1 = -3;$ $x + 1 - 1 = -3 - 1;$ $x = -4.$ Check: $-4 + 1 = -3?$
$-3 = -3\checkmark$ **30.** $-5y + 7 = -9y - 5;$ $-5y + 7 - 7 = -9y - 5 - 7;$ $-5y =$

$-9y - 12;$ $-5y + 9y = -9y - 12 + 9y;$ $4y = -12;$ $\frac{4y}{4} = \frac{-12}{4};$ $y = -3.$

Check: $-5(-3) + 7 = -9(-3) - 5?$ $15 + 7 = 27 - 5?$ $22 = 22\checkmark$

31. $7a - (a - 8) = a + 3;$ $7a - a + 8 = a + 3;$ $6a + 8 = a + 3;$ $6a + 8 - 8 =$
$a + 3 - 8;$ $6a = a - 5;$ $6a - a = a - 5 - a;$ $5a = -5;$ $a = -1.$ Check:
$7(-1) - (-1 - 8) = -1 + 3?$ $-7 - (-9) = 2?$ $-7 + 9 = 2?$ $2 = 2\checkmark$

32. $|-9| = 9$ and $|-1| = 1;$ $9 > 1,$ so $|-9| > |-1|;$ true

33. $|-4| = 4$ and $|3| = 3;$ $4 > 3,$ so $|-4| > |3|;$ false

34. $|-5| = 5$ and $|5| = 5;$ $5 = 5,$ so $|-5| = |5|;$ false

Pages 108–109 • MIXED REVIEW

1. $(-4)^2 = (-4)(-4) = 16$ **2.** $y \cdot 7 \cdot x \cdot 4 \cdot y \cdot y = (7 \cdot 4)(x)(y \cdot y \cdot y) = 28xy^3$

3. $-3x - (-x) + 4x = -3x + x + 4x = (-3 + 1 + 4)x = 2x$

4. $2(x - 5) - 3x = 2 \cdot x + 2(-5) - 3x = 2x - 10 - 3x = -x - 10$

5. $2 - (4 - a) = 2 + (-1)(4) + (-1)(-a) = 2 + (-4) + a = -2 + a$

6. $2r + 6s - 5s - 2r = (2 - 2)r + (6 - 5)s = s$ **7.** $\frac{x}{2} = 8;$ $2 \cdot \frac{x}{2} = 2 \cdot 8;$

$x = 16$ **8.** $y - 13 = 14;$ $y - 13 + 13 = 14 + 13;$ $y = 27$ **9.** $3(2a - 1) =$
$2(a + 3) + a;$ $6a - 3 = 2a + 6 + a;$ $6a - 3 = 3a + 6;$ $6a - 3 + 3 =$
$3a + 6 + 3;$ $6a = 3a + 9;$ $6a - 3a = 3a + 9 - 3a;$ $3a = 9;$ $a = 3$

10. $9x - 8 = 37$; $9x - 8 + 8 = 37 + 8$; $9a = 45$; $a = 5$

11. $2n + 5 = 9n - 2$; $2n + 5 + 2 = 9n - 2 + 2$; $2n + 7 = 9n$; $2n + 7 - 2n = 9n - 2n$; $7 = 7n$; $1 = n$ **12.** $7 - 3r = 6r - 11$; $7 - 3r + 11 = 6r - 11 + 11$; $18 - 3r = 6r$; $18 - 3r + 3r = 6r + 3r$; $18 = 9r$; $2 = r$ **13.** $5n - 8 = n$; $5n - 8 + 8 = n + 8$; $5n = n + 8$; $5n - n = n + 8 - n$; $4n = 8$; $n = 2$

14. Answers may vary. For example: $x + 6 = 4$.

15. If $x = 1$: $5 \cdot 1 - 3 = 5 - 3 = 2 < 14$; if $x = 2$: $5 \cdot 2 - 3 = 10 - 3 = 7 < 14$; if $x = 3$: $5 \cdot 3 - 3 = 15 - 3 = 12 < 14$; if $x = 4$: $5 \cdot 4 - 3 = 20 - 3 = 17 > 14$.   4

16. $d + 4$; $2(d + d + 4)$, or $4d + 8$ **17.** $>$ **18.**

19. $2 \cdot 14 + 9 = 28 + 9 = 37$ **20.** $2(14 + 9) = 2(23) = 46$ **21.** $(14 - 4)^2 = 10^2 = 100$

22. $(14 - 2) \div 4 = 12 \div 4 = 3$ **23.** $a(7a - 5b) = a \cdot 7a + a(-5b) = 7a^2 - 5ab$

24. **a.** $\dfrac{x}{x - x} = \dfrac{x}{0}$; impossible **b.** $\dfrac{2x}{x} = 2$ **c.** $\dfrac{x - x}{4} = \dfrac{0}{4} = 0$

25. $9 + x = 2$; $9 + x - 9 = 2 - 9$; $x = -7$ **26.** $-7y = 56$; $\dfrac{-7y}{-7} = \dfrac{56}{-7}$; $y = -8$

27. $6(y + 1) = 22 + 2y$; $6y + 6 = 22 + 2y$; $6y + 6 - 6 = 22 + 2y - 6$; $6y = 16 + 2y$; $6y - 2y = 16 + 2y - 2y$; $4y = 16$; $y = 4$

28. Let x = the brother's age now. Then $2x$ = Amos's age now, and last year their ages were $x - 1$ and $2x - 1$. $(x - 1) + (2x - 1) = 7$; $3x - 2 = 7$; $3x = 9$; $x = 3$; $2x = 6$; 6 years old

29. $4n + 6n = 50$; $10n = 50$; $n = 5$ **30.** $-5 < -2 < 1$

31. $5 - 8 = 5 + (-8) = -3$ **32.** $-24 \div (-4) = 6$ **33.** $(-2)(-9) = 18$

34. $-7 + (-4) = -11$ **35.** $63 \div (-9) = -7$ **36.** $3 - (-9) = 3 + 9 = 12$

37. $-19 + 7 = -12$ **38.** $-(-2)(-2) = -4$

39. Let x = the calories in a cup of orange juice. Then $x + 30$ = the calories in a cup of whole milk. $x + x + 30 = 270$; $2x + 30 = 270$; $2x = 240$; $x = 120$ and $x + 30 = 150$. orange juice: 120; whole milk: 150

40. $7(3a - 1) + 5(2 - 9a) = 21a - 7 + 10 - 45a = -24a + 3$

41. Steps: n; $3n$; $3n - 6$; $4(3n - 6)$ or $12n - 24$; $12n - 24 + 8$ or $12n - 16$; $6n - 8$. Thus, $6n - 8 = 34$; $6n = 42$; $n = 7$.

42. $2^5 = 32$ and $(3 + 1 \cdot 5)4 = (3 + 5)4 = 8 \cdot 4 = 32$; $=$

43. $w + 8$

44. Let n = the larger number. Then $n - 6$ = the smaller number. $3(n - 6) = n + 2$; $3n - 18 = n + 2$; $3n - 18 + 18 = n + 2 + 18$; $3n = n + 20$; $3n - n = n + 20 - n$; $2n = 20$; $n = 10$; $n - 6 = 4$; 10 and 4

45. $+9$ **46.** $x - 2 = 3$; $x = 5$

47. $7 + (-8) = -1$; $-1 - 4 = -1 + (-4) = -5$

48. $3(-5)(2) + (-9)(-8) = (-15)2 + 72 = -30 + 72 = 42$

Pages 113–114 · WRITTEN EXERCISES

A   1. $P = 2l + 2w$; $P = 2 \cdot 7 + 2 \cdot 4 = 14 + 8 = 22$ 2. $P = 5s$

      3. $P = x + 2x + 3x + 3x = 9x$ 4. $P = 2x + 2y$

      5. $P = x + x + 1 + x + 2 = 3x + 3$

      6. $P = 2 \cdot 2h + 4 \cdot h + 3 \cdot x = 4h + 4h + 3x = 8h + 3x$

      7. Each unlabeled side has length $2a + a = 3a$. Thus, $P = 2 \cdot 3a + 2 \cdot 2a + 2 \cdot a = 6a + 4a + 2a = 12a$

      8. The longer unlabeled side has length $x + x + x = 3x$; the shorter unlabeled sides have length x. Thus, $P = 3x + 7 \cdot x = 3x + 7x = 10x$

      9. The unlabeled side at the top has length $4x - x = 3x$; the unlabeled side at the right has length $3x - x = 2x$. Thus, $P = 2 \cdot 3x + 2 \cdot x + 2x + 4x = 6x + 2x + 2x + 4x = 14x$

    10. Let x = the width (in cm). Then $x + 2$ = the length. $P = 2l + 2w$; $20 = 2(x + 2) + 2x$; $20 = 2x + 4 + 2x$; $20 = 4x + 4$; $16 = 4x$; $4 = x$, and $x + 2 = 6$.   length: 6 cm;   width: 4 cm

    11. Let x = the width (in cm). Then $x + 5$ = the length. $P = 2l + 2w$; $50 = 2(x + 5) + 2x$; $50 = 2x + 10 + 2x$; $50 = 4x + 10$; $40 = 4x$; $10 = x$, and $x + 5 = 15$.   length: 15 cm;   width: 10 cm

    12. $48 = x + 2x + 2x + 3$; $48 = 5x + 3$; $45 = 5x$; $9 = x$

B   13. $C = 3.14d = 3.14 \cdot 4 = 12.56$;   12.56 cm

    14. $C = 3.14d = 3.14 \cdot 10 = 31.4$;   31.4 cm

    15. $C = 3.14d = 3.14 \cdot 18 = 56.52$;   56.52 cm

    16. $C = 3.14d = 3.14 \cdot 40 = 125.6$;   125.6 cm

    17. $C = 3.14d = 3.14 \cdot 3 = 9.42$;   9.42 m

    18. $C = 3.14d = 3.14 \cdot 16 = 50.24$;   50.24 cm

C   19. a. $C = 3.14d = 3.14 \cdot 56 = 175.84$;   175.84 cm

       b. $100 \cdot 175.84 = 17{,}584$;   17,584 cm

Pages 117–118 · WRITTEN EXERCISES

A   1. $A = s^2 = 7^2 = 49$;   49 cm²   2. $A = lw = 9 \cdot 6 = 54$;   54 m²

      3. $A = lw = 2 \cdot 8 = 16$;   16 m²   4. $A = lw = 9 \cdot 7 = 63$;   63 m²

      5. $A = \frac{1}{2}bh = \frac{1}{2} \cdot 8 \cdot 3 = 12$;   12 cm²   6. $A = \frac{1}{2}bh = \frac{1}{2} \cdot 14 \cdot 12 = 84$;   84 m²

      7. $A = \frac{1}{2}bh = \frac{1}{2} \cdot 8 \cdot 4 = 16$;   16 cm²   8. $A = \frac{1}{2}bh = \frac{1}{2} \cdot 6 \cdot 4 = 12$;   12 m²

      9. a. $A = (3x)(2x) = 6x^2$

        b. A = area rectangle − area cut-out square = $(3x)(2x) - x^2 = 6x^2 - x^2 = 5x^2$

    10. a. $A = (5x)(3x) = 15x^2$ b. A = area large rectangle − area cut-out rectangle = $(5x)(3x) - (x)(2x) = 15x^2 - 2x^2 = 13x^2$

    11. a. $A = (4y)(3x) = 12xy$ b. $A = (4y)(3x) - (2y)(x) = 12xy - 2xy = 10xy$

12. $A = a^2 - b^2$ **13.** $A = (3y)(2x) + (y)(x) = 6xy + xy = 7xy$

14. $A = xw - zy$ **15.** $A = (9y)(3y) - 2 \cdot y^2 = 27y^2 - 2y^2 = 25y^2$

16. $A = (6b)(3b) + 2 \cdot b^2 = 18b^2 + 2b^2 = 20b^2$

17. $A = \frac{1}{2}bh = \frac{1}{2} \cdot 4x \cdot 2x = 4x^2$ **18.** square centimeters   **19.** centimeters

20. Area of rug $= 5 \cdot 3 = 15$ m^2;   Cost $= 18 \cdot 15 = 270$;   \$270

21. Area of driveway $= 20 \cdot 4 = 80$ m^2;   Cost $= 24 \cdot 80 = 1920$;   \$1920

22. $10^2 = 100$

B   **23.** $A = \frac{1}{2}sh + s^2$ **24.** $A = \frac{1}{2}wh + lw + \frac{1}{2}wh = wh + lw$

25. $A = \frac{1}{2} \cdot 9b \cdot 2h + 9b \cdot 4h - (b \cdot h + b \cdot h + b \cdot 2h) = 9bh + 36bh - (4bh) = 41bh$

26. distributive

Pages 121–122 • WRITTEN EXERCISES

A   **1.**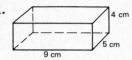

$V = Bh = (9 \cdot 5) \cdot 4 = 180$;
180 cm^3

2.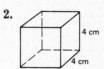

$V = Bh = (4 \cdot 4) \cdot 4 = 64$;
64 cm^3

3.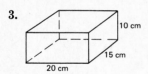

$V = 20 \cdot 15 \cdot 10 = 3000$;
3000 cm^3, or 3 liters

4.

$V = 40 \cdot 30 \cdot 50 = 60{,}000$;
60,000 cm^3, or 60 liters

5. $V = Bh = (4x \cdot 3x) \cdot 2x = 24x^3$ **6.** $V = Bh = (4x \cdot 4x) \cdot 6x = 96x^3$

7. $B = (6 \cdot 4) - (4 \cdot 2) = 24 - 8 = 16$; $V = Bh = 16 \cdot 3 = 48$

8. $B = (4x \cdot 2x) - x^2 = 8x^2 - x^2 = 7x^2$; $V = Bh = 7x^2 \cdot 2x = 14x^3$

9. $B = (3 \cdot 2) - 1^2 = 6 - 1 = 5$; $V = Bh = 5 \cdot 4 = 20$

10. $B = 1 \cdot 2 + 3 \cdot 1 = 2 + 3 = 5$; $V = Bh = 5 \cdot 2 = 10$

B   **11.** $B = 50 \cdot 20 = 1000$; $V = Bh = 1000 \cdot 40 = 40{,}000$;   40,000 cm^3

12. $40{,}000 \div 1000 = 40$;   40 liters   **13.** $40{,}000 \div 1 = 40{,}000$;   40,000 grams

14. $40{,}000 \div 1000 = 40$;   40 kilograms

15. $45 \cdot 30 \cdot 30 = 40{,}500$; $40{,}500 > 40{,}000$;   more

16. $B = (6x \cdot 2x) + (2x \cdot 4x) = 12x^2 + 8x^2 = 20x^2$; $V = Bh = 20x^2 \cdot 6x = 120x^3$

17. $B = (3x \cdot 3x) - (x \cdot x) = 9x^2 - x^2 = 8x^2$; $V = Bh = 8x^2 \cdot 2x = 16x^3$

18. $B = (3x \cdot 3x) - 4 \cdot x^2 = 9x^2 - 4x^2 = 5x^2$; $V = Bh = 5x^2 \cdot 4x = 20x^3$

Page 125 · WRITTEN EXERCISES

A   1. qk km   2. xp mi   3. $2t$ dollars   4. ry cents   5. $6t$ 6. $3f$

   7. $6t + 3f$ 8. $32p$ dollars   9. $6z$ dollars   10. dn dollars   11. $(s - b)$ dollars

   12. $(b - s)$ dollars

B   13. $(2u + 3w)$ dollars   14. $(8r + 12s)$ dollars   15. $(6d + s)$ cents

   16. $(a + 12b)$ dollars

Pages 128–129 · WRITTEN EXERCISES

A   1.

	rate	time	Distance
Maya	20	t	$20t$
Mike	8	t	$8t$
			56

$20t + 8t = 56$; $28t = 56$; $t = 2$.

2 hours

2.

	rate	time	Distance
Honolulu-San Francisco	550	t	$550t$
San Francisco-Honolulu	400	t	$400t$
			3800

$550t + 400t = 3800$;
$950t = 3800$; $t = 4$.
In 4 hours, or at 4:00 P.M.

3.

	rate	time	Distance
Kim	r	2	$2r$
Lucia	$r + 4$	2	$2(r + 4)$
			80

$2r + 2(r + 4) = 80$; $2r + 2r + 8 = 80$;
$4r + 8 = 80$; $4r = 72$; $r = 18$, and
$r + 4 = 22$.   Kim: 18 km/h;
Lucia: 22 km/h

4.

	rate	time	Distance
Maria	80	3	240
Tom	r	3	$3r$
			510

$240 + 3r = 510$; $3r = 270$; $r = 90$.
90 km/h

5.

	rate	time	Distance
Bike	25	$t + 2$	$25(t + 2)$
Car	75	t	$75t$

$25(t + 2) = 75t$; $25t + 50 = 75t$;
$50 = 50t$; $1 = t$, so Distance $= 75t =$
$75 \cdot 1 = 75$.   75 km

6.

	rate	time	Distance
Float	6	$t + 4$	$6(t + 4)$
Motorboat	18	t	$18t$

$6(t + 4) = 18t$; $6t + 24 = 18t$;
$24 = 12t$; $2 = t$, so Distance $= 18t =$
$18 \cdot 2 = 36$.   36 km

B **7.**

	rate	time	Distance
Alvin	x	$\dfrac{1}{2}$	$\dfrac{1}{2}x$
Curtis	$2x$	$\dfrac{1}{2}$	x

$x = \dfrac{1}{2}x + 6$; $\dfrac{1}{2}x = 6$; $x = 12$.

Alvin walks at 12 km/h.

Pages 131–133 · WRITTEN EXERCISES

A **1.**

	price	number	Cost
Sell	30	$n - 4$	$30(n - 4)$
Buy	20	n	$20n$
			240

$30(n - 4) - 20n = 240$; $30n - 120 -$
$20n = 240$; $10n - 120 = 240$;
$10n = 360$; $n = 36$.   36 bagels bought

2.

	price	number	Cost
Sell	290	$n - 1$	$290(n - 1)$
Buy	220	n	$220n$
			1250

$290(n - 1) - 220n = 1250$; $290n -$
$290 - 220n = 1250$; $70n - 290 =$
1250; $70n = 1540$; $n = 22$, so
$n - 1 = 21$.   21 recorders sold

3.

	price	number	Cost
Sell	60	$x - 2$	$60(x - 2)$
Buy	40	x	$40x$
			0

$60(x - 2) - 40x = 0$; $60x - 120 -$
$40x = 0$; $20x - 120 = 0$; $20x = 120$;
$x = 6$.   6 pens bought

4.

	pay/hour	hours	Weekly pay
Mrs. Wolfe	12	x	$12x$
Mr. Wolfe	12	$x + 4$	$12(x + 4)$
			912

$12x + 12(x + 4) = 912$; $12x + 12x + 48 = 912$; $24x + 48 = 912$; $24x = 864$;
$x = 36$ and $x + 4 = 40$. Mrs. Wolfe: 36 hours; Mr. Wolfe: 40 hours

5.

	pay/hour	hours	Weekly pay
Maureen	$x + 1$	40	$40(x + 1)$
Karl	x	40	$40x$
			840

$40(x + 1) + 40x = 840$; $40x + 40 + 40x = 840$; $80x + 40 = 840$; $80x = 800$; $x = 10$ and $x + 1 = 11$.   Maureen: \$11/hour;   Karl: \$10/hour

6.

	price	number	Cost
Student	1	$x + 80$	$x + 80$
Adult	2	x	$2x$
			980

$x + 80 + 2x = 980$; $3x + 80 = 980$; $3x = 900$; $x = 300$, so $x + 80 = 380$.   380 student tickets sold

B   **7.**

	price	number	Cost
Advance	3	$2x$	$6x$
Door	4	x	$4x$
			1540

$6x + 4x = 1540$; $10x = 1540$; $x = 154$ and $2x = 308$.   154 tickets at door, 308 advance tickets sold

8.

	price	number	Cost
\$25 pledge	25	$2x$	$50x$
\$40 pledge	40	x	$40x$
			15,300

$50x + 40x = 15,300$; $90x = 15,300$; $x = 170$, so $2x = 340$.   340 people pledged \$25.

Pages 136–137 · WRITTEN EXERCISES

A   **1. a.** $y - 5 = 7$; $y - 5 + 5 = 7 + 5$; $y = 12$ **b.** $y - x = 7$; $y - x + x = 7 + x$; $y = 7 + x$ **c.** $y - x = t$; $y - x + x = t + x$; $y = t + x$

2. a. $y + 5 = 7$; $y + 5 - 5 = 7 - 5$; $y = 2$ **b.** $y + x = 7$; $y = 7 - x$ **c.** $y + x = t$; $y = t - x$

3. $2 + y = 5$; $2 + y - 2 = 5 - 2$; $y = 3$

4. $2x + y = 4$; $2x + y - 2x = 4 - 2x$; $y = 4 - 2x$

5. $3 + y = 11$; $y = 8$ **6.** $4s + y = 9$; $y = 9 - 4s$

7. $y - 2x = 5$; $y - 2x + 2x = 5 + 2x$; $y = 5 + 2x$

8. $y + 3x = 7$; $y + 3x - 3x = 7 - 3x$; $y = 7 - 3x$

9. $7 - y = 5$; $7 - y - 7 = 5 - 7$; $-y = -2$; $(-1)(-y) = (-1)(-2)$; $y = 2$

10. $3x - y = 2$; $-y = 2 - 3x$; $(-1)(-y) = (-1)(2 - 3x)$; $y = -2 + 3x$

11. a. $8x = 24$; $\quad \dfrac{8x}{8} = \dfrac{24}{8}$; $\quad x = 3$ **b.** $8x = y$; $\quad x = \dfrac{y}{8}$ **c.** $ax = y$; $\quad x = \dfrac{y}{a}$

12. a. $3x = 45$; $\quad \dfrac{3x}{3} = \dfrac{45}{3}$; $\quad x = 15$ **b.** $3x = y$; $\quad x = \dfrac{y}{3}$ **c.** $kx = y$; $\quad x = \dfrac{y}{k}$

13. $3 + x = 11$; $\quad 3 + x - 3 = 11 - 3$; $\quad x = 8$

14. $y + x = 9$; $\quad x = 9 - y$ $\qquad\qquad$ **15.** $2y + x = 4$; $\quad x = 4 - 2y$

16. $x + 3y = 9$; $\quad x = 9 - 3y$ $\qquad\quad$ **17.** $4y + x = 6$; $\quad x = 6 - 4y$

18. $x - 3y = 5$; $\quad x - 3y + 3y = 5 + 3y$; $\quad x = 5 + 3y$

19. $y - x = 5$; $\quad y - x - y = 5 - y$; $\quad -x = 5 - y$; $\quad (-1)(-x) = (-1)(5 - y)$; $\quad x = -5 + y$

20. $5y - x = 10$; $\quad -x = 10 - 5y$; $\quad (-1)(-x) = (-1)(10 - 5y)$; $\quad x = -10 + 5y$

21. $A = 8w$; $\quad \dfrac{A}{8} = \dfrac{w}{8}$; $\quad \dfrac{A}{8} = w$ $\qquad\qquad$ **22.** $A = Pr$; $\quad \dfrac{A}{P} = \dfrac{Pr}{P}$; $\quad \dfrac{A}{P} = r$

23. $D = rt$; $\quad \dfrac{D}{r} = t$ $\qquad\qquad\qquad$ **24.** $E = 360n$; $\quad \dfrac{E}{360} = n$

25. $N = a + b$; $\quad N - a = a + b - a$; $\quad N - a = b$

26. $P = a + b + c$; $\quad P - (a + b) = a + b + c - (a + b)$; $\quad P - a - b = c$

27. $W = fd$; $\quad \dfrac{W}{f} = d$ $\qquad\qquad\qquad$ **28.** $W = AV$; $\quad \dfrac{W}{A} = V$

29. $\dfrac{H}{T} = A$; $\quad T \cdot \dfrac{H}{T} = T \cdot A$; $\quad H = TA$ $\qquad$ **30.** $h = \dfrac{A}{b} = \dfrac{84}{14} = 6$

31. $h = \dfrac{V}{lw} = \dfrac{100}{10 \cdot 5} = \dfrac{100}{50} = 2$ $\qquad\qquad$ **32.** $h = \dfrac{V}{B} = \dfrac{450}{50} = 9$

33. $D = rt$; $\quad \dfrac{D}{t} = r$; $\quad \dfrac{256}{2} = r$; $\quad 128 = r$; $\quad 128$ km/h

34. $D = rt$; $\quad \dfrac{D}{t} = r$; $\quad \dfrac{2208}{24} = r$; $\quad 92 = r$; $\quad 92$ km/h

B $\quad$ **35.** $D = rt$; $\quad \dfrac{D}{t} = r$; $\quad \dfrac{48000}{4} = r$; $\quad 12{,}000 = r$; $\quad 12{,}000$ km/h

Page 137 · PUZZLE PROBLEMS

25¢	10¢	5¢	1¢
1	0	0	2
0	2	1	2
0	2	0	7
0	1	3	2
0	1	2	7
0	1	1	12
0	1	0	17
0	0	5	2
0	0	4	7
0	0	3	12
0	0	2	17
0	0	1	22
0	0	0	27

Page 140 · WRITTEN EXERCISES

A   1. clockwise   2. counterclockwise

       3. A: 16 teeth; B: 8 teeth. Let s = speed of B; $2 \cdot 16 = s \cdot 8$; $32 = 8s$; $4 = s$; 4 turns/second.

       4. Let s = speed of first gear; $1 \cdot 52 = s \cdot 13$; $52 = 13s$; $4 = s$; 4 turns/second.

       5. $20 \cdot 6 = 10 \cdot d$; $120 = 10d$; $12 = d$; 12 m

       6. $20 \cdot 3 = 30 \cdot d$; $60 = 30d$; $2 = d$; 2 m

B   7. Let w = the weight of the rock; $75 \cdot 200 = w \cdot 20$; $15{,}000 = 20w$; $750 = w$; about 750 kg.

Page 141 · READING ALGEBRA

       1. They walked in opposite directions. Ben: 4 km/h; Christina: 5 km/h

       2. The distance apart was 27 km.

       3. Let n = number of hours each walked; Ben: $4n$ km; Christina: $5n$ km

       4. $4n + 5n = 27$; $9n = 27$; $n = 3$

       5. Yes; No; You must add 3 hours to 1:15 P.M. to get a time of 4:15 P.M.

       6. Yes, $(3 \times 4) + (3 \times 5) = 12 + 15 = 27$; Yes

Pages 142–143 · PROBLEM SOLVING STRATEGIES

1. c.

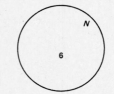

First box	Second box	Third box
2	2	1
3	1	1
1	2	2
1	3	1
1	1	3
2	1	2

2. a.

     b. A: $12 - x$; S: $14 - x$ c. $(12 - x) + (14 - x) + x + 6 = 30$

     d. $-x + 32 = 30$; $-x = -2$; $x = 2$; 2 students

3.

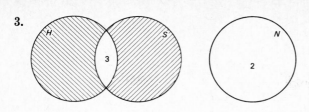

number taking only history:　$13 - 3 = 10$;　number taking only science:　$10 - 3 = 7$;
total = $10 + 7 + 3 + 2 = 22$.　**22 students**

Page 144 • SKILLS REVIEW

1. $3 \times 5 - 7 = 15 - 7 = 8$.　b　　　　　2. $4 + 12 \div 2 = 4 + 6 = 10$.　c

3. $20 - 12 \div (6 - 2) = 20 - 12 \div 4 = 20 - 3 = 17$.　b

4. $24 \div (-4) + 2 = -6 + 2 = -4$.　b

5. $-3(-5 + 1) + (-2) = -3(-4) + (-2) = 12 + (-2) = 10$.　b

6. $5 + 15 \div 3 + 2 = 5 + 5 + 2 = 12$.　b

7. $3 \cdot 4 + 18 \div 2 + 7 = 12 + 9 + 7 = 28$.　c

8. $-7 - (-15) \div 5 + 6 = -7 - (-3) + 6 = -7 + 3 + 6 = 2$.　c

9. $15 \div (-3) + 2 = -5 + 2 = -3$　　　　10. $-15 \div (-3 + 2) = -15 \div (-1) = 15$

11. $-(15 \div (-3)) + 2 = -(-5) + 2 = 5 + 2 = 7$

12. $12 \cdot 6 - 4 + 3 = 72 - 4 + 3 = 71$　　　13. $(12 \cdot 6) - (4 + 3) = 72 - 7 = 65$

14. $12(6 - 4) + 3 = 12(2) + 3 = 24 + 3 = 27$

15. $21 \div 7 - (4 - 1) = 3 - 3 = 0$　　　16. $4 \cdot 5 + 9 \div 3 = 20 + 3 = 23$

17. $-(15 \cdot 3 - 2) = -(45 - 2) = -43$　　　18. $7 - 8 + 24 \div 8 = 7 - 8 + 3 = 2$

19. $2 + 6 \cdot 10 \div 2 = 2 + 60 \div 2 = 2 + 30 = 32$

20. $84 - 16 \div 2 + 2 = 84 - 8 + 2 = 78$　　　21. $17 - 12 \cdot 2 + 3 = 17 - 24 + 3 = -4$

22. $6 \cdot 7 + 28 \div 4 = 42 + 7 = 49$

23. $-36 \div (-9) - 5 - (-4) = 4 - 5 + 4 = 3$

24. $216 \div 24 \div (-3) \div (-1) = 9 \div (-3) \div (-1) = -3 \div (-1) = 3$

Pages 145–146 • CHAPTER REVIEW EXERCISES

1. $P = x + x + 5 + 2x + 1 = 4x + 6$

2. $P = 2 \cdot y + 2 \cdot 2x + x = 2y + 4x + x = 2y + 5x$

3. The unlabeled side at the top has length $5x - 2x = 3x$; the other unlabeled side has length $3x - x = 2x$.　Thus, $P = 3x + 3x + x + 2x + 2x + 5x = 16x$.

4. $A = lw = 6x \cdot 2x = 12x^2$

5. $A = (2x)(3x) - x^2 = 6x^2 - x^2 = 5x^2$

6. $A = (4y)(3x) - (2y)(x) = 12xy - 2xy = 10xy$

7. Area of rug = $4 \cdot 3 = 12$ (m²);　$11 \cdot 12 = 132$;　$132

8. $B = (6 \cdot 2) + (2 \cdot 4) = 12 + 8 = 20$;　$V = Bh = 20 \cdot 2 = 40$

9. $B = (2x \cdot x) + (4x \cdot 2x) = 2x^2 + 8x^2 = 10x^2$; $V = Bh = 10x^2 \cdot 3x = 30x^3$

10. $V = 50 \cdot 30 \cdot 20 = 30{,}000$; $30{,}000$ cm³, or 30 liters

11. $90h$ km   **12.** st dollars

13.

	rate	time	Distance
Bike	20	3	60
Car	r	2	$2r$
			246

$60 + 2r = 246$; $2r = 186$; $r = 93$.
93 km/h

14.

	rate	time	Distance
Marsha	r	4	$4r$
Cammie	$r - 2$	4	$4(r - 2)$
			48

$4r + 4(r - 2) = 48$; $4r + 4r - 8 = 48$;
$8r - 8 = 48$; $8r = 56$; $r = 7$ and
$r - 2 = 5$.   Marsha: 7 km/h;   Cammie:
5 km/h

15.

	rate	number	Cost
Sell	210	$n - 2$	$210(n - 2)$
Buy	150	n	$150n$
			1440

$210(n - 2) - 150n = 1440$; $210n -$
$420 - 150n = 1440$; $60n - 420 = 1440$;
$60n = 1860$; $n = 31$, so $n - 2 = 29$.
29 monitors sold

16. a. $7x = 21$; $\dfrac{7x}{7} = \dfrac{21}{7}$; $x = 3$ **b.** $kx = 21$; $x = \dfrac{21}{k}$ **c.** $kx = a$; $x = \dfrac{a}{k}$

17. a. $x + y = 7$; $x + y - y = 7 - y$; $x = 7 - y$ **b.** $x - y = 7$; $x - y + y =$
$7 + y$; $x = 7 + y$ **c.** $y - x = 7$; $y - x - y = 7 - y$; $-x = 7 - y$;
$(-1)(-x) = (-1)(7 - y)$; $x = -7 + y$

18. $\dfrac{t}{3} = 8$; $3 \cdot \dfrac{t}{3} = 3 \cdot 8$; $t = 24$ **19.** $W = fd$; $\dfrac{W}{d} = \dfrac{fd}{d}$; $\dfrac{W}{d} = f$

Page 147 · CHAPTER TEST

1. $P = x + 2x + x + 3 = 4x + 3$ **2.** The unlabeled side at the top has length
$5x - 2x = 3x$;   the unlabeled side at the right has length $4x - 2x = 2x$.   Thus,
$P = 3 \cdot 2x + 3x + 4x + 5x = 6x + 3x + 4x + 5x = 18x$.   **3.** Let $w =$ the width (in
in.).   Then $3w =$ the length. $P = 2l + 2w$; $56 = 2 \cdot 3w + 2w$; $56 = 6w + 2w$;
$56 = 8w$; $7 = w$ and $3w = 21$.   length: 21 in.;   width: 7 in.   **4.** $A =$ area
rectangle $-$ area cut-out square $= (4x)(3x) - x^2 = 12x^2 - x^2 = 11x^2$

5. $A = \dfrac{1}{2}bh = \dfrac{1}{2} \cdot 6y \cdot 4y = 12y^2$ **6.** $A = 90 \cdot 5 = 450$ ft²; $2A = 900$ ft²;
$900 \div 300 = 3$;   3 cans of paint

7. $V = Bh = (10 \cdot 6) \cdot 8 = 480$;   480 cm^3

8. rx km   **9.** $(x - 5c)$ cents

10.

	rate	time	Distance
1st car	$x + 10$	4	$4(x + 10)$
2nd car	x	4	$4x$
			640

$4(x + 10) + 4x = 640$; $4x + 40 +$
$4x = 640$; $8x + 40 = 640$; $8x = 600$;
$x = 75$ and $x + 10 = 85$.
75 km/h and 85 km/h

11.

	price	number	Cost
Sell	35	x	$35x$
Buy	26	$x + 2$	$26(x + 2)$
			92

$35x - 26(x + 2) = 92$; $35x - 26x -$
$52 = 92$; $9x - 52 = 92$; $9x = 144$;
$x = 16$;   16 irons sold

12. $3y - x = 4$; $3y - x - 3y = 4 - 3y$; $-x = 4 - 3y$; $(-1)(-x) =$
$(-1)(4 - 3y)$; $x = -4 + 3y$ **13.** $C = pn$; $\dfrac{C}{p} = \dfrac{pn}{p}$; $\dfrac{C}{p} = n$

Page 148 · CUMULATIVE REVIEW

1. $>$ **2.** $<$ **3.** $>$ **4.** $>$ **5.** $x - 3 = 4$; $x = 7$

6. $y + 1 = 1$; $y = 0$

7.   **8.**

9. $3 + (-3) = 0$ **10.** $-3 + (-3) = -6$ **11.** $-7 + 5 = -2$

12. $-11 + (-3) = -14$ **13.** $8 - 4 = 4$ **14.** $-8 - 4 = -8 + (-4) = -12$

15. $8 - (-4) = 8 + 4 = 12$ **16.** $-8 - (-4) = -8 + 4 = -4$

17. $7 \cdot 20 = 140$ **18.** $-4 \cdot 16 = -64$ **19.** $-5(-9) = 45$

20. $-(-12)^2 = -144$ **21.** $3a(-7y) = (3)(-7)(ay) = -21ay$

22. $-(-1 - x) = (-1)(-1) + (-1)(-x) = 1 + x$ **23.** $t - (3t - 5) =$
$t - 3t + 5 = -2t + 5$ **24.** $(2z - y)(-8) = (2z)(-8) + (-y)(-8) = -16z + 8y$

25. $24 \div -8 = -3$ **26.** $-16 \div 2 = -8$ **27.** $-54 \div -9 = 6$ **28.** $26 \div -2 = -13$

29. $x + 5 = -1$; $x + 5 - 5 = -1 - 5$; $x = -6$

30. $4z = -60$; $\dfrac{4z}{4} = \dfrac{-60}{4}$; $z = -15$

31. $4t + 2 = t - 4$; $4t + 2 - 2 = t - 4 - 2$; $4t = t - 6$; $4t - t = t - 6 - t$;
$3t = -6$; $t = -2$ **32.** The unlabeled side has length $3x - x = 2x$.   Thus, $P =$
$3 \cdot x + 2 \cdot 2x + 3x = 3x + 4x + 3x = 10x$ **33.** $A = \dfrac{1}{2}bh = \dfrac{1}{2} \cdot 8 \cdot 4 = 16$;
16 cm^2 **34.** $V = Bh = (4y \cdot y) \cdot 2y = 8y^3$ **35.** $2w + t$

36.

	rate	time	Distance
A to B	750	t	$750t$
B to A	600	$t + 1$	$600(t + 1)$

$750t = 600(t + 1)$; $750t = 600t + 600$; $150t = 600$; $t = 4$.   Thus, Distance = $750t = 750 \cdot 4 = 3000$.   3000 km

37.

	price	number	Cost
Sell	3	$n - 4$	$3(n - 4)$
Buy	2	n	$2n$
			56

$3(n - 4) - 2n = 56$; $3n - 12 - 2n = 56$; $n - 12 = 56$; $n = 68$.
68 pies bought

Page 153 · WRITTEN EXERCISES

A

1. $3y + 11$

2. $9x^2 + 15$

3. $2m^2$

4. $-11x + 13y$

5. $5ab + 2b^2$

6. $3a^2b + 2ab^2$

7. $2x^2 + xy + 2y^2$

8. $10n - 13$

9. $-5y^2 + 1$

10. $\begin{array}{l} 2x - y \\ \underline{7x + y} \\ 9x \end{array}$

11. $\begin{array}{l} n + 1 \\ \underline{11n + 1} \\ 12n + 2 \end{array}$

12. $\begin{array}{l} 3y + z + 1 \\ \underline{18y + 9z + 16} \\ 21y + 10z + 17 \end{array}$

13. $\begin{array}{l} m + 3n + 8 \\ \underline{9m + 9n + 7} \\ 10m + 12n + 15 \end{array}$

14. $\begin{array}{l} n^2 + 2n + 1 \\ \underline{3n^2 + 4n - 3} \\ 4n^2 + 6n - 2 \end{array}$

15. $\begin{array}{l} -m^2 + 4m \\ \underline{m^2 - 8m + 8} \\ -4m + 8 \end{array}$

16. $\begin{array}{l} a^2 - 2ab + b^2 \\ \underline{-2a^2 + b^2} \\ {-a^2} - 2ab + 2b^2 \end{array}$

17. $\begin{array}{l} x^2 - 10x + 5 \\ \underline{9x^2 - 10x - 3} \\ 10x^2 - 20x + 2 \end{array}$

18. $\begin{array}{l} 3a^2 + 3a - 6 \\ \underline{a^2 - 1} \\ 4a^2 + 3a - 7 \end{array}$

19. $\begin{array}{l} x^2 + 2x + 4 \\ \underline{4x^2 - 5} \\ 5x^2 + 2x - 1 \end{array}$

20. $\begin{array}{l} n^3 - 4 \\ \underline{n^3 - n^2 + 10} \\ 2n^3 - n^2 + 6 \end{array}$

21. $\begin{array}{l} k^3 - 1 \\ \underline{ + k^2 - 2k + 7} \\ k^3 + k^2 - 2k + 6 \end{array}$

22. $(n^2 - n) + (n^3 - 2n) = n^3 + n^2 - n - 2n = n^3 + n^2 - 3n$

23. $(2x^2 + 3) + (x^3 + 4x^2 - 6) = x^3 + 2x^2 + 4x^2 + 3 - 6 = x^3 + 6x^2 - 3$

24. $(2a^3 - ab + 1) + (a^3 - 2ab) = 2a^3 + a^3 - ab - 2ab + 1 = 3a^3 - 3ab + 1$

25. $(4a^2 - 2ab) + (-3a^2 + ab - b^2) = 4a^2 - 3a^2 - 2ab + ab - b^2 = a^2 - ab - b^2$

26. **a.** $P = 3n + n + 6 + 2n - 8 + n + 7 + 2n - 4 = 9n + 1$

   b. $P = 9n + 1 = 9 \cdot 5 + 1 = 45 + 1 = 46$

Page 155 · WRITTEN EXERCISES

A

1. $\begin{array}{l} 2x + y \\ \underline{-x + 3y} \\ x + 4y \end{array}$

2. $\begin{array}{l} 3a - 4b \\ \underline{-a - 5b} \\ 2a - 9b \end{array}$

3. $\begin{array}{l} -x^2 + y^2 \\ \underline{-x^2 + y^2} \\ -2x^2 + 2y^2 \end{array}$

4. $\begin{array}{l} x^2 - 2x - 1 \\ \underline{-x^2 - 2x - 1} \\ -4x - 2 \end{array}$

5. $\begin{array}{l} 3x^2 - 4x + 5 \\ \underline{-x^2 + 4x - 1} \\ 2x^2 + 4 \end{array}$

6. $\begin{array}{l} n^2 - 4n + 5 \\ \underline{-3n^2 - 7n - 3} \\ -2n^2 - 11n + 2 \end{array}$

7. $\begin{array}{l} x^2 - 2xy + y^2 \\ \underline{-x^2 - 2xy - y^2} \\ -4xy \end{array}$

8. $\begin{array}{l} y^3 + y^2 + 6 \\ \underline{-y^3 - 4} \\ y^2 + 2 \end{array}$

9. $\begin{array}{l} x^2y + xy^2 + 4 \\ \underline{ 3xy^2 - 2} \\ x^2y + 4xy^2 + 2 \end{array}$

10. $\begin{array}{l} a + 1 \\ \underline{-2a + 4} \\ {-a} + 5 \end{array}$

11. $\begin{array}{l} 2b - 2 \\ \underline{-b + 4} \\ b + 2 \end{array}$

12. $\begin{array}{l} x^2 + 1 \\ \underline{-x^2 - 1} \\ 0 \end{array}$

13. $\begin{array}{l} 5m - 16 \\ \underline{-m - 2} \\ 4m - 18 \end{array}$

14. $\begin{array}{l} 3n - 2 \\ \underline{n + 2} \\ 4n \end{array}$

15. $\begin{array}{l} y + 6 \\ \underline{3y + 8} \\ 4y + 14 \end{array}$

55

16. $\quad x^2 + 3x + 2$
$\quad\ \underline{-x^2 + 4x - 1}$
$\qquad\qquad 7x + 1$

17. $\quad x^2 + 2x + 1$
$\quad\ \underline{-4x^2 + 3x - 7}$
$\quad -3x^2 + 5x - 6$

18. $\quad a^2 + 5ab - 2c$
$\quad\ \underline{-3a^2 - \ ab + 4c}$
$\quad -2a^2 + 4ab + 2c$

19. $\qquad\ \ 2ab - 4c$
$\quad\ \underline{-a^2 + 3ab - 6c}$
$\quad -a^2 + 5ab - 10c$

20. $\quad\ x^2 - 8x + 7$
$\quad\ \underline{-5x^2 \qquad - 9}$
$\quad -4x^2 - 8x - 2$

21. $\quad 4y^2 \qquad\ - 8$
$\quad\ \underline{-2y^2 + 5y - 10}$
$\quad\ 2y^2 + 5y - 18$

22. $3x - (-24 - x) = 60;\quad 3x + 24 + x = 60;\quad 4x + 24 = 60;\quad 4x = 36;\quad x = 9$

23. $(13n + 5) - (3n + 6) = 99;\quad 13n + 5 - 3n - 6 = 99;\quad 10n - 1 = 99;\quad 10n = 100;\quad n = 10$

24. $(2x - 4) - (x + 8) = 24;\quad 2x - 4 - x - 8 = 24;\quad x - 12 = 24;\quad x = 36$

25. $(12n - 40) - (10n + 30) = n + 10;\quad 12n - 40 + 10n - 30 = n + 10;\quad 2n - 70 = n + 10;\quad 2n = n + 80;\quad n = 80$

B   26. $(4n - 7) - (3n + 1) = 4n - 7 - 3n - 1 = n - 8;\quad (n - 8)$ cm

Page 157 · WRITTEN EXERCISES

A

1. $a \cdot a^2 = a^{1+2} = a^3$
2. $b^2 \cdot b^2 = b^{2+2} = b^4$
3. $x^2 \cdot 3x = 3x^{2+1} = 3x^3$
4. $x^4 \cdot x = x^{4+1} = x^5$
5. $n^3 \cdot n^4 = n^{3+4} = n^7$
6. $a^6 \cdot a = a^{6+1} = a^7$
7. $(2x)(2x^2) = (2 \cdot 2)(x \cdot x^2) = 4x^{1+2} = 4x^3$
8. $(3x)(-2x^4) = (3 \cdot -2)(x \cdot x^4) = -6x^{1+4} = -6x^5$
9. $(c^2)(-5c^3) = (1 \cdot -5)(c^2 \cdot c^3) = -5c^{2+3} = -5c^5$
10. $(ab)(a^2b) = (a \cdot a^2)(b \cdot b) = a^{1+2}b^{1+1} = a^3b^2$
11. $(3x^2)(-2x^5) = (3 \cdot -2)(x^2 \cdot x^5) = -6x^{2+5} = -6x^7$
12. $(-y^2)(-y^7) = (-1 \cdot -1)(y^2 \cdot y^7) = y^{2+7} = y^9$
13. $(3x^2)(4x^4) = (3 \cdot 4)(x^2 \cdot x^4) = 12x^{2+4} = 12x^6$
14. $(-x^2)(-4x) = (-1 \cdot -4)(x^2 \cdot x) = 4x^{2+1} = 4x^3$
15. $(5a)(-ab^2) = (5 \cdot -1)(a \cdot a)(b^2) = -5a^{1+1}b^2 = -5a^2b^2$
16. $(xy)(-2x) = (1 \cdot -2)(x \cdot x)(y) = -2x^{1+1}y = -2x^2y$
17. $(cd)(-3d^3) = (1 \cdot -3)(c)(d \cdot d^3) = -3cd^{1+3} = -3cd^4$
18. $(2mn)(-8m^2) = (2 \cdot -8)(m \cdot m^2)(n) = -16m^{1+2}n = -16m^3n$
19. $(5x^2y)(4xy^2) = (5 \cdot 4)(x^2 \cdot x)(y \cdot y^2) = 20x^{2+1}y^{1+2} = 20x^3y^3$
20. $(-5xy)(2xy^2) = (-5 \cdot 2)(x \cdot x)(y \cdot y^2) = -10x^{1+1}y^{1+2} = -10x^2y^3$
21. $(-r^2s)(-10r^2s^2) = (-1 \cdot -10)(r^2 \cdot r^2)(s \cdot s^2) = 10r^{2+2}s^{1+2} = 10r^4s^3$
22. $(-6a^2)(4ab^5) = (-6 \cdot 4)(a^2 \cdot a)(b^5) = -24a^{2+1}b^5 = -24a^3b^5$
23. $(-x^4)(-3xyz^2) = (-1 \cdot -3)(x^4 \cdot x)(yz^2) = 3x^{4+1}yz^2 = 3x^5yz^2$
24. $(-6a^2b^5)(abc^3) = (-6 \cdot 1)(a^2 \cdot a)(b^5 \cdot b)(c^3) = -6a^{2+1}b^{5+1}c^3 = -6a^3b^6c^3$
25. $(a^2b)(-5a^2b^2) = (1 \cdot -5)(a^2 \cdot a^2)(b \cdot b^2) = -5a^{2+2}b^{1+2} = -5a^4b^3$
26. $(-x^3)(-5x^2y) = (-1 \cdot -5)(x^3 \cdot x^2)(y) = 5x^{3+2}y = 5x^5y$
27. $(xy^3)(-2x^3y^2) = (1 \cdot -2)(x \cdot x^3)(y^3 \cdot y^2) = -2x^{1+3}y^{3+2} = -2x^4y^5$
28. $(-a^4b^3)(a^2bc^5) = -(a^4 \cdot a^2)(b^3 \cdot b)(c^5) = -a^{4+2}b^{3+1}c^5 = -a^6b^4c^5$

29. $(m^3n)(-4m^3n^2p^4) = (1 \cdot -4)(m^3 \cdot m^3)(n \cdot n^2)(p^4) = -4m^{3+3}n^{1+2}p^4 = -4m^6n^3p^4$

30. $(r^2s^3t^4)(r^5st^3) = (r^2 \cdot r^5)(s^3 \cdot s)(t^4 \cdot t^3) = r^{2+5}s^{3+1}t^{4+3} = r^7s^4t^7$

Page 157 · PUZZLE PROBLEMS

The word "nothing" has 2 different meanings here: in the first sentence, the absence of something; in the second, no thing. Thus, logic cannot be used to conclude the third sentence.

Page 159 · WRITTEN EXERCISES

A　1. $(x^2)^3 = x^{2 \cdot 3} = x^6$　　　　　　　2. $(a^3)^4 = a^{3 \cdot 4} = a^{12}$

3. $(b^6)^2 = b^{6 \cdot 2} = b^{12}$　　　　　　　4. $(x^2)^5 = x^{2 \cdot 5} = x^{10}$

5. $(c^3)^5 = c^{3 \cdot 5} = c^{15}$　　　　　　　6. $(n^4)^{10} = n^{4 \cdot 10} = n^{40}$

7. $(2x)^2 = 2^2 \cdot x^2 = 4x^2$　　　　　　8. $(4a)^2 = 4^2 \cdot a^2 = 16a^2$

9. $(ab)^4 = a^4b^4$　　　　　　　　　10. $(xy)^6 = x^6y^6$

11. $(6ax)^2 = 6^2 \cdot a^2x^2 = 36a^2x^2$　　　12. $(-2xy)^3 = (-2)^3(x^3y^3) = -8x^3y^3$

13. $A = (5a)^2 = 5^2 \cdot a^2 = 25a^2$　　　14. $A = (6x)^2 = 6^2 \cdot x^2 = 36x^2$

15. $A = (3x)^2 + x^2 = 3^2 \cdot x^2 + x^2 = 9x^2 + x^2 = 10x^2$

16. $(5x^3)^2 = 5^2(x^3)^2 = 25x^6$　　　　　17. $(3a^2)^2 = 3^2(a^2)^2 = 9a^4$

18. $(2b^3)^2 = 2^2(b^3)^2 = 4b^6$　　　　　19. $(5x^5)^2 = 5^2(x^5)^2 = 25x^{10}$

20. $(-2a^5)^2 = (-2)^2(a^5)^2 = 4a^{10}$　　　21. $(2a^2b)^2 = 2^2(a^2)^2b^2 = 4a^4b^2$

22. $(3xy^3)^2 = 3^2x^2(y^3)^2 = 9x^2y^6$　　　23. $(-x^3y^3)^2 = (-1)^2(x^3)^2(y^3)^2 = x^6y^6$

24. $2(ab^4)^2 = 2a^2(b^4)^2 = 2a^2b^8$　　　25. $3(x^2y)^2 = 3(x^2)^2y^2 = 3x^4y^2$

26. $-(2n^2)^3 = -(2)^3(n^2)^3 = -8n^6$　　　27. $-(4n^2)^2 = -(4)^2(n^2)^2 = -16n^4$

B　28. $(xy)^2(-xy) = (x^2y^2)(-xy) = -(x^2 \cdot x)(y^2 \cdot y) = -x^3y^3$

29. $(-xy^3)(-xy)^2 = (-xy^3)(-1)^2(x^2y^2) = (-xy^3)(x^2y^2) = -(x \cdot x^2)(y^3 \cdot y^2) = -x^3y^5$

30. $(3xy^4)^2(-4x^2)^2 = 3^2x^2(y^4)^2 \cdot (-4)^2(x^2)^2 = 9x^2y^8 \cdot 16x^4 = (9 \cdot 16)(x^2 \cdot x^4)(y^8) = 144x^6y^8$

31. $(2mn)^3(3n)^2 = 2^3m^3n^3 \cdot 3^2n^2 = 8m^3n^3 \cdot 9n^2 = (8 \cdot 9)(m^3)(n^3 \cdot n^2) = 72m^3n^5$

32. $-(x^2y^2)(x^2) = -(x^2 \cdot x^2)(y^2) = -x^4y^2$

33. $-(r^4s^3)^2(rs^2)^3 = -(r^4)^2(s^3)^2 \cdot (r^3)(s^2)^3 = -r^8s^6 \cdot r^3s^6 = -r^{11}s^{12}$

Page 161 · WRITTEN EXERCISES

A　1. $2(x + 4) = (2 \cdot x) + (2 \cdot 4) = 2x + 8$

2. $3(a - b) = (3 \cdot a) - (3 \cdot b) = 3a - 3b$

3. $4(x + y) = (4 \cdot x) + (4 \cdot y) = 4x + 4y$

4. $5(a^2 + b) = (5 \cdot a^2) + (5 \cdot b) = 5a^2 + 5b$

5. $-6(n + 2m) = (-6)(n) + (-6)(2m) = -6n - 12m$

6. $-1(5a + b^2) = (-1)(5a) + (-1)(b^2) = -5a - b^2$

7. $a(a - b) = (a \cdot a) - (a \cdot b) = a^2 - ab$

8. $x(x + 3y) = (x \cdot x) + (x \cdot 3y) = x^2 + 3xy$

9. $-c(a + b) = (-c)(a) + (-c)(b) = -ca - cb$

10. $-ab(2a - 4b) = (-ab)(2a) - (-ab)(4b) = -2a^2b + 4ab^2$

11. $-5x(3x + 2y) = (-5x)(3x) + (-5x)(2y) = -15x^2 - 10xy$

12. $2x(3x - 1) = (2x \cdot 3x) - (2x \cdot 1) = 6x^2 - 2x$

13. $4a(a + 2b + 3) = (4a \cdot a) + (4a \cdot 2b) + (4a \cdot 3) = 4a^2 + 8ab + 12a$

14. $-4(1 + 5x + x^2) = (-4)(1) + (-4)(5x) + (-4)(x^2) = -4 - 20x - 4x^2$

15. $-1(2x + y + z) = (-1)(2x) + (-1)(y) + (-1)(z) = -2x - y - z$

16. $2x(x^2 - 2x - 4) = (2x \cdot x^2) - (2x \cdot 2x) - (2x \cdot 4) = 2x^3 - 4x^2 - 8x$

17. $-4y(y^3 - 2y + 1) = (-4y)(y^3) - (-4y)(2y) + (-4y)(1) = -4y^4 + 8y^2 - 4y$

18. $ab(a^2 + 2ab - 1) = (ab \cdot a^2) + (ab \cdot 2ab) - (ab \cdot 1) = a^3b + 2a^2b^2 - ab$

19. $-x^2(x + 2x^2) = (-x^2)(x) + (-x^2)(2x^2) = -x^3 - 2x^4$

20. $-3c(2c^2 + 4c - 5) = (-3c)(2c^2) + (-3c)(4c) - (-3c)(5) = -6c^3 - 12c^2 + 15c$

21. $-y^2(y^3 - 2y^2 + 4y) = (-y^2)(y^3) - (-y^2)(2y^2) + (-y^2)(4y) = -y^5 + 2y^4 - 4y^3$

22. $A = 10n(n + 6) = (10n \cdot n) + (10n \cdot 6) = 10n^2 + 60n$; $(10n^2 + 60n)$ cm^2

B **23.** Value $= 10(3n + 1) = 10 \cdot 3n + 10 \cdot 1 = 30n + 10$; $(30n + 10)$ cents

24. $4(2n + 3) - 3(n - 1) = 0$; $(4 \cdot 2n) + (4 \cdot 3) - (3 \cdot n) - (3 \cdot -1) = 0$;
$8n + 12 - 3n + 3 = 0$; $5n + 15 = 0$; $5n = -15$; $n = -3$

25. $-(n + 3) + 2(n + 7) = 0$; $-n - 3 + 2n + 14 = 0$; $n + 11 = 0$; $n = -11$

26. $5x + 2 - 2(2x + 6) = 0$; $5x + 2 - 4x - 12 = 0$; $x - 10 = 0$; $x = 10$

27. $(2y - 3) - (y + 6) = 63$; $2y - 3 - y - 6 = 63$; $y - 9 = 63$; $y = 72$

28. $2(5x - 6) - 3(2x - 4) = 0$; $10x - 12 - 6x + 12 = 0$; $4x = 0$; $x = 0$

29. $3(1 - 2a) - (6 - 2a) = -7$; $3 - 6a - 6 + 2a = -7$; $-4a - 3 = -7$;
$-4a = -4$; $a = 1$

30. $3(x - 4) + 2(2x + 1) = 4$; $3x - 12 + 4x + 2 = 4$; $7x - 10 = 4$; $7x = 14$;
$x = 2$

31. $2(n - 6) + 5(2n + 4) = 32$; $2n - 12 + 10n + 20 = 32$; $12n + 8 = 32$;
$12n = 24$; $n = 2$

32. $6(1 - 3x) - 2(2x + 5) = 40$; $6 - 18x - 4x - 10 = 40$; $-22x - 4 = 40$;
$-22x = 44$; $x = -2$

33. $4(2a - 3) - 2(a - 8) = 22$; $8a - 12 - 2a + 16 = 22$; $6a + 4 = 22$; $6a = 18$;
$a = 3$

34. $7(m + 3) - 5(2 - m) = -1$; $7m + 21 - 10 + 5m = -1$; $12m + 11 = -1$;
$12m = -12$; $m = -1$

35. $2(2x - 1) + 3(x + 4) = 52$; $4x - 2 + 3x + 12 = 52$; $7x + 10 = 52$; $7x = 42$;
$x = 6$

Page 163 · WRITTEN EXERCISES

A

1.
$$\begin{array}{r} x + 2 \\ \underline{x + 1} \\ x^2 + 2x \\ \underline{ + x + 2} \\ x^2 + 3x + 2 \end{array}$$

2.
$$\begin{array}{r} y + 4 \\ \underline{y + 3} \\ y^2 + 4y \\ \underline{ + 3y + 12} \\ y^2 + 7y + 12 \end{array}$$

3.
$$\begin{array}{r} n + 6 \\ \underline{n + 4} \\ n^2 + 6n \\ \underline{ + 4n + 24} \\ n^2 + 10n + 24 \end{array}$$

4.
$$\begin{array}{r} a + 5 \\ \underline{a + 3} \\ a^2 + 5a \\ \underline{ + 3a + 15} \\ a^2 + 8a + 15 \end{array}$$

5.
$$\begin{array}{r} x + 7 \\ \underline{x + 1} \\ x^2 + 7x \\ \underline{ + x + 7} \\ x^2 + 8x + 7 \end{array}$$

6.
$$\begin{array}{r} y + 3 \\ \underline{y + 9} \\ y^2 + 3y \\ \underline{ + 9y + 27} \\ y^2 + 12y + 27 \end{array}$$

7.
$$\begin{array}{r} x + y \\ \underline{x + y} \\ x^2 + xy \\ \underline{ + xy + y^2} \\ x^2 + 2xy + y^2 \end{array}$$

8.
$$\begin{array}{r} a + c \\ \underline{a + b} \\ a^2 + ac \\ \underline{ + ab + bc} \\ a^2 + ac + ab + bc \end{array}$$

9.
$$\begin{array}{r} x + 4 \\ \underline{x + y} \\ x^2 + 4x \\ \underline{ + xy + 4y} \\ x^2 + 4x + xy + 4y \end{array}$$

10.
$$\begin{array}{r} x - 2 \\ \underline{x - 1} \\ x^2 - 2x \\ \underline{ - x + 2} \\ x^2 - 3x + 2 \end{array}$$

11.
$$\begin{array}{r} y - 4 \\ \underline{y - 3} \\ y^2 - 4y \\ \underline{ - 3y + 12} \\ y^2 - 7y + 12 \end{array}$$

12.
$$\begin{array}{r} n - 1 \\ \underline{n - 3} \\ n^2 - n \\ \underline{ - 3n + 3} \\ n^2 - 4n + 3 \end{array}$$

13.
$$\begin{array}{r} x + 2 \\ \underline{x - 1} \\ x^2 + 2x \\ \underline{ - x - 2} \\ x^2 + x - 2 \end{array}$$

14.
$$\begin{array}{r} y - 4 \\ \underline{y + 3} \\ y^2 - 4y \\ \underline{ + 3y - 12} \\ y^2 - y - 12 \end{array}$$

15.
$$\begin{array}{r} y + 4 \\ \underline{y - 5} \\ y^2 + 4y \\ \underline{ - 5y - 20} \\ y^2 - y - 20 \end{array}$$

16.
$$\begin{array}{r} x + 2y \\ \underline{2x + y} \\ 2x^2 + 4xy \\ \underline{ + xy + 2y^2} \\ 2x^2 + 5xy + 2y^2 \end{array}$$

17.
$$\begin{array}{r} 4x + 1 \\ \underline{4x + 1} \\ 16x^2 + 4x \\ \underline{ + 4x + 1} \\ 16x^2 + 8x + 1 \end{array}$$

18.
$$\begin{array}{r} 4x - 1 \\ \underline{4x + 1} \\ 16x^2 - 4x \\ \underline{ + 4x - 1} \\ 16x^2 - 1 \end{array}$$

19.
$$\begin{array}{r} a - b \\ \underline{a + b} \\ a^2 - ab \\ \underline{ + ab - b^2} \\ a^2 - b^2 \end{array}$$

20.
$$\begin{array}{r} a - 2b \\ \underline{a + 2b} \\ a^2 - 2ab \\ \underline{ + 2ab - 4b^2} \\ a^2 - 4b^2 \end{array}$$

21.
$$\begin{array}{r} a + b \\ \underline{2a - 1} \\ 2a^2 + 2ab \\ \underline{ - a - b} \\ 2a^2 + 2ab - a - b \end{array}$$

22.
$$\begin{array}{r} 6 - x \\ \underline{6 - x} \\ 36 - 6x \\ \underline{ - 6x + x^2} \\ 36 - 12x + x^2 \end{array}$$

23.
$$\begin{array}{r} 6x - 5 \\ \underline{6x - 5} \\ 36x^2 - 30x \\ \underline{ - 30x + 25} \\ 36x^2 - 60x + 25 \end{array}$$

24.
$$\begin{array}{r} 2y + 1 \\ \underline{3y - 2} \\ 6y^2 + 3y \\ \underline{ - 4y - 2} \\ 6y^2 - y - 2 \end{array}$$

B

25.
$$n^2 + 2n + 1$$
$$n + 1$$
$$n^3 + 2n^2 + n$$
$$+ n^2 + 2n + 1$$
$$n^3 + 3n^2 + 3n + 1$$

26.
$$x^2 + 2x + 1$$
$$x + 2$$
$$x^3 + 2x^2 + x$$
$$+ 2x^2 + 4x + 2$$
$$x^3 + 4x^2 + 5x + 2$$

27.
$$x^2 - 4x + 4$$
$$x - 2$$
$$x^3 - 4x^2 + 4x$$
$$- 2x^2 + 8x - 8$$
$$x^3 - 6x^2 + 12x - 8$$

28.
$$y^2 + 6y + 9$$
$$y + 3$$
$$y^3 + 6y^2 + 9y$$
$$+ 3y^2 + 18y + 27$$
$$y^3 + 9y^2 + 27y + 27$$

29.
$$m^2 - 2m + 4$$
$$m - 2$$
$$m^3 - 2m^2 + 4m$$
$$- 2m^2 + 4m - 8$$
$$m^3 - 4m^2 + 8m - 8$$

30.
$$n^2 - 8n + 16$$
$$n - 4$$
$$n^3 - 8n^2 + 16n$$
$$- 4n^2 + 32n - 64$$
$$n^3 - 12n^2 + 48n - 64$$

31.
$$a^2 + 2ab + b^2$$
$$a + b$$
$$a^3 + 2a^2b + ab^2$$
$$+ a^2b + 2ab^2 + b^3$$
$$a^3 + 3a^2b + 3ab^2 + b^3$$

32.
$$a^2 - b^2$$
$$a - b$$
$$a^3 - ab^2$$
$$- a^2b + b^3$$
$$a^3 - ab^2 - a^2b + b^3$$

C

33.
$$n^2 + 2n + 1$$
$$n^2 + 2n + 1$$
$$n^4 + 2n^3 + n^2$$
$$+ 2n^3 + 4n^2 + 2n$$
$$+ n^2 + 2n + 1$$
$$n^4 + 4n^3 + 6n^2 + 4n + 1$$

34.
$$x^2 - 3x - 1$$
$$x^2 - 7x + 12$$
$$x^4 - 3x^3 - x^2$$
$$- 7x^3 + 21x^2 + 7x$$
$$+ 12x^2 - 36x - 12$$
$$x^4 - 10x^3 + 32x^2 - 29x - 12$$

35.
$$5a^2 + ab + b^2$$
$$2a^2 - ab + 4b^2$$
$$10a^4 + 2a^3b + 2a^2b^2$$
$$- 5a^3b - a^2b^2 - ab^3$$
$$+ 20a^2b^2 + 4ab^3 + 4b^4$$
$$10a^4 - 3a^3b + 21a^2b^2 + 3ab^3 + 4b^4$$

36.
$$2x^4 - 5x^3 - 7x^2 + 10$$
$$x - 4$$
$$2x^5 - 5x^4 - 7x^3 + 10x$$
$$- 8x^4 + 20x^3 + 28x^2 - 40$$
$$2x^5 - 13x^4 + 13x^3 + 28x^2 + 10x - 40$$

37.
$$n^4 - 3n^3 + n^2 + 1$$
$$n - 1$$
$$n^5 - 3n^4 + n^3 + n$$
$$- n^4 + 3n^3 - n^2 - 1$$
$$n^5 - 4n^4 + 4n^3 - n^2 + n - 1$$

38.
$$a^4 + a^3b + a^2b^2 + ab^3 + b^4$$
$$a - b$$
$$a^5 + a^4b + a^3b^2 + a^2b^3 + ab^4$$
$$- a^4b - a^3b^2 - a^2b^3 - ab^4 - b^5$$
$$a^5 - b^5$$

Pages 166–167 · WRITTEN EXERCISES

A

1. $(x + 4)(x + 3) = x^2 + 3x + 4x + 12 = x^2 + 7x + 12$

2. $(x + 3)(x + 5) = x^2 + 5x + 3x + 15 = x^2 + 8x + 15$

3. $(x + 6)(x + 2) = x^2 + 2x + 6x + 12 = x^2 + 8x + 12$

4. $(x + 2)(x + 1) = x^2 + x + 2x + 2 = x^2 + 3x + 2$

5. $(x - 1)(x - 2) = x^2 - 3x + 2$ **6.** $(x - 6)(x - 3) = x^2 - 9x + 18$

7. $(x - 4)(x - 3) = x^2 - 7x + 12$ **8.** $(a - 3)(a - 5) = a^2 - 8a + 15$

9. $(y - 4)(y + 1) = y^2 - 3y - 4$ **10.** $(a - 1)(a + 3) = a^2 + 2a - 3$

11. $(n + 2)(n - 1) = n^2 + n - 2$ **12.** $(x + 5)(x - 2) = x^2 + 3x - 10$

13. $(y + 6)(y - 2) = y^2 + 4y - 12$ **14.** $(a + 5)(a - 3) = a^2 + 2a - 15$

15. $(2x + 1)(x + 4) = 2x^2 + 9x + 4$ 16. $(x + 2)(x - 2) = x^2 - 4$

17. $(a + b)(a - b) = a^2 - b^2$ 18. $(2x + 5)(2x - 5) = 4x^2 - 25$

19. $(6a + 7)(6a - 7) = 36a^2 - 49$ 20. $(2y - 3)(2y + 3) = 4y^2 - 9$

21. $(3x + 4)(3x - 4) = 9x^2 - 16$ 22. $(5n - 1)(2n + 3) = 10n^2 + 13n - 3$

23. $(3x - 1)(2x + 5) = 6x^2 + 13x - 5$

24. $(3x + 2y)(2x + 3y) = 6x^2 + 13xy + 6y^2$

25. $(3x + y)(7x + 2y) = 21x^2 + 13xy + 2y^2$

26. $(3a - b)(a - 3b) = 3a^2 - 10ab + 3b^2$

27. $(4x - 3y)(x - 5y) = 4x^2 - 23xy + 15y^2$

28. $(3a - b)(7a + 4b) = 21a^2 + 5ab - 4b^2$

29. $(6x - y)(4x + 2y) = 24x^2 + 8xy - 2y^2$

30. $(3n + 2p)(7n - p) = 21n^2 + 11np - 2p^2$

31. $A = (x + 6)(x + 4) = x^2 + 10x + 24$ 32. $A = (x + 3)(x + 3) = x^2 + 6x + 9$

33. $A = (a + 4)(2a - 3) = 2a^2 + 5a - 12$

34. $A = (17 - x)(x - 4) = 17x - 68 - x^2 + 4x = 21x - 68 - x^2;\quad (21x - 68 - x^2)\ \text{cm}^2$

35. $A = (2y + 3)(3y + 1) = 6y^2 + 11y + 3;\quad (6y^2 + 11y + 3)\ \text{cm}^2$

B   36. $A = \frac{1}{2}(3n - 1)(4n) = \frac{1}{2}(12n^2 - 4n) = 6n^2 - 2n$

37. $A = \frac{1}{2}(2n - 1)(n + 3) = \frac{1}{2}(2n^2 + 5n - 3)$

38. $A = \frac{1}{2}(2x + 3)(x - 3) + (2x + 3)(x - 1) = \frac{1}{2}(2x^2 - 3x - 9) + 2x^2 + x - 3$

C   39. $A = \frac{1}{2} \cdot 2x(3x - 1 + 3x + 1) = x(6x) = 6x^2$

40. Area (top) $= (5x + 1)(9x - 2) = 45x^2 - x - 2$;   area (bottom) $= 10x(9x - 2) = 90x^2 - 20x$;   area (2 rectangular sides) $= 2 \cdot (9x - 2)(6x) = 2(54x^2 - 12x) = 108x^2 - 24x$;   area (2 trapezoidal sides) $= 2 \cdot \frac{1}{2}(5x + 2)(5x + 1 + 10x) = (5x + 2)(15x + 1) = 75x^2 + 35x + 2$.   Thus, total surface area $= (45x^2 - x - 2) + (90x^2 - 20x) + (108x^2 - 24x) + (75x^2 + 35x + 2) = 318x^2 - 10x$

Page 169 · WRITTEN EXERCISES

A   1. $(x + 4)^2 = x^2 + 2 \cdot 4 \cdot x + 4^2 = x^2 + 8x + 16$

2. $(m + 1)^2 = m^2 + 2 \cdot 1 \cdot m + 1^2 = m^2 + 2m + 1$

3. $(r + 2)^2 = r^2 + 2 \cdot 2 \cdot r + 2^2 = r^2 + 4r + 4$

4. $(y - 3)^2 = y^2 - 2 \cdot 3 \cdot y + (-3)^2 = y^2 - 6y + 9$

5. $(x + 3)^2 = x^2 + 6x + 9$ 6. $(n + 6)^2 = n^2 + 12n + 36$

7. $(x - 2)^2 = x^2 - 4x + 4$ 8. $(m - 6)^2 = m^2 - 12m + 36$

9. $(x - 4)^2 = x^2 - 8x + 16$ 10. $(m + 5)^2 = m^2 + 10m + 25$

11. $(y - 7)^2 = y^2 - 14y + 49$ **12.** $(a + 10)^2 = a^2 + 20a + 100$

13. $(x + 9)^2 = x^2 + 18x + 81$ **14.** $(y - 8)^2 = y^2 - 16y + 64$

15. $(m + 7)^2 = m^2 + 14m + 49$ **16.** $(n + 8)^2 = n^2 + 16n + 64$

17. $(n - 5)^2 = n^2 - 10n + 25$ **18.** $(y - 1)^2 = y^2 - 2y + 1$

19. $(3a + 2)^2 = (3a)^2 + 2 \cdot 2 \cdot 3a + 2^2 = 9a^2 + 12a + 4$

20. $(2x + 3)^2 = (2x)^2 + 2 \cdot 3 \cdot 2x + 3^2 = 4x^2 + 12x + 9$

21. $(4x - 2)^2 = 16x^2 - 16x + 4$ **22.** $(2a + b)^2 = 4a^2 + 4ab + b^2$

23. $(3x + y)^2 = 9x^2 + 6xy + y^2$ **24.** $(6n - 1)^2 = 36n^2 - 12n + 1$

25. $(10a - b)^2 = 100a^2 - 20ab + b^2$ **26.** $(2m + 3n)^2 = 4m^2 + 12mn + 9n^2$

27. $(4x - 3y)^2 = 16x^2 - 24xy + 9y^2$ **28.** $(3x - 4y)^2 = 9x^2 - 24xy + 16y^2$

29. $2(a + 1)^2 = 2(a^2 + 2a + 1) = 2a^2 + 4a + 2$

30. $2(a + 5)^2 = 2(a^2 + 10a + 25) = 2a^2 + 20a + 50$

31. $3(x - 2)^2 = 3(x^2 - 4x + 4) = 3x^2 - 12x + 12$

32. $-1(x + 3)^2 = -1(x^2 + 6x + 9) = -x^2 - 6x - 9$

33. $-2(x - 2)^2 = -2(x^2 - 4x + 4) = -2x^2 + 8x - 8$

34. $-1(2n + 3)^2 = -1(4n^2 + 12n + 9) = -4n^2 - 12n - 9$

35. $2(2x + y)^2 = 2(4x^2 + 4xy + y^2) = 8x^2 + 8xy + 2y^2$

36. $-1(a - 2b)^2 = -1(a^2 - 4ab + 4b^2) = -a^2 + 4ab - 4b^2$

B   **37.** $(a + b)^2 = a^2 + ab + ab + b^2$

Page 171 · WRITTEN EXERCISES

A   **1.** $\dfrac{2a}{2a} = 1$ **2.** $\dfrac{-6a}{a} = \dfrac{-6 \cdot a}{a} = -6$ **3.** $\dfrac{a^2}{a} = \dfrac{a \cdot a}{a} = a$

4. $\dfrac{6c}{2} = \dfrac{6 \cdot c}{2} = 3c$ **5.** $\dfrac{-5ax}{a} = \dfrac{-5 \cdot a \cdot x}{a} = -5x$

6. $\dfrac{x^4}{x^3} = \dfrac{x \cdot x \cdot x \cdot x}{x \cdot x \cdot x} = x$

7. $\dfrac{c^9}{c^4} = \dfrac{c \cdot c \cdot c \cdot c \cdot c \cdot c \cdot c \cdot c \cdot c}{c \cdot c \cdot c \cdot c} = c^5$

8. $\dfrac{2x^2}{x} = \dfrac{2 \cdot x \cdot x}{x} = 2x$ **9.** $\dfrac{7n^3}{n^2} = \dfrac{7 \cdot n \cdot n \cdot n}{n \cdot n} = 7n$

10. $\dfrac{10a^5}{2a^3} = \dfrac{10 \cdot a \cdot a \cdot a \cdot a \cdot a}{2 \cdot a \cdot a \cdot a} = 5a^2$ **11.** $\dfrac{-6mn}{mn} = \dfrac{-6 \cdot m \cdot n}{m \cdot n} = -6$

12. $\dfrac{2xy}{2y} = \dfrac{2 \cdot x \cdot y}{2 \cdot y} = x$ **13.** $\dfrac{12rs}{-3} = \dfrac{12 \cdot r \cdot s}{-3} = -4rs$

14. $\dfrac{4ab}{-2b} = \dfrac{4 \cdot a \cdot b}{-2 \cdot b} = -2a$ **15.** $\dfrac{-9x^2y}{-3} = \dfrac{-9 \cdot x \cdot x \cdot y}{-3} = 3x^2y$

16. $\dfrac{-8ab^2}{-2a} = \dfrac{-8 \cdot a \cdot b \cdot b}{-2 \cdot a} = 4b^2$ **17.** $\dfrac{-56x^4}{8x^2} = \dfrac{-56 \cdot x \cdot x \cdot x \cdot x}{8 \cdot x \cdot x} = -7x^2$

18. $\dfrac{30m^3}{-6m} = \dfrac{30 \cdot m \cdot m \cdot m}{-6 \cdot m} = -5m^2$ **19.** $\dfrac{-63b^2}{9b} = \dfrac{-63 \cdot b \cdot b}{9 \cdot b} = -7b$

20. $\dfrac{36n^6}{6n^4} = \dfrac{36 \cdot n \cdot n \cdot n \cdot n \cdot n \cdot n}{6 \cdot n \cdot n \cdot n \cdot n} = 6n^2$

21. $\dfrac{-10cd^4}{-5d^2} = \dfrac{-10 \cdot c \cdot d \cdot d \cdot d \cdot d}{-5 \cdot d \cdot d} = 2cd^2$

22. $\dfrac{48bc^2}{-8c^2} = \dfrac{48 \cdot b \cdot c \cdot c}{-8 \cdot c \cdot c} = -6b$

23. $\dfrac{42a^2b^2}{-6ab} = \dfrac{42 \cdot a \cdot a \cdot b \cdot b}{-6 \cdot a \cdot b} = -7ab$

24. $\dfrac{-72x^4y^3}{9xy^2} = \dfrac{-72 \cdot x \cdot x \cdot x \cdot x \cdot y \cdot y \cdot y}{9 \cdot x \cdot y \cdot y} = -8x^3y$

25. $\dfrac{50m^3n^2}{2m^2n} = \dfrac{50 \cdot m \cdot m \cdot m \cdot n \cdot n}{2 \cdot m \cdot m \cdot n} = 25mn$

26. $\dfrac{48pq^2}{-3pq} = \dfrac{48 \cdot p \cdot q \cdot q}{-3 \cdot p \cdot q} = -16q$

27. $\dfrac{5a^5b^3}{ab^2} = \dfrac{5 \cdot a \cdot a \cdot a \cdot a \cdot a \cdot b \cdot b \cdot b}{a \cdot b \cdot b} = 5a^4b$

28. $\dfrac{64r^6s^4}{4r^4s^3} = \dfrac{64 \cdot r \cdot r \cdot r \cdot r \cdot r \cdot r \cdot s \cdot s \cdot s \cdot s}{4 \cdot r \cdot r \cdot r \cdot r \cdot s \cdot s \cdot s} = 16r^2s$

29. $\dfrac{-12x^2y^6}{-xy^4} = \dfrac{-12 \cdot x \cdot x \cdot y \cdot y \cdot y \cdot y \cdot y \cdot y}{-x \cdot y \cdot y \cdot y \cdot y} = 12xy^2$

30. $\dfrac{-32c^9d^6}{8c^3d^2} = \dfrac{-32 \cdot c \cdot c \cdot c \cdot c \cdot c \cdot c \cdot c \cdot c \cdot c \cdot d \cdot d \cdot d \cdot d \cdot d \cdot d}{8 \cdot c \cdot c \cdot c \cdot d \cdot d} = -4c^6d^4$

B 31. $\dfrac{4x^2}{x} - \dfrac{2x^3}{x^2} = \dfrac{4 \cdot x \cdot x}{x} - \dfrac{2 \cdot x \cdot x \cdot x}{x \cdot x} = 4x - 2x = 2x$

32. $\dfrac{2x^4}{x^3} - \dfrac{3x}{3} = \dfrac{2 \cdot x \cdot x \cdot x \cdot x}{x \cdot x \cdot x} - \dfrac{3 \cdot x}{3} = 2x - x = x$

33. $\dfrac{5a^2}{5a} + \dfrac{6a^5}{2a^4} = \dfrac{5 \cdot a \cdot a}{5 \cdot a} + \dfrac{6 \cdot a \cdot a \cdot a \cdot a \cdot a}{2 \cdot a \cdot a \cdot a \cdot a} = a + 3a = 4a$

34. $\dfrac{10xy}{5x} + \dfrac{4x^2y^3}{4x^2y^2} = 2y + y = 3y$

35. $\dfrac{18ab^3}{6b^2} + \dfrac{5a^2b}{5a} = 3ab + ab = 4ab$

36. $\dfrac{-4x^3y^4}{-2x^2y^2} - \dfrac{3x^2y^5}{xy^3} = 2xy^2 - 3xy^2 = -xy^2$

Page 171 · PUZZLE PROBLEMS

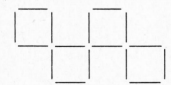

Page 173 · WRITTEN EXERCISES

A 1. $\dfrac{a^5}{a^2} = a^{5-2} = a^3$

2. $\dfrac{x^9}{x^4} = x^{9-4} = x^5$

3. $\dfrac{x^3}{x^7} = \dfrac{1}{x^{7-3}} = \dfrac{1}{x^4}$

4. $\dfrac{5a^7}{a^3} = 5a^{7-3} = 5a^4$

5. $\dfrac{9x^6}{x^3} = 9x^{6-3} = 9x^3$

6. $\dfrac{12x^3}{3x^5} = \dfrac{4}{x^{5-3}} = \dfrac{4}{x^2}$

7. $\dfrac{-8n^2}{4n^5} = \dfrac{-2}{n^{5-2}} = \dfrac{-2}{n^3}$

8. $\dfrac{14b^{12}}{2b^2} = 7b^{12-2} = 7b^{10}$

9. $\dfrac{-3a^2b^2}{ab^2} = -3a^{2-1}b^{2-2} = -3a \cdot 1 = -3a$

10. $\dfrac{-18a^3b}{-3a^2} = 6a^{3-2}b = 6ab$

11. $\dfrac{16x^4y^3}{-4xy} = -4x^{4-1}y^{3-1} = -4x^3y^2$

12. $\dfrac{25x^4y^2}{5xy} = 5x^{4-1}y^{2-1} = 5x^3y$

13. $\dfrac{-12x^5y^2}{-6x^2y^2} = 2x^{5-2}y^{2-2} = 2x^3 \cdot 1 = 2x^3$

14. $\dfrac{14a^3b^2}{-7a^2b^5} = \dfrac{-2a^{3-2}}{b^{5-2}} = \dfrac{-2a}{b^3}$

15. $\dfrac{-42m^6n^3}{-7m^4n^4} = \dfrac{6m^{6-4}}{n^{4-3}} = \dfrac{6m^2}{n}$

16. $\dfrac{56x^5y^5}{7xy^4} = 8x^{5-1}y^{5-4} = 8x^4y$

17. $\dfrac{-20x^6y^7}{5xy^6} = -4x^{6-1}y^{7-6} = -4x^5y$

18. $\dfrac{-12a^3b^8}{-3a^2b^5} = 4a^{3-2}b^{8-5} = 4ab^3$

19. $\dfrac{9x^9y^4z}{3xy} = 3x^{9-1}y^{4-1}z = 3x^8y^3z$

20. $\dfrac{15a^2b^4c^3}{-3ac^2} = -5a^{2-1}b^4c^{3-2} = -5ab^4c$

21. $\dfrac{-7a^9bc}{a^5bc^3} = \dfrac{-7a^{9-5}b^{1-1}}{c^{3-1}} = \dfrac{-7a^4}{c^2}$

22. $\dfrac{49c^2d^{10}x}{-7c^2d^5} = -7c^{2-2}d^{10-5}x = -7d^5x$

23. $\dfrac{-3r^4st^6}{-rst^4} = 3r^{4-1}s^{1-1}t^{6-4} = 3r^3t^2$

24. $\dfrac{36a^6b^2c^7}{-12abc^6} = -3a^{6-1}b^{2-1}c^{7-6} = -3a^5bc$

25. $\dfrac{12x^2y}{-2xy} = -6x^{2-1}y^{1-1} = -6x$

26. $\dfrac{21a^2b^3c}{-3a^2b} = -7a^{2-2}b^{3-1}c = -7b^2c$

27. $\dfrac{-40a^4b^2c^5}{-10a^3bc^3} = 4a^{4-3}b^{2-1}c^{5-3} = 4abc^2$

28. $\dfrac{30x^2y^3z}{-5xy^2} = -6x^{2-1}y^{3-2}z = -6xyz$

29. $\dfrac{-42a^5b^2}{6a^3b} = -7a^{5-3}b^{2-1} = -7a^2b$

30. $\dfrac{36x^4y^2z}{-6x^4} = -6x^{4-4}y^2z = -6y^2z$

31. $\dfrac{-81a^4b^2c^9}{-3ab^2c^5} = 27a^{4-1}b^{2-2}c^{9-5} = 27a^3c^4$

32. $\dfrac{54r^2s^4t^{10}}{-9s^3t^4} = -6r^2s^{4-3}t^{10-4} = -6r^2st^6$

33. $\dfrac{16ab^4}{2a^2b^3} = \dfrac{8b^{4-3}}{a^{2-1}} = \dfrac{8b}{a}$

34. $\dfrac{15m^2n^2s}{-5mn^3s^4} = \dfrac{-3m^{2-1}}{n^{3-2}s^{4-1}} = \dfrac{-3m}{ns^3}$

35. $\dfrac{-x^2y^2z^2}{x^4y^2z} = \dfrac{-y^{2-2}z^{2-1}}{x^{4-2}} = \dfrac{-z}{x^2}$

36. $\dfrac{-45cd^3}{-9c^3d^6} = \dfrac{5}{c^{3-1}d^{6-3}} = \dfrac{5}{c^2d^3}$

Page 173 • PUZZLE PROBLEMS

Since the number doubles every minute, and the basket is full after 1 hour, the basket was half full in 1 hour minus 1 minute, or 59 minutes.

Page 175 • WRITTEN EXERCISES

A

1. $\dfrac{5a + 15b}{5} = \dfrac{5a}{5} + \dfrac{15b}{5} = a + 3b$

2. $\dfrac{7a - 7b}{7} = \dfrac{7a}{7} - \dfrac{7b}{7} = a - b$

3. $\dfrac{8x - 4y}{4} = \dfrac{8x}{4} - \dfrac{4y}{4} = 2x - y$

4. $\dfrac{3a + 12c}{3} = \dfrac{3a}{3} + \dfrac{12c}{3} = a + 4c$

5. $\dfrac{8x - 4y}{2} = \dfrac{8x}{2} - \dfrac{4y}{2} = 4x - 2y$

6. $\dfrac{20x - 16}{4} = \dfrac{20x}{4} - \dfrac{16}{4} = 5x - 4$

7. $\dfrac{56m + 63n}{7} = \dfrac{56m}{7} + \dfrac{63n}{7} = 8m + 9n$

8. $\dfrac{42x + 54y}{6} = \dfrac{42x}{6} + \dfrac{54y}{6} = 7x + 9y$

9. $\dfrac{5a^2 + 45a}{5a} = \dfrac{5a^2}{5a} + \dfrac{45a}{5a} = a + 9$

10. $\dfrac{6x^2 - 48x}{6x} = \dfrac{6x^2}{6x} - \dfrac{48x}{6x} = x - 8$

11. $\dfrac{ab^2 - a^2b}{a} = \dfrac{ab^2}{a} - \dfrac{a^2b}{a} = b^2 - ab$

12. $\dfrac{xy - 4x^2y^2}{xy} = \dfrac{xy}{xy} - \dfrac{4x^2y^2}{xy} = 1 - 4xy$

13. $\dfrac{a^3 + a^2b + ab^2}{a} = \dfrac{a^3}{a} + \dfrac{a^2b}{a} + \dfrac{ab^2}{a} = a^2 + ab + b^2$

14. $\dfrac{x^3 + 3x^2 - x}{x} = \dfrac{x^3}{x} + \dfrac{3x^2}{x} - \dfrac{x}{x} = x^2 + 3x - 1$

15. $\dfrac{6m^2n^2 - 2m + 4m^2}{2m} = \dfrac{6m^2n^2}{2m} - \dfrac{2m}{2m} + \dfrac{4m^2}{2m} = 3mn^2 - 1 + 2m$

16. $\dfrac{3n^4 + 3n^3 - 6n^2}{3n} = \dfrac{3n^4}{3n} + \dfrac{3n^3}{3n} - \dfrac{6n^2}{3n} = n^3 + n^2 - 2n$

17. $\dfrac{5x^2 - 15x^3y + 10x}{5x} = \dfrac{5x^2}{5x} - \dfrac{15x^3y}{5x} + \dfrac{10x}{5x} = x - 3x^2y + 2$

18. $\dfrac{a^2b^3 - a^3b + 4a^2b^2}{a^2b} = \dfrac{a^2b^3}{a^2b} - \dfrac{a^3b}{a^2b} + \dfrac{4a^2b^2}{a^2b} = b^2 - a + 4b$

19. $\dfrac{6r^2s + 9rs^2 - 18r^3}{3r} = \dfrac{6r^2s}{3r} + \dfrac{9rs^2}{3r} - \dfrac{18r^3}{3r} = 2rs + 3s^2 - 6r^2$

20. $\dfrac{n^2 - 3n^4 + 5n^5}{-n^2} = \dfrac{n^2}{-n^2} - \dfrac{3n^4}{-n^2} + \dfrac{5n^5}{-n^2} = -1 + 3n^2 - 5n^3$

21. $\dfrac{24a^2b - 6a^2b^2 + 12a^3}{-6} = \dfrac{24a^2b}{-6} - \dfrac{6a^2b^2}{-6} + \dfrac{12a^3}{-6} = -4a^2b + a^2b^2 - 2a^3$

B **22.** $\dfrac{-4x^4 + 12x^3 - 48x^2 + 8x}{4x} = \dfrac{-4x^4}{4x} + \dfrac{12x^3}{4x} - \dfrac{48x^2}{4x} + \dfrac{8x}{4x} = -x^3 + 3x^2 - 12x + 2$

23. $\dfrac{32n - 16n + 16n^2 + 32n^2}{-8n} = \dfrac{16n + 48n^2}{-8n} = \dfrac{16n}{-8n} + \dfrac{48n^2}{-8n} = -2 - 6n$

24. $\dfrac{6a^4b^4 + 2a^3b^3 + 4a^2b^2 + 2ab}{-2ab} = \dfrac{6a^4b^4}{-2ab} + \dfrac{2a^3b^3}{-2ab} + \dfrac{4a^2b^2}{-2ab} + \dfrac{2ab}{-2ab} =$

$-3a^3b^3 - a^2b^2 - 2ab - 1$

25. $\dfrac{-18x^3y + 48x^2y^2 - 54xy^3}{-6xy} = \dfrac{-18x^3y}{-6xy} + \dfrac{48x^2y^2}{-6xy} - \dfrac{54xy^3}{-6xy} = 3x^2 - 8xy + 9y^2$

26. $\dfrac{-24a^6b^3 + 12a^4b^2 - 9a^2b^3}{-3a^2b^2} = \dfrac{-24a^6b^3}{-3a^2b^2} + \dfrac{12a^4b^2}{-3a^2b^2} - \dfrac{9a^2b^3}{-3a^2b^2} = 8a^4b - 4a^2 + 3b$

27. $\dfrac{21x^3b^4 - 28x^5b^3 + 42x^2b^5}{-7x^2b^3} = \dfrac{21x^3b^4}{-7x^2b^3} - \dfrac{28x^5b^3}{-7x^2b^3} + \dfrac{42x^2b^5}{-7x^2b^3} = -3xb + 4x^3 - 6b^2$

Page 177 • WRITTEN EXERCISES

A **1.**
$$\begin{array}{r} x + 2 \\ x + 2 \overline{)\, x^2 + 4x + 4} \\ \underline{x^2 + 2x} \\ 2x + 4 \\ \underline{2x + 4} \\ 0 \end{array}$$

Check: $(x + 2)^2 = x^2 + 4x + 4$

2.
$$\begin{array}{r} a + 2 \\ a + 4 \overline{)\, a^2 + 6a + 8} \\ \underline{a^2 + 4a} \\ 2a + 8 \\ \underline{2a + 8} \\ 0 \end{array}$$

Check: $(a + 4)(a + 2) =$
$a^2 + 6a + 8$

3.
$$\begin{array}{r} y + 4 \\ y + 6 \overline{)\, y^2 + 10y + 24} \\ \underline{y^2 + 6y} \\ 4y + 24 \\ \underline{4y + 24} \\ 0 \end{array}$$

Check: $(y + 6)(y + 4) = y^2 + 10y + 24$

4.
$$\begin{array}{r} x - 2 \\ x - 2 \overline{)\, x^2 - 4x + 4} \\ \underline{x^2 - 2x} \\ -2x + 4 \\ \underline{-2x + 4} \\ 0 \end{array}$$

Check: $(x - 2)^2 = x^2 - 4x + 4$

5.
$$\begin{array}{r} x + 2 \\ x - 6\,\overline{)\,x^2 - 4x - 12} \\ \underline{x^2 - 6x} \\ 2x - 12 \\ \underline{2x - 12} \\ 0 \end{array}$$

Check: $(x - 6)(x + 2) = x^2 - 4x - 12$

6.
$$\begin{array}{r} x + 1 \\ x - 2\,\overline{)\,x^2 - x - 2} \\ \underline{x^2 - 2x} \\ x - 2 \\ \underline{x - 2} \\ 0 \end{array}$$

Check: $(x - 2)(x + 1) = $
$x^2 - x - 2$

7.
$$\begin{array}{r} x - 4 \\ x + 3\,\overline{)\,x^2 - x - 12} \\ \underline{x^2 + 3x} \\ -4x - 12 \\ \underline{-4x - 12} \\ 0 \end{array}$$

Check: $(x + 3)(x - 4) = x^2 - x - 12$

8.
$$\begin{array}{r} n - 4 \\ n + 1\,\overline{)\,n^2 - 3n - 4} \\ \underline{n^2 + n} \\ -4n - 4 \\ \underline{-4n - 4} \\ 0 \end{array}$$

Check: $(n + 1)(n - 4) = $
$n^2 - 3n - 4$

9.
$$\begin{array}{r} a - 5 \\ a + 2\,\overline{)\,a^2 - 3a - 10} \\ \underline{a^2 + 2a} \\ -5a - 10 \\ \underline{-5a - 10} \\ 0 \end{array}$$

Check: $(a + 2)(a - 5) = a^2 - 3a - 10$

10.
$$\begin{array}{r} s - 2 \\ s - 5\,\overline{)\,s^2 - 7s + 10} \\ \underline{s^2 - 5s} \\ -2s + 10 \\ \underline{-2s + 10} \\ 0 \end{array}$$

Check: $(s - 5)(s - 2) = $
$s^2 - 7s + 10$

11.
$$\begin{array}{r} x - 6 \\ x - 3\,\overline{)\,x^2 - 9x + 18} \\ \underline{x^2 - 3x} \\ -6x + 18 \\ \underline{-6x + 18} \\ 0 \end{array}$$

Check: $(x - 3)(x - 6) = x^2 - 9x + 18$

12.
$$\begin{array}{r} c - 3 \\ c - 7\,\overline{)\,c^2 - 10c + 21} \\ \underline{c^2 - 7c} \\ -3c + 21 \\ \underline{-3c + 21} \\ 0 \end{array}$$

Check: $(c - 7)(c - 3) = $
$c^2 - 10c + 21$

B 13. Width $= \dfrac{x^2 + 4x - 12}{x + 6}$

$$\begin{array}{r} x - 2 \\ x + 6\,\overline{)\,x^2 + 4x - 12} \\ \underline{x^2 + 6x} \\ -2x - 12 \\ \underline{-2x - 12} \\ 0 \end{array}$$

Width $= x - 2$

14. $V = Bh$, and
$B = (x + 6)(x + 2)$
$\quad = x^2 + 8x + 12$, so
$h = \dfrac{x^3 + 11x^2 + 36x + 36}{x^2 + 8x + 12}$.

$$\begin{array}{r} x + 3 \\ x^2 + 8x + 12\,\overline{)\,x^3 + 11x^2 + 36x + 36} \\ \underline{x^3 + 8x^2 + 12x} \\ 3x^2 + 24x + 36 \\ \underline{3x^2 + 24x + 36} \\ 0 \end{array}$$

Height $= x + 3$

Page 177 · CALCULATOR ACTIVITIES

1. $(-b^3)(-b) = b^4 = 4^4 = 256$

2. $(ab^2)(a^2b) = a^3b^3 = 3^3 \cdot 4^3 = 1728$

3. $a^2b^4c = 3^2 \cdot 4^4 \cdot (-2) = -4608$

4. $(6a^3)^2 = 6^2 \cdot a^6 = 36 \cdot 3^6 = 26{,}244$

5. $(8ab^4)(a^2b) = 8a^3b^5 = 8 \cdot 3^3 \cdot 4^5 = 221{,}184$

6. $(-2a^2b)(3ab^4) = -6a^3b^5 = -6 \cdot 3^3 \cdot 4^5 = -165{,}888$

Page 178 · SKILLS REVIEW

1. Let n = the number. $2n + 7 = 11$ 2. Let n = the number. $2n + 11 = 51$

3. Let x = the number of tickets Clarisse sold. $x + x + 82 = 286$

4. Let x = the calories in a half cup of milk. $x + x + 50 = 210$

5. Let n = the larger number. $2n = 5(n - 8) + 4$

6. Let x = the width. $66 = 2(x + 9) + 2x$

7. Let r = André's rate. $\dfrac{1}{2}r + \dfrac{1}{2}(r + 12) = 13$

8. Let x = the number of regular tickets sold. $5x + 7(x + 425) = 12{,}575$

Pages 179–180 · CHAPTER REVIEW EXERCISES

1. $2x - 2$ 2. $3x^2 - 6$ 3. $7n + 2$ 4. $5a^2 - 1$

5. $\begin{array}{r} n + 3 \\ 2n + 5 \\ \hline 3n + 8 \end{array}$ 6. $\begin{array}{r} 2x - y \\ 9x - 7y \\ \hline 11x - 8y \end{array}$ 7. $\begin{array}{r} n^2 - 1 \\ 3n^2 + 1 \\ \hline 4n^2 \end{array}$

8. $\begin{array}{r} 3x - 7 \\ x + 14 \\ \hline 4x + 7 \end{array}$ 9. $\begin{array}{r} a^2 + b^2 \\ 5a^2 + 7b^2 \\ \hline 6a^2 + 8b^2 \end{array}$ 10. $\begin{array}{r} 1 - x^2 \\ 5 + 9x^2 \\ \hline 6 + 8x^2 \end{array}$

11. $\begin{array}{r} x + 2 \\ -x + 3 \\ \hline 5 \end{array}$ 12. $\begin{array}{r} 3y - 1 \\ -2y - 1 \\ \hline y - 2 \end{array}$ 13. $\begin{array}{r} 12x - 7y \\ -10x - 5y \\ \hline 2x - 12y \end{array}$

14. $\begin{array}{r} 15m^2 + n \\ -11m^2 + 2n \\ \hline 4m^2 + 3n \end{array}$ 15. $\begin{array}{r} 5m - n \\ -m - 4n \\ \hline 4m - 5n \end{array}$ 16. $\begin{array}{r} 6a - b \\ -7a - b \\ \hline -a - 2b \end{array}$

17. $\begin{array}{r} 9x + y \\ -2x - 3y \\ \hline 7x - 2y \end{array}$ 18. $\begin{array}{r} 6x - 3y \\ -5x + 3y \\ \hline x \end{array}$ 19. $\begin{array}{r} 2x + 1 \\ -9x - 1 \\ \hline -7x \end{array}$

20. $\begin{array}{r} 8a + b \\ -a + 5b \\ \hline 7a + 6b \end{array}$ 21. $a^2 \cdot a^4 = a^{2+4} = a^6$

22. $2a^2 \cdot a = 2a^{2+1} = 2a^3$ 23. $a \cdot a^3 = a^{1+3} = a^4$

24. $(5a)(-6a^2) = (5 \cdot -6)(a \cdot a^2) = -30a^{1+2} = -30a^3$

25. $(14x)(2x^3) = (14 \cdot 2)(x \cdot x^3) = 28x^{1+3} = 28x^4$

26. $(3x^2)(8x) = (3 \cdot 8)(x^2 \cdot x) = 24x^{2+1} = 24x^3$

27. $(-7y^2)(8y^3) = (-7 \cdot 8)(y^2 \cdot y^3) = -56y^{2+3} = -56y^5$

28. $(-2n^3)(-9n^2) = (-2 \cdot -9)(n^3 \cdot n^2) = 18n^{3+2} = 18n^5$

29. $(a^2)^3 = a^{2 \cdot 3} = a^6$

30. $-(a^2)^3 = -(a^{2 \cdot 3}) = -a^6$

31. $(-a^2)^3 = (-1)^3(a^2)^3 = -a^6$

32. $(-3x^2)^3 = (-3)^3(x^2)^3 = -27x^6$

33. $(mn^2)^4 = m^4(n^2)^4 = m^4n^8$

34. $(3mn^2)^3 = 3^3m^3(n^2)^3 = 27m^3n^6$

35. $(-x^2y^3)^3 = (-1)^3(x^2)^3(y^3)^3 = -x^6y^9$

36. $(3xy^4)^2 = 3^2x^2(y^4)^2 = 9x^2y^8$

37. $5(6 + n) = (5 \cdot 6) + (5 \cdot n) = 30 + 5n$

38. $7(6x + 3) = (7 \cdot 6x) + (7 \cdot 3) = 42x + 21$

39. $a(a + 4) = (a \cdot a) + (a \cdot 4) = a^2 + 4a$

40.
$$\begin{array}{r} n - 2 \\ n - 3 \\ \hline n^2 - 2n \\ - 3n + 6 \\ \hline n^2 - 5n + 6 \end{array}$$

41.
$$\begin{array}{r} 4n - 1 \\ 3n - 1 \\ \hline 12n^2 - 3n \\ - 4n + 1 \\ \hline 12n^2 - 7n + 1 \end{array}$$

42.
$$\begin{array}{r} x - y \\ 6x + y \\ \hline 6x^2 - 6xy \\ + xy - y^2 \\ \hline 6x^2 - 5xy - y^2 \end{array}$$

43.
$$\begin{array}{r} 2r + 8 \\ r + 4 \\ \hline 2r^2 + 8r \\ + 8r + 32 \\ \hline 2r^2 + 16r + 32 \end{array}$$

44.
$$\begin{array}{r} 5a + 2 \\ 2a - 7 \\ \hline 10a^2 + 4a \\ - 35a - 14 \\ \hline 10a^2 - 31a - 14 \end{array}$$

45.
$$\begin{array}{r} x - y \\ 3x + y \\ \hline 3x^2 - 3xy \\ + xy - y^2 \\ \hline 3x^2 - 2xy - y^2 \end{array}$$

46. $(n - 6)(n - 3) = n^2 - 9n + 18$

47. $(2x - 1)(7x + 2) = 14x^2 - 3x - 2$

48. $(3n + 1)(10n + 2) = 30n^2 + 16n + 2$

49. $(1 + 4n)(1 - 2n) = 1 + 2n - 8n^2$

50. $(y + 4)(3y + 1) = 3y^2 + 13y + 4$

51. $(x - 5)(x - 7) = x^2 - 12x + 35$

52. $(n + 4)^2 = n^2 + 2 \cdot 4 \cdot n + 4^2 = n^2 + 8n + 16$

53. $(n - 4)^2 = n^2 - 2 \cdot 4 \cdot n + (-4)^2 = n^2 - 8n + 16$

54. $(3m - 2n)^2 = 9m^2 - 12mn + 4n^2$

55. $(4a + 3b)^2 = 16a^2 + 24ab + 9b^2$

56. $\dfrac{6n^2}{2n} = \dfrac{6 \cdot n \cdot n}{2 \cdot n} = 3n$

57. $\dfrac{50x^3}{5x} = \dfrac{50 \cdot x \cdot x \cdot x}{5 \cdot x} = 10x^2$

58. $\dfrac{-10x^3}{x^2} = \dfrac{-10 \cdot x \cdot x \cdot x}{x \cdot x} = -10x$

59. $\dfrac{24xy^2}{-8xy} = \dfrac{24 \cdot x \cdot y \cdot y}{-8 \cdot x \cdot y} = -3y$

60. $\dfrac{18a^2b^4}{-3ab} = \dfrac{18 \cdot a \cdot a \cdot b \cdot b \cdot b \cdot b}{-3 \cdot a \cdot b} = -6ab^3$

61. $\dfrac{40a^4b^3}{8ab^2} = \dfrac{40 \cdot a \cdot a \cdot a \cdot a \cdot b \cdot b \cdot b}{8 \cdot a \cdot b \cdot b} = 5a^3b$

62. $\dfrac{121mn^2}{11m} = \dfrac{121 \cdot m \cdot n \cdot n}{11 \cdot m} = 11n^2$

63. $\dfrac{48x^2y^2}{-4x^2y} = \dfrac{48 \cdot x \cdot x \cdot y \cdot y}{-4 \cdot x \cdot x \cdot y} = -12y$

64. $\dfrac{6n^2 + n}{n} = \dfrac{6n^2}{n} + \dfrac{n}{n} = 6n + 1$

65. $\dfrac{4m^2 + 12m}{4m} = \dfrac{4m^2}{4m} + \dfrac{12m}{4m} = m + 3$

66. $\dfrac{6x^2 - 3x}{3x} = \dfrac{6x^2}{3x} - \dfrac{3x}{3x} = 2x - 1$

67. $\dfrac{12a^2b^2 - 9ab}{3ab} = \dfrac{12a^2b^2}{3ab} - \dfrac{9ab}{3ab} = 4ab - 3$

68. $\dfrac{4a^3 - 2a^2 + 2a}{2a} = \dfrac{4a^3}{2a} - \dfrac{2a^2}{2a} + \dfrac{2a}{2a} = 2a^2 - a + 1$

69. $\dfrac{6x^2y + 8xy^2 + 10xy}{2xy} = \dfrac{6x^2y}{2xy} + \dfrac{8xy^2}{2xy} + \dfrac{10xy}{2xy} = 3x + 4y + 5$

70. $\dfrac{y^5}{y^9} = \dfrac{1}{y^{9-5}} = \dfrac{1}{y^4}$

71. $\dfrac{15x^2y^3}{5xy^4} = \dfrac{3x^{2-1}}{y^{4-3}} = \dfrac{3x}{y}$

72. $\dfrac{18m^8n^6}{-9m^5n^{11}} = \dfrac{-2m^{8-5}}{n^{11-6}} = \dfrac{-2m^3}{n^5}$

73. $\dfrac{21a^3b^2c}{7ab^2c^3} = \dfrac{3a^{3-1}b^{2-2}}{c^{3-1}} = \dfrac{3a^2}{c^2}$

74.
$$
\begin{array}{r}
x - 5 \\
x - 7 \overline{) x^2 - 12x + 35} \\
\underline{x^2 - 7x} \\
-5x + 35 \\
\underline{-5x + 35} \\
0
\end{array}
$$

75.
$$
\begin{array}{r}
y + 7 \\
y + 4 \overline{) y^2 + 11y + 28} \\
\underline{y^2 + 4y} \\
7y + 28 \\
\underline{7y + 28} \\
0
\end{array}
$$

76.
$$
\begin{array}{r}
m + 3 \\
m - 5 \overline{) m^2 - 2m - 15} \\
\underline{m^2 - 5m} \\
3m - 15 \\
\underline{3m - 15} \\
0
\end{array}
$$

Page 181 · CHAPTER TEST

1.
$$
\begin{array}{l}
x^2 - xy + y^2 \\
\underline{x^2 - y^2} \\
2x^2 - xy
\end{array}
$$

2.
$$
\begin{array}{l}
-2m^2 + m - 1 \\
\underline{4m^2 - m + 5} \\
2m^2 + 4
\end{array}
$$

3.
$$
\begin{array}{l}
3s - 4t \\
\underline{-8s + 7t} \\
-5s + 3t
\end{array}
$$

4.
$$
\begin{array}{l}
6y^2 - y - 4 \\
\underline{-y^2 - 2y + 5} \\
5y^2 - 3y + 1
\end{array}
$$

5. $(4x + 9) - (5x - 8) = 11;\quad 4x + 9 - 5x + 8 = 11;\quad -x + 17 = 11;\quad -x = -6;$
$x = 6$

6. $(-4x^3)(-7x) = (-4 \cdot -7)(x^3 \cdot x) = 28x^{3+1} = 28x^4$

7. $(5b^2)(-2ab^2c^3) = (5 \cdot -2)(a)(b^2 \cdot b^2)(c^3) = -10ab^{2+2}c^3 = -10ab^4c^3$

8. $(-9xz^3)(6x^2y^4) = (-9 \cdot 6)(x \cdot x^2)(y^4z^3) = -54x^{1+2}y^4z^3 = -54x^3y^4z^3$

9. $(-3x)^4 = (-3)^4x^4 = 81x^4$

10. $(2x^2y^3)^3 = 2^3(x^2)^3(y^3)^3 = 8x^6y^9$

11. $-(xy^2)^2(-x^2y) = -(x^2)(y^2)^2(-x^2y) = (-x^2y^4)(-x^2y) =$
$(-1 \cdot -1)(x^2 \cdot x^2)(y^4 \cdot y) = x^4y^5$

12. $-4x(x^2 - 3x + 5) = (-4x)(x^2) - (-4x)(3x) + (-4x)(5) = -4x^3 + 12x^2 - 20x$

13. $rs(r^2 - 3s) = (rs \cdot r^2) - (rs \cdot 3s) = r^3s - 3rs^2$

14. $y^3(2y^4 - 7y^2 + 1) = (y^3 \cdot 2y^4) - (y^3 \cdot 7y^2) + (y^3 \cdot 1) = 2y^7 - 7y^5 + y^3$

15.
$$
\begin{array}{l}
a - b \\
\underline{a + 2b} \\
a^2 - ab \\
\underline{ + 2ab - 2b^2} \\
a^2 + ab - 2b^2
\end{array}
$$

16.
$$
\begin{array}{l}
x - 4 \\
\underline{3x - 1} \\
3x^2 - 12x \\
\underline{ - x + 4} \\
3x^2 - 13x + 4
\end{array}
$$

17.
$$
\begin{array}{l}
n^2 - 3n + 4 \\
\underline{n + 2} \\
n^3 - 3n^2 + 4n \\
\underline{ + 2n^2 - 6n + 8} \\
n^3 - n^2 - 2n + 8
\end{array}
$$

18. $(r + 7)(r + 2) = r^2 + 2r + 7r + 14 = r^2 + 9r + 14$

19. $(x + 4y)(x - 4y) = x^2 - 16y^2$

20. $(3a - 2b)(2a + 5b) = 6a^2 + 11ab - 10b^2$

21. $(n - 9)^2 = n^2 - 2 \cdot 9 \cdot n + (-9)^2 = n^2 - 18n + 81$

22. $(5x + 2y)^2 = 25x^2 + 20xy + 4y^2$

23. $-3(2m - 1)^2 = -3(4m^2 - 4m + 1) = -12m^2 + 12m - 3$

24. $\dfrac{16n^4}{-2n^2} = \dfrac{16 \cdot n \cdot n \cdot n \cdot n}{-2 \cdot n \cdot n} = -8n^2$ 25. $\dfrac{-60r^3s^2}{-6rs^2} = \dfrac{-60 \cdot r \cdot r \cdot r \cdot s \cdot s}{-6 \cdot r \cdot s \cdot s} = 10r^2$

26. $\dfrac{45c^3d^4}{5d^3} = \dfrac{45 \cdot c \cdot c \cdot c \cdot d \cdot d \cdot d \cdot d}{5 \cdot d \cdot d \cdot d} = 9c^3d$

27. $\dfrac{15x^4}{-5x^9} = \dfrac{-3}{x^{9-4}} = \dfrac{-3}{x^5}$ 28. $\dfrac{-8a^3b}{8ab^4} = \dfrac{-1a^{3-1}}{b^{4-1}} = \dfrac{-a^2}{b^3}$

29. $\dfrac{-35a^4b^2c}{-7a^2b^2c^2} = \dfrac{5a^{4-2}b^{2-2}}{c^{2-1}} = \dfrac{5a^2}{c}$

30. $\dfrac{2a^5 - a^3 + 7a^2}{a^2} = \dfrac{2a^5}{a^2} - \dfrac{a^3}{a^2} + \dfrac{7a^2}{a^2} = 2a^3 - a + 7$

31. $\dfrac{42x^2y + 14xy - 63xy^2}{7xy} = \dfrac{42x^2y}{7xy} + \dfrac{14xy}{7xy} - \dfrac{63xy^2}{7xy} = 6x + 2 - 9y$

32.
$$\begin{array}{r} a + 4 \\ a - 3 \overline{)\, a^2 + a - 12} \\ \underline{a^2 - 3a} \\ 4a - 12 \\ \underline{4a - 12} \\ 0 \end{array}$$

33.
$$\begin{array}{r} b - 8 \\ b - 4 \overline{)\, b^2 - 12b + 32} \\ \underline{b^2 - 4b} \\ -8b + 32 \\ \underline{-8b + 32} \\ 0 \end{array}$$

Pages 182–183 · MIXED REVIEW

1. $(2x^2 - 5x + 9) - (3x^2 - 7x - 4) = 2x^2 - 5x + 9 - 3x^2 + 7x + 4 =$
   $-x^2 + 2x + 13$

2.

	rate	time	Distance
Jannette	$r + 12$	4	$4(r + 12)$
Anne	r	4	$4r$
			640

$4(r + 12) + 4r = 640$; $4r + 48 + 4r = 640$; $8r + 48 = 640$; $8r = 592$; $r = 74$ and $r + 12 = 86$.   Jannette: 86 km/h;   Anne: 74 km/h

3. $27 - x = 3 - 5x$; $27 - x - 27 = 3 - 5x - 27$; $-x = -24 - 5x$;
   $-x + 5x = -24 - 5x + 5x$; $4x = -24$; $x = -6$

4. $y - 8 = -2$; $y - 8 + 8 = -2 + 8$; $y = 6$

5. $3(1 - 2n) = n - 4$; $3 - 6n = n - 4$; $3 - 6n + 4 = n - 4 + 4$;
   $-6n + 7 = n$; $-6n + 7 + 6n = n + 6n$; $7 = 7n$; $1 = n$

6. $5(-9) - 7(-3) = -45 + 21 = -24$ 7. $-9 + 5 + (-8) = -4 + (-8) = -12$

8. $(5x + 2y)(3x - 2y) = 15x^2 - 10xy + 6xy - 4y^2 = 15x^2 - 4xy - 4y^2$

9. Let n and $n + 1$ be the consecutive whole numbers. $n + n + 1 = 35$; $2n + 1 = 35$; $2n = 34$; $n = 17$ and $n + 1 = 18$.   17 and 18

10. $-5a + 3b - 9b - (-2a) = -5a - (-2a) + 3b - 9b = -3a - 6b$

11. $5x - y = 7$; $5x - y - 5x = 7 - 5x$; $-y = 7 - 5x$; $(-1)(-y) =$ $(-1)(7 - 5x)$; $y = -7 + 5x$

12.

	price	number	Cost
Sell	4	n	$4n$
Buy	3	$n + 3$	$3(n + 3)$
			8

$4n - 3(n + 3) = 8$; $4n - 3n - 9 = 8$; $n - 9 = 8$; $n = 17$.   17 sets sold

13. $\dfrac{16a^4b^4 - 12a^3b^3 + 14a^2b^2}{-2a^2b} = \dfrac{16a^4b^4}{-2a^2b} - \dfrac{12a^3b^3}{-2a^2b} + \dfrac{14a^2b^2}{-2a^2b} = -8a^2b^3 + 6ab^2 - 7b$

14. $A = 4x \cdot 3x + \dfrac{1}{2} \cdot 4x \cdot 2x = 12x^2 + 4x^2 = 16x^2$

15.

16. Let n = the number. $n = -6 - 17$; $n = -23$

17. $(5a - 1)^2 = (5a)^2 - 2(5a)(1) + 1^2 = 25a^2 - 10a + 1$

18. $x + 7y$

19. $4(a - 4) = 9 - 3(2a - 5)$; $4a - 16 = 9 - 6a + 15$; $4a - 16 = 24 - 6a$; $4a - 16 + 16 = 24 - 6a + 16$; $4a = 40 - 6a$; $4a + 6a = 40 - 6a + 6a$; $10a = 40$; $a = 4$

20. $(5m^2 - 7m - 6) + (4m^3 - 7m^2 + 7m - 2) = 4m^3 + 5m^2 - 7m^2 - 7m + 7m - 6 - 2 = 4m^3 - 2m^2 - 8$

21. $(-3xy^2)(4x^3y^4) = (-3 \cdot 4)(x \cdot x^3)(y^2 \cdot y^4) = -12x^{1+3}y^{2+4} = -12x^4y^6$

22. $-x(3 - x) = (-x)(3) + (-x)(-x) = -3x + x^2$, or $x^2 - 3x$

23.
$$x^2 + 2x - 4$$
$$\underline{x - 3}$$
$$x^3 + 2x^2 - 4x$$
$$\underline{- 3x^2 - 6x + 12}$$
$$x^3 - x^2 - 10x + 12$$

24. Let n = the number. $3n + 7 = 5n - 3$; $3n + 7 + 3 = 5n - 3 + 3$; $3n + 10 = 5n$; $3n + 10 - 3n = 5n - 3n$; $10 = 2n$; $5 = n$

25. $-7 < 0 < 2$ 26. $A = (2a + 3b)^2 = 4a^2 + 12ab + 9b^2$

27. $V = (x + 2)(x)(x) = (x + 2)(x^2) = (x^3 + 2x^2)$ cubic units

28. $-(4x^2)^3 = -(4^3)(x^2)^3 = -64x^6$

29. $(3m^3n)^2 \cdot 5mn^4 = (3^2)(m^3)^2(n^2) \cdot 5mn^4 = 9m^6n^2 \cdot 5mn^4 = (9 \cdot 5)m^{6+1}n^{2+4} = 45m^7n^6$

30. $\dfrac{9x^5}{x^3} - \dfrac{8x^3}{2x} = 9x^2 - 4x^2 = 5x^2$

31. $P = 2l + 2w = 2(4.5) + 2(3) = 9 + 6 = 15;$ 15 cm

32. $5x + 7 = 5(-2) + 7 = -10 + 7 = -3;$ $5(x + 7) = 5(-2 + 7) = 5(5) = 25$

33. If $a = 0$: $8 \cdot 0 < 3(0 + 6)$? $0 < 3(6)$? $0 < 18$?   Yes.

     If $a = 1$: $8 \cdot 1 < 3(1 + 6)$? $8 < 3(7)$? $8 < 21$?   Yes.

     If $a = 2$: $8 \cdot 2 < 3(2 + 6)$? $16 < 3(8)$? $16 < 24$?   Yes.

     If $a = 3$: $8 \cdot 3 < 3(3 + 6)$? $24 < 3(9)$? $24 < 27$?   Yes. $0, 1, 2, 3$

34. $(s + 4) - 6$, or $s - 2$

35. $\dfrac{-12}{-6} = 2$ and $(-1)^3 = -1$; $2 > -1$; $>$

36. $40x + 8(x + 2) = 256;$ $40x + 8x + 16 = 256;$ $48x + 16 = 256;$ $48x = 240;$
$x = 5$, so $x + 2 = 7;$ $7 \times 8 = 56.$ \$56

37. $7(2a + 5) - (3a - 1) = 58;$ $14a + 35 - 3a + 1 = 58;$ $11a + 36 = 58;$ $11a = 22;$ $a = 2$

38. **a.** $\dfrac{1 \cdot 0}{1 - 0} = \dfrac{0}{1} = 0$ **b.** $\dfrac{1 \div 0}{1 + 0};$ impossible

    c. $\dfrac{1 + 0}{1 \cdot 0} = \dfrac{1}{0};$ impossible   **d.** $\dfrac{1 - 0}{1 + 0} = \dfrac{1}{1} = 1$

39. $A = \dfrac{1}{2}bh;$ $18 = \dfrac{1}{2} \cdot 4 \cdot h;$ $18 = 2h;$ $h = 9$

40. **a.** $P = (3x - 1) + (2x + 7) + (x + 8) = 6x + 14$

    b. $P = 6 \cdot 6 + 14 = 36 + 14 = 50$

    c. $P = 6x + 14;$ $80 = 6x + 14;$ $80 - 14 = 6x + 14 - 14;$ $66 = 6x;$ $11 = x$

Page 187 · WRITTEN EXERCISES

A
1. $42 = 2 \cdot 3 \cdot 7$
2. $100 = 2 \cdot 2 \cdot 5 \cdot 5$
3. $36 = 2 \cdot 2 \cdot 3 \cdot 3$
4. $54 = 2 \cdot 3 \cdot 3 \cdot 3$
5. $310 = 2 \cdot 5 \cdot 31$
6. $125 = 5 \cdot 5 \cdot 5$
7. $120 = 2 \cdot 2 \cdot 2 \cdot 3 \cdot 5$
8. $625 = 5 \cdot 5 \cdot 5 \cdot 5$
9. $624 = 2 \cdot 2 \cdot 2 \cdot 2 \cdot 3 \cdot 13$
10. $93 = 3 \cdot 31$

Note: For Exs. 11–38, g.c.f. = greatest common factor.

11. $12 = 2 \cdot 2 \cdot 3$ and $15 = 3 \cdot 5$; g.c.f. $= 3$
12. $16 = 2 \cdot 2 \cdot 2 \cdot 2$ and $22 = 2 \cdot 11$; g.c.f. $= 2$
13. $14 = 2 \cdot 7$ and $21 = 3 \cdot 7$; g.c.f. $= 7$
14. $27 = 3 \cdot 3 \cdot 3$ and $36 = 2 \cdot 2 \cdot 3 \cdot 3$; g.c.f. $= 3 \cdot 3 = 9$
15. $11 = 11$ and $20 = 2 \cdot 2 \cdot 5$; g.c.f. $= 1$
16. $12 = 2 \cdot 2 \cdot 3$ and $18 = 2 \cdot 3 \cdot 3$; g.c.f. $= 2 \cdot 3 = 6$
17. $22 = 2 \cdot 11$ and $33 = 3 \cdot 11$; g.c.f. $= 11$
18. $15 = 3 \cdot 5$ and $40 = 2 \cdot 2 \cdot 2 \cdot 5$; g.c.f. $= 5$
19. $16 = 2 \cdot 2 \cdot 2 \cdot 2$ and $64 = 2 \cdot 2 \cdot 2 \cdot 2 \cdot 2 \cdot 2$; g.c.f. $= 2 \cdot 2 \cdot 2 \cdot 2 = 16$
20. $21 = 3 \cdot 7$ and $42 = 2 \cdot 3 \cdot 7$; g.c.f. $= 3 \cdot 7 = 21$
21. $36 = 2 \cdot 2 \cdot 3 \cdot 3$ and $48 = 2 \cdot 2 \cdot 2 \cdot 2 \cdot 3$; g.c.f. $= 2 \cdot 2 \cdot 3 = 12$
22. $14 = 2 \cdot 7$ and $49 = 7 \cdot 7$; g.c.f. $= 7$
23. $42 = 2 \cdot 3 \cdot 7$ and $28 = 2 \cdot 2 \cdot 7$; g.c.f. $= 2 \cdot 7 = 14$
24. $18 = 2 \cdot 3 \cdot 3$ and $24 = 2 \cdot 2 \cdot 2 \cdot 3$; g.c.f. $= 2 \cdot 3 = 6$
25. $100 = 2 \cdot 2 \cdot 5 \cdot 5$ and $125 = 5 \cdot 5 \cdot 5$; g.c.f. $= 5 \cdot 5 = 25$
26. $30 = 2 \cdot 3 \cdot 5$ and $100 = 2 \cdot 2 \cdot 5 \cdot 5$; g.c.f. $= 2 \cdot 5 = 10$
27. $16 = 2 \cdot 2 \cdot 2 \cdot 2$ and $36 = 2 \cdot 2 \cdot 3 \cdot 3$; g.c.f. $= 2 \cdot 2 = 4$
28. $15 = 3 \cdot 5$ and $24 = 2 \cdot 2 \cdot 2 \cdot 3$; g.c.f. $= 3$
29. $24 = 2 \cdot 2 \cdot 2 \cdot 3$ and $56 = 2 \cdot 2 \cdot 2 \cdot 7$; g.c.f. $= 2 \cdot 2 \cdot 2 = 8$
30. $19 = 19$ and $21 = 3 \cdot 7$; g.c.f. $= 1$
31. $12 = 2 \cdot 2 \cdot 3$ and $20 = 2 \cdot 2 \cdot 5$; g.c.f. $= 2 \cdot 2 = 4$
32. $28 = 2 \cdot 2 \cdot 7$ and $35 = 5 \cdot 7$; g.c.f. $= 7$
33. $40 = 2 \cdot 2 \cdot 2 \cdot 5$ and $100 = 2 \cdot 2 \cdot 5 \cdot 5$; g.c.f. $= 2 \cdot 2 \cdot 5 = 20$
34. $15 = 3 \cdot 5$ and $35 = 5 \cdot 7$; g.c.f. $= 5$
35. $42 = 2 \cdot 3 \cdot 7$ and $48 = 2 \cdot 2 \cdot 2 \cdot 2 \cdot 3$; g.c.f. $= 2 \cdot 3 = 6$
36. $50 = 2 \cdot 5 \cdot 5$ and $60 = 2 \cdot 2 \cdot 3 \cdot 5$; g.c.f. $= 2 \cdot 5 = 10$
37. $100 = 2 \cdot 2 \cdot 5 \cdot 5$ and $400 = 2 \cdot 2 \cdot 2 \cdot 2 \cdot 5 \cdot 5$; g.c.f. $= 2 \cdot 2 \cdot 5 \cdot 5 = 100$
38. $36 = 2 \cdot 2 \cdot 3 \cdot 3$ and $81 = 3 \cdot 3 \cdot 3 \cdot 3$; g.c.f. $= 3 \cdot 3 = 9$

Page 189 · WRITTEN EXERCISES

A
1. $9 + 3x = (3 \cdot 3) + (3 \cdot x) = 3(3 + x)$

2. $5 - 15n = (5 \cdot 1) - (5 \cdot 3n) = 5(1 - 3n)$

3. $2x - 10 = (2 \cdot x) - (2 \cdot 5) = 2(x - 5)$

4. $4y - 16 = (4 \cdot y) - (4 \cdot 4) = 4(y - 4)$

5. $3x^2 - x = (3x \cdot x) - (1 \cdot x) = x(3x - 1)$

6. $5n^2 - 2n = (5n \cdot n) - (2 \cdot n) = n(5n - 2)$

7. $2x^2 - 6x = (2x \cdot x) - (2x \cdot 3) = 2x(x - 3)$

8. $y^2 - 2y = (y \cdot y) - (2 \cdot y) = y(y - 2)$

9. $28n^2 - 7n = (7n \cdot 4n) - (7n \cdot 1) = 7n(4n - 1)$

10. $2xy - y^2 = (2x \cdot y) - (y \cdot y) = y(2x - y)$

11. $4x^2 - 8x = (4x \cdot x) - (4x \cdot 2) = 4x(x - 2)$

12. $21y^2 - 7xy = (7y \cdot 3y) - (7y \cdot x) = 7y(3y - x)$

13. $25mn - 5m^2n^2 = 5mn(5 - mn)$ **14.** $8a^2b - 24ab^2 = 8ab(a - 3b)$

15. $9x^2 - 27x^2y = 9x^2(1 - 3y)$ **16.** $12a^2 + 36a^2b = 12a^2(1 + 3b)$

17. $3x^2 - 6x + 21 = 3(x^2 - 2x + 7)$

18. $5n^3 + 15n^2 + 25n = 5n(n^2 + 3n + 5)$

19. $55y^2 + 22y + 44 = 11(5y^2 + 2y + 4)$

20. $2n^2 + 4mn + 80m^2 = 2(n^2 + 2mn + 40m^2)$

21. $4a^2 + 12ab - 16b^2 = 4(a^2 + 3ab - 4b^2)$

22. $3x^2 - 12xy + 9y^2 = 3(x^2 - 4xy + 3y^2)$

B   **23.** $6x^2 + 6x + 24xy + 42 = 6(x^2 + x + 4xy + 7)$

24. $-13x + 26x^2 + 39x^3 = 13x(-1 + 2x + 3x^2)$

25. $-50a^2 + 25b^2 + 75ab = 25(-2a^2 + b^2 + 3ab)$

26. $48mn + 72m^2n^2 + 60m^3n^3 = 12mn(4 + 6mn + 5m^2n^2)$

27. $56x^3y^3 - 72x^2y^2 - 64xy = 8xy(7x^2y^2 - 9xy - 8)$

28. $32a^2b^4 - 16ab^3 + 48a^3b^5 = 16ab^3(2ab - 1 + 3a^2b^2)$

C   **29.** $x(a - b) + 2(a - b) = (x + 2)(a - b)$

30. $y(a + d) - 4(a + d) = (y - 4)(a + d)$

31. $mn - nx + my - xy = n(m - x) + y(m - x) = (n + y)(m - x)$

32. $ab - 3ad + b^2 - 3bd = a(b - 3d) + b(b - 3d) = (a + b)(b - 3d)$

33. $2a - a^2 + 2b - ab = a(2 - a) + b(2 - a) = (a + b)(2 - a)$

34. $3x^2 - 2xy - 3x + 2y = x(3x - 2y) - 1(3x - 2y) = (x - 1)(3x - 2y)$

35. $2x^2 - 8 + x^2y - 4y = 2(x^2 - 4) + y(x^2 - 4) = (2 + y)(x^2 - 4)$

36. $a^3 - ab^2 - a^2b + b^3 = a(a^2 - b^2) - b(a^2 - b^2) = (a - b)(a^2 - b^2)$

Page 191 · WRITTEN EXERCISES

A   **1.** A = area of square − area of circle = $(2r)^2 - \pi r^2 = 4r^2 - \pi r^2 = r^2(4 - \pi)$

2. A = area of circle − area of square = $\pi \cdot (3s)^2 - (2s)^2 = \pi \cdot 9s^2 - 4s^2 = s^2(9\pi - 4)$

3. $A = \pi a^2 - \pi b^2 = \pi(a^2 - b^2)$

4. $A = (28 \cdot 10) + (28 \cdot 7) + (28 \cdot 3) = 28(10 + 7 + 3) = 28 \cdot 20 = 560; \quad 560 \text{ cm}^2$

5. $A = (4 \cdot 4) + (4 \cdot 5) + (4 \cdot 9) = 4(4 + 5 + 9) = 4 \cdot 18 = 72$

6. $A = \dfrac{1}{2} \cdot 26 \cdot 8 + \dfrac{1}{2} \cdot 26 \cdot 14 = \dfrac{1}{2} \cdot 26(8 + 14) = 13 \cdot 22 = 286$

B **7.** $A = [(x + 5)(x + 3) - 3 \cdot x] - \pi x^2 = (x^2 + 8x + 15 - 3x) - \pi x^2 =$
$x^2 + 5x + 15 - \pi x^2 = x^2(1 - \pi) + 5(x + 3)$

Pages 194–195 • WRITTEN EXERCISES

A **1.** 1 and 5   **2.** 2 and 3   **3.** 2 and 4

  4. 2 and 6   **5.** 3 and 4   **6.** 3 and 3

  7. 3 and 5   **8.** 2 and 7   **9.** 3 and 6

  10. 4 and 5   **11.** 3 and 7   **12.** 2 and 12

  13. 3 and 8   **14.** 4 and 9   **15.** 6 and 6

  16. $(x + 1)(x + 7)$ **17.** $(a + 1)(a + 6)$ **18.** $(n + 4)(n + 5)$

  19. $(x + 3)(x + 6)$ **20.** $(y + 2)(y + 7)$ **21.** $(r + 3)(r + 7)$

  22. $y^2 + 3y + 2 = (y + 1)(y + 2)$ **23.** $n^2 + 17n + 16 = (n + 1)(n + 16)$

  24. $y^2 + 16y + 15 = (y + 1)(y + 15)$ **25.** $a^2 + 5a + 4 = (a + 1)(a + 4)$

  26. $m^2 + 7m + 6 = (m + 1)(m + 6)$ **27.** $x^2 + 11x + 10 = (x + 1)(x + 10)$

  28. $x^2 + 11x + 18 = (x + 2)(x + 9)$ **29.** $n^2 + 6n + 8 = (n + 2)(n + 4)$

  30. $y^2 + 6y + 9 = (y + 3)(y + 3)$ **31.** $x^2 + 9x + 14 = (x + 2)(x + 7)$

  32. $a^2 + 9a + 18 = (a + 3)(a + 6)$ **33.** $y^2 + 12y + 11 = (y + 1)(y + 11)$

  34. $r^2 + 11r + 30 = (r + 5)(r + 6)$ **35.** $b^2 + 12b + 27 = (b + 3)(b + 9)$

  36. $m^2 + 11m + 28 = (m + 4)(m + 7)$ **37.** $x^2 + 12x + 32 = (x + 4)(x + 8)$

  38. $y^2 + 24y + 23 = (y + 1)(y + 23)$ **39.** $x^2 + 13x + 30 = (x + 3)(x + 10)$

  40. $a^2 + 9a + 20 = (a + 4)(a + 5)$ **41.** $a^2 + 12a + 20 = (a + 2)(a + 10)$

  42. $n^2 + 12n + 35 = (n + 5)(n + 7)$

B **43.** $x^2 + 27x + 50 = (x + 2)(x + 25)$ **44.** $x^2 + 17x + 72 = (x + 8)(x + 9)$

  45. $m^2 + 28m + 75 = (m + 3)(m + 25)$ **46.** $x^2 + 20x + 64 = (x + 4)(x + 16)$

  47. $a^2 + 24a + 63 = (a + 3)(a + 21)$ **48.** $y^2 + 24y + 44 = (y + 2)(y + 22)$

Page 197 • WRITTEN EXERCISES

A **1.** -2 and -3 **2.** -2 and -5 **3.** -3 and -6

  4. -3 and -4 **5.** -4 and -6

  6. $n^2 - 3n + 2 = (n - 1)(n - 2)$ **7.** $x^2 - 7x + 10 = (x - 2)(x - 5)$

  8. $y^2 - 8y + 12 = (y - 2)(y - 6)$ **9.** $x^2 - 5x + 6 = (x - 2)(x - 3)$

  10. $n^2 - 10n + 21 = (n - 3)(n - 7)$ **11.** $y^2 - 6y + 5 = (y - 1)(y - 5)$

  12. $b^2 - 2b + 1 = (b - 1)(b - 1)$ **13.** $x^2 - 11x + 24 = (x - 3)(x - 8)$

  14. $x^2 - 13x + 30 = (x - 3)(x - 10)$ **15.** $n^2 + 9n + 18 = (n + 3)(n + 6)$

16. $n^2 - 11n + 18 = (n - 2)(n - 9)$

17. $n^2 - 5n + 4 = (n - 1)(n - 4)$

18. $a^2 - 9a + 14 = (a - 2)(a - 7)$

19. $x^2 + 6x + 8 = (x + 2)(x + 4)$

20. $x^2 - 7x + 12 = (x - 3)(x - 4)$

21. $y^2 - 11y + 28 = (y - 4)(y - 7)$

22. $x^2 - 12x + 27 = (x - 3)(x - 9)$

23. $n^2 + 10n + 25 = (n + 5)(n + 5)$

24. $n^2 - 11n + 24 = (n - 3)(n - 8)$

25. $y^2 - 11y + 30 = (y - 5)(y - 6)$

26. $r^2 - 14r + 33 = (r - 3)(r - 11)$

27. $x^2 - 14x + 49 = (x - 7)(x - 7)$

28. $a^2 - 12a + 36 = (a - 6)(a - 6)$

29. $m^2 - 13m + 40 = (m - 5)(m - 8)$

30. $a^2 - 4a + 4 = (a - 2)(a - 2)$

31. $z^2 - 18z + 32 = (z - 2)(z - 16)$

32. $x^2 - 11x + 30 = (x - 5)(x - 6)$

33. $y^2 - 15y + 36 = (y - 3)(y - 12)$

34. $b^2 - 12b + 36 = (b - 6)(b - 6)$

35. $d^2 - 14d + 45 = (d - 5)(d - 9)$

36. $c^2 - 24c + 80 = (c - 4)(c - 20)$

37. $x^2 - 52x + 100 = (x - 2)(x - 50)$

38. $n^2 - 30n + 200 = (n - 10)(n - 20)$

Page 199 · WRITTEN EXERCISES

A

1. $x^2 + 2x + 1 = (x + 1)^2$; yes

2. $y^2 + 4y + 4 = (y + 2)^2$; yes

3. $a^2 + 6a + 9 = (a + 3)^2$; yes

4. $y \neq$ twice $2 \cdot y$; no

5. $4y \neq$ twice $1 \cdot y$; no

6. $2y \neq$ twice $2 \cdot y$; no

7. $a^2 + 10a + 25 = (a + 5)^2$

8. $n^2 + 4n + 4 = (n + 2)^2$

9. $x^2 + 8x + 16 = (x + 4)^2$

10. $n^2 + 12n + 36 = (n + 6)^2$

11. $x^2 - 4x + 4 = (x - 2)^2$

12. $a^2 - 9a + 20 = (a - 4)(a - 5)$

13. $y^2 + 14y + 49 = (y + 7)^2$

14. $x^2 - 20x + 100 = (x - 10)^2$

15. $n^2 + 18n + 81 = (n + 9)^2$

16. $x^2 + 15x + 56 = (x + 7)(x + 8)$

17. $x^2 - 19x + 90 = (x - 9)(x - 10)$

18. $b^2 - 15b + 36 = (b - 3)(b - 12)$

19. $a^2 - 2ab + b^2 = (a - b)^2$

20. $m^2n^2 + 18mn + 81 = (mn + 9)^2$

21. $1 - 20x + 100x^2 = (1 - 10x)^2$

Page 202 · WRITTEN EXERCISES

A

1. 2 and -5

2. -2 and 9

3. 1 and -4

4. 3 and -7

5. 1 and -2

6. -2 and 3

7. 3 and -10

8. 5 and -6

9. -4 and 7

10. -3 and 5

11. -1 and 14

12. 4 and -6

13. $x^2 + 4x - 21 = (x - 3)(x + 7)$

14. $x^2 - 3x - 4 = (x + 1)(x - 4)$

15. $x^2 + 7x - 18 = (x - 2)(x + 9)$

16. $x^2 - 6x - 16 = (x + 2)(x - 8)$

17. $b^2 + b - 12 = (b - 3)(b + 4)$

18. $b^2 - 4b - 12 = (b + 2)(b - 6)$

19. $n^2 - 3n - 18 = (n + 3)(n - 6)$

20. $y^2 - 9y - 10 = (y + 1)(y - 10)$

21. $x^2 - x - 20 = (x + 4)(x - 5)$

22. $a^2 + 3a - 10 = (a - 2)(a + 5)$

23. $y^2 + 14y - 15 = (y - 1)(y + 15)$

24. $n^2 - 3n - 28 = (n + 4)(n - 7)$

25. $b^2 + 2b - 24 = (b - 4)(b + 6)$

26. $x^2 - x - 30 = (x + 5)(x - 6)$

27. $b^2 - 7b - 30 = (b + 3)(b - 10)$

28. $a^2 - 9a - 22 = (a + 2)(a - 11)$

29. $x^2 + 12x - 28 = (x - 2)(x + 14)$ **30.** $a^2 + 11a - 26 = (a - 2)(a + 13)$

31. $y^2 + 4y - 32 = (y - 4)(y + 8)$ **32.** $x^2 - 5x - 36 = (x + 4)(x - 9)$

33. $y^2 + 6y - 27 = (y - 3)(y + 9)$ **34.** $x^2 + 9x - 36 = (x - 3)(x + 12)$

35. $m^2 + 12m - 64 = (m - 4)(m + 16)$ **36.** $x^2 - x - 72 = (x + 8)(x - 9)$

37. $n^2 - 2n - 63 = (n + 7)(n - 9)$ **38.** $b^2 + 10b - 24 = (b - 2)(b + 12)$

39. $y^2 + y - 42 = (y - 6)(y + 7)$ **40.** $x^2 - 4x - 45 = (x + 5)(x - 9)$

41. $y^2 - y - 56 = (y + 7)(y - 8)$ **42.** $z^2 + 8z - 65 = (z - 5)(z + 13)$

43. $c^2 + 16c - 80 = (c - 4)(c + 20)$ **44.** $m^2 - 24m - 81 = (m + 3)(m - 27)$

45. $a^2 + 21a - 100 = (a - 4)(a + 25)$

Page 203 · MIXED PRACTICE EXERCISES

1. $x^2 + 10x + 9 = (x + 1)(x + 9)$ **2.** $x^2 + 8x + 15 = (x + 3)(x + 5)$

3. $n^2 - 8n + 12 = (n - 2)(n - 6)$ **4.** $a^2 - 9a + 20 = (a - 4)(a - 5)$

5. $y^2 + y - 12 = (y - 3)(y + 4)$ **6.** $x^2 + 2x - 15 = (x - 3)(x + 5)$

7. $y^2 - 4y - 21 = (y + 3)(y - 7)$ **8.** $x^2 - 9x - 22 = (x + 2)(x - 11)$

9. $x^2 + 10x + 21 = (x + 3)(x + 7)$ **10.** $a^2 + 4a - 32 = (a - 4)(a + 8)$

11. $b^2 + 4b - 5 = (b - 1)(b + 5)$ **12.** $x^2 - 8x + 16 = (x - 4)^2$

13. $b^2 - 8b + 7 = (b - 1)(b - 7)$ **14.** $x^2 + 10x - 11 = (x - 1)(x + 11)$

15. $y^2 + y - 20 = (y - 4)(y + 5)$ **16.** $a^2 - 4a - 32 = (a + 4)(a - 8)$

17. $x^2 + 12x + 35 = (x + 5)(x + 7)$ **18.** $y^2 + 5y - 36 = (y - 4)(y + 9)$

19. $n^2 - 15n + 56 = (n - 7)(n - 8)$ **20.** $y^2 - y - 42 = (y + 6)(y - 7)$

21. $x^2 + 15x + 54 = (x + 6)(x + 9)$ **22.** $a^2 + 2a - 48 = (a - 6)(a + 8)$

23. $n^2 + 14n + 45 = (n + 5)(n + 9)$ **24.** $x^2 + x - 72 = (x - 8)(x + 9)$

25. $n^2 - 5n - 50 = (n + 5)(n - 10)$ **26.** $x^2 - 2x - 48 = (x + 6)(x - 8)$

27. $y^2 + 24y - 52 = (y - 2)(y + 26)$ **28.** $x^2 - 9x + 20 = (x - 4)(x - 5)$

29. $a^2 - 7a - 44 = (a + 4)(a - 11)$ **30.** $y^2 + 17y + 52 = (y + 4)(y + 13)$

31. $c^2 + 4c - 60 = (c - 6)(c + 10)$ **32.** $z^2 - 10z - 75 = (z + 5)(z - 15)$

33. $m^2 - 14m + 40 = (m - 4)(m - 10)$

34. $x^2 - 2x - 15 = (x + 3)(x - 5);$ $x + 3$ and $x - 5$

35. $P = 2(x + 3) + 2(x - 5) = 2x + 6 + 2x - 10 = 4x - 4$

36. $y^2 - 2y - 48 = (y + 6)(y - 8);$ $y + 6$ and $y - 8$

37. $P = 2(y + 6) + 2(y - 8) = 2y + 12 + 2y - 16 = 4y - 4$

Page 205 · WRITTEN EXERCISES

A   **1.** $(n - 7)(n + 7) = n^2 - 7^2 = n^2 - 49$ **2.** $(x - 8)(x + 8) = x^2 - 8^2 = x^2 - 64$

     3. $(a + 10)(a - 10) = a^2 - 100$ **4.** $(y - 9)(y + 9) = y^2 - 81$

     5. $(m + n)(m - n) = m^2 - n^2$ **6.** $(a + c)(a - c) = a^2 - c^2$

     7. $(x + y)(x - y) = x^2 - y^2$

     8. $(1 - 3b)(1 + 3b) = 1^2 - (3b)^2 = 1 - 9b^2$

9. $(1 - 2x)(1 + 2x) = 1^2 - (2x)^2 = 1 - 4x^2$

10. $(m + 4n)(m - 4n) = m^2 - (4n)^2 = m^2 - 16n^2$

11. $(3x - 2)(3x + 2) = (3x)^2 - 2^2 = 9x^2 - 4$

12. $(6x + 1)(6x - 1) = (6x)^2 - 1^2 = 36x^2 - 1$

13. $(5y - 2)(5y + 2) = 25y^2 - 4$ **14.** $(7x - 1)(7x + 1) = 49x^2 - 1$

15. $(9m - n)(9m + n) = 81m^2 - n^2$ **16.** $(4a - b)(4a + b) = 16a^2 - b^2$

17. $(5x + 6)(5x - 6) = 25x^2 - 36$ **18.** $(10x - 8)(10x + 8) = 100x^2 - 64$

B   **19.** $(5x + 3y)(5x - 3y) = (5x)^2 - (3y)^2 = 25x^2 - 9y^2$

20. $(2x - 6y)(2x + 6y) = (2x)^2 - (6y)^2 = 4x^2 - 36y^2$

21. $(10x + 5y)(10x - 5y) = (10x)^2 - (5y)^2 = 100x^2 - 25y^2$

22. $(x^2 - 1)(x^2 + 1) = (x^2)^2 - 1^2 = x^4 - 1$

23. $(a^2 - 2)(a^2 + 2) = (a^2)^2 - 2^2 = a^4 - 4$

24. $(b^3 - 5)(b^3 + 5) = (b^3)^2 - 5^2 = b^6 - 25$

C   **25.** $(21)(19) = (20 + 1)(20 - 1) = 400 - 1 = 399$

26. $(51)(49) = (50 + 1)(50 - 1) = 2500 - 1 = 2499$

27. $(52)(48) = (50 + 2)(50 - 2) = 2500 - 4 = 2496$

28. $(18)(22) = (20 - 2)(20 + 2) = 400 - 4 = 396$

29. $(32)(28) = (30 + 2)(30 - 2) = 900 - 4 = 896$

30. $(38)(42) = (40 - 2)(40 + 2) = 1600 - 4 = 1596$

31. $(93)(87) = (90 + 3)(90 - 3) = 8100 - 9 = 8091$

32. $(83)(77) = (80 + 3)(80 - 3) = 6400 - 9 = 6391$

33. $(24)(36) = (30 - 6)(30 + 6) = 900 - 36 = 864$

34. $(94)(86) = (90 + 4)(90 - 4) = 8100 - 16 = 8084$

35. $(55)(65) = (60 - 5)(60 + 5) = 3600 - 25 = 3575$

36. $(95)(85) = (90 + 5)(90 - 5) = 8100 - 25 = 8075$

37. $(84)(76) = (80 + 4)(80 - 4) = 6400 - 16 = 6384$

38. $(59)(61) = (60 - 1)(60 + 1) = 3600 - 1 = 3599$

39. $(48)(52) = (50 - 2)(50 + 2) = 2500 - 4 = 2496$

40. $(73)(67) = (70 + 3)(70 - 3) = 4900 - 9 = 4891$

Pages 206–207 · WRITTEN EXERCISES

A   **1.** Yes (n and 3)   **2.** Yes (x and 4)   **3.** No

4. No   **5.** Yes ($5m$ and n)   **6.** No

7. No   **8.** Yes ($6m$ and 1)

9. $x^2 - 16 = (x + 4)(x - 4)$ **10.** $x^2 - 64 = (x + 8)(x - 8)$

11. $n^2 - 9 = (n + 3)(n - 3)$ **12.** $y^2 - 81 = (y + 9)(y - 9)$

13. $a^2 - 9b^2 = (a + 3b)(a - 3b)$ **14.** $b^2 - 4a^2 = (b + 2a)(b - 2a)$

15. $x^2 - 4y^2 = (x + 2y)(x - 2y)$ **16.** $r^2 - 16t^2 = (r + 4t)(r - 4t)$

17. $b^2 - 64c^2 = (b + 8c)(b - 8c)$ 18. $9n^2 - 49 = (3n - 7)(3n + 7)$

19. $4a^2 - 25b^2 = (2a + 5b)(2a - 5b)$ 20. $49m^2 - n^2 = (7m + n)(7m - n)$

21. $x^2 - 121 = (x + 11)(x - 11)$ 22. $y^2 - 144 = (y + 12)(y - 12)$

23. $64x^2 - 9y^2 = (8x + 3y)(8x - 3y)$ 24. $81a^2 - 49b^2 = (9a + 7b)(9a - 7b)$

B   25. $-x^2 + 4 = 4 - x^2 = (2 + x)(2 - x)$

26. $-1 + 4y^2 = 4y^2 - 1 = (2y + 1)(2y - 1)$

27. $-x^2 + 64 = 64 - x^2 = (8 - x)(8 + x)$

28. $-4x^2 + 9 = 9 - 4x^2 = (3 + 2x)(3 - 2x)$

29. $-y^2 + x^2 = x^2 - y^2 = (x + y)(x - y)$ 30. $4y^4 - 1 = (2y^2 + 1)(2y^2 - 1)$

31. $4x^6 - 9 = (2x^3 + 3)(2x^3 - 3)$

32. $-1 + 49y^4 = 49y^4 - 1 = (7y^2 + 1)(7y^2 - 1)$

33. $144 - 121x^2 = (12 + 11x)(12 - 11x)$ 34. $225a^2 - 1 = (15a + 1)(15a - 1)$

35. $-4a^2 + 36b^2c^2 = 36b^2c^2 - 4a^2 = (6bc + 2a)(6bc - 2a)$

36. $-a^2 + 100b^2 = 100b^2 - a^2 = (10b + a)(10b - a)$

C   37. $y^4 - 81 = (y^2 + 9)(y^2 - 9) = (y^2 + 9)(y + 3)(y - 3)$

38. $16y^4 - 81z^4 = (4y^2 + 9z^2)(4y^2 - 9z^2) = (4y^2 + 9z^2)(2y + 3z)(2y - 3z)$

39. $16a^4 - 1 = (4a^2 + 1)(4a^2 - 1) = (4a^2 + 1)(2a + 1)(2a - 1)$

40. $-1 + 81a^4 = 81a^4 - 1 = (9a^2 + 1)(9a^2 - 1) = (9a^2 + 1)(3a + 1)(3a - 1)$

41. $256 - x^4y^4 = (16 + x^2y^2)(16 - x^2y^2) = (16 + x^2y^2)(4 + xy)(4 - xy)$

42. $a^8 - b^8 = (a^4 + b^4)(a^4 - b^4) = (a^4 + b^4)(a^2 + b^2)(a^2 - b^2) =$
$(a^4 + b^4)(a^2 + b^2)(a + b)(a - b)$

43. $x^8y^4 - 1 = (x^4y^2 + 1)(x^4y^2 - 1) = (x^4y^2 + 1)(x^2y + 1)(x^2y - 1)$

44. $81a^8b^4 - c^8 = (9a^4b^2 + c^4)(9a^4b^2 - c^4) = (9a^4b^2 + c^4)(3a^2b + c^2)(3a^2b - c^2)$

45.

46. If x and y are both even, let $x = 2a$ and $y = 2b$. Then $x^2 - y^2 = (2a)^2 -$
$(2b)^2 = 4a^2 - 4b^2 = 4(a^2 - b^2)$, and 4 is a factor. If x and y are both odd, let
$x = 2a + 1$ and $y = 2b + 1$. Then $x^2 - y^2 = (2a + 1)^2 - (2b + 1)^2 =$
$(4a^2 + 4a + 1) - (4b^2 + 4b + 1) = 4a^2 + 4a - 4b^2 - 4b = 4(a^2 + a - b^2 - b)$,
and 4 is a factor.

Page 207 • PUZZLE PROBLEMS

No, it is not possible.

Page 209 • WRITTEN EXERCISES

A   1. $3y^2 + 18y + 24 = 3(y^2 + 6y + 8) = 3(y + 2)(y + 4)$

2. $4x^2 + 24x - 64 = 4(x^2 + 6x - 16) = 4(x - 2)(x + 8)$

3. $2y^2 - 16y + 30 = 2(y^2 - 8y + 15) = 2(y - 3)(y - 5)$

4. $2b^2 - 4b - 48 = 2(b^2 - 2b - 24) = 2(b + 4)(b - 6)$

5. $5a^2 - 20a - 60 = 5(a^2 - 4a - 12) = 5(a + 2)(a - 6)$

6. $3x^2 + 30x + 27 = 3(x^2 + 10x + 9) = 3(x + 1)(x + 9)$

7. $4x^2 + 8x - 60 = 4(x^2 + 2x - 15) = 4(x - 3)(x + 5)$

8. $3y^2 - 12y - 63 = 3(y^2 - 4y - 21) = 3(y + 3)(y - 7)$

9. $2x^2 - 16x + 32 = 2(x^2 - 8x + 16) = 2(x - 4)^2$

10. $5a^2 - 5b^2 = 5(a^2 - b^2) = 5(a + b)(a - b)$

11. $8x^2 - 32 = 8(x^2 - 4) = 8(x + 2)(x - 2)$

12. $x^2y - y^3 = y(x^2 - y^2) = y(x + y)(x - y)$

13. $4x^2 - 36 = 4(x^2 - 9) = 4(x + 3)(x - 3)$

14. $3a^2 - 27b^2 = 3(a^2 - 9b^2) = 3(a + 3b)(a - 3b)$

15. $6x^2 - 24y^2 = 6(x^2 - 4y^2) = 6(x + 2y)(x - 2y)$

B 16. $-x^2 - 2xy - y^2 = -1(x^2 + 2xy + y^2) = -1(x + y)^2$

17. $-x^2 + 4x - 3 = -1(x^2 - 4x + 3) = -1(x - 1)(x - 3)$

18. $2a^3 - 8a = 2a(a^2 - 4) = 2a(a - 2)(a + 2)$

19. $-162 + 2x^2 = 2x^2 - 162 = 2(x^2 - 81) = 2(x + 9)(x - 9)$

20. $-48 + 3x^2 = 3x^2 - 48 = 3(x^2 - 16) = 3(x + 4)(x - 4)$

21. $4a - 4a^2 + 8 = -4a^2 + 4a + 8 = -4(a^2 - a - 2) = -4(a + 1)(a - 2)$

22. $-3x^2 + 30x - 75 = -3(x^2 - 10x + 25) = -3(x - 5)^2$

23. $-200b^2 + 2a^2 = 2a^2 - 200b^2 = 2(a^2 - 100b^2) = 2(a + 10b)(a - 10b)$

24. $-2x^2 + 24x - 72 = -2(x^2 - 12x + 36) = -2(x - 6)^2$

25. $-12z^2 + 27 = 27 - 12z^2 = 3(9 - 4z^2) = 3(3 + 2z)(3 - 2z)$

26. $-5x^2 + 30x + 35 = -5(x^2 - 6x - 7) = -5(x + 1)(x - 7)$

27. $25a^2 + 50a - 200 = 25(a^2 + 2a - 8) = 25(a - 2)(a + 4)$

28. $x^3 - 4x^2 + 4x = x(x^2 - 4x + 4) = x(x - 2)^2$

29. $a^2b^2 + 12a^2b + 36a^2 = a^2(b^2 + 12b + 36) = a^2(b + 6)^2$

30. $x^3y - xy^3 = xy(x^2 - y^2) = xy(x + y)(x - y)$

Page 211 · WRITTEN EXERCISES

A

1. $3x^2 + 4x + 1 = (3x + 1)(x + 1)$

2. $2x^2 + 5x + 2 = (2x + 1)(x + 2)$

3. $2a^2 + 9a + 9 = (2a + 3)(a + 3)$

4. $3y^2 + 7y + 2 = (3y + 1)(y + 2)$

5. $2b^2 + 5b + 3 = (2b + 3)(b + 1)$

6. $2x^2 + 7x + 3 = (2x + 1)(x + 3)$

7. $3a^2 - 8a + 5 = (3a - 5)(a - 1)$

8. $5a^2 + 4a - 1 = (5a - 1)(a + 1)$

9. $3x^2 - 10x + 8 = (3x - 4)(x - 2)$

10. $5x^2 - 7x + 2 = (5x - 2)(x - 1)$

11. $3y^2 - 5y - 2 = (3y + 1)(y - 2)$

12. $3x^2 + x - 2 = (3x - 2)(x + 1)$

13. $2x^2 + 5x - 3 = (2x - 1)(x + 3)$

14. $7x^2 + 13x - 2 = (7x - 1)(x + 2)$

15. $2x^2 - 3x - 9 = (2x + 3)(x - 3)$

16. $4x^2 - 7x - 2 = (4x + 1)(x - 2)$

17. $4a^2 + 5a - 9 = (4a + 9)(a - 1)$ 18. $6x^2 - 19x + 3 = (6x - 1)(x - 3)$

19. $6b^2 - 5b - 25 = (3b + 5)(2b - 5)$

20. $8x^2 - 14x - 4 = 2(4x^2 - 7x - 2) = 2(4x + 1)(x - 2)$

21. $4a^2 - a - 5 = (4a - 5)(a + 1)$ 22. $6a^2 - 7a + 2 = (3a - 2)(2a - 1)$

23. $4x^2 - 8x - 5 = (2x + 1)(2x - 5)$ 24. $6y^2 + y - 15 = (3y + 5)(2y - 3)$

25. $8r^2 - 6r - 9 = (2r - 3)(4r + 3)$ 26. $6r^2 - 23r + 7 = (3r - 1)(2r - 7)$

27. $8x^2 + 2x - 3 = (4x + 3)(2x - 1)$ 28. $8c^2 - 10c + 3 = (4c - 3)(2c - 1)$

29. $4c^2 + 6c + 2 = 2(2c^2 + 3c + 1) = 2(2c + 1)(c + 1)$

30. $6x^2 + 21x - 90 = 3(2x^2 + 7x - 30) = 3(2x - 5)(x + 6)$

31. $9y^2 + 15y + 6 = 3(3y^2 + 5y + 2) = 3(3y + 2)(y + 1)$

32. $6x^2 + 11x - 10 = (3x - 2)(2x + 5)$

33. $6x^2 + 15x + 6 = 3(2x^2 + 5x + 2) = 3(2x + 1)(x + 2)$

34. $10a^2 + 14a + 4 = 2(5a^2 + 7a + 2) = 2(5a + 2)(a + 1)$

35. $10y^2 - 29y + 10 = (5y - 2)(2y - 5)$ 36. $12n^2 - 11n + 2 = (3n - 2)(4n - 1)$

37. $6x^2 + 11x - 121 = (3x - 11)(2x + 11)$ 38. $10x^2 - 31x - 14 = (5x + 2)(2x - 7)$

39. $12x^2 + 7x - 12 = (4x - 3)(3x + 4)$ 40. $6x^2 + 19x + 15 = (3x + 5)(2x + 3)$

41. $15n^2 + 17n - 4 = (3n + 4)(5n - 1)$ 42. $20y^2 - 11y - 4 = (4y + 1)(5y - 4)$

43. $25x^2 - 5x - 2 = (5x + 1)(5x - 2)$

44. $30a^2 + 37a + 9 = (3a + 1)(10a + 9)$ 45. $10c^2 - 23c + 12 = (2c - 3)(5c - 4)$

46. $9y^2 + 14y - 8 = (y + 2)(9y - 4)$ 47. $18z^2 + 27z + 10 = (3z + 2)(6z + 5)$

48. $36m^2 - 35m - 1 = (m - 1)(36m + 1)$ 49. $4x^2 + 12x + 9 = (2x + 3)^2$

50. $9n^2 + 30n + 25 = (3n + 5)^2$ 51. $18b^2 + 13b - 21 = (2b + 3)(9b - 7)$

52. $16s^2 - 10s - 21 = (2s - 3)(8s + 7)$ 53. $16g^2 - 48g + 27 = (4g - 3)(4g - 9)$

54. $30x^2 + x - 8 = (2x - 1)(15x + 8)$ 55. $24x^2 + 2x - 15 = (4x - 3)(6x + 5)$

56. $27r^2 - 39r - 10 = (3r - 5)(9r + 2)$ 57. $48d^2 - 7d - 3 = (3d - 1)(16d + 3)$

58. $24y^2 - 73y + 24 = (3y - 8)(8y - 3)$

59. $36x^2 + 39x + 10 = (3x + 2)(12x + 5)$

60. $36a^2 - 61a + 20 = (4a - 5)(9a - 4)$

Page 211 · PUZZLE PROBLEMS

There is a built-in contradiction. Consider the following two cases.

(1) *The barber shaves himself.* Then (as the barber) he *cannot* shave himself.

(2) *The barber does not shave himself.* Since he does not shave himself, then (as the barber) he *must* shave himself.

Pages 212–213 · PROBLEM SOLVING STRATEGIES

1. **a.** $7^1 = 7, 7$; $7^2 = 49, 9$; $7^3 = 343, 3$; $7^4 = 2401, 1$; $7^5 = 16,807, 7$;

$7^6 = 117,649, 9$; $7^7 = 823,543, 3$; $7^8 = 5,764,801, 1$; $7^9 = 40,353,607, 7$

b. pattern of units digits: $7, 9, 3, 1, 7, 9, 3, 1, 7 \ldots$;   units digit of 7^{22} is 9.

2. pattern of units digits: $3, 9, 7, 1, 3, 9, 7, 1, 3 \ldots$;   units digit of 3^{15} is 7.

3. a. $1 + 4 + 9 + 16 = 30$; $1 + 4 + 9 + 16 + 25 = 55$

    b. The picture has 8 squares along each edge. $1 + 4 + 9 + 16 + 25 +$
       $36 + 49 + 64 = 204$;   204 squares

Pages 214–215 · COMPUTER ACTIVITIES

1. a.
```
ORIGINAL AMOUNT?500
WHAT IS THE GROWTH RATE
(EXAMPLE:FOR 5% TYPE 5)?8
HOW MANY YEARS?5
NEW AMOUNT:734.664039
```
$735

b.
```
ORIGINAL AMOUNT?10000
WHAT IS THE GROWTH RATE
(EXAMPLE:FOR 5% TYPE 5)?9.5
HOW MANY YEARS?8
NEW AMOUNT:20668.6902
```
$20,669

c.
```
ORIGINAL AMOUNT?2500
WHAT IS THE GROWTH RATE
(EXAMPLE:FOR 5% TYPE 5)?12
HOW MANY YEARS?10
NEW AMOUNT:7764.62053
```
$7765

2. a.
```
ORIGINAL AMOUNT?238000000
WHAT IS THE GROWTH RATE
(EXAMPLE:FOR 5% TYPE 5)?2.5
HOW MANY YEARS?25
NEW AMOUNT:441238699
```
441,000,000

b.
```
ORIGINAL AMOUNT?238000000
WHAT IS THE GROWTH RATE
(EXAMPLE:FOR 5% TYPE 5)?2
HOW MANY YEARS?25
NEW AMOUNT:390464233
```
390,000,000

c.
```
ORIGINAL AMOUNT?238000000
WHAT IS THE GROWTH RATE
(EXAMPLE:FOR 5% TYPE 5)?1.5
HOW MANY YEARS?25
NEW AMOUNT:345324999
```
345,000,000

d.
```
ORIGINAL AMOUNT?238000000
WHAT IS THE GROWTH RATE
(EXAMPLE:FOR 5% TYPE 5)?1
HOW MANY YEARS?25
NEW AMOUNT:305218814
```
305,000,000

3. a.
```
ORIGINAL AMOUNT?100
WHAT IS THE GROWTH RATE
(EXAMPLE:FOR 5% TYPE 5)?12
HOW MANY YEARS?6
NEW AMOUNT:197.382269
```
6 years

b.
```
ORIGINAL AMOUNT?100
WHAT IS THE GROWTH RATE
(EXAMPLE:FOR 5% TYPE 5)?10
HOW MANY YEARS?7
NEW AMOUNT:194.871711
```
7 years

c.
```
ORIGINAL AMOUNT?100
WHAT IS THE GROWTH RATE
(EXAMPLE:FOR 5% TYPE 5)?8
HOW MANY YEARS?9
NEW AMOUNT:199.900463
```
9 years

d.
```
ORIGINAL AMOUNT?100
WHAT IS THE GROWTH RATE
(EXAMPLE:FOR 5% TYPE 5)?6
HOW MANY YEARS?12
NEW AMOUNT:201.219649
```
12 years

4. a. ORIGINAL AMOUNT?8900
WHAT IS THE GROWTH RATE
(EXAMPLE:FOR 5% TYPE 5)?-20
HOW MANY YEARS?4
NEW AMOUNT:3645.44001

$3645

b. ORIGINAL AMOUNT?100000
WHAT IS THE GROWTH RATE
(EXAMPLE:FOR 5% TYPE 5)?-15
HOW MANY YEARS?4
NEW AMOUNT:52200.6251

4 years

5. a.

years	1	2	3	4	5	6	7	8	9	10
total value	$105	$110	$116	$122	$128	$134	$141	$148	$155	$163

b.

years	1	2	3	4	5	6	7	8	9	10
total value	$110	$121	$133	$146	$161	$177	$195	$214	$236	$259

6.

years	1	2	3	4	5	6	7	8	9	10
total value	$108	$117	$126	$136	$147	$159	$171	$185	$200	$216

7.

years	1	2	3	4	5	6	7	8	9	10
total value	$113	$128	$144	$163	$184	$208	$235	$266	$300	$339

8. The Rule of 72 claims that the number of years needed for an investment to double is 72 ÷ (interest rate). Example: An investment will double in 6 years at a rate of 12% since 72 ÷ 12 = 6.

Page 216 · SKILLS REVIEW

1. $\frac{2}{4} = \frac{2 \cdot 1}{2 \cdot 2} = \frac{1}{2}$

2. $\frac{3}{15} = \frac{3 \cdot 1}{3 \cdot 5} = \frac{1}{5}$

3. $\frac{4}{16} = \frac{4 \cdot 1}{4 \cdot 4} = \frac{1}{4}$

4. $\frac{6}{8} = \frac{2 \cdot 3}{2 \cdot 4} = \frac{3}{4}$

5. $\frac{5}{10} = \frac{5 \cdot 1}{5 \cdot 2} = \frac{1}{2}$

6. $\frac{3}{9} = \frac{3 \cdot 1}{3 \cdot 3} = \frac{1}{3}$

7. $\frac{8}{10} = \frac{2 \cdot 4}{2 \cdot 5} = \frac{4}{5}$

8. $\frac{12}{3} = \frac{3 \cdot 4}{3 \cdot 1} = \frac{4}{1} = 4$

9. $\frac{16}{12} = \frac{4 \cdot 4}{4 \cdot 3} = \frac{4}{3}$

10. $\frac{18}{9} = \frac{9 \cdot 2}{9 \cdot 1} = \frac{2}{1} = 2$

11. $\frac{1}{3} + \frac{1}{3} = \frac{2}{3}$

12. $\frac{3}{5} + \frac{1}{5} = \frac{4}{5}$

13. $\frac{2}{8} + \frac{4}{8} = \frac{6}{8} = \frac{3}{4}$

14. $\frac{2}{6} + \frac{3}{6} = \frac{5}{6}$

15. $\frac{3}{12} + \frac{5}{12} = \frac{8}{12} = \frac{2}{3}$

16. $\frac{2}{9} + \frac{4}{9} = \frac{6}{9} = \frac{2}{3}$

17. $\frac{3}{14} + \frac{5}{14} = \frac{8}{14} = \frac{4}{7}$

18. $\frac{3}{10} + \frac{4}{10} = \frac{7}{10}$

19. $\dfrac{2}{7} + \dfrac{3}{7} = \dfrac{5}{7}$

20. $\dfrac{1}{4} + \dfrac{2}{4} = \dfrac{3}{4}$

21. $\dfrac{5}{14} + \dfrac{2}{14} = \dfrac{7}{14} = \dfrac{1}{2}$

22. $\dfrac{1}{12} + \dfrac{3}{12} = \dfrac{4}{12} = \dfrac{1}{3}$

23. $\dfrac{3}{10} + \dfrac{2}{10} = \dfrac{5}{10} = \dfrac{1}{2}$

24. $\dfrac{3}{15} + \dfrac{7}{15} = \dfrac{10}{15} = \dfrac{2}{3}$

25. $\dfrac{4}{13} + \dfrac{6}{13} = \dfrac{10}{13}$

26. $\dfrac{3}{14} + \dfrac{6}{14} = \dfrac{9}{14}$

27. $\dfrac{4}{5} - \dfrac{2}{5} = \dfrac{2}{5}$

28. $\dfrac{7}{8} - \dfrac{3}{8} = \dfrac{4}{8} = \dfrac{1}{2}$

29. $\dfrac{7}{9} - \dfrac{2}{9} = \dfrac{5}{9}$

30. $\dfrac{9}{11} - \dfrac{5}{11} = \dfrac{4}{11}$

31. $\dfrac{3}{4} - \dfrac{1}{4} = \dfrac{2}{4} = \dfrac{1}{2}$

32. $\dfrac{9}{10} - \dfrac{3}{10} = \dfrac{6}{10} = \dfrac{3}{5}$

33. $\dfrac{11}{12} - \dfrac{4}{12} = \dfrac{7}{12}$

34. $\dfrac{9}{14} - \dfrac{7}{14} = \dfrac{2}{14} = \dfrac{1}{7}$

35. $\dfrac{8}{13} - \dfrac{5}{13} = \dfrac{3}{13}$

36. $\dfrac{7}{12} - \dfrac{4}{12} = \dfrac{3}{12} = \dfrac{1}{4}$

37. $\dfrac{5}{6} - \dfrac{2}{6} = \dfrac{3}{6} = \dfrac{1}{2}$

38. $\dfrac{6}{10} - \dfrac{3}{10} = \dfrac{3}{10}$

39. $\dfrac{5}{8} - \dfrac{3}{8} = \dfrac{2}{8} = \dfrac{1}{4}$

40. $\dfrac{10}{14} - \dfrac{5}{14} = \dfrac{5}{14}$

41. $\dfrac{5}{12} - \dfrac{1}{12} = \dfrac{4}{12} = \dfrac{1}{3}$

42. $\dfrac{11}{13} - \dfrac{7}{13} = \dfrac{4}{13}$

Pages 217–218 · CHAPTER REVIEW EXERCISES

Note: For Exs. 1–12, g.c.f. = greatest common factor.

1. $6 = 2 \cdot 3$ and $9 = 3 \cdot 3$; g.c.f. $= 3$

2. $16 = 2 \cdot 2 \cdot 2 \cdot 2$ and $4 = 2 \cdot 2$; g.c.f. $= 2 \cdot 2 = 4$

3. $15 = 3 \cdot 5$ and $3 = 3$; g.c.f. $= 3$

4. $24 = 2 \cdot 2 \cdot 2 \cdot 3$ and $32 = 2 \cdot 2 \cdot 2 \cdot 2 \cdot 2$; g.c.f. $= 2 \cdot 2 \cdot 2 = 8$

5. $56 = 2 \cdot 2 \cdot 2 \cdot 7$ and $72 = 2 \cdot 2 \cdot 2 \cdot 3 \cdot 3$; g.c.f. $= 2 \cdot 2 \cdot 2 = 8$

6. $30 = 2 \cdot 3 \cdot 5$ and $27 = 3 \cdot 3 \cdot 3$; g.c.f. $= 3$

7. $3y = 3 \cdot y$ and $9y = 3 \cdot 3 \cdot y$; g.c.f. $= 3 \cdot y = 3y$

8. $16n = 2 \cdot 2 \cdot 2 \cdot 2 \cdot n$ and $8n^3 = 2 \cdot 2 \cdot 2 \cdot n \cdot n \cdot n$; g.c.f. $= 2 \cdot 2 \cdot 2 \cdot n = 8n$

9. $4x^2 = 2 \cdot 2 \cdot x \cdot x$ and $x^3 = x \cdot x \cdot x$; g.c.f. $= x \cdot x = x^2$

10. $24x = 2 \cdot 2 \cdot 2 \cdot 3 \cdot x$ and $6x^2 = 2 \cdot 3 \cdot x \cdot x$; g.c.f. $= 2 \cdot 3 \cdot x = 6x$

11. $20x = 2 \cdot 2 \cdot 5 \cdot x$ and $4y = 2 \cdot 2 \cdot y$; g.c.f. $= 2 \cdot 2 = 4$

12. $5x^3 = 5 \cdot x \cdot x \cdot x$ and $30x^2 = 2 \cdot 3 \cdot 5 \cdot x \cdot x$; g.c.f. $= 5 \cdot x \cdot x = 5x^2$

13. $2n^2 + 4 = (2 \cdot n^2) + (2 \cdot 2) = 2(n^2 + 2)$

14. $3x^2 - 9 = (3 \cdot x^2) - (3 \cdot 3) = 3(x^2 - 3)$

15. $ab^2 + a^2b = (ab \cdot b) + (ab \cdot a) = ab(b + a)$

16. $7x^2 - 21x = 7x(x - 3)$

17. $2y^3 + 8y^2 = 2y^2(y + 4)$

18. $3xy - 27 = 3(xy - 9)$

19. $2x^2 - 8x^3 = 2x^2(1 - 4x)$

20. $x^2 + 4x = x(x + 4)$

21. $36x + 6 = 6(6x + 1)$

22. $n^2 + 6n + 9 = (n + 3)^2$

23. $x^2 + 8x + 7 = (x + 1)(x + 7)$

24. $x^2 - 10x + 25 = (x - 5)^2$

25. $x^2 - 9x + 14 = (x - 2)(x - 7)$

26. $n^2 + 9n + 20 = (n + 4)(n + 5)$

27. $n^2 + 12n + 20 = (n + 2)(n + 10)$

28. $y^2 - 9y + 8 = (y - 1)(y - 8)$

29. $y^2 - 11y + 30 = (y - 5)(y - 6)$

30. $n^2 - 13n + 22 = (n - 2)(n - 11)$

31. $y^2 + 17y + 30 = (y + 2)(y + 15)$

32. $n^2 + 12n + 35 = (n + 5)(n + 7)$

33. $y^2 + 12y + 32 = (y + 4)(y + 8)$

34. $y^2 - 13y + 40 = (y - 5)(y - 8)$

35. $n^2 - 18n + 32 = (n - 2)(n - 16)$

36. $y^2 + 9y + 18 = (y + 3)(y + 6)$

37. $y^2 + 11y + 18 = (y + 2)(y + 9)$

38. $x^2 + 11x + 30 = (x + 5)(x + 6)$

39. $h^2 - 24h + 23 = (h - 1)(h - 23)$

40. $n^2 + 20n + 64 = (n + 4)(n + 16)$

41. $b^2 - 25b + 100 = (b - 5)(b - 20)$

42. $x^2 + 12x + 27 = (x + 3)(x + 9)$

43. $x^2 + 3x - 18 = (x - 3)(x + 6)$

44. $n^2 - 7n - 18 = (n + 2)(n - 9)$

45. $y^2 - 2y - 24 = (y + 4)(y - 6)$

46. $m^2 + 6m - 16 = (m - 2)(m + 8)$

47. $x^2 - 5x - 24 = (x + 3)(x - 8)$

48. $y^2 - 10y - 24 = (y + 2)(y - 12)$

49. $x^2 - x - 12 = (x + 3)(x - 4)$

50. $y^2 - 12y + 36 = (y - 6)^2$

51. $n^2 - n - 20 = (n + 4)(n - 5)$

52. $a^2 + 3a - 10 = (a - 2)(a + 5)$

53. $a^2 - 9a - 10 = (a + 1)(a - 10)$

54. $b^2 + 5b - 6 = (b - 1)(b + 6)$

55. $(x - 2)(x + 2) = x^2 - 4$

56. $(y + 4)(y - 4) = y^2 - 16$

57. $(n + 5)(n - 5) = n^2 - 25$

58. $(1 + 2a)(1 - 2a) = 1^2 - (2a)^2 = 1 - 4a^2$

59. $(2y + 3)(2y - 3) = (2y)^2 - 3^2 = 4y^2 - 9$

60. $(4n + 1)(4n - 1) = (4n)^2 - 1^2 = 16n^2 - 1$

61. $(x + y)(x - y) = x^2 - y^2$

62. $(xy + 1)(xy - 1) = x^2y^2 - 1$

63. $(4x - 3y)(4x + 3y) = 16x^2 - 9y^2$

64. $a^2 - 1 = (a + 1)(a - 1)$

65. $a^2 - b^2 = (a + b)(a - b)$

66. $x^2 - 4y^2 = (x + 2y)(x - 2y)$

67. $25 - n^2 = (5 + n)(5 - n)$

68. $n^2 - 25 = (n + 5)(n - 5)$

69. $4x^2 - y^2 = (2x + y)(2x - y)$

70. $9n^2 - 1 = (3n + 1)(3n - 1)$

71. $16a^2 - b^2 = (4a + b)(4a - b)$

72. $x^2y^2 - 1 = (xy + 1)(xy - 1)$

73. $4x^2 - 16 = 4(x^2 - 4) = 4(x + 2)(x - 2)$

74. $3a^2 + 6a + 3 = 3(a^2 + 2a + 1) = 3(a + 1)^2$

75. $2n^2 + 6n - 20 = 2(n^2 + 3n - 10) = 2(n - 2)(n + 5)$

76. $2y^2 - 4y - 48 = 2(y^2 - 2y - 24) = 2(y + 4)(y - 6)$

77. $3x^2 - 3x - 36 = 3(x^2 - x - 12) = 3(x + 3)(x - 4)$

78. $x^3 - xy^2 = x(x^2 - y^2) = x(x + y)(x - y)$

79. $2n^2 + 8n - 42 = 2(n^2 + 4n - 21) = 2(n - 3)(n + 7)$

80. $4a^2 - 20a - 24 = 4(a^2 - 5a - 6) = 4(a + 1)(a - 6)$

81. $ab^2 - 4ac^2 = a(b^2 - 4c^2) = a(b + 2c)(b - 2c)$

82. $2x^2 + 3x + 1 = (2x + 1)(x + 1)$

83. $3y^2 + 5y - 2 = (3y - 1)(y + 2)$

84. $5z^2 - 3z - 2 = (5z + 2)(z - 1)$

85. $4t^2 + 8t + 3 = (2t + 3)(2t + 1)$

86. $10m^2 - 13m - 3 = (5m + 1)(2m - 3)$

87. $10k^2 + 17t + 6 = (5k + 6)(2k + 1)$

Page 219 · CHAPTER TEST

1. $126 = 2 \cdot 3 \cdot 3 \cdot 7$ **2.** $21 = 3 \cdot 7$ and $33 = 3 \cdot 11$;   g.c.f. $= 3$

3. $50 = 2 \cdot 5 \cdot 5$ and $63 = 3 \cdot 3 \cdot 7$;   g.c.f. $= 1$

4. $12 = 2 \cdot 2 \cdot 3$ and $24 = 2 \cdot 2 \cdot 2 \cdot 3$; g.c.f. $= 2 \cdot 2 \cdot 3 = 12$

5. $65 = 5 \cdot 13$ and $26 = 2 \cdot 13$; g.c.f. $= 13$

6. $5x^3 - 10x^2 = 5x^2(x - 2)$ 7. $7a^3b - 21a^2b^2 + 7ab^3 = 7ab(a^2 - 3ab + b^2)$

8. $40y^3 - 5y^4 + 15xy = 5y(8y^2 - y^3 + 3x)$

9. $A = (1 \cdot 3) + (3 \cdot 5) + (3 \cdot 2) = 3 + 15 + 6 = 24$

10. $A = $ area of rectangle $-$ area of 2 circles $= (5r \cdot 2r) - 2\pi r^2 = 10r^2 - 2\pi r^2 = 2r^2(5 - \pi)$

11. $x^2 + 7x + 10 = (x + 2)(x + 5)$

12. $n^2 + 13n + 12 = (n + 12)(n + 1)$

13. $a^2 + 15a + 36 = (a + 12)(a + 3)$

14. $y^2 - 12y + 20 = (y - 2)(y - 10)$

15. $x^2 - 52x + 100 = (x - 2)(x - 50)$

16. $m^2 - 12m + 35 = (m - 5)(m - 7)$

17. $x^2 + 16x + 64 = (x + 8)^2$

18. $r^2 - 8r + 16 = (r - 4)^2$

19. $a^2 - 18ab + 81b^2 = (a - 9b)^2$

20. $n^2 - n - 6 = (n + 2)(n - 3)$

21. $y^2 + 5y - 24 = (y - 3)(y + 8)$

22. $b^2 - 12b - 45 = (b + 3)(b - 15)$

23. $(2x + 1)(2x - 1) = (2x)^2 - 1^2 = 4x^2 - 1$

24. $(a + 3b)(a - 3b) = a^2 - (3b)^2 = a^2 - 9b^2$

25. $(rs - 7)(rs + 7) = (rs)^2 - 7^2 = r^2s^2 - 49$

26. $(x^2 - 5y)(x^2 + 5y) = (x^2)^2 - (5y)^2 = x^4 - 25y^2$

27. $x^2 - 4y^2 = (x + 2y)(x - 2y)$

28. $81 - 16a^2 = (9 + 4a)(9 - 4a)$

29. $-100 + a^4 = a^4 - 100 = (a^2 + 10)(a^2 - 10)$

30. $b^3 - 5b^2 + 4b = b(b^2 - 5b + 4) = b(b - 1)(b - 4)$

31. $98x^2 - 2y^2 = 2(49x^2 - y^2) = 2(7x + y)(7x - y)$

32. $3m^2 + 12m - 36 = 3(m^2 + 4m - 12) = 3(m + 6)(m - 2)$

33. $3a^2 + 7a + 4 = (3a + 4)(a + 1)$

34. $8c^2 + 22c + 9 = (2c + 1)(4c + 9)$

35. $7x^2 + 20x - 3 = (7x - 1)(x + 3)$

Page 220 · CUMULATIVE REVIEW

1. $7x - y$ 2. $-2a^2 - 5a$ 3. $-2m - 10n$ 4. $18c^2d - 10c$

5. $\begin{array}{r} 3x - 4 \\ -2x + 1 \\ \hline x - 3 \end{array}$ 6. $\begin{array}{r} 9z^2 + 2z \\ -3z^2 + 2z \\ \hline 6z^2 + 4z \end{array}$ 7. $\begin{array}{r} 2t - 5 \\ t + 3 \\ \hline 3t - 2 \end{array}$ 8. $\begin{array}{r} 4m^2n - 4n^2 \\ -6m^2n - 3n^2 \\ \hline -2m^2n - 7n^2 \end{array}$

9. $-3y^3(y^4) = -3y^{3+4} = -3y^7$

10. $(x^3y^2)^2 = (x^3)^2(y^2)^2 = x^6y^4$

11. $xy(2x^2 - 3xy + y^2) = xy(2x^2) + xy(-3xy) + xy(y^2) = 2x^3y - 3x^2y^2 + xy^3$

12. $(a + 2)(a + 3) = a^2 + 5a + 6$

13. $(y - 4)(y - 1) = y^2 - 5y + 4$

14. $(2z - 3)(3z + 2) = 6z^2 - 5z - 6$

15. $A = (2x - 3)(3x + 4) = 6x^2 - x - 12$; $(6x^2 - x - 12)$ cm^2

16. $(3d + f)^2 = (3d)^2 + 2 \cdot 3d \cdot f + f^2 = 9d^2 + 6df + f^2$

17. $\dfrac{26m^3}{2m} = \dfrac{26 \cdot m \cdot m \cdot m}{2 \cdot m} = 13m^2$

18. $\dfrac{-60z^5}{4z^4} = \dfrac{-60 \cdot z \cdot z \cdot z \cdot z \cdot z}{4 \cdot z \cdot z \cdot z \cdot z} = -15z$

19. $\dfrac{-56y^8z^4}{-14y^4z^2} = \dfrac{-56}{-14} \cdot y^{8-4}z^{4-2} = 4y^4z^2$

20. $\dfrac{8y^3 + 12y}{4y} = \dfrac{8y^3}{4y} + \dfrac{12y}{4y} = 2y^2 + 3$

21. $\dfrac{10x^5 + 15x^3 - 25x}{5x} = \dfrac{10x^5}{5x} + \dfrac{15x^3}{5x} - \dfrac{25x}{5x} = 2x^4 + 3x^2 - 5$

22. $\dfrac{3a^3b^3 - 9a^2b^4 + 9ab^2}{3ab} = \dfrac{3a^3b^3}{3ab} - \dfrac{9a^2b^4}{3ab} + \dfrac{9ab^2}{3ab} = a^2b^2 - 3ab^3 + 3b$

23. $27 = 3 \cdot 3 \cdot 3$ and $51 = 3 \cdot 17$; g.c.f. $= 3$

24. $42 = 2 \cdot 3 \cdot 7$ and $24 = 2 \cdot 2 \cdot 2 \cdot 3$; g.c.f. $= 2 \cdot 3 = 6$

25. $3x^3 = 3 \cdot x \cdot x \cdot x$ and $15x = 3 \cdot 5 \cdot x$; g.c.f. $= 3 \cdot x = 3x$

26. $42ab = 2 \cdot 3 \cdot 7 \cdot a \cdot b$ and $48a^2 = 2 \cdot 2 \cdot 2 \cdot 2 \cdot 3 \cdot a \cdot a$; g.c.f. $= 2 \cdot 3 \cdot a = 6a$

27. $7m^2 - 21 = 7(m^2 - 3)$

28. $15y^3 - 9y^2 + 3y = 3y(5y^2 - 3y + 1)$ **29.** $x^2 + 10x + 16 = (x + 2)(x + 8)$

30. $y^2 + 7y + 12 = (y + 3)(y + 4)$ **31.** $t^2 - 18t + 45 = (t - 3)(t - 15)$

32. $z^2 - 12z + 35 = (z - 5)(z - 7)$ **33.** $1 + 6n + 9n^2 = (1 + 3n)^2$

34. $b^2 - 16b + 64 = (b - 8)^2$ **35.** $x^2 - 8x - 9 = (x + 1)(x - 9)$

36. $m^2 + 2m - 63 = (m + 9)(m - 7)$ **37.** $9 - t^2 = (3 + t)(3 - t)$

38. $16x^3 - 25xy^2 = x(16x^2 - 25y^2) = x(4x + 5y)(4x - 5y)$

39. $2x^2 - 16x - 40 = 2(x^2 - 8x - 20) = 2(x + 2)(x - 10)$

40. $5r^2 + 50r + 125 = 5(r^2 + 10r + 25) = 5(r + 5)^2$

41. $-5a^2 + 20 = 20 - 5a^2 = 5(4 - a^2) = 5(2 + a)(2 - a)$

42. $-s^2 + 32s - 60 = -1(s^2 - 32s + 60) = -1(s - 2)(s - 30)$

43. $-2y^2 - 40y - 150 = -2(y^2 + 20y + 75) = -2(y + 5)(y + 15)$

44. $mn^2 - m^3 = m(n^2 - m^2) = m(n + m)(n - m)$

Pages 226–227 · WRITTEN EXERCISES

A **1.** July and August   **2.** No;   the scale from 0 to 29 is condensed.

3. 33,000 − 30,500 = 2500

4.

Time	Number of eggs
Two years ago	1,100,000
One year ago	1,250,000
This year	1,400,000
Next year	1,150,000

5.

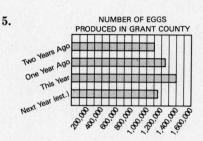

6. 35 people   **7.** Swimming;   about 23 people   **8.** Football;   10 people

9. Answers will vary.   **10.** $44 − $42 = $2;   Dry King is $2 less than Easy Dry.

11.

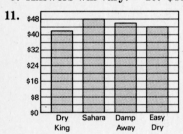

Pages 229–231 · WRITTEN EXERCISES

A **1.** 0.9   **2.** 8   **3.** 1992

4.

POPULATION OF LATIN AMERICA

5. and 6. Answers may vary considerably.
1990: about 451 million
1995: about 500 million

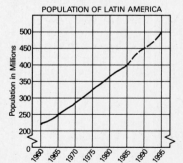

POPULATION OF LATIN AMERICA

7. about 70   **8.** about $8\frac{1}{2}$ years ago;   about $2\frac{1}{2}$ years ago

9. 40   **10.** 28   **11.** 10;   about 70   **12.** about 48°F   **13.** about 46°F   **14.** May

15. May   **16.** 110   **17.** 1905   **18.** about 215 million

19. about 370 million

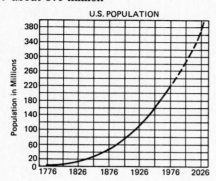

U.S. POPULATION

20. Transportation;   about $1280

21. $600 + $800 + $1100 = $2500

22. 4 years ago

23. 2 years ago

24. Heating

25. Since you average 25 miles per gallon, you use $\frac{1}{25}$ gallon for 1 mile.   Thus, it costs $\frac{1}{25}$ ($1.25), or 5¢ per mile.

26.

Miles traveled per year	10,000	20,000	30,000
Gasoline cost in dollars per year	$500	$1000	$1500

27.

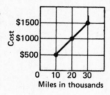

28. about $1200

29. $\frac{1}{30}$ ($1.20) = 4¢ for 1 mile;   2(4¢) = 8¢ for 2 miles

30.

Miles traveled per year	10,000	20,000	30,000
Gasoline cost in dollars per year	$400	$800	$1200

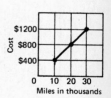

Pages 235–236 • WRITTEN EXERCISES

A   **1.** (−5, 4)   **2.** (5, 3)   **3.** (−3, −5)   **4.** (1, −2)   **5.** (0, 4)   **6.** (−4, 2)

7. (2, −4)   **8.** (−3, −3)   **9.** (−2, 3)   **10.** *P*   **11.** *M*   **12.** *O*

13. *R*   **14.** *J*   **15.** *N*   **16.** *L*   **17.** *Q*   **18.** *K*   **19.** 2   **20.** 1

21. (2, 4)   **22.** (2, −2)   **23.** (5, 1)   **24.** (−1, 1)

25.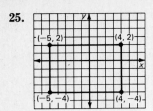

26.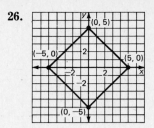

27.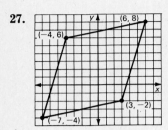

B 28.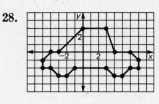

Pages 239–240 · WRITTEN EXERCISES

A   **1.** $(5, 3)$: $5 + 3 = 8$; $(-5, -3)$: $-5 + (-3) = -8 \neq 8$; $(-4, 12)$: $-4 + 12 = 8$.
   a and c

   2. $(10, 2)$: $10 - 2 = 8$; $(-4, 4)$: $-4 - 4 = -8 \neq 8$; $(-6, -2)$: $-6 - (-2) =$
   $-4 \neq 8$.   **a**

   3. $(0, 2)$: $3 \cdot 0 - 2 = -2 \neq 2$; $(-2, 0)$: $3(-2) - 2 = -8 \neq 0$; $(2, 4)$: $3 \cdot 2 -$
   $2 = 4$.   **c**

   4. $(3, 1)$: $5 - 2 \cdot 3 = -1 \neq 1$; $(1, 3)$: $5 - 2 \cdot 1 = 3$; $(-1, 7)$: $5 - 2(-1) = 7$.
   b and c

   5. $(0, 4)$: $4 - 3 \cdot 0 = 4 \neq 10$; $(-1, 7)$: $7 - 3(-1) = 10$; $(2, 10)$: $10 - 3 \cdot 2 =$
   $4 \neq 10$.   **b**

   6. $(1, -6)$: $2 \cdot 1 - (-6) = 8$; $(0, -8)$: $2 \cdot 0 - (-8) = 8$; $(-8, 8)$: $2(-8) - 8 =$
   $-24 \neq 8$.   **a and b**

7.

x	y
3	6
5	10
7	14
8	16

8.

x	y
2	−6
−4	12
3	−9
−3	9

9.

x	y
0	3
2	5
−4	−1
5	8

10.

x	y
0	−1
−1	−3
1	1
2	3

11.

x	y
0	2
1	1
2	0
−2	4

12.

x	y
0	9
2	5
3	3
−3	15

13.

t	s
0	6
−1	3
1	9
−2	0

14.

a	b
0	−1
1	3
−1	−5
2	7

15. $x + y = 9$; $x + y - x = 9 - x$; $y = 9 - x$

16. $y + 2x = 4$; $y + 2x - 2x = 4 - 2x$; $y = 4 - 2x$

17. $y - x = 6$; $y - x + x = 6 + x$; $y = 6 + x$

18. $3x + y = 7$; $3x + y - 3x = 7 - 3x$; $y = 7 - 3x$

19. $4x + y = 5$; $y = 5 - 4x$ **20.** $y - 3x = 7$; $y = 7 + 3x$

21. $y + 4x = 9$; $y = 9 - 4x$ **22.** $y + 5 = 4x$; $y = 4x - 5$

23. $7 + y = 9x$; $y = 9x - 7$

24. $x - y = 9$; $x = 9 + y$; $x - 9 = y$

25. $2x - y = 5$; $2x = 5 + y$; $2x - 5 = y$

26. $3x - y = 6$; $3x = 6 + y$; $3x - 6 = y$

Note: Answers to Exs. 27–38 may vary.

27. If $x = 1$, $y = 0$; if $x = 2$, $y = 1$; if $x = -1$, $y = -2$. $(1, 0)$, $(2, 1)$, $(-1, -2)$

28. If $x = 0$, $y = 5$; if $x = 1$, $y = 7$; if $x = 2$, $y = 9$. $(0, 5)$, $(1, 7)$, $(2, 9)$

29. $y = -x$. If $x = 1$, $y = -1$; if $x = 2$, $y = -2$; if $x = -1$, $y = 1$. $(1, -1)$, $(2, -2)$, $(-1, 1)$

30. $y = 7 - x$. If $x = 0$, $y = 7$; if $x = 1$, $y = 6$; if $x = -1$, $y = 8$. $(0, 7)$, $(1, 6)$, $(-1, 8)$

31. $y = 6 + x$. If $x = 0$, $y = 6$; if $x = 1$, $y = 7$; if $x = -1$, $y = 5$. $(0, 6)$, $(1, 7)$, $(-1, 5)$

32. $y = 8 + x$. If $x = 0$, $y = 8$; if $x = 2$, $y = 10$; if $x = -2$, $y = 6$. $(0, 8)$, $(2, 10)$, $(-2, 6)$

33. $y = 4 + 2x$. If $x = 0$, $y = 4$; if $x = 2$, $y = 8$; if $x = -2$, $y = 0$. $(0, 4)$, $(2, 8)$, $(-2, 0)$

34. $y = 2 + 3x$. If $x = 0$, $y = 2$; if $x = 1$, $y = 5$; if $x = -1$, $y = -1$. $(0, 2)$, $(1, 5)$, $(-1, -1)$

35. $y = 7 + 4x$. If $x = 0$, $y = 7$; if $x = 1$, $y = 11$; if $x = -1$, $y = 3$. $(0, 7)$, $(1, 11)$, $(-1, 3)$

36. $y = -5x$. If $x = 0$, $y = 0$; if $x = 1$, $y = -5$; if $x = -1$, $y = 5$. $(0, 0)$, $(1, -5)$, $(-1, 5)$

37. $y = 2 - 2x$. If $x = 0$, $y = 2$; if $x = 1$, $y = 0$; if $x = -1$, $y = 4$. $(0, 2)$, $(1, 0)$, $(-1, 4)$

38. $y = 9 + 3x$. If $x = 0$, $y = 9$; if $x = 1$, $y = 12$; if $x = -3$, $y = 0$. $(0, 9)$, $(1, 12)$, $(-3, 0)$

B 39. $y = 4x$ 40. $y = x + 1$

41. $y = 4x + 1$ 42. $y = 2x - 1$

43. **a.** $8 > 2 \cdot 2$? $8 > 4$? Yes   **b.** $8 > 2 \cdot 3$? $8 > 6$? Yes

   c. $8 > 2 \cdot 4$? $8 > 8$? No

44. **a.** $5 < 3 \cdot 2$? $5 < 6$? Yes   **b.** $6 < 3 \cdot 2$? $6 < 6$? No

   c. $7 < 3 \cdot 2$? $7 < 6$? No

45. **a.** $1 < 5 - 2$? $1 < 3$? Yes   **b.** $2 < 5 - 2$? $2 < 3$? Yes

   c. $3 < 5 - 2$? $3 < 3$? No

46. **a.** $1 > 2 \cdot 0 - 3$? $1 > -3$? Yes   **b.** $0 > 2 \cdot 0 - 3$? $0 > -3$? Yes

   c. $8 > 2 \cdot 15 - 3$? $8 > 27$? No

Page 242 · WRITTEN EXERCISES

A 1.
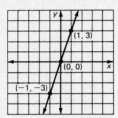

2.
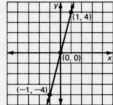

3.
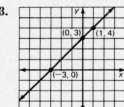

4.
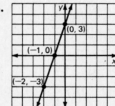

5.
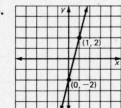

6.
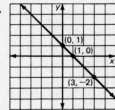

7.
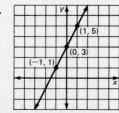

8.
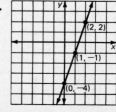

9.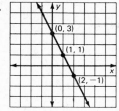

10.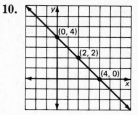

11.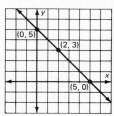

12.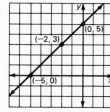

13.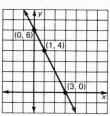

14.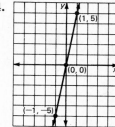

15.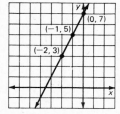

16.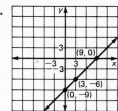

17.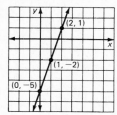

18.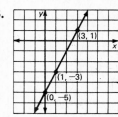

19.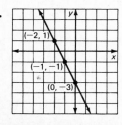

20.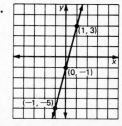

21.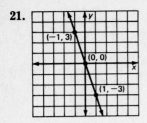

B **22.**

x	y
-2	-1
-1	0
0	1
1	2

$y = x + 1$

23.

x	y
-1	3
0	2
1	1
2	0

$y = -x + 2$

24.

x	y
-1	-2
0	0
1	2
2	4

$y = 2x$

Page 244 • WRITTEN EXERCISES

A **1.** $\dfrac{2}{3}$ **2.** $\dfrac{5}{1} = 5$ **3.** $\dfrac{3}{2}$ **4.** $\dfrac{1}{3}$ **5.** $\dfrac{-1}{3} = -\dfrac{1}{3}$ **6.** $\dfrac{-2}{1} = -2$ **7.** 3

8. 4 **9.** 2 **10.** 4 **11.** -1 **12.** -1 **13.** -3 **14.** -5

15. $\dfrac{1}{3}$ **16.** $\dfrac{3}{4}$ **17.** $-\dfrac{1}{5}$ **18.** $-\dfrac{2}{3}$

19. Answers may vary.   **20.** Answers may vary.

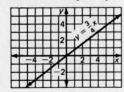   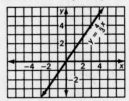

B **21.** 0 **22.** The run varies. For example, the run from $(0, 1)$ to $(2, 1)$ is 2.

23. 0 **24.** the y-axis

25. The rise varies. For example, the rise from $(1, 0)$ to $(1, 2)$ is 2.

26. 0 **27.** No **28.** the x-axis

Pages 247–248 • WRITTEN EXERCISES

A **1.** $y = -4$; $x = 5$ **2.** $2x$ **3.** $x = -9$ **4.** $x + 1$

5.

n	1	2	3	12	10
C	48	96	144	576	480

6. $48n$

7.

Rate	Distance
10	40
20	80
30	120
40	160

8.

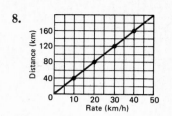

9. −17°C   **10.** −41°C   **11.** 40

B 12.

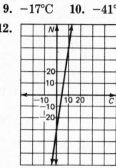

13. about 15°C

14. about 145 times per minute

15. about 4°C

Page 249 · CALCULATOR ACTIVITIES

1. a, c   **2.** b   **3.** a, b   **4.** c, d   **5.** b, d

Page 252 · WRITTEN EXERCISES

A   1.

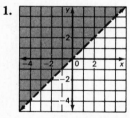

2.

3.

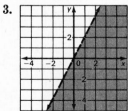

4.

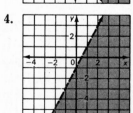

5.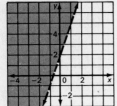

6.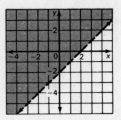

7.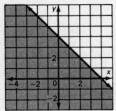

8.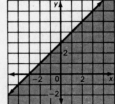

9.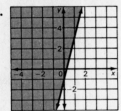

10.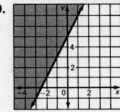

11.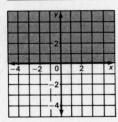

12.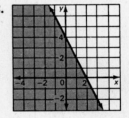

Page 253 · READING ALGEBRA

1. exponent   2. perimeter   3. trinomial

4. integers   5. coordinates   6. factors

Page 254 · SKILLS REVIEW

1. $\dfrac{3}{12}$ 2. $\dfrac{4}{20}$ 3. $\dfrac{3}{9}$ 4. $\dfrac{4}{8}$ 5. $\dfrac{3}{18}$ 6. $\dfrac{4}{6}$ 7. $\dfrac{18}{48}$ 8. $\dfrac{12}{16}$

9. $\dfrac{20}{35}$ 10. $\dfrac{20}{24}$ 11. $\dfrac{4}{12},\dfrac{3}{12}$ 12. $\dfrac{7}{14},\dfrac{2}{14}$ 13. $\dfrac{2}{8},\dfrac{1}{8}$ 14. $\dfrac{5}{15},\dfrac{3}{15}$

15. $\dfrac{5}{20},\dfrac{4}{20}$ 16. $\dfrac{6}{15},\dfrac{5}{15}$ 17. $\dfrac{3}{18},\dfrac{8}{18}$ 18. $\dfrac{8}{12},\dfrac{3}{12}$ 19. $\dfrac{6}{14},\dfrac{1}{14}$

20. $\dfrac{3}{24},\dfrac{20}{24}$ 21. $\dfrac{21}{28},\dfrac{12}{28}$ 22. $\dfrac{36}{63},\dfrac{14}{63}$ 23. $\dfrac{21}{30},\dfrac{20}{30}$ 24. $\dfrac{25}{30},\dfrac{12}{30}$

25. $\dfrac{32}{72},\dfrac{27}{72}$ 26. $\dfrac{1}{2}+\dfrac{1}{7}=\dfrac{7}{14}+\dfrac{2}{14}=\dfrac{7+2}{14}=\dfrac{9}{14}$

27. $\dfrac{1}{6} + \dfrac{1}{5} = \dfrac{5}{30} + \dfrac{6}{30} = \dfrac{5+6}{30} = \dfrac{11}{30}$

28. $\dfrac{1}{4} + \dfrac{1}{6} = \dfrac{3}{12} + \dfrac{2}{12} = \dfrac{3+2}{12} = \dfrac{5}{12}$

29. $\dfrac{1}{9} + \dfrac{1}{3} = \dfrac{1}{9} + \dfrac{3}{9} = \dfrac{1+3}{9} = \dfrac{4}{9}$

30. $\dfrac{1}{4} + \dfrac{1}{7} = \dfrac{7}{28} + \dfrac{4}{28} = \dfrac{11}{28}$

31. $\dfrac{1}{2} - \dfrac{1}{3} = \dfrac{3}{6} - \dfrac{2}{6} = \dfrac{1}{6}$

32. $\dfrac{1}{3} - \dfrac{1}{8} = \dfrac{8}{24} - \dfrac{3}{24} = \dfrac{5}{24}$

33. $\dfrac{1}{4} - \dfrac{1}{5} = \dfrac{5}{20} - \dfrac{4}{20} = \dfrac{1}{20}$

34. $\dfrac{1}{2} - \dfrac{1}{6} = \dfrac{3}{6} - \dfrac{1}{6} = \dfrac{2}{6} = \dfrac{1}{3}$

35. $\dfrac{1}{3} - \dfrac{1}{7} = \dfrac{7}{21} - \dfrac{3}{21} = \dfrac{4}{21}$

36. $\dfrac{2}{3} + \dfrac{1}{4} = \dfrac{8}{12} + \dfrac{3}{12} = \dfrac{11}{12}$

37. $\dfrac{5}{6} + \dfrac{1}{7} = \dfrac{35}{42} + \dfrac{6}{42} = \dfrac{41}{42}$

38. $\dfrac{3}{8} + \dfrac{1}{6} = \dfrac{9}{24} + \dfrac{4}{24} = \dfrac{13}{24}$

39. $\dfrac{4}{5} + \dfrac{3}{7} = \dfrac{28}{35} + \dfrac{15}{35} = \dfrac{43}{35}$

40. $\dfrac{2}{9} + \dfrac{3}{10} = \dfrac{20}{90} + \dfrac{27}{90} = \dfrac{47}{90}$

41. $\dfrac{3}{4} - \dfrac{1}{8} = \dfrac{6}{8} - \dfrac{1}{8} = \dfrac{5}{8}$

42. $\dfrac{7}{8} - \dfrac{2}{5} = \dfrac{35}{40} - \dfrac{16}{40} = \dfrac{19}{40}$

43. $\dfrac{9}{11} - \dfrac{1}{3} = \dfrac{27}{33} - \dfrac{11}{33} = \dfrac{16}{33}$

44. $\dfrac{5}{7} - \dfrac{2}{5} = \dfrac{25}{35} - \dfrac{14}{35} = \dfrac{11}{35}$

45. $\dfrac{2}{3} - \dfrac{5}{8} = \dfrac{16}{24} - \dfrac{15}{24} = \dfrac{1}{24}$

Pages 255–256 • CHAPTER REVIEW EXERCISES

1. about 7.5 cm   **2.** January   **3.** July   **4.** $11.5 - 3.5 = 8$;   8 cm

5. about 117 million   **6.** 1967

7.

x	y
0	0
1	−4
2	−8
−2	8

8.

x	y
0	3
1	7
2	11
3	15

9.

x	y
0	10
1	9
2	8
−1	11

10.

x	y
0	5
1	2
2	−1
−2	11

11.

12.

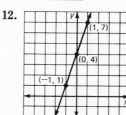

13.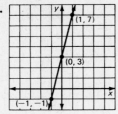

14.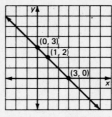

15.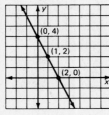

16.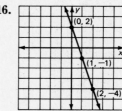

17. 3   **18.** −1   **19.** $\frac{3}{4}$ **20.** the number of yards purchased

21.

n	1	2	3	5	8
C	$4.20	$8.40	$12.60	$21.00	$33.60

22. $4.20n$ **23.** the number of bumper stickers sold

24.

n	10	15	20	50	110
A	$5	$7.50	$10	$25	$55

25.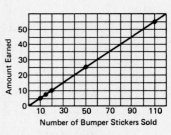

26.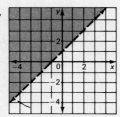

27.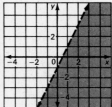

28.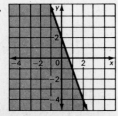

Page 257 · CHAPTER TEST

1. 20   2. 1981

3.

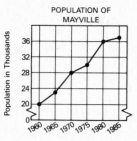

4.

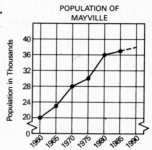

about 38 thousand

5.

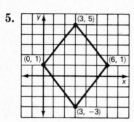

6. $3x - y = 6$; $3x = 6 + y$;
   $3x - 6 = y$

7. Answers may vary. $y = 2x + 2$.   If $x = 0$, $y = 2$;   if $x = 1$, $y = 4$;   if $x = -1$,
   $y = 0$. $(0, 2), (1, 4), (-1, 0)$

8.

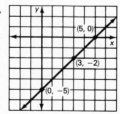

9.

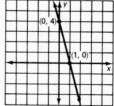

10. 7

11.

x	4	10	50	64	100
y	20	50	250	320	500

Pages 258–259 · MIXED REVIEW

1. Answers may vary.

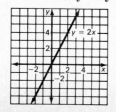

2. $\dfrac{7x^4 - 14x^3y + 35x^2}{-7x^2} = \dfrac{7x^4}{-7x^2} - \dfrac{14x^3y}{-7x^2} + \dfrac{35x^2}{-7x^2} = -x^2 + 2xy - 5$

3.

	rate	time	Distance
Jog	10	$x + 4$	$10(x + 4)$
Bus	50	x	$50x$

$10(x + 4) = 50x; \quad 10x + 40 = 50x;$
$40 = 40x; \quad 1 = x,$ so $50x = 50 \cdot 1 = 50.$
50 km

4. $2(1 - 4a) - 7(-2 - a) = 1 - 6a; \quad 2 - 8a + 14 + 7a = 1 - 6a; \quad 16 - a = 1 - 6a; \quad 16 - a - 1 = 1 - 6a - 1; \quad 15 - a = -6a; \quad 15 - a + a = -6a + a; \quad 15 = -5a; \quad -3 = a$

5. $(2r^3s^2)(-9rs^3) = (2 \cdot -9)(r^3 \cdot r)(s^2 \cdot s^3) = -18r^4s^5$

6. $(x^2 + 3)(x^2 - 3) = x^4 - 9$

7.
$$
\begin{array}{r}
n^2 + 2n\phantom{{}+4} + 4 \\
\underline{n\phantom{{}^2+2n} - 2} \\
n^3 + 2n^2 + 4n\phantom{{}-8} \\
\underline{- 2n^2 - 4n - 8} \\
n^3 \phantom{{}+2n^2+4n} - 8
\end{array}
$$

8. $75 = 3 \cdot 5^2; \quad 30 = 2 \cdot 3 \cdot 5;$
g.c.f. $= 3 \cdot 5 = 15$

9. 1980　　**10.** $110 - 45 = 65;$　$65 million

11.

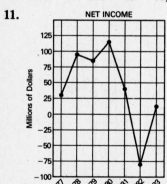

NET INCOME

12. $(-5, -3)$

13. a. $P = 2(2n + 3) + 2(n + 4) = 4n + 6 + 2n + 8 = 6n + 14$

　b. $A = (2n + 3)(n + 4) = 2n^2 + 11n + 12$

14. $b^2 + 2b - 35 = (b + 7)(b - 5)$　　**15.** $m^2 - 14m + 48 = (m - 6)(m - 8)$

16. $-9t^2 + r^2s^2 = r^2s^2 - 9t^2 = (rs + 3t)(rs - 3t)$

17. $7x^3 + 21x^2y + 28x = 7x(x^2 + 3xy + 4)$

18. $a^2 + 14a + 24 = (a + 2)(a + 12)$

19. $x^2 - 8xy + 16y^2 = (x - 4y)^2$

20. $\dfrac{8mn}{-2n} + \dfrac{-15m^3}{-5m^2} = -4m + 3m = -m$

21. a. $P = 13 + 14 + 15 = 42$　**b.** $A = \dfrac{1}{2} \cdot 14 \cdot 12 = 84$

22. $12 + 6 \div (-2) = 12 + (-3) = 9; \quad -(-3)^2 = -9; \quad 9 > -9; \quad >$

23.
```
  ←————•———————————————○——+——→
   -5  -4  -3  -2  -1   0   1
```

24. Answers may vary. $y = 5 + 3x.$ If $x = 0,$ $y = 5;$ if $x = 1,$ $y = 8;$ if $x = -1,$ $y = 2.$ $(0, 5), (1, 8), (-1, 2)$

25. $-3(3x - y)^2 = -3(9x^2 - 6xy + y^2) = -27x^2 + 18xy - 3y^2$

26. $2x^2 - 2x - 40 = 2(x^2 - x - 20) = 2(x - 5)(x + 4)$

27.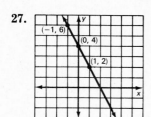

28. $k = \dfrac{C}{d};$ $kd = C;$ $d = \dfrac{C}{k}$

29. $(-x^4y^2)^3(2x)^2 = (-1)^3(x^4)^3(y^2)^3(2)^2(x^2) = (-1 \cdot 4)(x^{12} \cdot x^2)(y^6) = -4x^{14}y^6$

30. $(y^2 - 3y - 9) - (2y + 6) = y^2 - 3y - 9 - 2y - 6 = y^2 - 5y - 15$

31. $(3a^2 - 5ab) + (ab - b^2) = 3a^2 - 5ab + ab - b^2 = 3a^2 - 4ab - b^2$

32. $C = 36 + 22(n - 1)$

33. $C = 36 + 22(6 - 1) = 36 + 22 \cdot 5 = 36 + 110 = 146;$ 146¢, or \$1.46

34. $300 = 36 + 22(n - 1);$ $300 = 36 + 22n - 22;$ $300 = 14 + 22n;$ $286 = 22n;$ $13 = n;$ 13 minutes

35. $8(3x - 1) - 5(7x + 2) = 4;$ $24x - 8 - 35x - 10 = 4;$ $-11x - 18 = 4;$ $-11x = 22;$ $x = -2$

36.

	price	number	Cost
Sell	14	$n - 8$	$14(n - 8)$
Buy	10	n	$10n$
			288

$14(n - 8) - 10n = 288;$ $14n - 112 - 10n = 288;$ $4n - 112 = 288;$ $4n = 400;$ $n = 100,$ so $n - 8 = 92.$ 92 pairs sold

Pages 264–265 · WRITTEN EXERCISES

A **1.** $(1, 2)$ **2.** $(-2, 3)$ **3.** $(2, -1)$

Note: Answers to Exs. 4–7 may vary.

4. $(0, 4)$ and $(2, 2)$ **5.** $y = 12 - x$; $(0, 12)$ and $(6, 6)$

6. $y = 6 + x$; $(0, 6)$ and $(-6, 0)$ **7.** $y = 1 - 2x$; $(0, 1)$ and $(1, -1)$

8. $(1, -1)$ **9.** $(1, 2)$

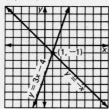   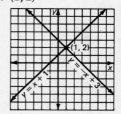

10. $(-2, 1)$ **11.** $(-1, -2)$

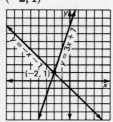   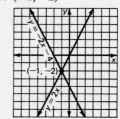

12. $(1, -1)$ **13.** $(-3, 0)$

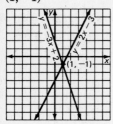   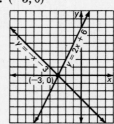

14. $y = x$ and $y = -x + 4$; $(2, 2)$ **15.** $y = x + 3$ and $y = -x - 1$; $(-2, 1)$

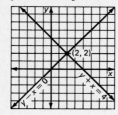   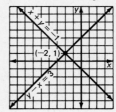

16. $y = -x - 4$ and $y = 2x + 5$; $(-3, -1)$

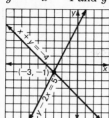

17. $y = x$ and $y = -3x - 4$; $(-1, -1)$

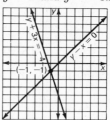

18. $y = 3x - 1$ and $y = 2x + 1$; $(2, 5)$

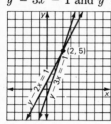

19. $y = 2x - 5$ and $y = x - 3$; $(2, -1)$

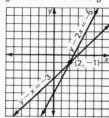

B   **20.** about $\left(-2\frac{1}{2}, 1\frac{1}{2}\right)$

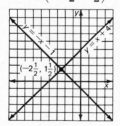

21. $y = -2x + 2$ and $y = 3x + 9$

about $\left(-1\frac{2}{5}, 4\frac{4}{5}\right)$

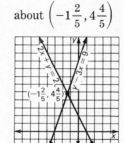

22. $y = 2x + 4$ and $y = -x + 5$

about $\left(\frac{1}{3}, 4\frac{2}{3}\right)$

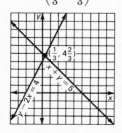

23. $y = 2x + 8$ and $y = -x - 2$

about $\left(-3\frac{1}{3}, 1\frac{1}{3}\right)$

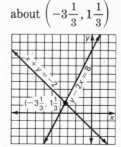

24. about $\left(-2\frac{1}{2}, 4\frac{1}{2}\right)$

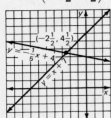

25. about $\left(2, -2\frac{1}{2}\right)$

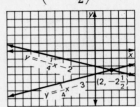

26. a.

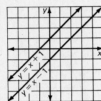

b. No   **c.** No

C   27.

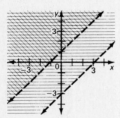

28.

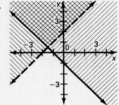

29.

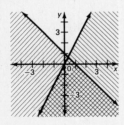

Pages 268–269 • WRITTEN EXERCISES

A   **1.** 4   **2.** 4   **3.** 1   **4.** $\frac{1}{3}$ **5.** $y = 3x + 6$;   3

6. $y = -2x + 4$; -2 **7.** $y = 4x$;   4   **8.** $y = -4x$; -4 **9.** $y = 2x - 3$;   2

10. $y = -5x + 2$; -5 **11.** -7 **12.** 1   **13.** Slope of both $= 2$;   no pair

14. Slope of both $= \frac{1}{3}$;   no pair

15. $y = -x + 8$ and $y = -2x + 8$;   slopes $= -1$ and -2;   one pair

16. $y = x - 4$ and $y = x - 6$;   slope of both $= -1$;   no pair

17. $y = x - 8$ and $y = -x + 6$;   slopes $= 1$ and -1;   one pair

18. $y = -2x + 9$ and $y = -4x + 6$;   slopes $= -2$ and -4;   one pair

19. $y = -3x + 4$ and $y = 3x + 2$;   slopes $= -3$ and 3;   one pair
20. $y = 4x - 6$ and $y = 4x + 7$;   slope of both $= 4$;   no pair
21. $y = -3x + 7$ and $y = -3x + 7$;   same equation;   all pairs
22. $y = -2x + 4$ and $y = -2x + 8$;   slope of both $= -2$;   no pair
23. $y = 4x + 9$ and $y = 4x + 9$;   same equation;   all pairs
24. $y = -x - 3$ and $y = 2x - 6$;   slopes $= -1$ and 2;   one pair
25. $y = -2x + 9$ and $y = -2x + 7$;   slope of both $= -2$;   no pair
26. $y = -3x + 3$ and $y = x + 5$;   slopes $= -3$ and 1;   one pair
27. $y = 3x - 4$ and $y = -3x - 4$;   slopes $= 3$ and -3;   one pair
28. $y = -2x + 4$ and $y = 2x - 4$;   slopes $= -2$ and 2;   one pair

B   29. **a.** No, since the slope of both is 2.

   b. Yes, it will intersect both since its slope is -1.

   c. No, since the first 2 graphs do not intersect.

30. **a.** No **b.** Yes **c.** Yes
31. **a.** Yes;   slope of both $= -4$. **b.** No **c.** No

Page 271 · WRITTEN EXERCISES

A   1. $y = 2x$ and $x + y = 9$; $x + (2x) = 9$; $3x = 9$; $x = 3$; $y = 2x = 2 \cdot 3 = 6$.
   $(3, 6)$

2. $y = 5x$ and $y - x = 8$; $(5x) - x = 8$; $4x = 8$; $x = 2$; $y = 5x = 5 \cdot 2 = 10$.
   $(2, 10)$

3. $x = y + 3$ and $x + y = 7$; $(y + 3) + y = 7$; $2y + 3 = 7$; $2y = 4$; $y = 2$;
   $x = y + 3 = 2 + 3 = 5$. $(5, 2)$

4. $y = x + 1$ and $x + y = 5$; $x + (x + 1) = 5$; $2x + 1 = 5$; $2x = 4$; $x = 2$;
   $y = x + 1 = 2 + 1 = 3$. $(2, 3)$

5. $y = x + 4$ and $x + y = 22$; $x + (x + 4) = 22$; $2x + 4 = 22$; $2x = 18$; $x = 9$;
   $y = x + 4 = 9 + 4 = 13$. $(9, 13)$

6. $x = 2 - y$ and $2y + x = 9$; $2y + (2 - y) = 9$; $y + 2 = 9$; $y = 7$;
   $x = 2 - y = 2 - 7 = -5$. $(-5, 7)$

7. $y = 3x + 1$ and $4x + y = 8$; $4x + (3x + 1) = 8$; $7x + 1 = 8$; $7x = 7$; $x = 1$;
   $y = 3x + 1 = 3 \cdot 1 + 1 = 3 + 1 = 4$. $(1, 4)$

8. $y = 2x - 3$ and $4x + y = 9$; $4x + (2x - 3) = 9$; $6x - 3 = 9$; $6x = 12$;
   $x = 2$; $y = 2x - 3 = 2 \cdot 2 - 3 = 4 - 3 = 1$. $(2, 1)$

9. $x + y = 2$ and $3x + y = 8$; $y = -x + 2$ and $3x + y = 8$; $3x + (-x + 2) = 8$;
   $2x + 2 = 8$; $2x = 6$; $x = 3$; $y = -x + 2 = -3 + 2 = -1$. $(3, -1)$

10. $2x + y = 5$ and $4x - y = 1$; $2x + y = 5$ and $y = 4x - 1$; $2x + (4x - 1) = 5$;
   $6x - 1 = 5$; $6x = 6$; $x = 1$; $y = 4x - 1 = 4 \cdot 1 - 1 = 4 - 1 = 3$. $(1, 3)$

11. $3x - y = 13$ and $2x + 3y = 16$; $\quad y = 3x - 13$ and $2x + 3y = 16$;

$2x + 3(3x - 13) = 16$; $\quad 2x + 9x - 39 = 16$; $\quad 11x - 39 = 16$; $\quad 11x = 55$;

$x = 5$; $\quad y = 3x - 13 = 3 \cdot 5 - 13 = 15 - 13 = 2$; $\quad (5, 2)$

12. $2x - y = 9$ and $5x + 2y = 27$; $\quad y = 2x - 9$ and $5x + 2y = 27$;

$5x + 2(2x - 9) = 27$; $\quad 5x + 4x - 18 = 27$; $\quad 9x - 18 = 27$; $\quad 9x = 45$; $\quad x = 5$;

$y = 2x - 9 = 2 \cdot 5 - 9 = 10 - 9 = 1$ $\quad (5, 1)$

13. $x = 5y$ and $3x = 7y + 16$; $\quad 3(5y) = 7y + 16$; $\quad 15y = 7y + 16$; $\quad 8y = 16$; $\quad y = 2$;

$x = 5y = 5 \cdot 2 = 10$. $\quad (10, 2)$

14. $x - 5y = 8$ and $4x + 2y = 10$; $\quad x = 5y + 8$ and $4x + 2y = 10$; $\quad 4(5y + 8) +$

$2y = 10$; $\quad 20y + 32 + 2y = 10$; $\quad 22y + 32 = 10$; $\quad 22y = -22$; $\quad y = -1$;

$x = 5y + 8 = 5(-1) + 8 = -5 + 8 = 3$. $\quad (3, -1)$

15. $a + 2b = 7$ and $2a = 3b$; $\quad a = -2b + 7$ and $2a = 3b$; $\quad 2(-2b + 7) = 3b$;

$-4b + 14 = 3b$; $\quad 14 = 7b$; $\quad 2 = b$; $\quad a = -2b + 7 = -2 \cdot 2 + 7 = -4 + 7 = 3$.

$(3, 2)$

16. $p - 5q = 6$ and $3p - 2q = 5$; $\quad p = 5q + 6$ and $3p - 2q = 5$; $\quad 3(5q + 6) - 2q =$

5; $\quad 15q + 18 - 2q = 5$; $\quad 13q + 18 = 5$; $\quad 13q = -13$; $\quad q = -1$; $\quad p = 5q + 6 =$

$5(-1) + 6 = -5 + 6 = 1$. $\quad (1, -1)$

B   **17.** $A = bh$ and $b = 2h$; $\quad A = (2h)h = 2h^2$

       18. $V = Bh$ and $B = 2h^2$; $\quad V = (2h^2)h = 2h^3$

       19. $V = lwh$, $l = 3h$, and $w = 2h$; $\quad V = (3h)(2h)h = 6h^3$

C   **20.** $x + y + z = 180$, $y = 3x$, and $z = 5x$; $\quad x + (3x) + (5x) = 180$; $\quad 9x = 180$;

$x = 20$; $\quad y = 3x = 3 \cdot 20 = 60$; $\quad z = 5x = 5 \cdot 20 = 100$

       21. $x + y + z = 62$, $x = 2z - 5$, and $y = 3z - 5$; $\quad (2z - 5) + (3z - 5) + z = 62$;

$6z - 10 = 62$; $\quad 6z = 72$; $\quad z = 12$; $\quad x = 2z - 5 = 2 \cdot 12 - 5 = 19$; $\quad y = 3z -$

$5 = 3 \cdot 12 - 5 = 31$

       22. $x + 2y + 3z = 0$, $2x + y = 6$, and $3x + z = 8$; $\quad x + 2y + 3z = 0$, $y = -2x + 6$,

and $z = -3x + 8$; $\quad x + 2(-2x + 6) + 3(-3x + 8) = 0$; $\quad x - 4x + 12 - 9x +$

$24 = 0$; $\quad -12x + 36 = 0$; $\quad 36 = 12x$; $\quad 3 = x$; $\quad y = -2x + 6 = -2 \cdot 3 + 6 = 0$;

$z = -3x + 8 = -3 \cdot 3 + 8 = -1$

Pages 273–274 · WRITTEN EXERCISES

A   **1.** $\quad x + y = 2$

            $\underline{x - y = 10}$

          $2x \quad\quad = 12$; $\cdot$ $x = 6$

          $6 + y = 2$; $\quad y = -4$. $\quad (6, -4)$

2. $\quad 5x + 4y = 1$

    $\underline{3x - 4y = 7}$

  $8x \quad\quad = 8$; $\quad x = 1$

  $5 \cdot 1 + 4y = 1$; $\quad 5 + 4y = 1$;

   $4y = -4$; $\quad y = -1$. $\quad (1, -1)$

3. $4x - y = 8$
$\underline{2x + y = -2}$
$6x \quad = 6; \quad x = 1$
$2 \cdot 1 + y = -2; \quad 2 + y = -2;$
$y = -4. \quad (1, -4)$

4. $5p + 3q = 10$
$\underline{2p - 3q = 4}$
$7p \quad = 14; \quad p = 2$
$5 \cdot 2 + 3q = 10; \quad 10 + 3q = 10;$
$3q = 0; \quad q = 0. \quad (2, 0)$

5. $2a - b = 3$
$\underline{4a + b = 9}$
$6a \quad = 12; \quad a = 2$
$4 \cdot 2 + b = 9; \quad 8 + b = 9;$
$b = 1. \quad (2, 1)$

6. $2x + y = 10$
$\underline{3x - y = 5}$
$5x \quad = 15; \quad x = 3$
$2 \cdot 3 + y = 10; \quad 6 + y = 10;$
$y = 4. \quad (3, 4)$

7. $3x - 2y = 8$
$\underline{x + 2y = 8}$
$4x \quad = 16; \quad x = 4$
$4 + 2y = 8; \quad 2y = 4; \quad y = 2.$
$(4, 2)$

8. $4r - 7s = 13$
$\underline{4r + 7s = -29}$
$8r \quad = -16; \quad r = -2$
$4(-2) + 7s = -29; \quad -8 + 7s = -29;$
$7s = -21; \quad s = -3. \quad (-2, -3)$

9. $x + 6y = 10$
$\underline{-x - 2y = -2}$
$\qquad 4y = 8; \quad y = 2$
$x + 6 \cdot 2 = 10; \quad x + 12 = 10;$
$x = -2. \quad (-2, 2)$

10. $x - y = 20$
$\underline{-x + 3y = -10}$
$\qquad 2y = 10; \quad y = 5$
$x - 5 = 20; \quad x = 25. \quad (25, 5)$

11. $3x + y = 7$
$\underline{2x - y = 8}$
$5x \quad = 15; \quad x = 3$
$3 \cdot 3 + y = 7; \quad 9 + y = 7;$
$y = -2. \quad (3, -2)$

12. $3x - 4y = 21$
$\underline{-2x + 4y = -18}$
$x \quad = 3$
$3 \cdot 3 - 4y = 21; \quad 9 - 4y = 21;$
$-4y = 12; \quad y = -3. \quad (3, -3)$

13. $3x + 2y = 18$
$\underline{-x - 2y = -14}$
$2x \quad = 4; \quad x = 2$
$3 \cdot 2 + 2y = 18; \quad 6 + 2y = 18;$
$2y = 12; \quad y = 6. \quad (2, 6)$

14. $2x - 5y = 14$
$\underline{-2x + 3y = -10}$
$\qquad -2y = 4; \quad y = -2$
$2x - 5(-2) = 14; \quad 2x + 10 = 14;$
$2x = 4; \quad x = 2. \quad (2, -2)$

15. Subtract. $\quad 2a + 3b = 7$
$\underline{-2a + b = 5}$
$\qquad\qquad 4b = 12; \quad b = 3$
$2a + 3 \cdot 3 = 7; \quad 2a + 9 = 7;$
$2a = -2; \quad a = -1. \quad (-1, 3)$

16. Subtract. $\quad 2s - 5r = 17$
$\underline{-6s + 5r = -1}$
$\qquad -4s \quad = 16; \quad s = -4$
$2(-4) - 5r = 17; \quad -8 - 5r = 17;$
$-5r = 25; \quad r = -5. \quad (-4, -5)$

17. Add. $\quad x + y = 9$
$\underline{x - y = 5}$
$2x \quad = 14; \quad x = 7$
$7 + y = 9; \quad y = 2. \quad (7, 2)$

18. Add. $\quad x + 2y = 7$
$\underline{3x - 2y = 5}$
$4x \quad = 12; \quad x = 3$
$3 + 2y = 7; \quad 2y = 4; \quad y = 2. \quad (3, 2)$

19. Subtract. $2x + 3y = 8$
                    $-2x + y = 8$
                    $4y = 16;\ y = 4$

$2x + 3 \cdot 4 = 8;\ 2x + 12 = 8;$

$2x = -4;\ x = -2.\ (-2, 4)$

21. Add. $3x - 2y = 13$
               $4x + 2y = 8$
               $7x\ \ \ \ \ \ \ = 21;\ x = 3$

$4 \cdot 3 + 2y = 8;\ 12 + 2y = 8;$

$2y = -4;\ y = -2.\ (3, -2)$

23. Add. $3x + y = 0$
               $6x - y = 18$
               $9x\ \ \ \ \ \ = 18;\ x = 2$

$3 \cdot 2 + y = 0;\ 6 + y = 0;$

$y = -6.\ (2, -6)$

25. Subtract. $4x - 2y = 10$
                    $x + 2y = 0$
                    $5x\ \ \ \ \ \ \ = 10;\ x = 2$

$4 \cdot 2 - 2y = 10;\ 8 - 2y = 10;$

$-2y = 2;\ y = -1.\ (2, -1)$

27. Add. $8x + 3y = 5$
               $x - 3y = 4$
               $9x\ \ \ \ \ \ = 9;\ x = 1$

$1 - 3y = 4;\ -3y = 3;\ y = -1.$

$(1, -1)$

29. Subtract. $3a - 12b = 9$
                    $-a + 12b = -11$
                    $2a\ \ \ \ \ \ \ \ = -2;\ a = -1$

$3(-1) - 12b = 9;\ -3 - 12b = 9;$

$-12b = 12;\ b = -1.\ (-1, -1)$

31. Subtract. $6x - 2y = 8$
                    $-5x + 2y = -5$
                    $x\ \ \ \ \ \ \ = 3$

$6 \cdot 3 - 2y = 8;\ 18 - 2y = 8;$

$-2y = -10;\ y = 5.\ (3, 5)$

20. Subtract. $a + 3b = 6$
                    $-2a - 3b = -9$
                    $-a\ \ \ \ \ \ = -3;\ a = 3$

$3 + 3b = 6;\ 3b = 3;\ b = 1.\ (3, 1)$

22. Subtract. $4x - 3y = 9$
                    $-2x + 3y = -3$
                    $2x\ \ \ \ \ \ \ = 6;\ x = 3$

$4 \cdot 3 - 3y = 9;\ 12 - 3y = 9;$

$-3y = -3;\ y = 1.\ (3, 1)$

24. Subtract. $4a - 7b = 13$
                    $-2a + 7b = -3$
                    $2a\ \ \ \ \ \ \ = 10;\ a = 5$

$4 \cdot 5 - 7b = 13;\ 20 - 7b = 13;$

$-7b = -7;\ b = 1.\ (5, 1)$

26. Add. $-4x + y = 7$
               $4x + 3y = 5$
               $4y = 12;\ y = 3$

$-4x + 3 = 7;\ -4x = 4;\ x = -1.$

$(-1, 3)$

28. Subtract. $5x - 2y = 7$
                    $-5x - 3y = -2$
                    $-5y = 5;$
                    $y = -1$

$5x - 2(-1) = 7;\ 5x + 2 = 7;$

$5x = 5;\ x = 1.\ (1, -1)$

30. Subtract. $2x - 11y = 18$
                    $-6x + 11y = -10$
                    $-4x\ \ \ \ \ \ = 8;$
                    $x = -2$

$2(-2) - 11y = 18;\ -4 - 11y = 18;$

$-11y = 22;\ y = -2.\ (-2, -2)$

32. Add. $4x - y = 15$
               $-4x + 3y = -5$
               $2y = 10;\ y = 5$

$4x - 5 = 15;\ 4x = 20;\ x = 5.$

$(5, 5)$

B   **33.** Subtract. $6x + 9y = 4$
$$\underline{-6x - 3y = 0}$$
$$6y = 4; \quad y = \frac{4}{6} = \frac{2}{3}$$

$$6x + 9\left(\frac{2}{3}\right) = 4; \quad 6x + 6 = 4;$$

$$6x = -2; \quad x = -\frac{2}{6} = -\frac{1}{3}. \quad \left(-\frac{1}{3}, \frac{2}{3}\right)$$

34. Add. $5a - b = 3$
$$\underline{10a + b = 3}$$
$$15a \quad = 6; \quad a = \frac{6}{15} = \frac{2}{5}$$

$$5\left(\frac{2}{5}\right) - b = 3; \quad 2 - b = 3;$$

$$-b = 1; \quad b = -1. \quad \left(\frac{2}{5}, -1\right)$$

35. Subtract. $8a + 6b = -1$
$$\underline{-8a + 4b = -4}$$
$$10b = -5;$$

$$b = -\frac{5}{10} = -\frac{1}{2}$$

$$8a + 6\left(-\frac{1}{2}\right) = -1; \quad 8a - 3 = -1;$$

$$8a = 2; \quad a = \frac{1}{4}. \quad \left(\frac{1}{4}, -\frac{1}{2}\right)$$

36. Add. $3x + 4y = 5$
$$\underline{6x - 4y = 1}$$
$$9x \quad = 6; \quad x = \frac{6}{9} = \frac{2}{3}$$

$$3\left(\frac{2}{3}\right) + 4y = 5; \quad 2 + 4y = 5;$$

$$4y = 3; \quad y = \frac{3}{4}. \quad \left(\frac{2}{3}, \frac{3}{4}\right)$$

37. Subtract. $x + 6y = 0$
$$\underline{-x + 3y = -3}$$
$$9y = -3;$$

$$y = -\frac{3}{9} = -\frac{1}{3}$$

$$x + 6\left(-\frac{1}{3}\right) = 0; \quad x - 2 = 0; \quad x = 2.$$

$$\left(2, -\frac{1}{3}\right)$$

38. Add. $4m - 3n = 2$
$$\underline{8m + 3n = 13}$$
$$12m \quad = 15;$$

$$m = \frac{15}{12} = \frac{5}{4}$$

$$4\left(\frac{5}{4}\right) - 3n = 2; \quad 5 - 3n = 2;$$

$$-3n = -3; \quad n = 1. \quad \left(\frac{5}{4}, 1\right)$$

Page 274 · PUZZLE PROBLEMS

Pass the center of one cord under the string circling the other person's wrist, over the person's hand, then back under the string again.

Page 275 · MIXED PRACTICE EXERCISES

1. about $\left(1\frac{1}{3}, 1\frac{2}{3}\right)$

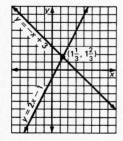

2. $y = -x$ and $y = -3x - 8.$ $(-4, 4)$

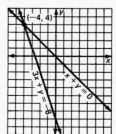

3. $y = -3x - 7$ and $y = 4x + 7$.

(-2, -1)

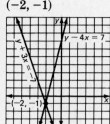

4. $y = -x + 4$ and

$y = -2x + 4$.   (0, 4)

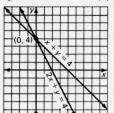

5. $x = 3y$ and $x + y = 12$; $(3y) + y = 12$; $4y = 12$; $y = 3$; $x = 3y =$
$3 \cdot 3 = 9$.   (9, 3)

6. $y = x + 1$ and $x + y = 5$; $x + (x + 1) = 5$; $2x + 1 = 5$; $2x = 4$; $x = 2$;
$y = x + 1 = 2 + 1 = 3$.   (2, 3)

7. $x + 2y = 2$ and $x + 3y = 13$; $x = -2y + 2$ and $x + 3y = 13$; $(-2y + 2) +$
$3y = 13$; $y + 2 = 13$; $y = 11$; $x = -2y + 2 = -2 \cdot 11 + 2 = -22 + 2 = -20$.
(-20, 11)

8. $y - 2x = 4$ and $y + 4x = 16$; $y = 2x + 4$ and $y + 4x = 16$; $(2x + 4) + 4x =$
16; $6x + 4 = 16$; $6x = 12$; $x = 2$; $y = 2x + 4 = 2 \cdot 2 + 4 = 4 + 4 = 8$.
(2, 8)

9. Add. $\begin{aligned} x + y &= 7 \\ 2x - y &= 5 \\ \hline 3x &= 12; \quad x = 4 \end{aligned}$

$4 + y = 7$; $y = 3$.   (4, 3)

10. Subtract. $\begin{aligned} 2a + 3b &= 5 \\ -2a - b &= -3 \\ \hline 2b &= 2; \quad b = 1 \end{aligned}$

$2a + 3 \cdot 1 = 5$; $2a + 3 = 5$;
$2a = 2$; $a = 1$.   (1, 1)

11. Add. $\begin{aligned} 3x - 2y &= 1 \\ -3x + 4y &= 7 \\ \hline 2y &= 8; \quad y = 4 \end{aligned}$

$3x - 2 \cdot 4 = 1$; $3x - 8 = 1$;
$3x = 9$; $x = 3$.   (3, 4)

12. Subtract. $\begin{aligned} 4x - 7y &= 9 \\ -6x + 7y &= -3 \\ \hline -2x &= 6; \\ x &= -3 \end{aligned}$

$4(-3) - 7y = 9$; $-12 - 7y = 9$;
$-7y = 21$; $y = -3$.   (-3, -3)

13. $y = x + 2$ and $2x + y = 11$; $2x + (x + 2) = 11$; $3x + 2 = 11$; $3x = 9$;
$x = 3$; $y = x + 2 = 3 + 2 = 5$.   (3, 5)

14. Subtract. $\begin{aligned} x + y &= 8 \\ -x + 3y &= -4 \\ \hline 4y &= 4; \\ y &= 1 \end{aligned}$

$x + 1 = 8$; $x = 7$.   (7, 1)

15. Add. $\begin{aligned} 2x + y &= 9 \\ 3x - y &= 6 \\ \hline 5x &= 15; \quad x = 3 \end{aligned}$

$2 \cdot 3 + y = 9$; $6 + y = 9$; $y = 3$.
(3, 3)

16. $x + 2y = 5$ and $3x + y = 10$; $x = -2y + 5$ and $3x + y = 10$; $3(-2y + 5) +$ $y = 10$; $-6y + 15 + y = 10$; $-5y + 15 = 10$; $-5y = -5$; $y = 1$; $x = -2y + 5 = -2 \cdot 1 + 5 = -2 + 5 = 3$.   (3, 1)

17. Add. $\begin{array}{r} 2x - 3y = 8 \\ x + 3y = 7 \\ \hline 3x = 15; \end{array}$ $x = 5$

     $5 + 3y = 7$; $3y = 2$; $y = \dfrac{2}{3}$. $\left(5, \dfrac{2}{3}\right)$

18. $x = 2y$ and $x + 3y = 5$;

     $(2y) + 3y = 5$; $5y = 5$; $y = 1$;

     $x = 2y = 2 \cdot 1 = 2$.   (2, 1)

19. Subtract. $\begin{array}{r} y - 2x = 7 \\ -y - 3x = -2 \\ \hline -5x = 5; \end{array}$ $x = -1$

     $y - 2(-1) = 7$; $y + 2 = 7$; $y = 5$. $(-1, 5)$

20. $y - x = 3$ and $y = 2x + 4$;

     $(2x + 4) - x = 3$; $x + 4 = 3$;

     $x = -1$; $y = 2x + 4 = 2(-1) +$ $4 = -2 + 4 = 2$. $(-1, 2)$

21. $3x - y = 8$ and $x + 2y = -2$; $y = 3x - 8$ and $x + 2y = -2$; $x +$ $2(3x - 8) = -2$; $x + 6x - 16 = -2$; $7x - 16 = -2$; $7x = 14$; $x = 2$; $y = 3x - 8 = 3 \cdot 2 - 8 = 6 - 8 = -2$. $(2, -2)$

22. $2a - 4b = 6$ and $-a - 3b = 7$; $2a - 4b = 6$ and $a = -3b - 7$; $2(-3b - 7) -$ $4b = 6$; $-6b - 14 - 4b = 6$; $-10b - 14 = 6$; $-10b = 20$; $b = -2$; $a =$ $-3b - 7 = -3(-2) - 7 = -1$; $(-1, -2)$

23. Subtract. $\begin{array}{r} x + y = -2 \\ -2x - y = -4 \\ \hline -x = -6; \end{array}$ $x = 6$

     $6 + y = -2$; $y = -8$. $(6, -8)$

24. $x - 5y = 2$ and $2x + y = 4$;

     $x = 5y + 2$ and $2x + y = 4$;

     $2(5y + 2) + y = 4$; $10y + 4 +$ $y = 4$; $11y + 4 = 4$; $11y = 0$;

     $y = 0$; $x = 5y + 2 = 5 \cdot 0 + 2 =$ 2.   (2, 0)

Pages 278–279 • WRITTEN EXERCISES

A

1. Let x and y = the numbers. $x + y = 35$ and $x - y = 13$; $x = -y + 35$ and $x - y = 13$; $(-y + 35) - y = 13$; $-2y + 35 = 13$; $-2y = -22$; $y = 11$; $x = -y + 35 = -11 + 35 = 24$.   24 and 11

2. Let x and y = the numbers. $x + y = 48$ and $x = 3y$; $(3y + y) = 48$; $4y = 48$; $y = 12$; $x = 3y = 3 \cdot 12 = 36$.   36 and 12

3. Let x and y = the numbers. $x = y + 9$ and $x + y = 53$; $(y + 9) + y = 53$; $2y + 9 = 53$; $2y = 44$; $y = 22$; $x = y + 9 = 22 + 9 = 31$.   31 and 22

4. Let x and y = the numbers. $x = 6y$ and $x = 2y + 12$; $2y + 12 = 6y$; $12 = 4y$; $3 = y$; $x = 6y = 6 \cdot 3 = 18$.   18 and 3

5. Let x = the cost of the coat and y = the cost of the pants. $x = 3y$ and $x + y = 180$; $(3y) + y = 180$; $4y = 180$; $y = 45$; $x = 3y = 3 \cdot 45 = 135$. coat: $135;   pants: $45

6. Let x = Arlene's age and y = Todd's age. $x = y + 7$ and $x + y = 43$; $(y + 7) + y = 43$; $2y + 7 = 43$; $2y = 36$; $y = 18$; $x = y + 7 = 18 + 7 = 25$. Arlene: 25 years old; Todd: 18 years old

7. Let x = rainfall (in cm) in 1986 and y = rainfall (in cm) in 1985. $x = y + 18$ and $x + y = 150$; $(y + 18) + y = 150$; $2y + 18 = 150$; $2y = 132$; $y = 66$; $x = y + 18 = 66 + 18 = 84$. 1986: 84 cm; 1985: 66 cm

8. Let x = the number of climbers and y = the number of skiers. $x = y + 3$ and $x + y = 35$; $(y + 3) + y = 35$; $2y + 3 = 35$; $2y = 32$; $y = 16$; $x = y + 3 = 16 + 3 = 19$. 19 climbers

B

9. *One variable:* Let x = the width (in cm). Then $3x$ = the length (in cm). $2 \cdot x + 2 \cdot 3x = 48$; $2x + 6x = 48$; $8x = 48$; $x = 6$ and $3x = 18$. length: 18 cm; width: 6 cm
Two variables: Let l = the length (in cm) and w = the width (in cm). $l = 3w$ and $2l + 2w = 48$; $2(3w) + 2w = 48$; $6w + 2w = 48$; $8w = 48$; $w = 6$; $l = 3w = 3 \cdot 6 = 18$. length: 18 cm; width: 6 cm

10. *One variable:* Let x = the Astros' score. Then $x + 3$ = the Kickers' score. $x + x + 3 = 13$; $2x + 3 = 13$; $2x = 10$; $x = 5$ and $x + 3 = 8$. Astros: 5 goals; Kickers: 8 goals
Two variables: Let a = the Astros' score and k = the Kickers' score. $k = a + 3$ and $a + k = 13$; $a + (a + 3) = 13$; $2a + 3 = 13$; $2a = 10$; $a = 5$; $k = a + 3 = 5 + 3 = 8$. Astros: 5 goals; Kickers: 8 goals

11. *One variable:* Let x = the number of books Sam read. Then $x + 12$ = the number of books Norm read. $x + x + 12 = 56$; $2x + 12 = 56$; $2x = 44$; $x = 22$ and $x + 12 = 34$. Norm: 34 books
Two variables: Let s = the number of books Sam read and n = the number of books Norm read. $n = 12 + s$ and $n + s = 56$; $(12 + s) + s = 56$; $12 + 2s = 56$; $2s = 44$; $s = 22$; $n = 12 + s = 12 + 22 = 34$. Norm: 34 books

12. *One variable:* Let x = the larger number. Then $x - 50$ = the smaller number. $2x = 3(x - 50)$; $2x = 3x - 150$; $-x = -150$; $x = 150$ and $x - 50 = 100$. 150 and 100
Two variables: Let x and y = the numbers. $x - y = 50$ and $2x = 3y$; $x = y + 50$ and $2x = 3y$; $2(y + 50) = 3y$; $2y + 100 = 3y$; $100 = y$; $x = y + 50 = 150$. 150 and 100

13. *One variable:* Let x = the grams of carbohydrates. Then $x - 18$ = the grams of protein. $x + x - 18 = 32$; $2x - 18 = 32$; $2x = 50$; $x = 25$ and $x - 18 = 25 - 18 = 7$. 7 grams of protein
Two variables: Let x = the grams of protein and y = the grams of carbohydrates. $y = x + 18$ and $x + y = 32$; $x + (x + 18) = 32$; $2x + 18 = 32$; $2x = 14$; $x = 7$. 7 grams of protein

Page 279 · PUZZLE PROBLEMS

Let b = the number of balls, c = the number of cups, and p = the number of pennies. Then, $b + c = 12p$ and $4b + 2p = c$; $b = 12p - c$, so $4(12p - c) + 2p = c$; $48p - 4c + 2p = c$; $50p = 5c$; $10p = c$. 10 pennies

Page 281 · WRITTEN EXERCISES

A

1. $a + b = 9 \longrightarrow 3a + 3b = 27$

$2a - 3b = 3 \longrightarrow \underline{2a - 3b = 3}$ Add

$5a = 30; \quad a = 6$

$6 + b = 9; \quad b = 3. \quad (6, 3)$

2. $x - 4y = 5 \longrightarrow 2x - 8y = 10$

$2x - 7y = 9 \longrightarrow \underline{2x - 7y = 9}$ Subt.

$-y = 1; \quad y = -1$

$x - 4(-1) = 5; \quad x + 4 = 5;$

$x = 1. \quad (1, -1)$

3. $y - 3x = 9 \longrightarrow y - 3x = 9$

$2y + x = 4 \longrightarrow \underline{6y + 3x = 12}$ Add

$7y = 21; \quad y = 3$

$3 - 3x = 9; \quad -3x = 6;$

$x = -2. \quad (-2, 3)$

4. $a + 3b = 16 \longrightarrow a + 3b = 16$

$2a - b = 4 \longrightarrow \underline{6a - 3b = 12}$ Add

$7a = 28; \quad a = 4$

$4 + 3b = 16; \quad 3b = 12;$

$b = 4. \quad (4, 4)$

5. $3m + n = 10 \longrightarrow 3m + n = 10$

$m + 2n = 10 \longrightarrow \underline{3m + 6n = 30}$ Subt.

$-5n = -20; \quad n = 4$

$3m + 4 = 10; \quad 3m = 6;$

$m = 2. \quad (2, 4)$

6. $3x - y = 5 \longrightarrow 6x - 2y = 10$

$x + 2y = 11 \longrightarrow \underline{x + 2y = 11}$ Add

$7x = 21; \quad x = 3$

$3 + 2y = 11; \quad 2y = 8;$

$y = 4. \quad (3, 4)$

7. $3x - 4y = 1 \longrightarrow 3x - 4y = 1$

$-x - 2y = 3 \longrightarrow \underline{-3x - 6y = 9}$ Add

$-10y = 10; \quad y = -1$

$3x - 4(-1) = 1; \quad 3x + 4 = 1;$

$3x = -3; \quad x = -1. \quad (-1, -1)$

8. $x - 4y = 3 \longrightarrow 2x - 8y = 6$

$-2x + y = 8 \longrightarrow \underline{-2x + y = 8}$ Add

$-7y = 14; \quad y = -2$

$x - 4(-2) = 3; \quad x + 8 = 3;$

$x = -5. \quad (-5, -2)$

9. $2x - 3y = 6 \longrightarrow 2x - 3y = 6$

$x - 2y = 3 \longrightarrow 2x - 4y = 6$ Subt.

$y = 0$

$x - 2 \cdot 0 = 3; \quad x = 3. \quad (3, 0)$

10. $6x - 5y = 7 \longrightarrow 6x - 5y = 7$

$2x + 2y = -16 \longrightarrow \underline{6x + 6y = -48}$ Subt.

$-11y = 55; \quad y = -5$

$2x + 2(-5) = -16; \quad 2x - 10 = -16; \quad 2x = -6; \quad x = -3. \quad (-3, -5)$

11. $9x + 6y = 0 \longrightarrow 9x + 6y = 0$

$3x + 5y = -9 \longrightarrow \underline{9x + 15y = -27}$ Subt.

$-9y = 27; \quad y = -3$

$9x + 6(-3) = 0; \quad 9x - 18 = 0; \quad 9x = 18; \quad x = 2. \quad (2, -3)$

12. $5m - 2n = 14 \longrightarrow 10m - 4n = 28$

$3m + 4n = -2 \longrightarrow \underline{3m + 4n = -2}$ Add

$13m = 26; \quad m = 2$

$3 \cdot 2 + 4n = -2; \quad 6 + 4n = -2; \quad 4n = -8; \quad n = -2. \quad (2, -2)$

13. $3a + 2b = 6 \longrightarrow 9a + 6b = 18$

$-4a + 6b = 44 \longrightarrow \underline{-4a + 6b = 44}$ Subt.

$13a = -26; \quad a = -2$

$3(-2) + 2b = 6; \quad -6 + 2b = 6; \quad 2b = 12; \quad b = 6. \quad (-2, 6)$

14. $5a + 4b = 7 \longrightarrow 5a + 4b = 7$
$\quad\ 3a + 2b = 3 \longrightarrow \underline{6a + 4b = 6}$ Subt.
$\qquad\qquad\qquad\qquad\ \ -a \quad\ = 1; \quad a = -1$

$3(-1) + 2b = 3; \quad -3 + 2b =$
$3; \quad 2b = 6; \quad b = 3. \quad (-1, 3)$

15. $5x - 9y = -3 \longrightarrow \ \ 5x - 9y = \ \ -3$
$\quad\ 4x - 3y = \ \ 6 \longrightarrow \underline{12x - 9y = \ \ 18}$ Subt.
$\qquad\qquad\qquad\qquad\quad -7x \qquad = -21; \quad x = 3$

$4 \cdot 3 - 3y = 6; \quad 12 - 3y = 6;$
$-3y = -6; \quad y = 2. \quad (3, 2)$

B   **16.** $2c - 3d = -1 \longrightarrow 6c - 9d = -3$
$\qquad\quad 3c - 4d = -3 \longrightarrow \underline{6c - 8d = -6}$ Subt.
$\qquad\qquad\qquad\qquad\qquad\ -d = \ \ 3; \quad d = -3$

$2c - 3(-3) = -1; \quad 2c + 9 =$
$-1; \quad 2c = -10; \quad c = -5.$
$(-5, -3)$

17. $3r - 2s = 15 \longrightarrow \ \ 9r - 6s = 45$
$\quad\ 7r - 3s = 15 \longrightarrow \underline{14r - 6s = 30}$ Subt.
$\qquad\qquad\qquad\qquad\ \ -5r \qquad = 15; \quad r = -3$

$3(-3) - 2s = 15; \quad -9 - 2s =$
$15; \quad -2s = 24; \quad s = -12.$
$(-3, -12)$

18. $2p + 3q = -1 \longrightarrow 6p + \ \ 9q = -3$
$\quad\ 3p + 5q = -2 \longrightarrow \underline{6p + 10q = -4}$ Subt.
$\qquad\qquad\qquad\qquad\qquad -q = \ \ 1; \quad q = -1$

$2p + 3(-1) = -1; \quad 2p - 3 =$
$-1; \quad 2p = 2; \quad p = 1. \quad (1, -1)$

19. $\quad 5a - 2b = 1 \longrightarrow \ \ 10a - \ \ 4b = \ \ 2$
$\quad -2a + 3b = 4 \longrightarrow \underline{-10a + 15b = 20}$ Add
$\qquad\qquad\qquad\qquad\qquad 11b = 22; \quad b = 2$

$5a - 2 \cdot 2 = 1; \quad 5a - 4 = 1;$
$5a = 5; \quad a = 1. \quad (1, 2)$

20. $2x + 4y = 2 \longrightarrow 6x + 12y = 6$
$\quad\ 3x + 5y = 2 \longrightarrow \underline{6x + 10y = 4}$ Subt.
$\qquad\qquad\qquad\qquad\quad 2y = 2; \quad y = 1$

$2x + 4 \cdot 1 = 2; \quad 2x + 4 = 2;$
$2x = -2; \quad x = -1. \quad (-1, 1)$

21. $\quad 3x - 2y = -5 \longrightarrow \ \ 9x - 6y = -15$
$\quad -4x + 3y = \ \ 8 \longrightarrow \underline{-8x + 6y = \ \ 16}$ Add
$\qquad\qquad\qquad\qquad\qquad x \qquad = \ \ 1$

$3 \cdot 1 - 2y = -5; \quad 3 - 2y =$
$-5; \quad -2y = -8; \quad y = 4. \quad (1, 4)$

Pages 282–283 · WRITTEN EXERCISES

A   **1.** $2p + \ \ g = 370 \longrightarrow 4p + 2g = 740$
$\qquad\ \ p + 2g = 335 \longrightarrow \underline{\ \ p + 2g = 335}$ Subt.
$\qquad\qquad\qquad\qquad\qquad 3p \qquad = 405; \quad p = 135$

$135 + 2g = 335; \quad 2g = 200;$
$g = 100. \quad$ plate: \$1.35;
glass: \$1.00

2. $3c + t = 149$
$\quad \underline{4c + t = 177}$ Subt.
$\quad -c \qquad = -28; \quad c = 28$

$3(28) + t = 149; \quad 84 + t = 149;$
$t = 65. \quad$ chair: \$28;   table: \$65

3. $2a + 3s = 12 \longrightarrow 2a + 3s = 12$
$\quad\ a + 2s = \ \ 7 \longrightarrow \underline{2a + 4s = 14}$ Subt.
$\qquad\qquad\qquad\qquad\ \ -s = -2; \quad s = 2$

$a + 2(2) = 7; \quad a + 4 = 7;$
$a = 3. \quad$ adult: \$3;   student: \$2

4. $3s + 2t = 405 \longrightarrow 9s + 6t = 1215$
$\quad\ 2s + 3t = 520 \longrightarrow \underline{4s + 6t = 1040}$ Subt.
$\qquad\qquad\qquad\qquad\quad 5s \qquad = \ \ 175; \quad s = 35$

$3(35) + 2t = 405; \quad 105 + 2t =$
$405; \quad 2t = 300; \quad t = 150.$
soap: 35¢;   toothpaste: \$1.50

5. $\quad 4t + 2r = \ \ 740$
$\quad \underline{6t + 2r = \ \ 970}$ Subt.
$\quad -2t \qquad = -230; \quad t = 115$

$4(115) + 2r = 740; \quad 460 + 2r = 740;$
$2r = 280; \quad r = 140. \quad$ tuna: \$1.15;
rice: \$1.40

B   **6.** $40r + 5o = 285$ $40r + 5(9) = 285;$ $40r + 45 = 285;$
$\underline{40r + 7o = 303}$ Subt. $40r = 240;$ $r = 6.$ regular: \$6 per hour;
$-2o = -18;$ $o = 9$ overtime: \$9 per hour

7.

Expensive	28	x	$28x$
Cheaper	16	y	$16y$
		20	380

$x +$ $y = 20 \rightarrow$ $16x + 16y = 320$
$28x + 16y = 380 \rightarrow$ $\underline{28x + 16y = 380}$ Subt.
$-12x$ $= -60;$ $x = 5$
$5 + y = 20;$ $y = 15.$ number of expensive
balls (x): 5;   number of cheaper balls (y): 15

Page 285 · WRITTEN EXERCISES

A   **1.** Let x = the speed of the boat in still water and y = the speed of the flowing water.
$2(x + y) = 12 \longrightarrow 2x + 2y = 12 \longrightarrow 6x + 6y = 36$
$3(x - y) = 12 \longrightarrow 3x - 3y = 12 \longrightarrow \underline{6x - 6y = 24}$ Add
$12x \qquad = 60;$ $x = 5.$ 5 km/h

2. Let x = the speed of the boat in still water and y = the speed of the flowing water.
$3(x + y) = 30 \longrightarrow 3x + 3y = 30 \longrightarrow 15x + 15y = 150$
$5(x - y) = 30 \longrightarrow 5x - 5y = 30 \longrightarrow \underline{15x - 15y = 90}$ Add
$30x \qquad = 240;$ $x = 8.$ 8 km/h

3.

Flying with the wind	$x + y$	2	1200
Flying into the wind	$x - y$	3	1200

$2(x + y) = 1200$
$3(x - y) = 1200$

$2x + 2y = 1200 \longrightarrow 6x + 6y = 3600$
$3x - 3y = 1200 \longrightarrow \underline{6x - 6y = 2400}$ Subt.
$12y = 1200;$ $y = 100.$ 100 km/h

4.

Flying with the wind	$x + y$	4	1560
Flying into the wind	$x - y$	5	1560

$4(x + y) = 1560$
$5(x - y) = 1560$

$4x + 4y = 1560 \longrightarrow 20x + 20y = 7800$
$5x - 5y = 1560 \longrightarrow \underline{20x - 20y = 6240}$ Subt.
$40y = 1560;$ $y = 39.$ 39 km/h

5.

Flying with the wind	$x + y$	3	1656
Flying into the wind	$x - y$	4	1656

$3(x + y) = 1656$
$4(x - y) = 1656$

$3x + 3y = 1656 \longrightarrow 12x + 12y = 6624$
$4x - 4y = 1656 \longrightarrow \underline{12x - 12y = 4968}$ Add
$24x \qquad = 11{,}592;$ $x = 483.$ 483 km/h

Page 287 · READING ALGEBRA

1. Answers may vary. At a pet show, 34 adult's tickets and 85 child's tickets were sold. An adult's ticket cost twice as much as a child's ticket. If a total of $153 in ticket sales was collected, what was the cost of each kind of ticket?

2. The graph shows that the two lines intersect at $(-1, 6)$.

Page 289 · PROBLEM SOLVING STRATEGIES

1. If the store bought 60 coats, the profit would be $(60 - 10)(\$95) - (60)(\$70) = \$4750 - \$4200 = \$550 \neq \$500$.

2. If the volleyball team played 24 games, the team would win $24 - 9 = 15$ games.
$$\frac{15}{24} = \frac{5}{8} \neq \frac{3}{4}.$$

3. If the cup cost $1.75, then the saucer cost $1.75 - \$1.20 = \$.55$. $\$1.75 + \$.55 = \$2.30 \neq \$2.50$.

4. If Andrew is 44 years old, his son is $44 - 30 = 14$ years old. Four years ago Andrew was 40 years old and his son was 10 years old. $40 \neq 3(10)$.

5. **a.** Let t = the number of tapes and a = the number of albums. Since there are 23 more tapes than albums, a cannot be greater than $51 - 23$, or 28. Thus $23 \leq t \leq 51$ and $0 \leq a \leq 28$. **b.** Answers may vary. **c.** $a + t = 51$ and $t = a + 23$; $a + (a + 23) = 51$; $2a + 23 = 51$; $2a = 28$; $a = 14$; $t = 14 + 23 = 37$; check: $23 \leq 37 \leq 51$ and $0 \leq 14 \leq 28$. 37 tapes

6. **a.** Let w = the width and l = the length. Since l and w must both be positive, $w > 5$ (if not, $l < 0$). Thus $5 < w < 50$ and $0 < l < 50$. **b.** Answers may vary. **c.** $2w + 2l = 100$ and $l = 2w - 10$; $2w + 2(2w - 10) = 100$; $2w + 4w - 20 = 100$; $6w - 20 = 100$; $6w = 120$; $w = 20$; $l = 2(20) - 10 = 30$; check: $5 < 20 < 50$ and $0 < 30 < 50$. width: 20 m

7. Let c = the number of cars and m = the number of motorcycles. **a.** $42 \div 4 = 10.5$; $c \leq 10$ **b.** $42 \div 2 = 21$; $m \leq 13$ **c.** Answers may vary. **d.** $4c + 2m = 42$ and $c + m = 13$; $m = 13 - c$; $4c + 2(13 - c) = 42$; $4c + 26 - 2c = 42$; $2c + 26 = 42$; $2c = 16$; $c = 8$; $m = 13 - 8 = 5$; check: $8 \leq 10$ and $5 \leq 13$. 8 cars

Page 290 · SKILLS REVIEW

1. $\dfrac{1}{5} \cdot \dfrac{2}{3} = \dfrac{1 \cdot 2}{5 \cdot 3} = \dfrac{2}{15}$

2. $\dfrac{1}{3} \cdot \dfrac{1}{2} = \dfrac{1 \cdot 1}{3 \cdot 2} = \dfrac{1}{6}$

3. $\dfrac{1}{2} \cdot \dfrac{5}{6} = \dfrac{1 \cdot 5}{2 \cdot 6} = \dfrac{5}{12}$

4. $\dfrac{3}{7} \cdot \dfrac{1}{2} = \dfrac{3 \cdot 1}{7 \cdot 2} = \dfrac{3}{14}$

5. $\dfrac{2}{5} \cdot \dfrac{3}{4} = \dfrac{\overset{1}{\cancel{2}} \cdot 3}{5 \cdot \underset{2}{\cancel{4}}} = \dfrac{3}{10}$

6. $\dfrac{9}{10} \cdot \dfrac{1}{4} = \dfrac{9 \cdot 1}{10 \cdot 4} = \dfrac{9}{40}$

7. $\dfrac{4}{5} \cdot \dfrac{2}{3} = \dfrac{4 \cdot 2}{5 \cdot 3} = \dfrac{8}{15}$

8. $\dfrac{2}{7} \cdot \dfrac{2}{3} = \dfrac{2 \cdot 2}{7 \cdot 3} = \dfrac{4}{21}$

9. $\dfrac{4}{5} \cdot \dfrac{1}{2} = \dfrac{\overset{2}{\cancel{4}} \cdot 1}{5 \cdot \underset{1}{\cancel{2}}} = \dfrac{2}{5}$

10. $\dfrac{8}{9} \cdot \dfrac{1}{4} = \dfrac{\overset{2}{\cancel{8}} \cdot 1}{9 \cdot \underset{1}{\cancel{4}}} = \dfrac{2}{9}$

11. $\dfrac{5}{7} \cdot \dfrac{4}{15} = \dfrac{\overset{1}{\cancel{5}} \cdot 4}{7 \cdot \underset{3}{\cancel{15}}} = \dfrac{4}{21}$

12. $\dfrac{7}{10} \cdot \dfrac{5}{6} = \dfrac{7 \cdot \overset{1}{\cancel{5}}}{\underset{2}{\cancel{10}} \cdot 6} = \dfrac{7}{12}$

13. $\dfrac{4}{9} \cdot \dfrac{3}{4} = \dfrac{\overset{1}{\cancel{4}} \cdot \overset{1}{\cancel{3}}}{\underset{3}{\cancel{9}} \cdot \underset{1}{\cancel{4}}} = \dfrac{1}{3}$

14. $\dfrac{5}{6} \cdot \dfrac{2}{3} = \dfrac{5 \cdot \overset{1}{\cancel{2}}}{\underset{3}{\cancel{6}} \cdot 3} = \dfrac{5}{9}$

15. $\dfrac{3}{8} \cdot \dfrac{4}{9} = \dfrac{\overset{1}{\cancel{3}} \cdot \overset{1}{\cancel{4}}}{\underset{2}{\cancel{8}} \cdot \underset{3}{\cancel{9}}} = \dfrac{1}{6}$

16. $\dfrac{6}{7} \cdot \dfrac{14}{15} = \dfrac{\overset{2}{\cancel{6}} \cdot \overset{2}{\cancel{14}}}{\underset{1}{\cancel{7}} \cdot \underset{5}{\cancel{15}}} = \dfrac{4}{5}$

17. $\dfrac{1}{3} \div \dfrac{1}{2} = \dfrac{1}{3} \cdot \dfrac{2}{1} = \dfrac{1 \cdot 2}{3 \cdot 1} = \dfrac{2}{3}$

18. $\dfrac{1}{2} \div \dfrac{1}{3} = \dfrac{1}{2} \cdot \dfrac{3}{1} = \dfrac{1 \cdot 3}{2 \cdot 1} = \dfrac{3}{2}$

19. $\dfrac{2}{5} \div \dfrac{3}{4} = \dfrac{2}{5} \cdot \dfrac{4}{3} = \dfrac{2 \cdot 4}{5 \cdot 3} = \dfrac{8}{15}$

20. $\dfrac{5}{8} \div \dfrac{1}{5} = \dfrac{5}{8} \cdot \dfrac{5}{1} = \dfrac{5 \cdot 5}{8 \cdot 1} = \dfrac{25}{8}$

21. $\dfrac{4}{5} \div \dfrac{1}{2} = \dfrac{4}{5} \cdot \dfrac{2}{1} = \dfrac{4 \cdot 2}{5 \cdot 1} = \dfrac{8}{5}$

22. $\dfrac{4}{5} \div 2 = \dfrac{4}{5} \cdot \dfrac{1}{2} = \dfrac{\overset{2}{\cancel{4}} \cdot 1}{5 \cdot \underset{1}{\cancel{2}}} = \dfrac{2}{5}$

23. $\dfrac{3}{8} \div \dfrac{1}{5} = \dfrac{3}{8} \cdot \dfrac{5}{1} = \dfrac{3 \cdot 5}{8 \cdot 1} = \dfrac{15}{8}$

24. $\dfrac{3}{8} \div 5 = \dfrac{3}{8} \cdot \dfrac{1}{5} = \dfrac{3 \cdot 1}{8 \cdot 5} = \dfrac{3}{40}$

25. $\dfrac{3}{5} \div \dfrac{3}{4} = \dfrac{3}{5} \cdot \dfrac{4}{3} = \dfrac{\overset{1}{\cancel{3}} \cdot 4}{5 \cdot \underset{1}{\cancel{3}}} = \dfrac{4}{5}$

26. $\dfrac{7}{10} \div \dfrac{14}{15} = \dfrac{7}{10} \cdot \dfrac{15}{14} = \dfrac{7 \cdot \overset{3}{\cancel{15}}}{\underset{2}{\cancel{10}} \cdot \underset{2}{\cancel{14}}} = \dfrac{3}{4}$

27. $\dfrac{5}{12} \div \dfrac{2}{3} = \dfrac{5}{12} \cdot \dfrac{3}{2} = \dfrac{5 \cdot \overset{1}{\cancel{3}}}{\underset{4}{\cancel{12}} \cdot 2} = \dfrac{5}{8}$

28. $\dfrac{4}{7} \div \dfrac{8}{14} = \dfrac{4}{7} \cdot \dfrac{14}{8} = \dfrac{\overset{1}{\cancel{4}} \cdot \overset{2}{\cancel{14}}}{\underset{1}{\cancel{7}} \cdot \underset{2}{\cancel{8}}} = \dfrac{2}{2} = 1$

29. $\dfrac{7}{8} \div \dfrac{3}{8} = \dfrac{7}{8} \cdot \dfrac{8}{3} = \dfrac{7 \cdot \overset{1}{\cancel{8}}}{\underset{1}{\cancel{8}} \cdot 3} = \dfrac{7}{3}$

30. $\dfrac{9}{10} \div \dfrac{3}{5} = \dfrac{9}{10} \cdot \dfrac{5}{3} = \dfrac{\overset{3}{\cancel{9}} \cdot \overset{1}{\cancel{5}}}{\underset{2}{\cancel{10}} \cdot \underset{1}{\cancel{3}}} = \dfrac{3}{2}$

31. $\dfrac{7}{9} \div \dfrac{2}{3} = \dfrac{7}{9} \cdot \dfrac{3}{2} = \dfrac{7 \cdot \overset{1}{\cancel{3}}}{\underset{3}{\cancel{9}} \cdot 2} = \dfrac{7}{6}$

32. $\dfrac{15}{16} \div \dfrac{5}{8} = \dfrac{15}{16} \cdot \dfrac{8}{5} = \dfrac{\overset{3}{\cancel{15}} \cdot \overset{1}{\cancel{8}}}{\underset{2}{\cancel{16}} \cdot \underset{1}{\cancel{5}}} = \dfrac{3}{2}$

Pages 291–292 • CHAPTER REVIEW EXERCISES

1. $(2, 4)$

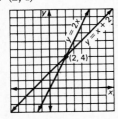

2. $y = -x + 3$ and $y = x + 1$. $(1, 2)$

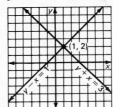

3. $(-2, 1)$

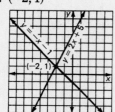

4. $y = -3x$ and $y = 2x + 5$. $(-1, 3)$

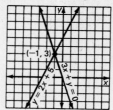

5. $y = -x - 2$ and $y = x - 6$. $(2, -4)$

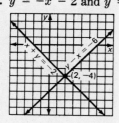

6. $y = 4x + 1$ and $y = 2x + 1$. $(0, 1)$

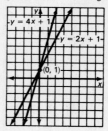

7. 9   **8.** -1 **9.** $y = x + 9$;   1   **10.** $y = -2x$; -2 **11.** $\dfrac{1}{2}$

12. $y = \dfrac{2}{3}x - 2$; $\dfrac{2}{3}$ **13.** Slope of both $= 4$;   no pair

14. $y = x + 4$ and $y = -x - 4$;   slopes $= 1$ and -1;   one pair

15. $y = -5x + 3$ and $y = -5x + 3$;   same equation;   all pairs

16. $y = 7x + 2$ and $y = -7x - 2$;   slopes $= 7$ and -7;   one pair

17. $y = 6x + 2$ and $y = 6x + 5$;   slope of both $= 6$;   no pair

18. $y = -2x + 6$ and $y = 2x + 6$;   slopes $= -2$ and 2;   one pair

19. $y = 3x$ and $x + y = 12$; $x + (3x) = 12$; $4x = 12$; $x = 3$; $y = 3x = 3 \cdot 3 = 9$.
$(3, 9)$

20. $y = -4x$ and $x - y = -10$; $x - (-4x) = -10$; $x + 4x = -10$; $5x = -10$;
$x = -2$; $y = -4x = -4(-2) = 8$. $(-2, 8)$

21. $y = 2x + 1$ and $3x + y = 6$; $3x + (2x + 1) = 6$; $5x + 1 = 6$; $5x = 5$; $x = 1$;
$y = 2x + 1 = 2 \cdot 1 + 1 = 2 + 1 = 3$. $(1, 3)$

22. $5x - y = 0$ and $x + 2y = 22$; $5x = y$ and $x + 2y = 22$; $x + 2(5x) = 22$;
$x + 10x = 22$; $11x = 22$; $x = 2$; $y = 5x = 5 \cdot 2 = 10$. $(2, 10)$

23. $3x - y = 1$ and $2x + 3y = 8$; $y = 3x - 1$ and $2x + 3y = 8$; $2x + 3(3x - 1) =$
8; $2x + 9x - 3 = 8$; $11x - 3 = 8$; $11x = 11$; $x = 1$; $y = 3x - 1 = 3 \cdot 1 -$
$1 = 3 - 1 = 2$. $(1, 2)$

24. $y + 4x = 5$ and $5x + y = 0$; $y + 4x = 5$ and $y = -5x$; $(-5x) + 4x = 5$;
$-x = 5$; $x = -5$; $y = -5x = -5(-5) = 25$. $(-5, 25)$

25. $2x + y = 0$
  $\underline{x - y = 6}$
  $3x \quad = 6; \quad x = 2$
  $2 - y = 6; \quad -y = 4; \quad y = -4.$
  $(2, -4)$

26. $-3x + 2y = 5$
  $\underline{3x + y = 7}$
  $3y = 12; \quad y = 4$
  $3x + 4 = 7; \quad 3x = 3;$
  $x = 1. \quad (1, 4)$

27. $x + 4y = 16$
  $\underline{-x + 3y = -2}$
  $7y = 14; \quad y = 2$
  $x + 4 \cdot 2 = 16; \quad x + 8 = 16;$
  $x = 8. \quad (8, 2)$

28. $x + 3y = -1$
  $\underline{-2x - 3y = 5}$
  $-x \quad = 4; \quad x = -4$
  $-4 + 3y = -1; \quad 3y = 3; \quad y = 1.$
  $(-4, 1)$

29. $5x + 2y = -18$
  $\underline{-3x - 2y = 6}$
  $2x \quad = -12; \quad x = -6$
  $5(-6) + 2y = -18; \quad -30 + 2y = -18;$
  $2y = 12; \quad y = 6. \quad (-6, 6)$

30. $3x + 2y = 5$
  $\underline{-3x + 2y = -1}$
  $4y = 4; \quad y = 1$
  $3x + 2 \cdot 1 = 5; \quad 3x + 2 = 5;$
  $3x = 3; \quad x = 1. \quad (1, 1)$

31. $x + 2y = 7$
  $\underline{-x + y = -1}$
  $3y = 6; \quad y = 2$
  $x + 2 \cdot 2 = 7; \quad x + 4 = 7; \quad x = 3.$
  $(3, 2)$

32. $2x + 3y = 7$
  $\underline{-x - 3y = -2}$
  $x \quad = 5$
  $2 \cdot 5 + 3y = 7; \quad 10 + 3y = 7;$
  $3y = -3; \quad y = -1. \quad (5, -1)$

33. $2x - y = 8$
  $\underline{-2x - 2y = -14}$
  $-3y = -6; \quad y = 2$
  $2x - 2 = 8; \quad 2x = 10; \quad x = 5. \quad (5, 2)$

34. $2x + 3y = 12 \longrightarrow \quad 2x + 3y = 12$
  $4x + y = 14 \longrightarrow \quad \underline{12x + 3y = 42}$ Subt.
  $\qquad\qquad\qquad -10x \quad = -30; \quad x = 3$
  $4 \cdot 3 + y = 14; \quad 12 + y = 14;$
  $y = 2. \quad (3, 2)$

35. $5x - 2y = 11 \longrightarrow 15x - 6y = 33$
  $3x - 6y = -3 \longrightarrow \underline{3x - 6y = -3}$ Subt.
  $\qquad\qquad\qquad 12x \quad = 36; \quad x = 3$
  $5 \cdot 3 - 2y = 11; \quad 15 - 2y =$
  $11; \quad -2y = -4; \quad y = 2. \quad (3, 2)$

36. $3x + 4y = 1 \longrightarrow 6x + 8y = 2$
  $6x + 5y = -1 \longrightarrow \underline{6x + 5y = -1}$ Subt.
  $\qquad\qquad\qquad 3y = 3; \quad y = 1$
  $3x + 4 \cdot 1 = 1; \quad 3x + 4 = 1;$
  $3x = -3; \quad x = -1. \quad (-1, 1)$

37. $3x - 2y = 0 \longrightarrow 12x - 8y = 0$
  $7x - 8y = 10 \longrightarrow \underline{7x - 8y = 10}$ Subt.
  $\qquad\qquad\qquad 5x \quad = -10; \quad x = -2$
  $3(-2) - 2y = 0; \quad -6 - 2y =$
  $0; \quad -6 = 2y; \quad -3 = y.$
  $(-2, -3)$

38. $3x - 4y = -1 \longrightarrow 6x - 8y = -2$
  $5x + 8y = 13 \longrightarrow \underline{5x + 8y = 13}$ Add
  $\qquad\qquad\qquad 11x \quad = 11; \quad x = 1$
  $5 \cdot 1 + 8y = 13; \quad 5 + 8y =$
  $13; \quad 8y = 8; \quad y = 1. \quad (1, 1)$

39. $9x - 2y = -3 \longrightarrow 9x - 2y = -3$
  $-3x + 5y = 27 \longrightarrow \underline{-9x + 15y = 81}$ Add
  $\qquad\qquad\qquad 13y = 78; \quad y = 6$
  $9x - 2 \cdot 6 = -3; \quad 9x - 12 =$
  $-3; \quad 9x = 9; \quad x = 1. \quad (1, 6)$

40. Let x and y = the numbers. $x + y = 27$ and $x - y = 13$; $x + y = 27$ and $y = x - 13$; $x + (x - 13) = 27$; $2x - 13 = 27$; $2x = 40$; $x = 20$; $y = x - 13 = 20 - 13 = 7$. **20 and 7**

41. Let x and y = the numbers. $x = y + 14$ and $x + y = 20$; $(y + 14) + y = 20$; $2y + 14 = 20$; $2y = 6$; $y = 3$; $x = y + 14 = 3 + 14 = 17$. **17 and 3**

42. Let s = the cost of a saw and h = the cost of a hammer.

$$
\begin{aligned}
s + 4h &= 72 \longrightarrow 2s + 8h = 144 \\
2s + 6h &= 114 \longrightarrow \underline{2s + 6h = 114} \quad \text{Subt.} \\
& \qquad\qquad\qquad 2h = 30; \quad h = 15
\end{aligned}
$$

$s + 4(15) = 72$; $s + 60 = 72$; $s = 12$. **saw: \$12; hammer: \$15**

43. Let a = the cost of an adult ticket (in cents) and s = the cost of a student ticket.

$$
\begin{aligned}
a + 3s &= 1150 \longrightarrow 3a + 9s = 3450 \\
3a + 2s &= 1700 \longrightarrow \underline{3a + 2s = 1700} \quad \text{Subt.} \\
& \qquad\qquad\qquad 7s = 1750; \quad s = 250
\end{aligned}
$$

$a + 3(250) = 1150$; $a + 750 = 1150$; $a = 400$. **adult: \$4.00; student: \$2.50**

44. Let x = the speed of the boat in still water and y = the speed of the flowing water.

$$
\begin{aligned}
2(x + y) &= 36 \longrightarrow 2x + 2y = 36 \longrightarrow 6x + 6y = 108 \\
3(x - y) &= 36 \longrightarrow 3x - 3y = 36 \longrightarrow \underline{6x - 6y = 72} \quad \text{Add} \\
& \qquad\qquad\qquad\qquad\qquad\qquad 12x = 180; \quad x = 15. \quad \textbf{15 km/h}
\end{aligned}
$$

Page 293 • CHAPTER TEST

1. $(-1, 3)$

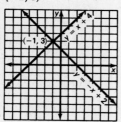

2. $y = 4x$ and $y = -2x - 6$; $(-1, -4)$

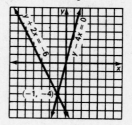

3. $y = 2x + 2$ and $y = -2x + 8$; about $\left(1\frac{1}{2}, 5\right)$

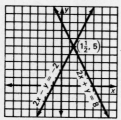

4. $y = 3x$ and $y = 3x + 9$; slopes of both = 3; **no pair**

5. $y = -x + 8$ and $y = x - 3$; slopes = -1 and 1; **one pair**

6. $y = 4x - 1$ and $y = 4x - 1$; same equation; **all pairs**

7. $y = x - 4$ and $3x + y = 12$; $3x + (x - 4) = 12$; $4x - 4 = 12$; $4x = 16$;
   $x = 4$; $y = x - 4 = 4 - 4 = 0$. $(4, 0)$

8. $3x - y = 1$ and $2x - 3y = 10$; $y = 3x - 1$ and $2x - 3y = 10$; $2x -$
   $3(3x - 1) = 10$; $2x - 9x + 3 = 10$; $-7x + 3 = 10$; $-7x = 7$; $x = -1$;
   $y = 3x - 1 = 3(-1) - 1 = -3 - 1 = -4$. $(-1, -4)$

9. $a = 2b + 1$ and $6a - 7b = 16$; $6(2b + 1) - 7b = 16$; $12b + 6 - 7b = 16$;
   $5b + 6 = 16$; $5b = 10$; $b = 2$; $a = 2b + 1 = 2 \cdot 2 + 1 = 4 + 1 = 5$. $(5, 2)$

10. $\begin{aligned} 3x + 2y &= 7 \\ -3x - y &= -11 \\ \hline y &= -4 \end{aligned}$

    $3x + 2(-4) = 7$; $3x - 8 = 7$;
    $3x = 15$; $x = 5$. $(5, -4)$

11. $\begin{aligned} 2x - 3y &= 5 \\ x + 3y &= 16 \\ \hline 3x &= 21; \quad x = 7 \end{aligned}$

    $7 + 3y = 16$; $3y = 9$;
    $y = 3$. $(7, 3)$

12. $\begin{aligned} 3x + 4y &= -2 \\ -5x - 4y &= -2 \\ \hline -2x &= -4; \quad x = 2 \end{aligned}$

    $3(2) + 4y = -2$; $6 + 4y = -2$;
    $4y = -8$; $y = -2$. $(2, -2)$

13. Let x and y = the numbers. $x + y = 16$ and $x - y = 8$; $x + y = 16$ and
    $x = y + 8$; $(y + 8) + y = 16$; $2y + 8 = 16$; $2y = 8$; $y = 4$; $x = y + 8 =$
    $4 + 8 = 12$.   12 and 4

14. Let p = the number of pennies and n = the number of nickels. $p + n = 16$ and
    $p + 5n = 40$; $p = 16 - n$ and $p + 5n = 40$; $(16 - n) + 5n = 40$; $16 + 4n =$
    40; $4n = 24$; $n = 6$.   6 nickels

15. $\begin{aligned} 5x - 3y = 7 &\longrightarrow 10x - 6y = 14 \\ 7x - 6y = 8 &\longrightarrow \underline{7x - 6y = 8} \quad \text{Subt.} \\ &\qquad\quad\ 3x = 6; \quad x = 2 \end{aligned}$

    $5(2) - 3y = 7$; $10 - 3y = 7$;
    $-3y = -3$; $y = 1$. $(2, 1)$

16. $\begin{aligned} 3x + 2y = 2 &\longrightarrow 3x + 2y = 2 \\ -x + 3y = 25 &\longrightarrow \underline{-3x + 9y = 75} \quad \text{Add} \\ &\qquad\quad 11y = 77; \quad y = 7 \end{aligned}$

    $-x + 3(7) = 25$; $-x + 21 =$
    25; $-x = 4$; $x = -4$. $(-4, 7)$

17. $\begin{aligned} 2x - 5y = 9 &\longrightarrow 10x - 25y = 45 \\ 5x - 7y = -5 &\longrightarrow \underline{10x - 14y = -10} \quad \text{Subt.} \\ &\qquad\quad -11y = 55; \quad y = -5 \end{aligned}$

    $2x - 5(-5) = 9$; $2x + 25 = 9$;
    $2x = -16$; $x = -8$. $(-8, -5)$

18. Let a = the cost of an album and t = the cost of a tape.

    $\begin{aligned} 5a + 2t = 57 &\longrightarrow 5a + 2t = 57 \\ 4a + t = 42 &\longrightarrow \underline{8a + 2t = 84} \quad \text{Subt.} \\ &\qquad\ -3a = -27; \quad a = 9 \end{aligned}$

    $4(9) + t = 42$; $36 + t = 42$;
    $t = 6$.   album: \$9;   tape: \$6

19. Let a = the cost of an adult ticket and s = the cost of a student ticket.

    $\begin{aligned} 2a + 3s = 14 &\longrightarrow 4a + 6s = 28 \\ 4a + 5s = 26 &\longrightarrow \underline{4a + 5s = 26} \quad \text{Subt.} \\ &\qquad\qquad\quad s = 2 \end{aligned}$

    $2a + 3(2) = 14$; $2a + 6 = 14$;
    $2a = 8$; $a = 4$.   adult: \$4;
    student: \$2

20. Let x = the speed of the boat in still water and y = the speed of the flowing water.

    $\begin{aligned} 3(x + y) = 48 &\longrightarrow 3x + 3y = 48 \longrightarrow 12x + 12y = 192 \\ 4(x - y) = 48 &\longrightarrow 4x - 4y = 48 \longrightarrow \underline{12x - 12y = 144} \quad \text{Subt.} \\ &\qquad\qquad\qquad\qquad\qquad\ 24y = 48; \quad y = 2. \quad \text{2 mph} \end{aligned}$

Page 294 · CUMULATIVE REVIEW

1. $37°C$ **2.** 3 A.M.   **3.** between 2 A.M. and 3 A.M.   **4.** No

5. $3x - y = 2$; $3x - y - 3x = 2 - 3x$; $-y = 2 - 3x$; $y = 3x - 2$

x	-2	-1	0	1	2
y	-8	-5	-2	1	4

6.

n	1	2	5	8
C	40	80	200	320

7. $40n$

8. $(-2, 1)$

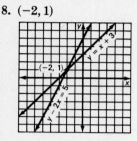

9. $y = -2x + 5$; -2

10. $y = x + 1$ and $y = -x + 3$;

slopes $= 1$ and -1;   one pair

11. $y = 2x + 1$ and $x + 3y = 10$; $x + 3(2x + 1) = 10$; $x + 6x + 3 = 10$;

$7x + 3 = 10$; $7x = 7$; $x = 1$; $y = 2x + 1 = 2(1) + 1 = 2 + 1 = 3$. $(1, 3)$

12.
$$x + y = 9$$
$$\underline{x - y = 1}$$
$$2x \quad\quad = 10; \quad x = 5$$
$$5 + y = 9; \quad y = 4. \quad (5, 4)$$

13.
$$5x - 2y = 7$$
$$\underline{-5x - 3y = -2}$$
$$-5y = 5; \quad y = -1$$
$$5x - 2(-1) = 7; \quad 5x + 2 = 7;$$
$$5x = 5; \quad x = 1. \quad (1, -1)$$

14.
$$5x - 2y = 2 \longrightarrow 15x - 6y = 6$$
$$2x - 3y = 3 \longrightarrow \underline{-4x + 6y = -6}$$
$$11x \quad\quad\quad = 0; \quad x = 0$$
$$2(0) - 3y = 3; \quad 0 - 3y = 3;$$
$$-3y = 3; \quad y = -1. \quad (0, -1)$$

15. Let x and $y =$ the numbers. $x + y = 56$ and $x = y + 8$; $(y + 8) + y = 56$;

$2y + 8 = 56$; $2y = 48$; $y = 24$; $x = y + 8 = 24 + 8 = 32$.   24 and 32

16. Let $a =$ the cost (in cents) of an apple and $p =$ the cost of a pear.

$$3a + 2p = 150 \longrightarrow 3a + 2p = 150$$
$$2a + p = 85 \longrightarrow \underline{4a + 2p = 170} \quad \text{Subt.}$$
$$-a \quad\quad = -20; \quad a = 20$$

$2(20) + p = 85$; $40 + p = 85$; $p = 45$.   apple: 20¢;   pear: 45¢

Page 299 · WRITTEN EXERCISES

A

1. $\dfrac{10}{15} = \dfrac{2 \cdot 5}{3 \cdot 5} = \dfrac{2}{3}$

2. $\dfrac{9}{12} = \dfrac{3 \cdot 3}{4 \cdot 3} = \dfrac{3}{4}$

3. $\dfrac{3}{18a} = \dfrac{3}{3 \cdot 6 \cdot a} = \dfrac{1}{6a}$

4. $\dfrac{7}{21g} = \dfrac{7}{7 \cdot 3 \cdot g} = \dfrac{1}{3g}$

5. $\dfrac{-5x}{10x} = \dfrac{-5 \cdot x}{5 \cdot 2 \cdot x} = \dfrac{-1}{2} = -\dfrac{1}{2}$

6. $\dfrac{3xy}{9y} = \dfrac{3 \cdot x \cdot y}{3 \cdot 3 \cdot y} = \dfrac{x}{3}$

7. $\dfrac{7ab}{14ab} = \dfrac{7 \cdot a \cdot b}{2 \cdot 7 \cdot a \cdot b} = \dfrac{1}{2}$

8. $\dfrac{4x}{12xy} = \dfrac{4 \cdot x}{3 \cdot 4 \cdot x \cdot y} = \dfrac{1}{3y}$

9. $\dfrac{-8x^2}{6x} = \dfrac{-4 \cdot 2 \cdot x \cdot x}{3 \cdot 2 \cdot x} = \dfrac{-4x}{3} = -\dfrac{4x}{3}$

10. $\dfrac{12abc}{9bcd} = \dfrac{3 \cdot 4 \cdot a \cdot b \cdot c}{3 \cdot 3 \cdot b \cdot c \cdot d} = \dfrac{4a}{3d}$

11. $\dfrac{28r}{21r^2} = \dfrac{4 \cdot 7 \cdot r}{3 \cdot 7 \cdot r \cdot r} = \dfrac{4}{3r}$

12. $\dfrac{6a}{18a^2} = \dfrac{6 \cdot a}{3 \cdot 6 \cdot a \cdot a} = \dfrac{1}{3a}$

13. $\dfrac{4ab^2}{12b} = \dfrac{4 \cdot a \cdot b \cdot b}{3 \cdot 4 \cdot b} = \dfrac{ab}{3}$

14. $\dfrac{-9st}{-6s^2} = \dfrac{-3 \cdot 3 \cdot s \cdot t}{-3 \cdot 2 \cdot s \cdot s} = \dfrac{3t}{2s}$

15. $-\dfrac{3a^2b}{15b} = -\dfrac{3 \cdot a \cdot a \cdot b}{3 \cdot 5 \cdot b} = -\dfrac{a^2}{5}$

16. $\dfrac{5x^2y}{10xy^2} = \dfrac{5 \cdot x \cdot x \cdot y}{2 \cdot 5 \cdot x \cdot y \cdot y} = \dfrac{x}{2y}$

17. $\dfrac{8ab}{4a^2b} = \dfrac{4 \cdot 2 \cdot a \cdot b}{4 \cdot a \cdot a \cdot b} = \dfrac{2}{a}$

18. $\dfrac{9x^2y}{15x} = \dfrac{3 \cdot 3 \cdot x \cdot x \cdot y}{5 \cdot 3 \cdot x} = \dfrac{3xy}{5}$

19. $\dfrac{8a^2x}{4ax^2} = \dfrac{2 \cdot 4 \cdot a \cdot a \cdot x}{4 \cdot a \cdot x \cdot x} = \dfrac{2a}{x}$

20. $\dfrac{9mn^2}{7n^2} = \dfrac{9 \cdot m \cdot n \cdot n}{7 \cdot n \cdot n} = \dfrac{9m}{7}$

21. $\dfrac{10x^2y}{25y^2} = \dfrac{2 \cdot 5 \cdot x \cdot x \cdot y}{5 \cdot 5 \cdot y \cdot y} = \dfrac{2x^2}{5y}$

22. $\dfrac{-6rs}{15r^2s} = \dfrac{-2 \cdot 3 \cdot r \cdot s}{5 \cdot 3 \cdot r \cdot r \cdot s} = \dfrac{-2}{5r} = -\dfrac{2}{5r}$

23. $\dfrac{10m^2k}{8mk} = \dfrac{2 \cdot 5 \cdot m \cdot m \cdot k}{2 \cdot 4 \cdot m \cdot k} = \dfrac{5m}{4}$

24. $\dfrac{9n^2r}{24nr^2} = \dfrac{3 \cdot 3 \cdot n \cdot n \cdot r}{3 \cdot 8 \cdot n \cdot r \cdot r} = \dfrac{3n}{8r}$

25. $\dfrac{4pq^2}{32qr} = \dfrac{4 \cdot p \cdot q \cdot q}{4 \cdot 8 \cdot q \cdot r} = \dfrac{pq}{8r}$

26. $\dfrac{7a}{28ab^2c} = \dfrac{7 \cdot a}{4 \cdot 7 \cdot a \cdot b \cdot b \cdot c} = \dfrac{1}{4b^2c}$

27. $\dfrac{6y^2z}{-3y} = \dfrac{2 \cdot 3 \cdot y \cdot y \cdot z}{-3 \cdot y} = \dfrac{2yz}{-1} = -2yz$

28. $\dfrac{12a^2b^4}{4a^3b} = \dfrac{3 \cdot 4 \cdot a \cdot a \cdot b \cdot b \cdot b \cdot b}{4 \cdot a \cdot a \cdot a \cdot b} = \dfrac{3b^3}{a}$

29. $\dfrac{10x^2w^3}{35xw^4} = \dfrac{5 \cdot 2 \cdot x \cdot x \cdot w \cdot w \cdot w}{5 \cdot 7 \cdot x \cdot w \cdot w \cdot w \cdot w} = \dfrac{2x}{7w}$

30. $\dfrac{12y^3t^3}{19y^2t^4} = \dfrac{12 \cdot y \cdot y \cdot y \cdot t \cdot t \cdot t}{19 \cdot y \cdot y \cdot t \cdot t \cdot t \cdot t} = \dfrac{12y}{19t}$

B

31. $\dfrac{125a^2b^3c}{-25a^3bc^2} = \dfrac{25 \cdot 5 \cdot a \cdot a \cdot b \cdot b \cdot b \cdot c}{-25 \cdot a \cdot a \cdot a \cdot b \cdot c \cdot c} = \dfrac{5b^2}{-ac} = -\dfrac{5b^2}{ac}$

32. $\dfrac{-76xy^2z^3}{19x^2z} = \dfrac{-4 \cdot 19 \cdot x \cdot y \cdot y \cdot z \cdot z \cdot z}{19 \cdot x \cdot x \cdot z} = \dfrac{-4y^2z^2}{x} = -\dfrac{4y^2z^2}{x}$

33. $\dfrac{12rst^3}{144r^2s^3t} = \dfrac{12 \cdot r \cdot s \cdot t \cdot t \cdot t}{12 \cdot 12 \cdot r \cdot r \cdot s \cdot s \cdot s \cdot t} = \dfrac{t^2}{12rs^2}$

34. $\dfrac{-21x^2yz^3}{-84xy^2z^2} = \dfrac{-21 \cdot x \cdot x \cdot y \cdot z \cdot z \cdot z}{-21 \cdot 4 \cdot x \cdot y \cdot y \cdot z \cdot z} = \dfrac{xz}{4y}$

35. $\dfrac{121ab^2c^3}{-11a^2b^3} = \dfrac{11 \cdot 11 \cdot a \cdot b \cdot b \cdot c \cdot c \cdot c}{-11 \cdot a \cdot a \cdot b \cdot b \cdot b} = \dfrac{11c^3}{-ab} = -\dfrac{11c^3}{ab}$

36. $\dfrac{-13r^3s^2t}{104rs^3t^2} = \dfrac{-13 \cdot r \cdot r \cdot r \cdot s \cdot s \cdot t}{13 \cdot 8 \cdot r \cdot s \cdot s \cdot s \cdot t \cdot t} = \dfrac{-r^2}{8st} = -\dfrac{r^2}{8st}$

37. $\dfrac{102x^3y^2}{-18x^2y^3z} = \dfrac{6 \cdot 17 \cdot x \cdot x \cdot x \cdot y \cdot y}{-6 \cdot 3 \cdot x \cdot x \cdot y \cdot y \cdot y \cdot z} = \dfrac{17x}{-3yz} = -\dfrac{17x}{3yz}$

38. $\dfrac{-24a^3b^2c^2}{-168a^2b^3c} = \dfrac{-24 \cdot a \cdot a \cdot a \cdot b \cdot b \cdot c \cdot c}{-24 \cdot 7 \cdot a \cdot a \cdot b \cdot b \cdot b \cdot c} = \dfrac{ac}{7b}$

39. $x = 0$ 40. $x - 1 = 0;\ \ x = 1$ 41. $x - 2 = 0;\ \ x = 2$

42. $x + 1 = 0;\ \ x = -1$ 43. $3x - 6 = 0;\ \ 3x = 6;\ \ x = 2$

Page 301 · WRITTEN EXERCISES

A

1. $\dfrac{x + 1}{3x + 3} = \dfrac{x + 1}{3(x + 1)} = \dfrac{1}{3}$

2. $\dfrac{x - 1}{2x - 2} = \dfrac{x - 1}{2(x - 1)} = \dfrac{1}{2}$

3. $\dfrac{3p + 3q}{7p + 7q} = \dfrac{3(p + q)}{7(p + q)} = \dfrac{3}{7}$

4. $\dfrac{2a - 2b}{2a + 2b} = \dfrac{2(a - b)}{2(a + b)} = \dfrac{a - b}{a + b}$

5. $\dfrac{x^2 + 2x}{x + 2} = \dfrac{x(x + 2)}{x + 2} = x$

6. $\dfrac{a + 1}{a^2 + a} = \dfrac{a + 1}{a(a + 1)} = \dfrac{1}{a}$

7. $\dfrac{3y}{9y^2 - 6y} = \dfrac{3y}{3y(3y - 2)} = \dfrac{1}{3y - 2}$

8. $\dfrac{6a + 6b}{9a + 9b} = \dfrac{2 \cdot 3(a + b)}{3 \cdot 3(a + b)} = \dfrac{2}{3}$

9. $\dfrac{2c + 6}{c^2 - 9} = \dfrac{2(c + 3)}{(c + 3)(c - 3)} = \dfrac{2}{c - 3}$

10. $\dfrac{a^2 - 4}{a - 2} = \dfrac{(a + 2)(a - 2)}{a - 2} = a + 2$

11. $\dfrac{2x + 10}{x^2 - 25} = \dfrac{2(x + 5)}{(x + 5)(x - 5)} = \dfrac{2}{x - 5}$

12. $\dfrac{x^2 - y^2}{x + y} = \dfrac{(x + y)(x - y)}{x + y} = x - y$

13. $\dfrac{n^2 - 9}{n^2 + 3n} = \dfrac{(n + 3)(n - 3)}{n(n + 3)} = \dfrac{n - 3}{n}$

14. $\dfrac{y^2 - 4y}{2y - 8} = \dfrac{y(y - 4)}{2(y - 4)} = \dfrac{y}{2}$

15. $\dfrac{2xy - 6x}{y^2 - 3y} = \dfrac{2x(y - 3)}{y(y - 3)} = \dfrac{2x}{y}$

16. $\dfrac{2x - 4xy}{1 - 2y} = \dfrac{2x(1 - 2y)}{1 - 2y} = 2x$

17. $\dfrac{5k + 10}{(k + 2)^2} = \dfrac{5(k + 2)}{(k + 2)(k + 2)} = \dfrac{5}{k + 2}$

18. $\dfrac{(m - 4)^2}{3m - 12} = \dfrac{(m - 4)(m - 4)}{3(m - 4)} = \dfrac{m - 4}{3}$

19. $\dfrac{6u + 6v}{u^2 - v^2} = \dfrac{6(u + v)}{(u + v)(u - v)} = \dfrac{6}{u - v}$

20. $\dfrac{g^2 - h^2}{5g - 5h} = \dfrac{(g + h)(g - h)}{5(g - h)} = \dfrac{g + h}{5}$

21. $\dfrac{12x^2y}{3xy^2 + 6x^2y} = \dfrac{3xy \cdot 4x}{3xy(y + 2x)} = \dfrac{4x}{y + 2x}$

22. $\dfrac{3x - 9xy}{1 - 3y} = \dfrac{3x(1 - 3y)}{1 - 3y} = 3x$

23. $\dfrac{x^2 + 6x + 9}{x^2 - 9} = \dfrac{(x + 3)(x + 3)}{(x + 3)(x - 3)} = \dfrac{x + 3}{x - 3}$

24. $\dfrac{y^2 - 4y + 4}{y^2 - 4} = \dfrac{(y - 2)(y - 2)}{(y + 2)(y - 2)} = \dfrac{y - 2}{y + 2}$

B

25. $\dfrac{2x^2 - 50}{2x + 10} = \dfrac{2(x^2 - 25)}{2(x + 5)} = \dfrac{2(x + 5)(x - 5)}{2(x + 5)} = x - 5$

26. $\dfrac{2ab^2 + 2a^2b}{4a^2 - 4b^2} = \dfrac{2ab(b + a)}{4(a^2 - b^2)} = \dfrac{2ab(a + b)}{4(a + b)(a - b)} = \dfrac{ab}{2(a - b)} = \dfrac{ab}{2a - 2b}$

27. $\dfrac{3x^2y}{6x^2y + 3xy^2} = \dfrac{3xy \cdot x}{3xy(2x + y)} = \dfrac{x}{2x + y}$

28. $\dfrac{x^2 - 8x + 15}{x^2 - 2x - 15} = \dfrac{(x - 3)(x - 5)}{(x + 3)(x - 5)} = \dfrac{x - 3}{x + 3}$

29. $\dfrac{x^2 - x - 6}{x^2 + 5x + 6} = \dfrac{(x + 2)(x - 3)}{(x + 2)(x + 3)} = \dfrac{x - 3}{x + 3}$

30. $\dfrac{x^2 - 6x + 8}{x^2 - x - 2} = \dfrac{(x - 2)(x - 4)}{(x - 2)(x + 1)} = \dfrac{x - 4}{x + 1}$

31. $\dfrac{y^2 - 4y - 5}{y^2 - 2y - 3} = \dfrac{(y + 1)(y - 5)}{(y + 1)(y - 3)} = \dfrac{y - 5}{y - 3}$

32. $\dfrac{t^2 + 7t + 10}{t^2 - t - 6} = \dfrac{(t + 2)(t + 5)}{(t + 2)(t - 3)} = \dfrac{t + 5}{t - 3}$

33. $\dfrac{z^2 - 12z + 20}{z^2 + 4z - 12} = \dfrac{(z - 2)(z - 10)}{(z - 2)(z + 6)} = \dfrac{z - 10}{z + 6}$

Page 303 · WRITTEN EXERCISES

A   1. $-2x = -1(2x)$ 2. $-3y = -1(3y)$ 3. $-a - b = -1(a + b)$

4. $-7 - 4x = -1(7 + 4x)$, or $-1(4x + 7)$

5. $2 - 3y = -1(-2 + 3y)$, or $-1(3y - 2)$

6. $6 - 5b = -1(-6 + 5b)$, or $-1(5b - 6)$

7. $3 - 8m = -1(-3 + 8m)$, or $-1(8m - 3)$

8. $16 - r = -1(-16 + r)$, or $-1(r - 16)$

9. $25 - v = -1(-25 + v)$, or $-1(v - 25)$

10. $3 - x^2 = -1(-3 + x^2)$, or $-1(x^2 - 3)$

11. $-4 - y^2 = -1(4 + y^2)$, or $-1(y^2 + 4)$

12. $a^2 - b^2 = -1(-a^2 + b^2)$, or $-1(b^2 - a^2)$

13. $\dfrac{a - b}{b - a} = \dfrac{a - b}{-1(a - b)} = \dfrac{1}{-1} = -1$ 14. $\dfrac{c - 2}{2 - c} = \dfrac{c - 2}{-1(c - 2)} = \dfrac{1}{-1} = -1$

15. $\dfrac{x - y}{y - x} = \dfrac{x - y}{-1(x - y)} = \dfrac{1}{-1} = -1$ 16. $\dfrac{a - c}{c - a} = \dfrac{a - c}{-1(a - c)} = \dfrac{1}{-1} = -1$

17. $\dfrac{1 - a}{a - 1} = \dfrac{-1(a - 1)}{a - 1} = \dfrac{-1}{1} = -1$ 18. $\dfrac{b - 1}{1 - b} = \dfrac{b - 1}{-1(b - 1)} = \dfrac{1}{-1} = -1$

19. $\dfrac{m - n}{n - m} = \dfrac{m - n}{-1(m - n)} = \dfrac{1}{-1} = -1$ 20. $\dfrac{a - 6}{6 - a} = \dfrac{a - 6}{-1(a - 6)} = \dfrac{1}{-1} = -1$

21. $\dfrac{2x + 2y}{-y - x} = \dfrac{2(x + y)}{-1(x + y)} = \dfrac{2}{-1} = -2$ 22. $\dfrac{3b + 6a}{-b - 2a} = \dfrac{3(b + 2a)}{-1(b + 2a)} = \dfrac{3}{-1} = -3$

23. $\dfrac{5x - 5y}{5y - 5x} = \dfrac{5(x - y)}{-5(x - y)} = \dfrac{1}{-1} = -1$ 24. $\dfrac{1 - b}{2b - 2} = \dfrac{-1(b - 1)}{2(b - 1)} = \dfrac{-1}{2} = -\dfrac{1}{2}$

25. $\dfrac{a^2 - 1}{1 - a} = \dfrac{(a + 1)(a - 1)}{-1(a - 1)} = \dfrac{a + 1}{-1} = -a - 1$

26. $\dfrac{m^2 - 1}{1 - m} = \dfrac{(m + 1)(m - 1)}{-1(m - 1)} = \dfrac{m + 1}{-1} = -m - 1$

27. $\dfrac{1 - r}{r^2 - 1} = \dfrac{-1(r - 1)}{(r + 1)(r - 1)} = \dfrac{-1}{r + 1} = -\dfrac{1}{r + 1}$

28. $\dfrac{a^2 - 9}{3 - a} = \dfrac{(a + 3)(a - 3)}{-1(a - 3)} = \dfrac{a + 3}{-1} = -a - 3$

29. $\dfrac{2 - c}{c^2 - 4} = \dfrac{-1(c - 2)}{(c + 2)(c - 2)} = \dfrac{-1}{c + 2} = -\dfrac{1}{c + 2}$

30. $\dfrac{3y^2 - 3}{3 - 3y} = \dfrac{3(y^2 - 1)}{3(1 - y)} = \dfrac{3(y + 1)(y - 1)}{-3(y - 1)} = \dfrac{y + 1}{-1} = -y - 1$

31. $\dfrac{2x^2 - 50}{10 - 2x} = \dfrac{2(x^2 - 25)}{2(5 - x)} = \dfrac{2(x + 5)(x - 5)}{-2(x - 5)} = \dfrac{x + 5}{-1} = -x - 5$

32. $\dfrac{-a - b}{b^2 - a^2} = \dfrac{-1(a + b)}{-1(a^2 - b^2)} = \dfrac{-1(a + b)}{-1(a + b)(a - b)} = \dfrac{1}{a - b}$

B   33. $\dfrac{8 - 4t}{t^2 + t - 6} = \dfrac{-4(t - 2)}{(t - 2)(t + 3)} = \dfrac{-4}{t + 3} = -\dfrac{4}{t + 3}$

34. $\dfrac{5 - 5x}{x^2 + 3x - 4} = \dfrac{-5(x - 1)}{(x - 1)(x + 4)} = \dfrac{-5}{x + 4} = -\dfrac{5}{x + 4}$

35. $\dfrac{p^2 - 4p - 21}{14 - 2p} = \dfrac{(p+3)(p-7)}{-2(p-7)} = \dfrac{p+3}{-2} = -\dfrac{p+3}{2}$

36. $\dfrac{z^2 - 6z + 9}{6 - 2z} = \dfrac{(z-3)(z-3)}{-2(z-3)} = \dfrac{z-3}{-2} = -\dfrac{z-3}{2}$

37. $\dfrac{r^2 - 7r + 12}{9 - 3r} = \dfrac{(r-3)(r-4)}{-3(r-3)} = \dfrac{r-4}{-3} = -\dfrac{r-4}{3}$

38. $\dfrac{8 - 4a}{a^2 - 5a + 6} = \dfrac{-4(a-2)}{(a-2)(a-3)} = \dfrac{-4}{a-3} = -\dfrac{4}{a-3}$

Page 305 · WRITTEN EXERCISES

A 1. $\dfrac{25}{40} = \dfrac{5}{8}$ 2. $\dfrac{30}{20} = \dfrac{3}{2}$ 3. $\dfrac{50}{75} = \dfrac{2}{3}$

4. $\dfrac{4}{12} = \dfrac{1}{3}$ 5. $\dfrac{12}{20} = \dfrac{3}{5}$ 6. $\dfrac{6}{8} = \dfrac{3}{4}$

7. $\dfrac{2}{1}$ 8. $\dfrac{40}{1000} = \dfrac{1}{25}$ 9. $\dfrac{9}{6} = \dfrac{3}{2}$

10. Number of boys = 2100 − 1200 = 900; $\dfrac{1200}{900} = \dfrac{4}{3}$

11. Distance on city streets = 40 − 24 = 16; $\dfrac{24}{16} = \dfrac{3}{2}$

12. **a.** rise;   run   **b.** $\dfrac{3}{5}$

B 13. $\dfrac{3 \text{ days}}{3 \text{ weeks}} = \dfrac{3 \text{ days}}{21 \text{ days}} = \dfrac{1 \text{ day}}{7 \text{ days}}$, or $\dfrac{1}{7}$

14. $\dfrac{30 \text{ min}}{3 \text{ h}} = \dfrac{30 \text{ min}}{180 \text{ min}} = \dfrac{1 \text{ min}}{6 \text{ min}}$, or $\dfrac{1}{6}$

15. $\dfrac{400 \text{ g}}{2 \text{ kg}} = \dfrac{400 \text{ g}}{2000 \text{ g}} = \dfrac{1 \text{ g}}{5 \text{ g}}$, or $\dfrac{1}{5}$

16. $\dfrac{9 \text{ months}}{2 \text{ years}} = \dfrac{9 \text{ mo}}{24 \text{ mo}} = \dfrac{3 \text{ mo}}{8 \text{ mo}}$, or $\dfrac{3}{8}$

17. $\dfrac{2 \text{ min}}{40 \text{ sec}} = \dfrac{120 \text{ sec}}{40 \text{ sec}} = \dfrac{3 \text{ sec}}{1 \text{ sec}}$, or $\dfrac{3}{1}$

18. $\dfrac{1 \text{ century}}{4 \text{ decades}} = \dfrac{100 \text{ years}}{40 \text{ years}} = \dfrac{5 \text{ years}}{2 \text{ years}}$, or $\dfrac{5}{2}$

Page 307 · WRITTEN EXERCISES

A 1. Let $4x$ = the number of boys and $3x$ = the number of girls. $4x + 3x = 35$; $7x = 35$; $x = 5$, so $4x = 4 \cdot 5 = 20$.   20 boys

2. Let $2x$ = number playing wind instruments and $3x$ = all others. $2x + 3x = 80$; $5x = 80$; $x = 16$, so $2x = 2 \cdot 16 = 32$.   32 members

3. Let $13x$ = number of heads and $12x$ = number of tails. $13x + 12x = 100$; $25x = 100$; $x = 4$, so $12x = 12 \cdot 4 = 48$.   48 tails

4. Let $5x$ = book sales (in dollars) and $3x$ = other sales. $5x + 3x = 4{,}000{,}000$; $8x = 4{,}000{,}000$; $x = 500{,}000$, so $5x = 5 \cdot 500{,}000 = 2{,}500{,}000$.   \$2,500,000 in book sales

5. Let $11x$ and $5x$ = the lengths of the pieces (in m). $11x + 5x = 80$; $16x = 80$; $x = 5$, so $5x = 5 \cdot 5 = 25$.   25 m

6. Let $7x$ = dollars spent on TV ads and $2x$ = dollars spent on newspaper ads. $7x + 2x = 360{,}000$; $9x = 360{,}000$; $x = 40{,}000$, so $2x = 2 \cdot 40{,}000 = 80{,}000$. $80{,}000$ on newspaper ads

7. Let $26x$ = dollars in taxes.   Then $100 - 26 = 74$, so $74x$ = dollars left. $26x + 74x = 30{,}000$; $100x = 30{,}000$; $x = 300$, so $74x = 74 \cdot 300 = 22{,}200$. $22{,}200$ left

8. Let $5x$ = number of full-time teachers and $3x$ = number of part-time teachers. $5x + 3x = 72$; $8x = 72$; $x = 9$, so $3x = 3 \cdot 9 = 27$.   27 part-time teachers

9. Let $4x$ = milliliters of vinegar and $3x$ = milliliters of oil. $4x + 3x = 735$; $7x = 735$; $x = 105$, so $4x = 4 \cdot 105 = 420$ and $3x = 3 \cdot 105 = 315$.   420 milliliters of vinegar and 315 milliliters of oil

B 10. Let $4x$ = dollars for streets, $3x$ = dollars for parks, and x = dollars for library. $4x + 3x + x = 64{,}000$; $8x = 64{,}000$; $x = 8000$, so $4x = 4 \cdot 8000 = 32{,}000$ and $3x = 3 \cdot 8000 = 24{,}000$. $32{,}000$ for streets, $24{,}000$ for parks, 8000 for library

11. Let $6x$, $5x$, and $5x$ = number of votes for each. $6x + 5x + 5x = 56{,}000$; $16x = 56{,}000$; $x = 3500$, so $6x = 6 \cdot 3500 = 21{,}000$ and $5x = 5 \cdot 3500 = 17{,}500$.   21,000 votes, 17,500 votes, and 17,500 votes

12. Let $7x$ = number of daffodils, $5x$ = number of hyacinths, and $3x$ = number of tulips. $7x + 5x + 3x = 60$; $15x = 60$; $x = 4$, so $3x = 3 \cdot 4 = 12$.   12 tulips

Page 309 • WRITTEN EXERCISES

A   1. $\dfrac{3}{5} = \dfrac{x}{15}$; $3 \cdot 15 = 5x$; $45 = 5x$; $9 = x$

2. $\dfrac{2}{9} = \dfrac{6}{x}$; $2x = 9 \cdot 6$; $2x = 54$; $x = 27$

3. $\dfrac{8}{6} = \dfrac{x}{9}$; $8 \cdot 9 = 6x$; $72 = 6x$; $12 = x$

4. $\dfrac{10}{a} = \dfrac{5}{10}$; $10 \cdot 10 = a \cdot 5$; $100 = 5a$; $20 = a$

5. $\dfrac{3x}{4} = \dfrac{9}{6}$; $3x \cdot 6 = 4 \cdot 9$; $18x = 36$; $x = 2$

6. $\dfrac{2x}{7} = \dfrac{4}{1}$; $2x \cdot 1 = 7 \cdot 4$; $2x = 28$; $x = 14$

7. $\dfrac{6}{1} = \dfrac{48}{x}$; $6x = 1 \cdot 48$; $6x = 48$; $x = 8$

8. $\dfrac{3a}{2} = \dfrac{15}{5}$; $3a \cdot 5 = 2 \cdot 15$; $15a = 30$; $a = 2$

9. $\dfrac{3a}{5} = \dfrac{12}{10}$; $3a \cdot 10 = 5 \cdot 12$; $30a = 60$; $a = 2$

10. $\dfrac{15}{5} = \dfrac{3a}{4}$; $15 \cdot 4 = 5 \cdot 3a$; $60 = 15a$; $4 = a$

11. $\dfrac{4}{5} = \dfrac{2x}{25}$; $4 \cdot 25 = 5 \cdot 2x$; $100 = 10x$; $10 = x$

12. $\dfrac{4r}{8} = \dfrac{3}{6}$; $4r \cdot 6 = 8 \cdot 3$; $24r = 24$; $r = 1$

13. $\dfrac{x}{4} = \dfrac{x + 2}{5}$; $x \cdot 5 = 4(x + 2)$; $5x = 4x + 8$; $x = 8$

14. $\dfrac{x}{5 - x} = \dfrac{2}{3}$; $x \cdot 3 = (5 - x)2$; $3x = 10 - 2x$; $5x = 10$; $x = 2$

15. $\dfrac{a - 2}{a + 1} = \dfrac{1}{2}$; $(a - 2)2 = (a + 1)1$; $2a - 4 = a + 1$; $2a = a + 5$; $a = 5$

16. $\dfrac{x + 4}{x} = \dfrac{5}{3}$; $(x + 4)3 = x \cdot 5$; $3x + 12 = 5x$; $12 = 2x$; $6 = x$

17. $\dfrac{x - 1}{3} = \dfrac{x + 1}{9}$; $(x - 1)9 = 3(x + 1)$; $9x - 9 = 3x + 3$; $9x = 3x + 12$; $6x = 12$; $x = 2$

18. $\dfrac{x + 1}{3} = \dfrac{x}{2}$; $(x + 1)2 = 3x$; $2x + 2 = 3x$; $2 = x$

19. $\dfrac{2a + 3}{4} = \dfrac{a - 6}{8}$; $(2a + 3)8 = 4(a - 6)$; $16a + 24 = 4a - 24$; $16a = 4a - 48$; $12a = -48$; $a = -4$

20. $\dfrac{y + 2}{y - 2} = \dfrac{5}{3}$; $(y + 2)3 = (y - 2)5$; $3y + 6 = 5y - 10$; $3y + 16 = 5y$; $16 = 2y$; $8 = y$

21. $\dfrac{3n - 5}{8n} = \dfrac{1}{4}$; $(3n - 5)4 = 8n \cdot 1$; $12n - 20 = 8n$; $-20 = -4n$; $5 = n$

22. $\dfrac{x + 3}{5} = \dfrac{x + 1}{4}$; $(x + 3)4 = 5(x + 1)$; $4x + 12 = 5x + 5$; $4x + 7 = 5x$; $7 = x$

23. $\dfrac{y + 2}{9} = \dfrac{y - 1}{18}$; $(y + 2)18 = 9(y - 1)$; $18y + 36 = 9y - 9$; $18y = 9y - 45$; $9y = -45$; $y = -5$

24. $\dfrac{a + 4}{6} = \dfrac{a + 6}{4}$; $(a + 4)4 = 6(a + 6)$; $4a + 16 = 6a + 36$; $4a - 20 = 6a$; $-20 = 2a$; $-10 = a$

25. $\dfrac{2x + 3}{4x} = \dfrac{3}{4}$; $(2x + 3)4 = 4x \cdot 3$; $8x + 12 = 12x$; $12 = 4x$; $3 = x$

26. $\dfrac{3a - 4}{1} = \dfrac{2a}{2}$; $(3a - 4)2 = 1 \cdot 2a$; $6a - 8 = 2a$; $-8 = -4a$; $2 = a$

27. $\dfrac{5m + 2}{3} = \dfrac{m - 1}{2}$; $(5m + 2)2 = 3(m - 1)$; $10m + 4 = 3m - 3$; $10m = 3m - 7$; $7m = -7$; $m = -1$

B   28. $\dfrac{x}{x - 2} = \dfrac{x + 2}{x + 1}$; $x(x + 1) = (x - 2)(x + 2)$; $x^2 + x = x^2 - 4$; $x = -4$

29. $\dfrac{y}{y - 2} = \dfrac{y + 2}{y + 4}$; $y(y + 4) = (y - 2)(y + 2)$; $y^2 + 4y = y^2 - 4$; $4y = -4$; $y = -1$

30. $\dfrac{a-1}{a} = \dfrac{a+1}{a+3}$;　$(a-1)(a+3) = a(a+1)$;　$a^2 + 2a - 3 = a^2 + a$;

$2a - 3 = a$;　$-3 = -a$;　$3 = a$

31. $\dfrac{x-3}{x} = \dfrac{x+3}{x+8}$;　$(x-3)(x+8) = x(x+3)$;　$x^2 + 5x - 24 = x^2 + 3x$;

$5x - 24 = 3x$;　$-24 = -2x$;　$12 = x$

32. $\dfrac{n-3}{n+5} = \dfrac{n-4}{n+3}$;　$(n-3)(n+3) = (n+5)(n-4)$;　$n^2 - 9 = n^2 + n - 20$;

$-9 = n - 20$;　$11 = n$

33. $\dfrac{c+5}{c-1} = \dfrac{c-3}{c-5}$;　$(c+5)(c-5) = (c-1)(c-3)$;　$c^2 - 25 = c^2 - 4c + 3$;

$-25 = -4c + 3$;　$-28 = -4c$;　$7 = c$

34. $\dfrac{x}{10} = \dfrac{36}{12}$;　$x \cdot 12 = 10 \cdot 36$;　$12x = 360$;　$x = 30$;　30 m

Pages 311–313 · WRITTEN EXERCISES

A　**1.**

Number	3	5
Cost	93¢	x

$\dfrac{3}{93} = \dfrac{5}{x}$;　$3x = 93 \cdot 5$;　$3x = 465$;　$x = 155$.　$1.55

2.

Number	6	9
Cost	90¢	x

$\dfrac{6}{90} = \dfrac{9}{x}$;　$6x = 90 \cdot 9$;　$6x = 810$;　$x = 135$.　$1.35

3.

Number	4	10
Cost	$2.76	x

$\dfrac{4}{276} = \dfrac{10}{x}$;　$4x = 276 \cdot 10$;　$4x = 2760$;　$x = 690$.
$6.90

4.

Number	3	5
Cost	$2.67	x

$\dfrac{3}{267} = \dfrac{5}{x}$;　$3x = 267 \cdot 5$;　$3x = 1335$;　$x = 445$.
$4.45

5.

Distance	1800 km	3150 km
Time	4	x

$\dfrac{1800}{4} = \dfrac{3150}{x}$;　$1800x = 12{,}600$;　$x = 7$.
7 hours

6.

Centimeters	160	x
Kilograms	4	7

$\dfrac{160}{4} = \dfrac{x}{7}$;　$4x = 1120$;　$x = 280$.　280 cm

7.

Tea	30 g	20 g
Chamomile	12 g	x

$\dfrac{30}{12} = \dfrac{20}{x}$; $30x = 240$; $x = 8$.   8 g

8.

Quantity	2000 L	1500 L
Time	80 min	x

$\dfrac{2000}{80} = \dfrac{1500}{x}$; $2000x = 120{,}000$; $x = 60$.

60 minutes, or 1 hour

9.

Larger measure	20 cm	28 cm
Shorter measure	15 cm	x

$\dfrac{20}{15} = \dfrac{28}{x}$; $20x = 420$; $x = 21$.   21 cm

10.

Grams	600 g	800 g
Servings	21	x

$\dfrac{600}{21} = \dfrac{800}{x}$; $600x = 16{,}800$; $x = 28$.   28 servings

11.

Map distance	5 cm	20 cm
Actual distance	2 km	x

$\dfrac{5}{2} = \dfrac{20}{x}$; $5x = 40$; $x = 8$.   8 km

12.

Car's worth	$8400	$9600
Tax	$476	x

$\dfrac{8400}{476} = \dfrac{9600}{x}$; $8400x = 4{,}569{,}600$; $x = 544$.

$544

13.

	Girl	Pole
Height of object	160 cm	x
Length of shadow	120 cm	660 cm

$\dfrac{160}{120} = \dfrac{x}{660}$; $120x = 105{,}600$;

$x = 880$.   880 cm

14.

	Man	Pole
Height of object	180 cm	x
Length of shadow	150 cm	540 cm

$\dfrac{180}{150} = \dfrac{x}{540}$; $150x = 97{,}200$; $x = 648$.

648 cm

B **15.**

Insurance	$660	x
Time	12 mo	9 mo

$\dfrac{660}{12} = \dfrac{x}{9}$; $12x = 5940$; $x = 495$. $495

16.

Tax	$1452	x
Time	12 mo	9 mo

$\dfrac{1452}{12} = \dfrac{x}{9}$; $12x = 13{,}068$; $x = 1089$. $1089

17.

Cost	$12,000	x
Area	18 m²	27 m²

$\dfrac{12,000}{18} = \dfrac{x}{27}; \quad 18x = 324,000; \quad x = 18,000.$

$18,000

Page 313 · PUZZLE PROBLEMS

There are 19 different ways as shown:

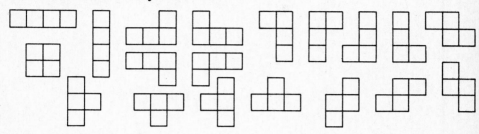

Page 315 · WRITTEN EXERCISES

A

1. $\dfrac{1}{3} \cdot \dfrac{1}{4} = \dfrac{1 \cdot 1}{3 \cdot 4} = \dfrac{1}{12}$

2. $\dfrac{1}{5} \cdot \dfrac{1}{8} = \dfrac{1 \cdot 1}{5 \cdot 8} = \dfrac{1}{40}$

3. $\dfrac{2}{3} \cdot \dfrac{4}{7} = \dfrac{2 \cdot 4}{3 \cdot 7} = \dfrac{8}{21}$

4. $\dfrac{x}{3} \cdot \dfrac{x}{4} = \dfrac{x \cdot x}{3 \cdot 4} = \dfrac{x^2}{12}$

5. $y \cdot \dfrac{y}{2} = \dfrac{y \cdot y}{1 \cdot 2} = \dfrac{y^2}{2}$

6. $m \cdot \dfrac{m}{8} = \dfrac{m \cdot m}{1 \cdot 8} = \dfrac{m^2}{8}$

7. $\dfrac{a}{b} \cdot \dfrac{2b}{3a} = \dfrac{a \cdot 2 \cdot b}{b \cdot 3 \cdot a} = \dfrac{2}{3}$

8. $\dfrac{xy}{5} \cdot \dfrac{10}{x} = \dfrac{x \cdot y \cdot 10}{5 \cdot x} = 2y$

9. $\dfrac{6x}{5y} \cdot \dfrac{y^2}{3x} = \dfrac{6 \cdot x \cdot y \cdot y}{5 \cdot y \cdot 3 \cdot x} = \dfrac{2y}{5}$

10. $\dfrac{5a}{3b^2} \cdot \dfrac{6b^3}{a^2} = \dfrac{5 \cdot a \cdot 6 \cdot b \cdot b \cdot b}{3 \cdot b \cdot b \cdot a \cdot a} = \dfrac{10b}{a}$

11. $\dfrac{4x}{3y} \cdot \dfrac{3}{8xy} = \dfrac{4 \cdot x \cdot 3}{3 \cdot y \cdot 8 \cdot x \cdot y} = \dfrac{1}{2y^2}$

12. $\dfrac{16x^2}{5} \cdot \dfrac{15}{4y^2} = \dfrac{16 \cdot x \cdot x \cdot 15}{5 \cdot 4 \cdot y \cdot y} = \dfrac{12x^2}{y^2}$

13. $\dfrac{9r}{7s} \cdot \dfrac{14rs}{3r} = \dfrac{9 \cdot r \cdot 14 \cdot r \cdot s}{7 \cdot s \cdot 3 \cdot r} = 6r$

14. $\dfrac{x^2}{3y} \cdot \dfrac{5xy}{x} = \dfrac{x \cdot x \cdot 5 \cdot x \cdot y}{3 \cdot y \cdot x} = \dfrac{5x^2}{3}$

15. $\dfrac{3a^2}{4} \cdot \dfrac{20b}{12b^2} = \dfrac{3 \cdot a \cdot a \cdot 20 \cdot b}{4 \cdot 12 \cdot b \cdot b} = \dfrac{5a^2}{4b}$

16. $\dfrac{12a}{5a^2} \cdot \dfrac{a^2}{6a^2} = \dfrac{12 \cdot a \cdot a \cdot a}{5 \cdot a \cdot a \cdot 6 \cdot a \cdot a} = \dfrac{2}{5a}$

17. $\dfrac{3}{7} \cdot \dfrac{7x - 7}{6} = \dfrac{3 \cdot 7(x - 1)}{7 \cdot 6} = \dfrac{x - 1}{2}$

18. $\dfrac{5}{4} \cdot \dfrac{4x - 4}{15} = \dfrac{5 \cdot 4(x - 1)}{4 \cdot 15} = \dfrac{x - 1}{3}$

19. $\dfrac{2}{3} \cdot \dfrac{3x - 12}{4} = \dfrac{2 \cdot 3(x - 4)}{3 \cdot 4} = \dfrac{x - 4}{2}$

20. $\dfrac{5}{4x + 16} \cdot \dfrac{4}{25} = \dfrac{5 \cdot 4}{4(x + 4) \cdot 25} = \dfrac{1}{5(x + 4)} = \dfrac{1}{5x + 20}$

21. $\dfrac{a^2 - a}{5} \cdot \dfrac{25}{a} = \dfrac{a(a - 1) \cdot 25}{5 \cdot a} = 5(a - 1) = 5a - 5$

22. $\dfrac{x}{b} \cdot \dfrac{b^2 - b}{x^2y} = \dfrac{x \cdot b(b - 1)}{b \cdot x \cdot x \cdot y} = \dfrac{b - 1}{xy}$

23. $\dfrac{a^2b}{2a - 2} \cdot \dfrac{2}{ab^2} = \dfrac{a \cdot a \cdot b \cdot 2}{2(a - 1) \cdot a \cdot b \cdot b} = \dfrac{a}{b(a - 1)} = \dfrac{a}{ab - b}$

24. $\dfrac{9}{x + 2} \cdot \dfrac{x^2 + 2x}{36} = \dfrac{9 \cdot x(x + 2)}{36(x + 2)} = \dfrac{x}{4}$

25. $\dfrac{a^2 - b^2}{10} \cdot \dfrac{5}{a - b} = \dfrac{5(a + b)(a - b)}{10(a - b)} = \dfrac{a + b}{2}$

26. $\dfrac{x^2 - 1}{4} \cdot \dfrac{12}{x - 1} = \dfrac{12(x + 1)(x - 1)}{4(x - 1)} = 3(x + 1) = 3x + 3$

27. $\dfrac{r^2 - 2r}{4} \cdot \dfrac{2}{r - 2} = \dfrac{2 \cdot r(r - 2)}{4(r - 2)} = \dfrac{r}{2}$

28. $\dfrac{3x - 3}{6} \cdot \dfrac{4}{1 - x} = \dfrac{4 \cdot 3(x - 1)}{6(-1)(x - 1)} = \dfrac{2}{-1} = -2$

29. $\dfrac{6}{x^2 - 4} \cdot \dfrac{2 - x}{18} = \dfrac{6(-1)(x - 2)}{18(x + 2)(x - 2)} = \dfrac{-1}{3(x + 2)} = -\dfrac{1}{3x + 6}$

B **30.** $\dfrac{x^2 - y^2}{x^2 - 49} \cdot \dfrac{x + 7}{x - y} = \dfrac{(x + y)(x - y)(x + 7)}{(x + 7)(x - 7)(x - y)} = \dfrac{x + y}{x - 7}$

31. $\dfrac{x^2 - x - 6}{14} \cdot \dfrac{7}{x - 3} = \dfrac{7(x + 2)(x - 3)}{14(x - 3)} = \dfrac{x + 2}{2}$

32. $\dfrac{2x}{x + 1} \cdot \dfrac{2x + 2}{6x^2} = \dfrac{2x \cdot 2(x + 1)}{6x^2(x + 1)} = \dfrac{2}{3x}$

33. $\dfrac{x^2 - 3x - 10}{2} \cdot \dfrac{4}{x^2 + x - 2} = \dfrac{4(x + 2)(x - 5)}{2(x + 2)(x - 1)} = \dfrac{2(x - 5)}{(x - 1)} = \dfrac{2x - 10}{x - 1}$

34. $\dfrac{2a - 4}{3a + 6} \cdot \dfrac{2a + 4}{a - 2} = \dfrac{2(a - 2) \cdot 2(a + 2)}{3(a + 2)(a - 2)} = \dfrac{4}{3}$

35. $\dfrac{3x - 9}{x - 3} \cdot \dfrac{x + 2}{3x + 12} = \dfrac{3(x - 3)(x + 2)}{(x - 3) \cdot 3(x + 4)} = \dfrac{x + 2}{x + 4}$

36. $\dfrac{4a - 1}{a^2} \cdot \dfrac{a^3 - a^2}{a - 1} = \dfrac{(4a - 1) \cdot a^2(a - 1)}{a^2(a - 1)} = 4a - 1$

37. $\dfrac{n^2 - 1}{3} \cdot \dfrac{21}{1 - n} = \dfrac{21(n + 1)(n - 1)}{3(-1)(n - 1)} = \dfrac{7(n + 1)}{-1} = -7n - 7$

C **38.** $\dfrac{3 - 3x}{x^2 - 2x - 3} \cdot \dfrac{x + 1}{x - 1} = \dfrac{-3(x - 1)(x + 1)}{(x + 1)(x - 3)(x - 1)} = \dfrac{-3}{x - 3} = -\dfrac{3}{x - 3}$

39. $\dfrac{m^2 - 1}{m^2 + 4m + 3} \cdot \dfrac{m^2 + m - 6}{m^2 + m - 2} = \dfrac{(m + 1)(m - 1)(m - 2)(m + 3)}{(m + 1)(m + 3)(m - 1)(m + 2)} = \dfrac{m - 2}{m + 2}$

Pages 317–318 · WRITTEN EXERCISES

A **1.** $\dfrac{5}{6} \div \dfrac{1}{3} = \dfrac{5}{6} \cdot \dfrac{3}{1} = \dfrac{5}{2}$ **2.** $\dfrac{x}{2} \div \dfrac{1}{4} = \dfrac{x}{2} \cdot \dfrac{4}{1} = 2x$

3. $\dfrac{a}{5} \div \dfrac{a}{6} = \dfrac{a}{5} \cdot \dfrac{6}{a} = \dfrac{6}{5}$ **4.** $\dfrac{a}{4} \div \dfrac{a^2}{2} = \dfrac{a}{4} \cdot \dfrac{2}{a^2} = \dfrac{1}{2a}$

5. $\dfrac{x}{10} \div \dfrac{6}{5x} = \dfrac{x}{10} \cdot \dfrac{5x}{6} = \dfrac{x^2}{12}$ **6.** $\dfrac{2x^2}{9} \div \dfrac{4x}{3} = \dfrac{2x^2}{9} \cdot \dfrac{3}{4x} = \dfrac{x}{6}$

7. $\dfrac{2a^2}{5b^2} \div \dfrac{4a^3}{10b} = \dfrac{2a^2}{5b^2} \cdot \dfrac{10b}{4a^3} = \dfrac{2 \cdot a \cdot a \cdot 10 \cdot b}{5 \cdot b \cdot b \cdot 4 \cdot a \cdot a \cdot a} = \dfrac{1}{ab}$

8. $\dfrac{5r}{12t^2} \div \dfrac{15r^2}{6s^3} = \dfrac{5r}{12t^2} \cdot \dfrac{6s^3}{15r^2} = \dfrac{5 \cdot r \cdot 6 \cdot s \cdot s \cdot s}{12 \cdot t \cdot t \cdot 15 \cdot r \cdot r} = \dfrac{s^3}{6t^2r}$

9. $3m^2 \div \dfrac{8m}{7} = \dfrac{3m^2}{1} \cdot \dfrac{7}{8m} = \dfrac{3 \cdot m \cdot m \cdot 7}{8 \cdot m} = \dfrac{21m}{8}$

10. $\dfrac{3km}{8} \div 18k^2 = \dfrac{3km}{8} \cdot \dfrac{1}{18k^2} = \dfrac{3 \cdot k \cdot m}{8 \cdot 18 \cdot k \cdot k} = \dfrac{m}{48k}$

11. $\dfrac{6}{a^2 b} \div \dfrac{a}{a^3 b^2} = \dfrac{6}{a^2 b} \cdot \dfrac{a^3 b^2}{a} = \dfrac{6 \cdot a \cdot a \cdot a \cdot b \cdot b}{a \cdot a \cdot b \cdot a} = 6b$

12. $\dfrac{(xy)^2}{x} \div \dfrac{xy}{y} = \dfrac{(xy)^2}{x} \cdot \dfrac{y}{xy} = \dfrac{(xy)(xy)y}{x(xy)} = y^2$

13. $\dfrac{(ab)^2}{4} \div 2ab = \dfrac{(ab)^2}{4} \cdot \dfrac{1}{2ab} = \dfrac{(ab)(ab)}{4 \cdot 2(ab)} = \dfrac{ab}{8}$

14. $\dfrac{a-b}{7} \div \dfrac{a-b}{6} = \dfrac{a-b}{7} \cdot \dfrac{6}{a-b} = \dfrac{6(a-b)}{7(a-b)} = \dfrac{6}{7}$

15. $\dfrac{h-g}{4} \div \dfrac{h-g}{8} = \dfrac{h-g}{4} \cdot \dfrac{8}{h-g} = \dfrac{8(h-g)}{4(h-g)} = 2$

16. $\dfrac{2x-2}{4} \div \dfrac{2}{9} = \dfrac{2x-2}{4} \cdot \dfrac{9}{2} = \dfrac{2(x-1) \cdot 9}{4 \cdot 2} = \dfrac{9(x-1)}{4} = \dfrac{9x-9}{4}$

17. $\dfrac{a^3}{3} \div \dfrac{a^4}{3a^2+3} = \dfrac{a^3}{3} \cdot \dfrac{3a^2+3}{a^4} = \dfrac{a^3 \cdot 3(a^2+1)}{3 \cdot a^4} = \dfrac{a^2+1}{a}$

18. $\dfrac{2x+2}{5} \div \dfrac{2}{3} = \dfrac{2x+2}{5} \cdot \dfrac{3}{2} = \dfrac{2(x+1) \cdot 3}{5 \cdot 2} = \dfrac{3(x+1)}{5} = \dfrac{3x+3}{5}$

19. $\dfrac{6}{y^2-9} \div \dfrac{3}{y-3} = \dfrac{6}{y^2-9} \cdot \dfrac{y-3}{3} = \dfrac{6(y-3)}{3(y+3)(y-3)} = \dfrac{2}{y+3}$

20. $\dfrac{a^2-b^2}{4} \div \dfrac{a-b}{8} = \dfrac{a^2-b^2}{4} \cdot \dfrac{8}{a-b} = \dfrac{8(a+b)(a-b)}{4(a-b)} = 2(a+b) = 2a+2b$

21. $\dfrac{2a-2}{2} \div (a-1) = \dfrac{2a-2}{2} \cdot \dfrac{1}{a-1} = \dfrac{2(a-1)}{2(a-1)} = 1$

22. $\dfrac{x^2-1}{2-x} \div (x-1) = \dfrac{x^2-1}{2-x} \cdot \dfrac{1}{x-1} = \dfrac{(x+1)(x-1)}{(2-x)(x-1)} = \dfrac{x+1}{2-x}$

23. $\dfrac{x^2-16}{2x} \div (x-4) = \dfrac{x^2-16}{2x} \cdot \dfrac{1}{x-4} = \dfrac{(x+4)(x-4)}{2x(x-4)} = \dfrac{x+4}{2x}$

24. $\dfrac{r-7}{r^2-4} \div \dfrac{1}{r-2} = \dfrac{r-7}{r^2-4} \cdot \dfrac{r-2}{1} = \dfrac{(r-7)(r-2)}{(r+2)(r-2)} = \dfrac{r-7}{r+2}$

B **25.** $\dfrac{a^2-a-6}{a^2-9} \div \dfrac{a+2}{a-3} = \dfrac{(a+2)(a-3)}{(a+3)(a-3)} \cdot \dfrac{a-3}{a+2} = \dfrac{a-3}{a+3}$

26. $\dfrac{x^2-16}{2x} \div \dfrac{x+4}{2x^2-2x} = \dfrac{(x+4)(x-4)}{2x} \cdot \dfrac{2x(x-1)}{x+4} = (x-4)(x-1) = x^2-5x+4$

27. $\dfrac{6y+12}{3y-9} \div \dfrac{4y-8}{3} = \dfrac{6(y+2)}{3(y-3)} \cdot \dfrac{3}{4(y-2)} = \dfrac{3(y+2)}{2(y-3)(y-2)} = \dfrac{3y+6}{2y^2-10y+12}$

28. $\dfrac{a-b}{4ab} \div \dfrac{b-a}{8a^2} = \dfrac{a-b}{4ab} \cdot \dfrac{8a^2}{-1(a-b)} = \dfrac{2a}{-b} = -\dfrac{2a}{b}$

29. $\dfrac{y+2}{y+1} \div \dfrac{2y+4}{y^2-2y-3} = \dfrac{y+2}{y+1} \cdot \dfrac{(y+1)(y-3)}{2(y+2)} = \dfrac{y-3}{2}$

30. $\dfrac{u^2+u^3}{36-u^2} \div \dfrac{u-u^3}{6u+u^2} = \dfrac{u^2(1+u)}{(6+u)(6-u)} \cdot \dfrac{u(6+u)}{u(1+u)(1-u)} = \dfrac{u^2}{(6-u)(1-u)} = \dfrac{u^2}{6-7u+u^2}$

C **31.** $\dfrac{(d-5)^2}{d+3} \div \dfrac{d^2-d-20}{d^2+7d+12} = \dfrac{(d-5)(d-5)}{d+3} \cdot \dfrac{(d+3)(d+4)}{(d+4)(d-5)} = d-5$

32. $\dfrac{h^2 + h - 6}{h^2 - 2h - 15} \div \dfrac{h^2 + 4h - 5}{h^2 - 25} = \dfrac{(h-2)(h+3)}{(h+3)(h-5)} \cdot \dfrac{(h+5)(h-5)}{(h-1)(h+5)} = \dfrac{h-2}{h-1}$

33. $\dfrac{y^2 - 7y + 12}{y^2 + y - 20} \div \dfrac{y^2 - 9y + 18}{y^2 + 7y + 10} = \dfrac{(y-3)(y-4)}{(y-4)(y+5)} \cdot \dfrac{(y+2)(y+5)}{(y-3)(y-6)} = \dfrac{y+2}{y-6}$

34. $\dfrac{x^2 + 5x + 6}{x^2 - 4} \div \dfrac{x^2 + 2x - 3}{x^2 - 3x + 2} = \dfrac{(x+2)(x+3)}{(x+2)(x-2)} \cdot \dfrac{(x-1)(x-2)}{(x-1)(x+3)} = 1$

Page 320 · WRITTEN EXERCISES

A

1. $\dfrac{4}{5} + \dfrac{3}{5} = \dfrac{4+3}{5} = \dfrac{7}{5}$

2. $\dfrac{4x}{5} + \dfrac{3x}{5} = \dfrac{4x+3x}{5} = \dfrac{7x}{5}$

3. $\dfrac{6}{2a} + \dfrac{3}{2a} = \dfrac{6+3}{2a} = \dfrac{9}{2a}$

4. $\dfrac{8}{9k} + \dfrac{10}{9k} = \dfrac{18}{9k} = \dfrac{2}{k}$

5. $\dfrac{8n}{9} - \dfrac{2n}{9} = \dfrac{8n-2n}{9} = \dfrac{6n}{9} = \dfrac{2n}{3}$

6. $\dfrac{a}{8} - \dfrac{4a}{8} = \dfrac{a-4a}{8} = \dfrac{-3a}{8} = -\dfrac{3a}{8}$

7. $\dfrac{2x}{14} + \dfrac{5x}{14} = \dfrac{2x+5x}{14} = \dfrac{7x}{14} = \dfrac{x}{2}$

8. $\dfrac{2s}{5} + \dfrac{8s}{5} = \dfrac{2s+8s}{5} = \dfrac{10s}{5} = 2s$

9. $\dfrac{4y}{3} - \dfrac{7y}{3} = \dfrac{4y-7y}{3} = \dfrac{-3y}{3} = -y$

10. $\dfrac{8p}{9} - \dfrac{5p}{9} = \dfrac{8p-5p}{9} = \dfrac{3p}{9} = \dfrac{p}{3}$

11. $\dfrac{-3x}{4} - \dfrac{x}{4} = \dfrac{-3x-x}{4} = \dfrac{-4x}{4} = -x$

12. $\dfrac{-a}{5} - \dfrac{3a}{5} = \dfrac{-a-3a}{5} = \dfrac{-4a}{5} = -\dfrac{4a}{5}$

13. $\dfrac{x}{y} + \dfrac{1}{y} = \dfrac{x+1}{y}$

14. $\dfrac{3}{a} + \dfrac{b}{a} = \dfrac{3+b}{a}$

15. $\dfrac{x}{3x} + \dfrac{1}{3x} = \dfrac{x+1}{3x}$

16. $\dfrac{a}{2a} - \dfrac{1}{2a} = \dfrac{a-1}{2a}$

17. $\dfrac{3}{a+b} + \dfrac{2}{a+b} = \dfrac{3+2}{a+b} = \dfrac{5}{a+b}$

18. $\dfrac{7}{x-1} + \dfrac{2}{x-1} = \dfrac{7+2}{x-1} = \dfrac{9}{x-1}$

19. $\dfrac{3}{y-2} - \dfrac{1}{y-2} = \dfrac{3-1}{y-2} = \dfrac{2}{y-2}$

20. $\dfrac{4}{a+3} - \dfrac{5}{a+3} = \dfrac{4-5}{a+3} = \dfrac{-1}{a+3} = -\dfrac{1}{a+3}$

21. $\dfrac{2x}{x+1} + \dfrac{2}{x+1} = \dfrac{2x+2}{x+1} = \dfrac{2(x+1)}{x+1} = 2$

22. $\dfrac{3a}{a-1} - \dfrac{3}{a-1} = \dfrac{3a-3}{a-1} = \dfrac{3(a-1)}{a-1} = 3$

23. $\dfrac{v}{v-4} - \dfrac{4}{v-4} = \dfrac{v-4}{v-4} = 1$

24. $\dfrac{x+y}{3xy} + \dfrac{2x-y}{3xy} = \dfrac{x+y+2x-y}{3xy} = \dfrac{3x}{3xy} = \dfrac{1}{y}$

25. $\dfrac{3a+2b}{4ab} + \dfrac{a+2b}{4ab} = \dfrac{3a+2b+a+2b}{4ab} = \dfrac{4a+4b}{4ab} = \dfrac{4(a+b)}{4ab} = \dfrac{a+b}{ab}$

26. $\dfrac{3x}{x+2} - \dfrac{x+2}{x+2} = \dfrac{3x-(x+2)}{x+2} = \dfrac{3x-x-2}{x+2} = \dfrac{2x-2}{x+2}$

27. $\dfrac{2c+1}{7c} - \dfrac{4c-3}{7c} = \dfrac{2c+1-(4c-3)}{7c} = \dfrac{2c+1-4c+3}{7c} = \dfrac{-2c+4}{7c}$

28. $\dfrac{3w+2}{w+3} - \dfrac{w-4}{w+3} = \dfrac{3w+2-(w-4)}{w+3} = \dfrac{3w+2-w+4}{w+3} = \dfrac{2w+6}{w+3} = \dfrac{2(w+3)}{w+3} = 2$

Page 322 · WRITTEN EXERCISES

A
1. $2 \cdot 4 = 8$
2. $6 \cdot 2 = 12$
3. $4 \cdot 4 = 16$
4. $3 \cdot y = 3y$
5. $4 \cdot x = 4x$
6. $3 \cdot x = 3x$
7. $3 \cdot y = 3y$
8. $3 \cdot 4a = 12a$
9. $2 \cdot 3x = 6x$
10. $2 \cdot 4 = 8$
11. $3 \cdot 3 = 9$
12. $3 \cdot 2 = 6$
13. $2a \cdot 3 = 6a$
14. $3x \cdot 5 = 15x$
15. $2y \cdot 2 = 4y$
16. $2b \cdot 4 = 8b$
17. $3y \cdot 5 = 15y$
18. $4b \cdot 3 = 12b$
19. $4m \cdot 2 = 8m$
20. $4s \cdot 1 = 4s$
21. $x \cdot 5 = 5x$
22. $b \cdot 3a = 3ab$
23. $b \cdot 2 = 2b$
24. $2x \cdot 3 = 6x$

Page 322 · PUZZLE PROBLEMS

The man borrowed one horse from a neighbor so that he then had 18 horses. The ratios $\frac{1}{2}, \frac{1}{3}$, and $\frac{1}{9}$ can be expressed as $\frac{9}{18}, \frac{6}{18}$, and $\frac{2}{18}$ respectively.   Floyd then received 9 horses, Denise 6 horses, Harriet 2 horses, and the borrowed horse was returned to the neighbor.

Page 324 · WRITTEN EXERCISES

A
1. $\dfrac{1}{2} + \dfrac{1}{3} = \dfrac{3}{6} + \dfrac{2}{6} = \dfrac{5}{6}$

2. $\dfrac{x}{5} + \dfrac{x}{3} = \dfrac{3x}{15} + \dfrac{5x}{15} = \dfrac{8x}{15}$

3. $\dfrac{y}{4} + \dfrac{y}{8} = \dfrac{2y}{8} + \dfrac{y}{8} = \dfrac{3y}{8}$

4. $\dfrac{4x}{9} - \dfrac{x}{6} = \dfrac{8x}{18} - \dfrac{3x}{18} = \dfrac{5x}{18}$

5. $\dfrac{2x}{3} + \dfrac{x}{2} = \dfrac{4x}{6} + \dfrac{3x}{6} = \dfrac{7x}{6}$

6. $\dfrac{3x}{2} + \dfrac{x}{4} = \dfrac{6x}{4} + \dfrac{x}{4} = \dfrac{7x}{4}$

7. $\dfrac{4x}{3} - \dfrac{x}{6} = \dfrac{8x}{6} - \dfrac{x}{6} = \dfrac{7x}{6}$

8. $\dfrac{a}{2} + \dfrac{2a}{3} = \dfrac{3a}{6} + \dfrac{4a}{6} = \dfrac{7a}{6}$

9. $\dfrac{3}{2a} + \dfrac{1}{a} = \dfrac{3}{2a} + \dfrac{2}{2a} = \dfrac{5}{2a}$

10. $\dfrac{5}{2x} - \dfrac{1}{4x} = \dfrac{10}{4x} - \dfrac{1}{4x} = \dfrac{9}{4x}$

11. $\dfrac{4}{3y} - \dfrac{7}{2y} = \dfrac{8}{6y} - \dfrac{21}{6y} = -\dfrac{13}{6y}$

12. $\dfrac{4}{3a} + \dfrac{1}{4a} = \dfrac{16}{12a} + \dfrac{3}{12a} = \dfrac{19}{12a}$

13. $\dfrac{3x}{4} + \dfrac{x}{3} = \dfrac{9x}{12} + \dfrac{4x}{12} = \dfrac{13x}{12}$

14. $\dfrac{a}{3} - \dfrac{2a}{7} = \dfrac{7a}{21} - \dfrac{6a}{21} = \dfrac{a}{21}$

15. $\dfrac{5}{x} + \dfrac{1}{2x} = \dfrac{10}{2x} + \dfrac{1}{2x} = \dfrac{11}{2x}$

16. $\dfrac{4}{a} + \dfrac{3}{2a} = \dfrac{8}{2a} + \dfrac{3}{2a} = \dfrac{11}{2a}$

17. $\dfrac{5}{n} + \dfrac{3}{4n} = \dfrac{20}{4n} + \dfrac{3}{4n} = \dfrac{23}{4n}$

18. $\dfrac{7}{3n} - \dfrac{4}{9n} = \dfrac{21}{9n} - \dfrac{4}{9n} = \dfrac{17}{9n}$

19. $\dfrac{3x}{8} - \dfrac{x}{12} = \dfrac{9x}{24} - \dfrac{2x}{24} = \dfrac{7x}{24}$

20. $\dfrac{1}{9y} - \dfrac{1}{15y} = \dfrac{5}{45y} - \dfrac{3}{45y} = \dfrac{2}{45y}$

21. $\dfrac{1}{5} - \dfrac{3}{x} = \dfrac{x}{5x} - \dfrac{15}{5x} = \dfrac{x - 15}{5x}$

22. $\dfrac{3}{4} + \dfrac{1}{a} = \dfrac{3a}{4a} + \dfrac{4}{4a} = \dfrac{3a + 4}{4a}$

23. $\dfrac{4}{x} + \dfrac{1}{2} = \dfrac{8}{2x} + \dfrac{x}{2x} = \dfrac{8 + x}{2x}$

24. $\dfrac{1}{a} + \dfrac{3}{2ab} = \dfrac{2b}{2ab} + \dfrac{3}{2ab} = \dfrac{2b + 3}{2ab}$

25. $\dfrac{2}{3x} + \dfrac{1}{6xy} = \dfrac{4y}{6xy} + \dfrac{1}{6xy} = \dfrac{4y + 1}{6xy}$

26. $\dfrac{1}{2x} - \dfrac{1}{3xy} = \dfrac{3y}{6xy} - \dfrac{2}{6xy} = \dfrac{3y - 2}{6xy}$

27. $\dfrac{2}{x} + \dfrac{3}{y} = \dfrac{2y}{xy} + \dfrac{3x}{xy} = \dfrac{2y + 3x}{xy}$

28. $\dfrac{3}{b} - \dfrac{1}{2b^2} = \dfrac{6b}{2b^2} - \dfrac{1}{2b^2} = \dfrac{6b - 1}{2b^2}$

B **29.** $\dfrac{3x}{4} - \dfrac{2x}{3} + \dfrac{x}{6} = \dfrac{9x}{12} - \dfrac{8x}{12} + \dfrac{2x}{12} = \dfrac{3x}{12} = \dfrac{x}{4}$

30. $\dfrac{a}{3} + \dfrac{2a}{6} - \dfrac{5a}{2} = \dfrac{2a}{6} + \dfrac{2a}{6} - \dfrac{15a}{6} = -\dfrac{11a}{6}$

31. $\dfrac{2x}{5} - \dfrac{1}{4} + \dfrac{3x}{5} = \dfrac{8x}{20} - \dfrac{5}{20} + \dfrac{12x}{20} = \dfrac{20x - 5}{20} = \dfrac{5(4x - 1)}{20} = \dfrac{4x - 1}{4}$

32. $\dfrac{3a}{a} + \dfrac{a}{a^2} - \dfrac{2a}{3a} = \dfrac{9a^2}{3a^2} + \dfrac{3a}{3a^2} - \dfrac{2a^2}{3a^2} = \dfrac{7a^2 + 3a}{3a^2} = \dfrac{a(7a + 3)}{3a^2} = \dfrac{7a + 3}{3a}$

33. $\dfrac{x}{2x} + \dfrac{3x}{x} - \dfrac{5}{3x} = \dfrac{3x}{6x} + \dfrac{18x}{6x} - \dfrac{10}{6x} = \dfrac{21x - 10}{6x}$

34. $\dfrac{6}{y^2} - \dfrac{5x}{3y} + \dfrac{4x}{6xy^2} = \dfrac{36x}{6xy^2} - \dfrac{10x^2y}{6xy^2} + \dfrac{4x}{6xy^2} = \dfrac{40x - 10x^2y}{6xy^2} = \dfrac{10x(4 - xy)}{6xy^2} =$

$\dfrac{5(4 - xy)}{3y^2} = \dfrac{20 - 5xy}{3y^2}$

Page 326 • WRITTEN EXERCISES

A **1.** $\dfrac{x + 1}{8} + \dfrac{3}{4} = \dfrac{x + 1}{8} + \dfrac{6}{8} = \dfrac{x + 1 + 6}{8} = \dfrac{x + 7}{8}$

2. $\dfrac{2k - 1}{6} - \dfrac{k + 1}{9} = \dfrac{3(2k - 1)}{18} - \dfrac{2(k + 1)}{18} = \dfrac{6k - 3 - 2k - 2}{18} = \dfrac{4k - 5}{18}$

3. $\dfrac{3t - 2}{6} + \dfrac{2t - 1}{4} = \dfrac{2(3t - 2)}{12} + \dfrac{3(2t - 1)}{12} = \dfrac{6t - 4 + 6t - 3}{12} = \dfrac{12t - 7}{12}$

4. $\dfrac{x + 2}{3} + \dfrac{x}{5} = \dfrac{5(x + 2)}{15} + \dfrac{3x}{15} = \dfrac{5x + 10 + 3x}{15} = \dfrac{8x + 10}{15}$

5. $\dfrac{x - 5}{4} + \dfrac{2x}{3} = \dfrac{3(x - 5)}{12} + \dfrac{4(2x)}{12} = \dfrac{3x - 15 + 8x}{12} = \dfrac{11x - 15}{12}$

6. $\dfrac{y - 9}{4} + \dfrac{3y}{8} = \dfrac{2(y - 9)}{8} + \dfrac{3y}{8} = \dfrac{2y - 18 + 3y}{8} = \dfrac{5y - 18}{8}$

7. $\dfrac{a - 2}{4} - \dfrac{a}{12} = \dfrac{3(a - 2)}{12} - \dfrac{a}{12} = \dfrac{3a - 6 - a}{12} = \dfrac{2a - 6}{12} = \dfrac{2(a - 3)}{12} = \dfrac{a - 3}{6}$

8. $\dfrac{a + 3b}{8} - \dfrac{a}{2} = \dfrac{a + 3b}{8} - \dfrac{4a}{8} = \dfrac{a + 3b - 4a}{8} = \dfrac{3b - 3a}{8}$

9. $\dfrac{x - 3}{4} - \dfrac{5x}{6} = \dfrac{3(x - 3)}{12} - \dfrac{2(5x)}{12} = \dfrac{3x - 9 - 10x}{12} = \dfrac{-7x - 9}{12}$

10. $\dfrac{x + 2}{3} + \dfrac{x + 1}{2} = \dfrac{2(x + 2)}{6} + \dfrac{3(x + 1)}{6} = \dfrac{2x + 4 + 3x + 3}{6} = \dfrac{5x + 7}{6}$

11. $\dfrac{x - 1}{2} + \dfrac{x + 3}{4} = \dfrac{2(x - 1)}{4} + \dfrac{x + 3}{4} = \dfrac{2x - 2 + x + 3}{4} = \dfrac{3x + 1}{4}$

12. $\dfrac{a - 5}{3} + \dfrac{a - 1}{6} = \dfrac{2(a - 5)}{6} + \dfrac{a - 1}{6} = \dfrac{2a - 10 + a - 1}{6} = \dfrac{3a - 11}{6}$

13. $\dfrac{x+5}{3} + 1 = \dfrac{x+5}{3} + \dfrac{3}{3} = \dfrac{x+5+3}{3} = \dfrac{x+8}{3}$

14. $\dfrac{2a+1}{5} - 2 = \dfrac{2a+1}{5} - \dfrac{10}{5} = \dfrac{2a+1-10}{5} = \dfrac{2a-9}{5}$

15. $\dfrac{4y-4}{3} + 8 = \dfrac{4y-4}{3} + \dfrac{24}{3} = \dfrac{4y-4+24}{3} = \dfrac{4y+20}{3}$

16. $\dfrac{4y+4}{4} - \dfrac{y-1}{2} = \dfrac{4y+4}{4} - \dfrac{2(y-1)}{4} = \dfrac{4y+4-2y+2}{4} = \dfrac{2y+6}{4} = \dfrac{2(y+3)}{4} = \dfrac{y+3}{2}$

17. $\dfrac{4x+2}{2} - \dfrac{x-2}{3} = \dfrac{3(4x+2)}{6} - \dfrac{2(x-2)}{6} = \dfrac{12x+6-2x+4}{6} = \dfrac{10x+10}{6} =$

$\dfrac{10(x+1)}{6} = \dfrac{5(x+1)}{3} = \dfrac{5x+5}{3}$

18. $\dfrac{2b-2}{4} - \dfrac{b-4}{3} = \dfrac{3(2b-2)}{12} - \dfrac{4(b-4)}{12} = \dfrac{6b-6-4b+16}{12} = \dfrac{2b+10}{12} =$

$\dfrac{2(b+5)}{12} = \dfrac{b+5}{6}$

B 19. $\dfrac{r}{4} + \dfrac{r-2}{8} - \dfrac{r-4}{16} = \dfrac{4r}{16} + \dfrac{2(r-2)}{16} - \dfrac{r-4}{16} = \dfrac{4r+2r-4-r+4}{16} = \dfrac{5r}{16}$

20. $\dfrac{4+3a}{6} - \dfrac{5a}{18} + \dfrac{3-a}{12} = \dfrac{6(4+3a)}{36} - \dfrac{2(5a)}{36} + \dfrac{3(3-a)}{36} = \dfrac{24+18a-10a+9-3a}{36} =$

$\dfrac{5a+33}{36}$

21. $\dfrac{10x+4}{5} - \dfrac{6x+2}{3} - \dfrac{2}{15} = \dfrac{3(10x+4)}{15} - \dfrac{5(6x+2)}{15} - \dfrac{2}{15} = \dfrac{30x+12-30x-10-2}{15} =$

$\dfrac{0}{15} = 0$

22. $\dfrac{x+1}{8x} + \dfrac{3x-1}{12x} - \dfrac{1}{6x} = \dfrac{3(x+1)}{24x} + \dfrac{2(3x-1)}{24x} - \dfrac{4}{24x} = \dfrac{3x+3+6x-2-4}{24x} =$

$\dfrac{9x-3}{24x} = \dfrac{3(3x-1)}{24x} = \dfrac{3x-1}{8x}$

Page 326 • PUZZLE PROBLEMS

Let x = his age at death.   Then $\dfrac{1}{6}x + \dfrac{1}{12}x + \dfrac{1}{7}x + 5 + \dfrac{1}{2}x + 4 = x$;

$\dfrac{14x}{84} + \dfrac{7x}{84} + \dfrac{12x}{84} + \dfrac{420}{84} + \dfrac{42x}{84} + \dfrac{336}{84} = x$; $\dfrac{75x+756}{84} = \dfrac{x}{1}$; $75x + 756 = 84x$;

$756 = 9x$; $84 = x$.   He lived to be 84 years old.

Page 328 • WRITTEN EXERCISES

A 1. $\dfrac{x}{3} - \dfrac{x}{5} = 4$; $\dfrac{5x}{15} - \dfrac{3x}{15} = 4$; $\dfrac{2x}{15} = \dfrac{4}{1}$; $2x = 60$; $x = 30$

2. $\dfrac{x}{6} + \dfrac{x}{7} = 13$; $\dfrac{7x}{42} + \dfrac{6x}{42} = 13$; $\dfrac{13x}{42} = \dfrac{13}{1}$; $13x = 546$; $x = 42$

3. $\dfrac{x}{6} + \dfrac{x}{3} = 6$; $\dfrac{x}{6} + \dfrac{2x}{6} = 6$; $\dfrac{3x}{6} = \dfrac{6}{1}$; $3x = 36$; $x = 12$

4. $\dfrac{1}{4} + \dfrac{1}{12} = \dfrac{x}{3}$; $\dfrac{3}{12} + \dfrac{1}{12} = \dfrac{x}{3}$; $\dfrac{4}{12} = \dfrac{x}{3}$; $12x = 12$; $x = 1$

5. $\dfrac{2}{3} - \dfrac{1}{6} = \dfrac{x}{4}$; $\dfrac{4}{6} - \dfrac{1}{6} = \dfrac{x}{4}$; $\dfrac{3}{6} = \dfrac{x}{4}$; $6x = 12$; $x = 2$

6. $\dfrac{4}{5} - \dfrac{2}{3} = \dfrac{n}{15}$; $\dfrac{12}{15} - \dfrac{10}{15} = \dfrac{n}{15}$; $\dfrac{2}{15} = \dfrac{n}{15}$; $15n = 30$; $n = 2$

7. $\dfrac{2}{5} + \dfrac{3}{10} = \dfrac{7}{n}$; $\dfrac{4}{10} + \dfrac{3}{10} = \dfrac{7}{n}$; $\dfrac{7}{10} = \dfrac{7}{n}$; $7n = 70$; $n = 10$

8. $\dfrac{4}{5} - \dfrac{2}{15} = \dfrac{6}{x}$; $\dfrac{12}{15} - \dfrac{2}{15} = \dfrac{6}{x}$; $\dfrac{10}{15} = \dfrac{6}{x}$; $10x = 90$; $x = 9$

9. $\dfrac{3}{4} - \dfrac{5}{6} = \dfrac{5}{y}$; $\dfrac{9}{12} - \dfrac{10}{12} = \dfrac{5}{y}$; $\dfrac{-1}{12} = \dfrac{5}{y}$; $-y = 60$; $y = -60$

10. $\dfrac{n}{2} + \dfrac{2n}{3} = 7$; $\dfrac{3n}{6} + \dfrac{4n}{6} = 7$; $\dfrac{7n}{6} = \dfrac{7}{1}$; $7n = 42$; $n = 6$

11. $\dfrac{a}{6} - \dfrac{2a}{9} = \dfrac{1}{2}$; $\dfrac{3a}{18} - \dfrac{4a}{18} = \dfrac{1}{2}$; $\dfrac{-a}{18} = \dfrac{1}{2}$; $-2a = 18$; $a = -9$

12. $\dfrac{1}{c} + \dfrac{3}{2c} = \dfrac{1}{6}$; $\dfrac{2}{2c} + \dfrac{3}{2c} = \dfrac{1}{6}$; $\dfrac{5}{2c} = \dfrac{1}{6}$; $2c = 30$; $c = 15$

13. $\dfrac{3}{4n} + \dfrac{1}{n} = \dfrac{7}{8}$; $\dfrac{3}{4n} + \dfrac{4}{4n} = \dfrac{7}{8}$; $\dfrac{7}{4n} = \dfrac{7}{8}$; $28n = 56$; $n = 2$

14. $\dfrac{3n}{7} - \dfrac{n}{3} = 4$; $\dfrac{9n}{21} - \dfrac{7n}{21} = 4$; $\dfrac{2n}{21} = \dfrac{4}{1}$; $2n = 84$; $n = 42$

15. $\dfrac{2}{x} - \dfrac{4}{3x} = \dfrac{2}{9}$; $\dfrac{6}{3x} - \dfrac{4}{3x} = \dfrac{2}{9}$; $\dfrac{2}{3x} = \dfrac{2}{9}$; $6x = 18$; $x = 3$

B 16. $\dfrac{a+2}{3} + \dfrac{a-1}{6} = 5$; $\dfrac{2(a+2)}{6} + \dfrac{a-1}{6} = 5$; $\dfrac{2a+4+a-1}{6} = 5$; $\dfrac{3a+3}{6} = \dfrac{5}{1}$;

$3a + 3 = 30$; $3a = 27$; $a = 9$

17. $\dfrac{x-1}{4} - \dfrac{2x-3}{4} = 5$; $\dfrac{x-1-2x+3}{4} = 5$; $\dfrac{-x+2}{4} = \dfrac{5}{1}$; $-x + 2 = 20$;

$-x = 18$; $x = -18$

18. $\dfrac{x}{4} - \dfrac{x+5}{3} = 6$; $\dfrac{3x}{12} - \dfrac{4(x+5)}{12} = 6$; $\dfrac{3x-4x-20}{12} = 6$; $\dfrac{-x-20}{12} = \dfrac{6}{1}$;

$-x - 20 = 72$; $-x = 92$; $x = -92$

19. $\dfrac{x+3}{8} - \dfrac{x-2}{6} = 1$; $\dfrac{3(x+3)}{24} - \dfrac{4(x-2)}{24} = 1$; $\dfrac{3x+9-4x+8}{24} = 1$; $\dfrac{-x+17}{24} =$

$\dfrac{1}{1}$; $-x + 17 = 24$; $-x = 7$; $x = -7$

20. $\dfrac{c-1}{2c} + \dfrac{c+3}{4c} = \dfrac{5}{8}$; $\dfrac{2(c-1)}{4c} + \dfrac{c+3}{4c} = \dfrac{5}{8}$; $\dfrac{2c-2+c+3}{4c} = \dfrac{5}{8}$; $\dfrac{3c+1}{4c} = \dfrac{5}{8}$;

$8(3c + 1) = 5 \cdot 4c$; $24c + 8 = 20c$; $8 = -4c$; $-2 = c$

21. $\dfrac{2-x}{x} - \dfrac{x+3}{3x} = \dfrac{-1}{3}$; $\dfrac{3(2-x)}{3x} - \dfrac{x+3}{3x} = \dfrac{-1}{3}$; $\dfrac{6-3x-x-3}{3x} = \dfrac{-1}{3}$; $\dfrac{-4x+3}{3x} =$

$\dfrac{-1}{3}$; $3(-4x + 3) = -3x$; $-12x + 9 = -3x$; $9 = 9x$; $1 = x$

C **22.** $\dfrac{6x - 4}{3} - 2 = \dfrac{18 - 4x}{3} + x$; $\dfrac{6x - 4}{3} - \dfrac{6}{3} = \dfrac{18 - 4x}{3} + \dfrac{3x}{3}$; $\dfrac{6x - 10}{3} = \dfrac{18 - x}{3}$;

$3(6x - 10) = 3(18 - x)$; $6x - 10 = 18 - x$; $6x = 28 - x$; $7x = 28$; $x = 4$

23. $2 - \dfrac{7x - 1}{6} = 3x - \dfrac{19x + 3}{4}$; $\dfrac{12}{6} - \dfrac{7x - 1}{6} = \dfrac{12x}{4} - \dfrac{19x + 3}{4}$; $\dfrac{-7x + 13}{6} = \dfrac{-7x - 3}{4}$;

$4(-7x + 13) = 6(-7x - 3)$; $-28x + 52 = -42x - 18$; $14x + 52 = -18$;

$14x = -70$; $x = -5$

Pages 330–332 · WRITTEN EXERCISES

A **1.** $\dfrac{1}{10} + \dfrac{1}{15} = \dfrac{1}{n}$; $\dfrac{3}{30} + \dfrac{2}{30} = \dfrac{1}{n}$; $\dfrac{5}{30} = \dfrac{1}{n}$; $5n = 30$; $n = 6$.   6 hours

2.

	Milo	Jean	Together
Hours needed	3	2	h
Part done in one hour	$\dfrac{1}{3}$	$\dfrac{1}{2}$	$\dfrac{1}{h}$

$\dfrac{1}{3} + \dfrac{1}{2} = \dfrac{1}{h}$; $\dfrac{2}{6} + \dfrac{3}{6} = \dfrac{1}{h}$; $\dfrac{5}{6} =$

$\dfrac{1}{h}$; $5h = 6$; $h = 1\dfrac{1}{5}$. $1\dfrac{1}{5}$ hours

3.

	Carpenter	Apprentice	Together
Hours needed	2	4	x
Part done in one hour	$\dfrac{1}{2}$	$\dfrac{1}{4}$	$\dfrac{1}{x}$

$\dfrac{1}{2} + \dfrac{1}{4} = \dfrac{1}{x}$; $\dfrac{2}{4} +$

$\dfrac{1}{4} = \dfrac{1}{x}$; $\dfrac{3}{4} = \dfrac{1}{x}$;

$3x = 4$; $x = 1\dfrac{1}{3}$.

$1\dfrac{1}{3}$ hours

4.

	Cold	Hot	Together
Minutes needed	6	9	m
Part done in one minute	$\dfrac{1}{6}$	$\dfrac{1}{9}$	$\dfrac{1}{m}$

$\dfrac{1}{6} + \dfrac{1}{9} = \dfrac{1}{m}$; $\dfrac{3}{18} + \dfrac{2}{18} = \dfrac{1}{m}$;

$\dfrac{5}{18} = \dfrac{1}{m}$; $5m = 18$;

$m = 3\dfrac{3}{5}$. $3\dfrac{3}{5}$ minutes

5.

	Corrie	Gary	Together
Hours needed	8	10	h
Part done in one hour	$\dfrac{1}{8}$	$\dfrac{1}{10}$	$\dfrac{1}{h}$

$\dfrac{1}{8} + \dfrac{1}{10} = \dfrac{1}{h}$; $\dfrac{5}{40} + \dfrac{4}{40} = \dfrac{1}{h}$;

$\dfrac{9}{40} = \dfrac{1}{h}$; $9h = 40$; $h = 4\dfrac{4}{9}$.

$4\dfrac{4}{9}$ hours

6.

	Eugene	Robot	Together
Hours needed	3	4	h
Part done in one hour	$\dfrac{1}{3}$	$\dfrac{1}{4}$	$\dfrac{1}{h}$

$\dfrac{1}{3} + \dfrac{1}{4} = \dfrac{1}{h}$; $\dfrac{4}{12} + \dfrac{3}{12} =$

$\dfrac{1}{h}$; $\dfrac{7}{12} = \dfrac{1}{h}$; $7h = 12$;

$h = 1\dfrac{5}{7}$. $1\dfrac{5}{7}$ hours

7. $\frac{1}{10} + \frac{1}{10} + \frac{1}{8} = \frac{1}{x}$; $\frac{4}{40} + \frac{4}{40} + \frac{5}{40} = \frac{1}{x}$; $\frac{13}{40} = \frac{1}{x}$; $13x = 40$; $x = 3\frac{1}{13}$.

$3\frac{1}{13}$ hours

8.

	Celia	Al	Julio	Together
Hours needed	4	4	3	n
Part done in one hour	$\frac{1}{4}$	$\frac{1}{4}$	$\frac{1}{3}$	$\frac{1}{n}$

$\frac{1}{4} + \frac{1}{4} + \frac{1}{3} = \frac{1}{n}$; $\frac{3}{12} +$
$\frac{3}{12} + \frac{4}{12} = \frac{1}{n}$; $\frac{10}{12} = \frac{1}{n}$;
$10n = 12$; $n = 1\frac{2}{10} = 1\frac{1}{5}$.

$1\frac{1}{5}$ hours

B 9.

	Pat	Temporary	Together
Hours needed	6	x	4
Part done in one hour	$\frac{1}{6}$	$\frac{1}{x}$	$\frac{1}{4}$

$\frac{1}{6} + \frac{1}{x} = \frac{1}{4}$; $\frac{x}{6x} + \frac{6}{6x} = \frac{1}{4}$;
$\frac{x + 6}{6x} = \frac{1}{4}$; $4(x + 6) = 6x$;
$4x + 24 = 6x$; $24 = 2x$;
$12 = x$. 12 hours

10.

	Frank	Paula	Together
Hours needed	10	n	6
Part done in one hour	$\frac{1}{10}$	$\frac{1}{n}$	$\frac{1}{6}$

$\frac{1}{10} + \frac{1}{n} = \frac{1}{6}$; $\frac{n}{10n} + \frac{10}{10n} = \frac{1}{6}$;
$\frac{n + 10}{10n} = \frac{1}{6}$; $6(n + 10) = 10n$;
$6n + 60 = 10n$; $60 = 4n$;
$15 = n$. 15 hours

11.

	Dana	Jack	Together
Hours needed	$2x$	x	4
Part done in one hour	$\frac{1}{2x}$	$\frac{1}{x}$	$\frac{1}{4}$

$\frac{1}{2x} + \frac{1}{x} = \frac{1}{4}$; $\frac{1}{2x} + \frac{2}{2x} = \frac{1}{4}$;
$\frac{3}{2x} = \frac{1}{4}$; $2x = 12$; $x = 6$ and
$2x = 12$. Dana: 12 hours;
Jack: 6 hours

C 12. The boss alone can complete $\frac{1}{6}$ of the job in one hour. Thus, in 2 hours she

completed $2 \cdot \frac{1}{6} = \frac{1}{3}$ of the job. For the *remainder* of the job $\left(\frac{2}{3}\right)$, we have the

following:

	Boss	Assistant	Together
Hours needed	4	x	3
Part done in one hour	$\frac{1}{4}$	$\frac{1}{x}$	$\frac{1}{3}$

Note: The boss needs only
4 hours to complete $\frac{2}{3}$ of the
job, not 6.

$$\frac{1}{4} + \frac{1}{x} = \frac{1}{3}; \quad \frac{x}{4x} + \frac{4}{4x} = \frac{1}{3}; \quad \frac{x + 4}{4x} = \frac{1}{3}; \quad 3(x + 4) = 4x; \quad 3x + 12 = 4x;$$

$12 = x$. Thus, it takes the assistant alone 12 hours to complete $\frac{2}{3}$ of the job.

If t = the time to complete the whole job, then $\frac{2}{3}t = 12$, and $t = 18$.   18 hours

Page 333 · CALCULATOR ACTIVITIES

1. $x = \dfrac{10 \cdot 13}{20} = 6.5$

2. $x = \dfrac{4 \cdot 45}{10} = 18$

3. $x = \dfrac{5 \cdot 18}{2 \cdot 15} = 3$

4. $x = \dfrac{15 \cdot 8}{10 \cdot 3} = 4$

5. $6(x + 1) = 3 \cdot 16; \quad x + 1 = \dfrac{3 \cdot 16}{6}; \quad x = \dfrac{3 \cdot 16}{6} - 1 = 7$

6. $12(4 - x) = 4 \cdot 3; \quad 4 - x = \dfrac{4 \cdot 3}{12}; \quad 4 - \dfrac{4 \cdot 3}{12} = x; \quad x = 3$

Page 335 · WRITTEN EXERCISES

A

1. $\dfrac{5}{k} + \dfrac{3}{k + 1} = \dfrac{5(k + 1)}{k(k + 1)} + \dfrac{3k}{k(k + 1)} = \dfrac{5k + 5 + 3k}{k(k + 1)} = \dfrac{8k + 5}{k(k + 1)}$

2. $\dfrac{2}{t} + \dfrac{3}{t + 5} = \dfrac{2(t + 5)}{t(t + 5)} + \dfrac{3t}{t(t + 5)} = \dfrac{2t + 10 + 3t}{t(t + 5)} = \dfrac{5t + 10}{t(t + 5)} = \dfrac{5(t + 2)}{t(t + 5)}$

3. $\dfrac{2}{x} - \dfrac{5}{x - 3} = \dfrac{2(x - 3)}{x(x - 3)} - \dfrac{5x}{x(x - 3)} = \dfrac{2x - 6 - 5x}{x(x - 3)} = \dfrac{-3x - 6}{x(x - 3)} = \dfrac{-3(x + 2)}{x(x - 3)}$

4. $\dfrac{r}{r + 5} - \dfrac{r}{r + 2} = \dfrac{r(r + 2)}{(r + 5)(r + 2)} - \dfrac{r(r + 5)}{(r + 5)(r + 2)} = \dfrac{r^2 + 2r - r^2 - 5r}{(r + 5)(r + 2)} = \dfrac{-3r}{(r + 5)(r + 2)}$

5. $\dfrac{a}{a + 4} - \dfrac{a}{a - 4} = \dfrac{a(a - 4)}{(a + 4)(a - 4)} - \dfrac{a(a + 4)}{(a + 4)(a - 4)} = \dfrac{a^2 - 4a - a^2 - 4a}{(a + 4)(a - 4)} =$

   $\dfrac{-8a}{(a + 4)(a - 4)}$

6. $\dfrac{9}{b + 3} - \dfrac{9}{b - 3} = \dfrac{9(b - 3)}{(b + 3)(b - 3)} - \dfrac{9(b + 3)}{(b + 3)(b - 3)} = \dfrac{9b - 27 - 9b - 27}{(b + 3)(b - 3)} = \dfrac{-54}{(b + 3)(b - 3)}$

7. $\dfrac{6}{y + 3} - \dfrac{y}{y - 2} = \dfrac{6(y - 2)}{(y + 3)(y - 2)} - \dfrac{y(y + 3)}{(y + 3)(y - 2)} = \dfrac{6y - 12 - y^2 - 3y}{(y + 3)(y - 2)} =$

   $\dfrac{-y^2 + 3y - 12}{(y + 3)(y - 2)}$

8. $\dfrac{c}{c - 4} - \dfrac{5}{c + 5} = \dfrac{c(c + 5)}{(c - 4)(c + 5)} - \dfrac{5(c - 4)}{(c - 4)(c + 5)} = \dfrac{c^2 + 5c - 5c + 20}{(c - 4)(c + 5)} = \dfrac{c^2 + 20}{(c - 4)(c + 5)}$

9. $\dfrac{7}{x + 3} + \dfrac{x}{x - 3} = \dfrac{7(x - 3)}{(x + 3)(x - 3)} + \dfrac{x(x + 3)}{(x + 3)(x - 3)} = \dfrac{7x - 21 + x^2 + 3x}{(x + 3)(x - 3)} =$

   $\dfrac{x^2 + 10x - 21}{(x + 3)(x - 3)}$

10. $\dfrac{6}{a+4} + \dfrac{3}{a-1} = \dfrac{6(a-1)}{(a+4)(a-1)} + \dfrac{3(a+4)}{(a+4)(a-1)} = \dfrac{6a-6+3a+12}{(a+4)(a-1)} =$

$\dfrac{9a+6}{(a+4)(a-1)} = \dfrac{3(3a+2)}{(a+4)(a-1)}$

11. $\dfrac{b}{b+6} + \dfrac{2b}{(b-1)} = \dfrac{b(b-1)}{(b+6)(b-1)} + \dfrac{2b(b+6)}{(b+6)(b-1)} = \dfrac{b^2-b+2b^2+12b}{(b+6)(b-1)} =$

$\dfrac{3b^2+11b}{(b+6)(b-1)} = \dfrac{b(3b+11)}{(b+6)(b-1)}$

12. $\dfrac{a}{4-a} - \dfrac{a}{a-4} = \dfrac{a}{4-a} - \dfrac{a}{-(4-a)} = \dfrac{a}{4-a} + \dfrac{a}{4-a} = \dfrac{2a}{4-a}$

13. $\dfrac{7}{x+y} - \dfrac{x}{x-y} = \dfrac{7(x-y)}{(x+y)(x-y)} - \dfrac{x(x+y)}{(x+y)(x-y)} = \dfrac{7x-7y-x^2-xy}{(x+y)(x-y)}$

14. $\dfrac{s}{r-s} + \dfrac{r}{r+2} = \dfrac{s(r+2)}{(r-s)(r+2)} + \dfrac{r(r-s)}{(r-s)(r+2)} = \dfrac{rs+2s+r^2-rs}{(r-s)(r+2)} = \dfrac{r^2+2s}{(r-s)(r+2)}$

15. $\dfrac{4}{a+b} + \dfrac{2a}{a-b} = \dfrac{4(a-b)}{(a+b)(a-b)} + \dfrac{2a(a+b)}{(a+b)(a-b)} = \dfrac{4a-4b+2a^2+2ab}{(a+b)(a-b)}$

16. $\dfrac{3}{x+6} - \dfrac{11}{x^2-36} = \dfrac{3(x-6)}{(x+6)(x-6)} - \dfrac{11}{(x+6)(x-6)} = \dfrac{3x-18-11}{(x+6)(x-6)} = \dfrac{3x-29}{(x+6)(x-6)}$

17. $\dfrac{8}{(c+d)^2} + \dfrac{5}{(c+d)} = \dfrac{8}{(c+d)^2} + \dfrac{5(c+d)}{(c+d)^2} = \dfrac{8+5c+5d}{(c+d)^2}$

18. $\dfrac{12}{t+s} - \dfrac{4}{t^2+ts} = \dfrac{12t}{t(t+s)} - \dfrac{4}{t(t+s)} = \dfrac{12t-4}{t(t+s)} = \dfrac{4(3t-1)}{t(t+s)}$

19. $\dfrac{1}{x^2-y^2} + \dfrac{3}{x-y} = \dfrac{1}{(x+y)(x-y)} + \dfrac{3(x+y)}{(x+y)(x-y)} = \dfrac{1+3x+3y}{(x+y)(x-y)}$

20. $\dfrac{c}{c^2-9} + \dfrac{1}{c+3} = \dfrac{c}{(c+3)(c-3)} + \dfrac{c-3}{(c+3)(c-3)} = \dfrac{c+c-3}{(c+3)(c-3)} = \dfrac{2c-3}{(c+3)(c-3)}$

21. $\dfrac{2a+5}{4-a^2} + \dfrac{5}{a+2} = \dfrac{-(2a+5)}{a^2-4} + \dfrac{5}{a+2} = \dfrac{-2a-5}{(a+2)(a-2)} + \dfrac{5(a-2)}{(a+2)(a-2)} =$

$\dfrac{-2a-5+5a-10}{(a+2)(a-2)} = \dfrac{3a-15}{(a+2)(a-2)} = \dfrac{3(a-5)}{(a+2)(a-2)}$

B 22. $\dfrac{2}{a+1} - \dfrac{1}{a+3} = \dfrac{1}{a}$; $\dfrac{2(a+3)}{(a+1)(a+3)} - \dfrac{a+1}{(a+1)(a+3)} = \dfrac{1}{a}$; $\dfrac{2a+6-a-1}{(a+1)(a+3)} = \dfrac{1}{a}$;

$\dfrac{a+5}{(a+1)(a+3)} = \dfrac{1}{a}$; $a(a+5) = (a+1)(a+3)$; $a^2+5a = a^2+4a+3$;

$5a = 4a + 3$; $a = 3$

23. $\dfrac{3}{x+1} - \dfrac{2}{x-1} = \dfrac{1}{x}$; $\dfrac{3(x-1)}{(x+1)(x-1)} - \dfrac{2(x+1)}{(x+1)(x-1)} = \dfrac{1}{x}$; $\dfrac{3x-3-2x-2}{(x+1)(x-1)} = \dfrac{1}{x}$;

$\dfrac{x-5}{(x+1)(x-1)} = \dfrac{1}{x}$; $x(x-5) = (x+1)(x-1)$; $x^2-5x = x^2-1$; $-5x =$

-1 ; $x = \dfrac{1}{5}$

24. $\dfrac{2}{s+6} + \dfrac{1}{s-3} = \dfrac{3}{s}$; $\dfrac{2(s-3)}{(s+6)(s-3)} + \dfrac{s+6}{(s+6)(s-3)} = \dfrac{3}{s}$; $\dfrac{2s-6+s-6}{(s+6)(s-3)} =$

$\dfrac{3}{s}$; $\dfrac{3s}{(s+6)(s-3)} = \dfrac{3}{s}$; $s(3s) = 3(s+6)(s-3)$; $3s^2 = 3s^2 + 9s - 54$;

$0 = 9s - 54$; $-9s = -54$; $s = 6$

25. $\dfrac{1}{m-1} + \dfrac{1}{m+2} = \dfrac{2}{m}$; $\quad \dfrac{m+2}{(m-1)(m+2)} + \dfrac{m-1}{(m-1)(m+2)} = \dfrac{2}{m}$; $\quad \dfrac{m+2+m-1}{(m-1)(m+2)} =$

$\dfrac{2}{m}$; $\quad \dfrac{2m+1}{(m-1)(m+2)} = \dfrac{2}{m}$; $\quad m(2m+1) = 2(m-1)(m+2)$; $\quad 2m^2 + m = 2m^2 +$

$2m - 4$; $\quad m = 2m - 4$; $\quad -m = -4$; $\quad m = 4$

Page 337 · COMPUTER ACTIVITIES

1. WHAT COUNTRY? FRANCE
EXCHANGE RATE
PER DOLLAR? 6.88

U.S.	FRANCE
1	6.88
2	13.76
3	20.64
4	27.52
5	34.4
6	41.28
7	48.16
8	55.04
9	61.92
10	68.8
15	103.2
20	137.6
25	172

2. Change line 110 to: 110 FOR C = 25 TO 100 STEP 25
WHAT COUNTRY? MEXICO
EXCHANGE RATE
PER DOLLAR? 468

U.S.	MEXICO
1	468
2	936
3	1404
4	1872
5	2340
6	2808
7	3276
8	3744
9	4212
10	4680
25	11700
50	23400
75	35100
100	46800

3. Change line 110 to: 110 FOR C = 10 TO 100 STEP 10
WHAT COUNTRY? CANADA
EXCHANGE RATE
PER DOLLAR? 1.39

U.S.	CANADA
10	13.9
20	27.8
30	41.7
40	55.6
50	69.5
60	83.4
70	97.3
80	111.2
90	125.1
100	139

Page 339 · SKILLS REVIEW

1. $<$ 2. $>$ 3. $<$ 4. $>$ 5. $<$ 6. $>$ 7. $<$ 8. $<$ 9. 6.7   10. 5.8

11. 9.3   12. 7.7   13. 4.3   14. 0.65   15. 9.21   16. 3.42   17. 0.08   18. 5.92

Pages 340–341 · CHAPTER REVIEW EXERCISES

1. $\dfrac{8}{24} = \dfrac{1 \cdot 8}{3 \cdot 8} = \dfrac{1}{3}$

2. $\dfrac{3}{15x} = \dfrac{3}{3 \cdot 5 \cdot x} = \dfrac{1}{5x}$

3. $\dfrac{11x}{22} = \dfrac{11 \cdot x}{11 \cdot 2} = \dfrac{x}{2}$

4. $\dfrac{-4a^2 b}{-ab^2} = \dfrac{-4 \cdot a \cdot a \cdot b}{-1 \cdot a \cdot b \cdot b} = \dfrac{4a}{b}$

5. $\dfrac{16x^2 y^3}{8x^2 y} = \dfrac{2 \cdot 8 \cdot x \cdot x \cdot y \cdot y \cdot y}{8 \cdot x \cdot x \cdot y} = 2y^2$

6. $\dfrac{3a - 15}{3} = \dfrac{3(a - 5)}{3} = a - 5$

7. $\dfrac{x - 4}{3x - 12} = \dfrac{x - 4}{3(x - 4)} = \dfrac{1}{3}$

8. $\dfrac{n^2 + 4n + 4}{n + 2} = \dfrac{(n + 2)^2}{n + 2} = n + 2$

9. $\dfrac{(3 + x)^2}{9 - x^2} = \dfrac{(3 + x)(3 + x)}{(3 + x)(3 - x)} = \dfrac{3 + x}{3 - x}$

10. $\dfrac{3 - a}{a - 3} = \dfrac{-1(a - 3)}{a - 3} = \dfrac{-1}{1} = -1$

11. $\dfrac{x - y}{y - x} = \dfrac{x - y}{-1(x - y)} = \dfrac{1}{-1} = -1$

12. $\dfrac{1 - x}{2x - 2} = \dfrac{-1(x - 1)}{2(x - 1)} = \dfrac{-1}{2} = -\dfrac{1}{2}$

13. $\dfrac{7 - a}{a^2 - 49} = \dfrac{-1(a - 7)}{(a + 7)(a - 7)} = \dfrac{-1}{a + 7} = -\dfrac{1}{a + 7}$

14. Let $3x$ = cans of blue and $2x$ = cans of yellow. $3x + 2x = 10$; $5x = 10$; $x = 2$, so $3x = 3 \cdot 2 = 6$.   6 cans of blue

15. Let $2x$ = number of coaches and $9x$ = number of players. $2x + 9x = 22$; $11x = 22$; $x = 2$, so $9x = 9 \cdot 2 = 18$.   18 players

16. $\dfrac{x}{14} = \dfrac{2}{7}$; $x \cdot 7 = 14 \cdot 2$; $7x = 28$; $x = 4$

17. $\dfrac{2n}{6} = \dfrac{12}{9}$; $2n \cdot 9 = 6 \cdot 12$; $18n = 72$; $n = 4$

18. $\dfrac{x - 2}{x + 2} = \dfrac{1}{3}$; $(x - 2)3 = (x + 2)1$; $3x - 6 = x + 2$; $3x = x + 8$; $2x = 8$;
$x = 4$

19. $\dfrac{2b}{b-7} = \dfrac{1}{4}$; $2b \cdot 4 = (b-7)1$; $8b = b-7$; $7b = -7$; $b = -1$

20.

Number	18	36
Cost	$2.70	x

$\dfrac{18}{270} = \dfrac{36}{x}$; $18x = 270 \cdot 36$; $18x = 9720$; $x = 540$. $5.40

21.

Number	3	2
Cost	$1.95	x

$\dfrac{3}{195} = \dfrac{2}{x}$; $3x = 390$; $x = 130$. $1.30

22. $\dfrac{9}{21} \cdot \dfrac{7}{63} = \dfrac{9 \cdot 7}{21 \cdot 63} = \dfrac{1}{21}$

23. $\dfrac{x}{4} \cdot \dfrac{12}{x^2} = \dfrac{x \cdot 12}{4 \cdot x \cdot x} = \dfrac{3}{x}$

24. $\dfrac{10m^2}{3m} \cdot \dfrac{m}{5} = \dfrac{10 \cdot m \cdot m \cdot m}{3 \cdot m \cdot 5} = \dfrac{2m^2}{3}$

25. $\dfrac{5}{3x-3} \cdot \dfrac{3}{75} = \dfrac{5 \cdot 3}{3(x-1) \cdot 75} = \dfrac{1}{15(x-1)} = \dfrac{1}{15x-15}$

26. $\dfrac{3}{4} \div \dfrac{1}{12} = \dfrac{3}{4} \cdot \dfrac{12}{1} = 9$

27. $\dfrac{n}{3} \div \dfrac{n}{9} = \dfrac{n}{3} \cdot \dfrac{9}{n} = 3$

28. $\dfrac{5}{a^2b^2} \div \dfrac{25}{ab} = \dfrac{5}{a^2b^2} \cdot \dfrac{ab}{25} = \dfrac{5 \cdot a \cdot b}{a \cdot a \cdot b \cdot b \cdot 25} = \dfrac{1}{5ab}$

29. $\dfrac{6}{mn^2} \div \dfrac{3}{m^2n} = \dfrac{6}{mn^2} \cdot \dfrac{m^2n}{3} = \dfrac{6 \cdot m \cdot m \cdot n}{m \cdot n \cdot n \cdot 3} = \dfrac{2m}{n}$

30. $\dfrac{x^2-4}{25} \cdot \dfrac{5}{x+2} = \dfrac{5(x+2)(x-2)}{25(x+2)} = \dfrac{x-2}{5}$

31. $\dfrac{a+b}{28} \div \dfrac{a-b}{7} = \dfrac{a+b}{28} \cdot \dfrac{7}{a-b} = \dfrac{a+b}{4(a-b)} = \dfrac{a+b}{4a-4b}$

32. $\dfrac{3a-3}{15} \div \dfrac{a-1}{3} = \dfrac{3a-3}{15} \cdot \dfrac{3}{a-1} = \dfrac{3 \cdot 3(a-1)}{15(a-1)} = \dfrac{3}{5}$

33. $\dfrac{5x}{3} + \dfrac{x}{3} = \dfrac{5x+x}{3} = \dfrac{6x}{3} = 2x$

34. $\dfrac{n}{2} - \dfrac{4}{2} = \dfrac{n-4}{2}$

35. $\dfrac{7}{a+1} + \dfrac{8}{a+1} = \dfrac{7+8}{a+1} = \dfrac{15}{a+1}$

36. $\dfrac{6a}{a+2} - \dfrac{3a}{a+2} = \dfrac{6a-3a}{a+2} = \dfrac{3a}{a+2}$

37. $\dfrac{1}{3} + \dfrac{1}{5} = \dfrac{5}{15} + \dfrac{3}{15} = \dfrac{8}{15}$

38. $\dfrac{n}{3} - \dfrac{n}{5} = \dfrac{5n}{15} - \dfrac{3n}{15} = \dfrac{2n}{15}$

39. $\dfrac{2x}{7} + \dfrac{x}{2} = \dfrac{2 \cdot 2x}{14} + \dfrac{7x}{14} = \dfrac{4x+7x}{14} = \dfrac{11x}{14}$

40. $\dfrac{3a}{4} - \dfrac{a}{3} = \dfrac{3 \cdot 3a}{12} - \dfrac{4a}{12} = \dfrac{9a-4a}{12} = \dfrac{5a}{12}$

41. $\dfrac{1}{x} + \dfrac{1}{4} = \dfrac{4}{4x} + \dfrac{x}{4x} = \dfrac{4+x}{4x}$

42. $\dfrac{1}{3} - \dfrac{3}{x} = \dfrac{x}{3x} - \dfrac{3 \cdot 3}{3x} = \dfrac{x-9}{3x}$

43. $\dfrac{1}{2b} - \dfrac{4}{3b^2} = \dfrac{3b}{6b^2} - \dfrac{2 \cdot 4}{6b^2} = \dfrac{3b-8}{6b^2}$

44. $\dfrac{1}{x} - \dfrac{1}{y} = \dfrac{y}{xy} - \dfrac{x}{xy} = \dfrac{y-x}{xy}$

45. $\dfrac{x+3}{6} + \dfrac{1}{2} = \dfrac{x+3}{6} + \dfrac{3}{6} = \dfrac{x+3+3}{6} = \dfrac{x+6}{6}$

46. $\dfrac{2y-3}{5} - \dfrac{y+1}{10} = \dfrac{2(2y-3)}{10} - \dfrac{y+1}{10} = \dfrac{2(2y-3)-(y+1)}{10} = \dfrac{4y-6-y-1}{10} = \dfrac{3y-7}{10}$

47. $\dfrac{3z-2}{4} + \dfrac{z+5}{8} = \dfrac{2(3z-2)}{8} + \dfrac{z+5}{8} = \dfrac{2(3z-2)+(z+5)}{8} = \dfrac{6z-4+z+5}{8} = \dfrac{7z+1}{8}$

48. $\dfrac{1}{6} + \dfrac{1}{2} = \dfrac{x}{3}$; $\dfrac{1}{6} + \dfrac{3}{6} = \dfrac{x}{3}$; $\dfrac{4}{6} = \dfrac{x}{3}$; $6x = 12$; $x = 2$

49. $\dfrac{n}{3} - \dfrac{n}{4} = 1$; $\dfrac{4n}{12} - \dfrac{3n}{12} = 1$; $\dfrac{n}{12} = \dfrac{1}{1}$; $n = 12$

50. $\dfrac{3}{4x} - \dfrac{5}{2x} = 7$; $\dfrac{3}{4x} - \dfrac{2 \cdot 5}{4x} = 7$; $\dfrac{3 - 10}{4x} = 7$; $\dfrac{-7}{4x} = \dfrac{7}{1}$; $7 \cdot 4x = -7$; $28x = -7$;

$x = -\dfrac{7}{28} = -\dfrac{1}{4}$

51. $\dfrac{1}{n} + \dfrac{3}{4n} = \dfrac{1}{4}$; $\dfrac{4}{4n} + \dfrac{3}{4n} = \dfrac{1}{4}$; $\dfrac{7}{4n} = \dfrac{1}{4}$; $4n = 28$; $n = 7$

52.

	Elton	Paulette	Together
Hours needed	6	5	x
Part done in one hour	$\dfrac{1}{6}$	$\dfrac{1}{5}$	$\dfrac{1}{x}$

$\dfrac{1}{6} + \dfrac{1}{5} = \dfrac{1}{x}$; $\dfrac{5}{30} + \dfrac{6}{30} = \dfrac{1}{x}$;

$\dfrac{11}{30} = \dfrac{1}{x}$; $11x = 30$; $x = 2\dfrac{8}{11}$.

$2\dfrac{8}{11}$ hours

53. $\dfrac{3}{t} + \dfrac{2}{t - 2} = \dfrac{3(t - 2)}{t(t - 2)} + \dfrac{2t}{t(t - 2)} = \dfrac{3(t - 2) + 2t}{t(t - 2)} = \dfrac{3t - 6 + 2t}{t(t - 2)} = \dfrac{5t - 6}{t(t - 2)}$

54. $\dfrac{2}{x} + \dfrac{3}{x - 1} = \dfrac{2(x - 1)}{x(x - 1)} + \dfrac{3x}{x(x - 1)} = \dfrac{2(x - 1) + 3x}{x(x - 1)} = \dfrac{2x - 2 + 3x}{x(x - 1)} = \dfrac{5x - 2}{x(x - 1)}$

55. $\dfrac{4z}{z + 1} - \dfrac{z}{z - 1} = \dfrac{4z(z - 1)}{(z + 1)(z - 1)} - \dfrac{z(z + 1)}{(z + 1)(z - 1)} = \dfrac{4z(z - 1) - z(z + 1)}{(z + 1)(z - 1)} =$

$\dfrac{4z^2 - 4z - z^2 - z}{(z + 1)(z - 1)} = \dfrac{3z^2 - 5z}{(z + 1)(z - 1)} = \dfrac{z(3z - 5)}{(z + 1)(z - 1)}$

Pages 342–343 · CHAPTER TEST

1. $\dfrac{18x^3y}{-6xy^2} = \dfrac{3 \cdot 6 \cdot x \cdot x \cdot x \cdot y}{-6 \cdot x \cdot y \cdot y} = \dfrac{3x^2}{-y} = -\dfrac{3x^2}{y}$

2. $\dfrac{12ab^2}{16bc^2} = \dfrac{3 \cdot 4 \cdot a \cdot b \cdot b}{4 \cdot 4 \cdot b \cdot c \cdot c} = \dfrac{3ab}{4c^2}$

3. $\dfrac{-85a^3b^2c^2}{-34ab^3c^2} = \dfrac{-5 \cdot 17 \cdot a \cdot a \cdot a \cdot b \cdot b \cdot c \cdot c}{-2 \cdot 17 \cdot a \cdot b \cdot b \cdot b \cdot c \cdot c} = \dfrac{-5a^2}{-2b} = \dfrac{5a^2}{2b}$

4. $\dfrac{x^2 - 1}{6x + 6} = \dfrac{(x + 1)(x - 1)}{6(x + 1)} = \dfrac{x - 1}{6}$ **5.** $\dfrac{k^2 + 4k + 4}{14k + 28} = \dfrac{(k + 2)(k + 2)}{14(k + 2)} = \dfrac{k + 2}{14}$

6. $\dfrac{x^2 - 7x - 8}{x^2 - 10x + 16} = \dfrac{(x - 8)(x + 1)}{(x - 8)(x - 2)} = \dfrac{x + 1}{x - 2}$

7. $\dfrac{1 - a}{7a - 7} = \dfrac{-1(a - 1)}{7(a - 1)} = \dfrac{-1}{7} = -\dfrac{1}{7}$

8. $\dfrac{4y - 4x}{x^2 - y^2} = \dfrac{-4(x - y)}{(x + y)(x - y)} = \dfrac{-4}{x + y} = -\dfrac{4}{x + y}$

9. $\dfrac{t^2 + 7t - 18}{-18 - 2t} = \dfrac{(t + 9)(t - 2)}{-2(t + 9)} = \dfrac{t - 2}{-2} = -\dfrac{t - 2}{2}$

10. $\dfrac{24}{40} = \dfrac{3}{5}$ **11.** $\dfrac{54}{42} = \dfrac{9}{7}$

12. $\dfrac{40 \text{ seconds}}{2 \text{ minutes}} = \dfrac{40 \text{ seconds}}{120 \text{ seconds}} = \dfrac{1 \text{ second}}{3 \text{ seconds}}$, or $\dfrac{1}{3}$

13. $\dfrac{1 \text{ quarter}}{55 \text{ cents}} = \dfrac{25 \text{ cents}}{55 \text{ cents}} = \dfrac{5 \text{ cents}}{11 \text{ cents}}$, or $\dfrac{5}{11}$

14. Let $5x$ = the number of boys and $4x$ = the number of girls. $5x + 4x = 306$; $9x = 306$; $x = 34$, so $4x = 4 \cdot 34 = 136$.   **136 girls**

15. $\dfrac{3}{14} = \dfrac{9}{2x}$; $3 \cdot 2x = 9 \cdot 14$; $6x = 126$; $x = 21$

16. $\dfrac{y + 1}{5y + 2} = \dfrac{2}{9}$; $(y + 1)9 = (5y + 2)2$; $9y + 9 = 10y + 4$; $9y + 5 = 10y$; $5 = y$

17. $\dfrac{a - 7}{a - 5} = \dfrac{a + 1}{a + 5}$; $(a - 7)(a + 5) = (a - 5)(a + 1)$; $a^2 - 2a - 35 = a^2 - 4a - 5$; $-2a - 35 = -4a - 5$; $-2a - 30 = -4a$; $-30 = -2a$; $15 = a$

18.

Number	8	12
Cost	$1.50	x

$2.25

$\dfrac{8}{150} = \dfrac{12}{x}$; $8 \cdot x = 150 \cdot 12$; $8x = 1800$; $x = 225$.

19.

Distance	170 km	x
Gas	16 L	24 L

$\dfrac{170}{16} = \dfrac{x}{24}$; $16x = 4080$; $x = 255$.   **255 km**

20. $\dfrac{6x^2}{25} \cdot \dfrac{10}{9xy^2} = \dfrac{6 \cdot x \cdot x \cdot 10}{25 \cdot 9 \cdot x \cdot y \cdot y} = \dfrac{4x}{15y^2}$

21. $\dfrac{12}{a^2 + a} \cdot \dfrac{a}{18} = \dfrac{12 \cdot a}{a(a + 1) \cdot 18} = \dfrac{2}{3(a + 1)} = \dfrac{2}{3a + 3}$

22. $\dfrac{r - 5}{r + 5} \cdot \dfrac{r^2 - 25}{5 - r} = \dfrac{(r - 5)(r + 5)(r - 5)}{(r + 5)(-1)(r - 5)} = \dfrac{r - 5}{-1} = -r + 5$

23. $\dfrac{(2x)^3}{xy} \div \dfrac{4x}{3y} = \dfrac{(2x)^3}{xy} \cdot \dfrac{3y}{4x} = \dfrac{2x \cdot 2x \cdot 2x \cdot 3 \cdot y}{x \cdot y \cdot 4x} = 6x$

24. $\dfrac{3b + 9}{9} \div (b + 3) = \dfrac{3b + 9}{9} \cdot \dfrac{1}{b + 3} = \dfrac{3(b + 3)}{9(b + 3)} = \dfrac{1}{3}$

25. $\dfrac{x^2 - x - 2}{2x - 4} \div \dfrac{x + 1}{8} = \dfrac{x^2 - x - 2}{2x - 4} \cdot \dfrac{8}{x + 1} = \dfrac{(x - 2)(x + 1)8}{2(x - 2)(x + 1)} = 4$

26. $\dfrac{5y}{8} + \dfrac{7y}{8} = \dfrac{5y + 7y}{8} = \dfrac{12y}{8} = \dfrac{3y}{2}$

27. $\dfrac{2x + 1}{x + 1} + \dfrac{1}{x + 1} = \dfrac{2x + 1 + 1}{x + 1} = \dfrac{2x + 2}{x + 1} = \dfrac{2(x + 1)}{x + 1} = 2$

28. $\dfrac{x - y}{2xy} - \dfrac{6x - y}{2xy} = \dfrac{x - y - (6x - y)}{2xy} = \dfrac{x - y - 6x + y}{2xy} = \dfrac{-5x}{2xy} = -\dfrac{5}{2y}$

29. $5x \cdot 4 = 20x$ **30.** $7 \cdot 9s = 63s$ **31.** $3 \cdot a = 3a$

32. $\dfrac{2}{3} - \dfrac{7}{a} = \dfrac{2a}{3a} - \dfrac{21}{3a} = \dfrac{2a - 21}{3a}$ **33.** $\dfrac{1}{2x} + \dfrac{1}{6x} = \dfrac{3}{6x} + \dfrac{1}{6x} = \dfrac{4}{6x} = \dfrac{2}{3x}$

34. $\dfrac{4}{9y} - \dfrac{3}{y^2} + \dfrac{1}{2y} = \dfrac{8y}{18y^2} - \dfrac{54}{18y^2} + \dfrac{9y}{18y^2} = \dfrac{17y - 54}{18y^2}$

35. $\dfrac{2x + 1}{5} - \dfrac{x - 4}{7} = \dfrac{7(2x + 1)}{35} - \dfrac{5(x - 4)}{35} = \dfrac{14x + 7 - 5x + 20}{35} = \dfrac{9x + 27}{35}$

36. $\dfrac{a - 3}{6a} - \dfrac{a - 2}{4a} = \dfrac{2(a - 3)}{12a} - \dfrac{3(a - 2)}{12a} = \dfrac{2a - 6 - 3a + 6}{12a} = \dfrac{-a}{12a} = -\dfrac{1}{12}$

37. $\dfrac{x}{6} + \dfrac{3x - 1}{8} + \dfrac{x}{12} = \dfrac{4x}{24} + \dfrac{3(3x - 1)}{24} + \dfrac{2x}{24} = \dfrac{4x + 9x - 3 + 2x}{24} = \dfrac{15x - 3}{24} =$

$\dfrac{3(5x - 1)}{24} = \dfrac{5x - 1}{8}$

38. $\dfrac{2b}{3} - \dfrac{b}{2} = 3; \quad \dfrac{4b}{6} - \dfrac{3b}{6} = 3; \quad \dfrac{b}{6} = 3; \quad b = 18$

39. $\dfrac{4}{x} + \dfrac{8}{3x} = \dfrac{5}{9}; \quad \dfrac{12}{3x} + \dfrac{8}{3x} = \dfrac{5}{9}; \quad \dfrac{20}{3x} = \dfrac{5}{9}; \quad 15x = 180; \quad x = 12$

40. $\dfrac{3}{c} - \dfrac{5c + 2}{3c} = \dfrac{8c - 2}{9c}; \quad \dfrac{27}{9c} - \dfrac{3(5c + 2)}{9c} = \dfrac{8c - 2}{9c}; \quad 27 - 15c - 6 = 8c - 2;$

$21 - 15c = 8c - 2; \quad 23 - 15c = 8c; \quad 23 = 23c; \quad 1 = c$

41.

	Rae	Yvette	Together
Minutes needed	60	40	m
Part done in one minute	$\dfrac{1}{60}$	$\dfrac{1}{40}$	$\dfrac{1}{m}$

$\dfrac{1}{60} + \dfrac{1}{40} = \dfrac{1}{m}; \quad \dfrac{2}{120} + \dfrac{3}{120} = \dfrac{1}{m};$

$\dfrac{5}{120} = \dfrac{1}{m}; \quad 5m = 120;$

$m = 24. \quad$ 24 min

42.

	Small pipe	Large pipe	Together
Hours needed	18	12	x
Part done in one hour	$\dfrac{1}{18}$	$\dfrac{1}{12}$	$\dfrac{1}{x}$

$\dfrac{1}{18} + \dfrac{1}{12} = \dfrac{1}{x}; \quad \dfrac{2}{36} +$

$\dfrac{3}{36} = \dfrac{1}{x}; \quad \dfrac{5}{36} = \dfrac{1}{x};$

$5x = 36; \quad x = 7\dfrac{1}{5}.$

$7\dfrac{1}{5}$ hours

43. $\dfrac{r}{r + 2} + \dfrac{3}{r - 3} = \dfrac{r(r - 3)}{(r + 2)(r - 3)} + \dfrac{3(r + 2)}{(r + 2)(r - 3)} = \dfrac{r^2 - 3r + 3r + 6}{(r + 2)(r - 3)} = \dfrac{r^2 + 6}{(r + 2)(r - 3)}$

44. $\dfrac{1}{t - 4} - \dfrac{t}{(t - 4)^2} = \dfrac{t - 4}{(t - 4)^2} - \dfrac{t}{(t - 4)^2} = \dfrac{t - 4 - t}{(t - 4)^2} = \dfrac{-4}{(t - 4)^2} = -\dfrac{4}{(t - 4)^2}$

45. $\dfrac{2}{1 - x} + \dfrac{4}{x^2 - 1} = \dfrac{-2}{x - 1} + \dfrac{4}{x^2 - 1} = \dfrac{-2(x + 1)}{(x + 1)(x - 1)} + \dfrac{4}{(x + 1)(x - 1)} =$

$\dfrac{-2x - 2 + 4}{(x + 1)(x - 1)} = \dfrac{-2x + 2}{(x + 1)(x - 1)} = \dfrac{-2(x - 1)}{(x + 1)(x - 1)} = \dfrac{-2}{x + 1} = -\dfrac{2}{x + 1}$

Pages 344–345 • MIXED REVIEW

1. $\begin{aligned} 7x + 3y &= -4 \\ 9x + 3y &= 0 \quad \text{Subt.} \\ \hline -2x &= -4; \quad x = 2 \end{aligned}$

$7(2) + 3y = -4; \quad 14 + 3y = -4;$

$3y = -18; \quad y = -6. \quad (2, -6)$

2. $4x + 3y = 15$ and $3x - y = 8$; $4x + 3y = 15$ and $y = 3x - 8$; $4x +$
   $3(3x - 8) = 15$; $4x + 9x - 24 = 15$; $13x - 24 = 15$; $13x = 39$; $x = 3$;
   $y = 3x - 8 = 3 \cdot 3 - 8 = 9 - 8 = 1$. $(3, 1)$

3. $2x + 5y = 20 \longrightarrow 6x + 15y = 60$
   $3x + 2y = 8 \longrightarrow \underline{6x + 4y = 16}$ Subt.
   $11y = 44$; $y = 4$

   $2x + 5 \cdot 4 = 20$; $2x + 20 = 20$; $2x = 0$; $x = 0$. $(0, 4)$

4. $\dfrac{x}{10} + \dfrac{x}{8} = \dfrac{4x}{40} + \dfrac{5x}{40} = \dfrac{9x}{40}$

5. $\dfrac{z^2 + 3z - 28}{z^2 - 5z + 4} = \dfrac{(z + 7)(z - 4)}{(z - 1)(z - 4)} = \dfrac{z + 7}{z - 1}$

6. $\dfrac{a^2 - b^2}{3ab - 3a^2} = \dfrac{(a + b)(a - b)}{-3a(a - b)} = \dfrac{a + b}{-3a} = -\dfrac{a + b}{3a}$

7. Answers may vary. $y = 4 - 3x$; if $x = 0$, $y = 4 - 3 \cdot 0 = 4 - 0 = 4$; if $x = 1$,
   $y = 4 - 3 \cdot 1 = 4 - 3 = 1$; if $x = 2$, $y = 4 - 3(2) = 4 - 6 = -2$. $(0, 4)$, $(1, 1)$,
   $(2, -2)$

8. Let $x =$ the cost (in cents) of a roll of tape and $y =$ the cost of a notebook.
   $4x + 3y = 530 \longrightarrow 8x + 6y = 1060$
   $6x + 2y = 570 \longrightarrow \underline{18x + 6y = 1710}$ Subt.
   $-10x = -650$; $x = 65$

   $6x + 2y = 570$; $6(65) + 2y = 570$; $390 + 2y = 570$; $2y = 180$; $y = 90$.

   tape: \$.65; notebook: \$.90

9.

Interest	\$12.50	x
Account	\$2000	\$4200

$\dfrac{12.50}{2000} = \dfrac{x}{4200}$; $2000x = 52{,}500$;

$x = 26.25$. \$26.25

10. $V = (2 \cdot 3 \cdot 5) + (6 \cdot 4 \cdot 5) = 30 + 120 = 150$

11. $x^2 - 16x + 63 = (x - 9)(x - 7)$

12. $6a^3 + 42a^2b + 12ab^2 = 6a(a^2 + 7ab + 2b^2)$

13. $-(-4)^3 = -(-64) = 64$ and $(2 - 9)8 - 7 = (-7)8 - 7 = -56 - 7 = -63$;
    $64 > -63$; $>$

14. **a.** $1 > 7 - 2 \cdot 1$? $1 > 7 - 2$? $1 > 5$? No
    b. $0 > 7 - 2 \cdot 4$? $0 > 7 - 8$? $0 > -1$? Yes
    c. $4 > 7 - 2 \cdot 0$? $4 > 7 - 0$? $4 > 7$? No
    d. $-1 > 7 - 2 \cdot 5$? $-1 > 7 - 10$? $-1 > -3$? Yes

15. $\dfrac{x - 7}{x - 3} = \dfrac{x - 1}{x + 6}$; $(x - 7)(x + 6) = (x - 3)(x - 1)$; $x^2 - x - 42 = x^2 - 4x + 3$;
    $-x - 42 = -4x + 3$; $-x = -4x + 45$; $3x = 45$; $x = 15$

16. $\dfrac{7}{3n} - \dfrac{1}{n} = \dfrac{2}{3}$; $\dfrac{7}{3n} - \dfrac{3}{3n} = \dfrac{2}{3}$; $\dfrac{4}{3n} = \dfrac{2}{3}$; $4 \cdot 3 = 3n \cdot 2$; $12 = 6n$; $2 = n$

17. $5(2x + 1) = -3(1 - 4x)$; $10x + 5 = -3 + 12x$; $10x + 8 = 12x$; $8 = 2x$;
$4 = x$

18.

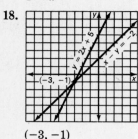

$(-3, -1)$

19. Let $3x$ = number of counselors and
$14x$ = number of campers.
$3x + 14x = 136$; $17x = 136$;
$x = 8$, so $3x = 3 \cdot 8 = 24$.
24 counselors

20. $P = 4s$, so $68 = 4(x + 9)$; $68 = 4x + 36$; $32 = 4x$; $8 = x$.

21. $(3y^3 - 7y + 4) - (-y^4 + 8y^3 - y^2 + 3y - 5) = 3y^3 - 7y + 4 + y^4 - 8y^3 + y^2 - 3y + 5 = y^4 - 5y^3 + y^2 - 10y + 9$

22. $(5a - 2b)^2 = (5a)^2 + 2(5a)(-2b) + (-2b)^2 = 25a^2 - 20ab + 4b^2$

23.

n	40	37	35.5	22.5
W	400	370	355	225

24. $\dfrac{10r^2}{r^2 + 3r} \cdot \dfrac{2r + 6}{5} = \dfrac{10r^2 \cdot 2(r + 3)}{r(r + 3) \cdot 5} = 4r$

25. $\dfrac{(2ab^2)^2}{3} \div 6a^3b^3 = \dfrac{(2ab^2)^2}{3} \cdot \dfrac{1}{6a^3b^3} = \dfrac{2 \cdot a \cdot b \cdot b \cdot 2 \cdot a \cdot b \cdot b}{3 \cdot 6 \cdot a \cdot a \cdot a \cdot b \cdot b \cdot b} = \dfrac{2b}{9a}$

26. $\dfrac{-6x^2y}{-8y^3z} = \dfrac{-6 \cdot x \cdot x \cdot y}{-8 \cdot y \cdot y \cdot y \cdot z} = \dfrac{3x^2}{4y^2z}$

27. $\dfrac{4a^2 + b^2}{2ab} - \dfrac{b^2 - 5b^3}{2ab} = \dfrac{4a^2 + b^2 - (b^2 - 5b^3)}{2ab} = \dfrac{4a^2 + b^2 - b^2 + 5b^3}{2ab} = \dfrac{4a^2 + 5b^3}{2ab}$

28. $(-5xy^3)(4x^2y)^2 = (-5xy^3)(16x^4y^2) = (-5 \cdot 16)(x \cdot x^4)(y^3 \cdot y^2) = -80x^5y^5$

29. $\dfrac{4a^2}{a - 1} \div \dfrac{2a^3}{1 - a} = \dfrac{4a^2}{a - 1} \cdot \dfrac{1 - a}{2a^3} = \dfrac{4a^2(-1)(a - 1)}{(a - 1)2a^3} = \dfrac{-2}{a} = -\dfrac{2}{a}$

30. Let x and y = the numbers. $x + y = 32$ and $x - y = 8$; $x + y = 32$ and $y = x - 8$; $x + (x - 8) = 32$; $2x - 8 = 32$; $2x = 40$; $x = 20$; $y = x - 8$ = $20 - 8 = 12$.   20 and 12

31. $(2a - b)(a^2 + ab) = 2a(a^2 + ab) - b(a^2 + ab) = 2a^3 + 2a^2b - a^2b - ab^2 = 2a^3 + a^2b - ab^2$

32. $-3x^2(x^2 + 5x - 1) = -3x^4 - 15x^3 + 3x^2$

33. $(5y - 7)(3y + 8) = 15y^2 + 19y - 56$

34. $2x + 3(x + 5)$, or $5x + 15$

35.

	Tony	Friend	Together
Hours needed	12	15	h
Part done in one hour	$\dfrac{1}{12}$	$\dfrac{1}{15}$	$\dfrac{1}{h}$

$\dfrac{1}{12} + \dfrac{1}{15} = \dfrac{1}{h}$; $\dfrac{5}{60} + \dfrac{4}{60} = \dfrac{1}{h}$; $\dfrac{9}{60} = \dfrac{1}{h}$; $9h = 60$; $h = 6\dfrac{6}{9} = 6\dfrac{2}{3}$. $6\dfrac{2}{3}$ hours

36.

37.

	rate	time	Distance
Leaving Toronto	92 km/h	t	$92t$
Leaving Detroit	97 km/h	t	$97t$
			378

$92t + 97t = 378$; $189t = 378$; $t = 2$. 10:00

38. $y = x - 2$ and $y = x + 2$; slopes of both $= 1$. no solution pair

CHAPTER 10 · Decimals and Percents

Page 349 · WRITTEN EXERCISES

A   **1.** 0.66   **2.** 8.07   **3.** 2.93   **4.** 2.86   **5.** 2.64   **6.** 3.702

     7. 5.12   **8.** 8.213   **9.** 7.765   **10.** 3.237   **11.** 12.817   **12.** −3.167

     13. 14.53   **14.** 11.86   **15.** 20.511   **16.** 15.704   **17.** 11.056

     18. 20.9204   **19.** 12.6   **20.** 23.5   **21.** 0.82   **22.** 56   **23.** 24.3

     24. 671   **25.** 610   **26.** 0.912   **27.** 4.1208   **28.** 0.316   **29.** 2.500

     30. 3.7842   **31.** 0.423526   **32.** 1.53807   **33.** 0.06018   **34.** 0.12586

     35. 0.004182   **36.** 0.0242   **37.** 0.206244   **38.** 15.10518   **39.** 16.1671

     40. $9.5 - 7.9 = 1.6$

Page 351 · WRITTEN EXERCISES

A   **1.** $4\overline{)24} = 6$ **2.** $7\overline{)21.7} = 3.1$ **3.** $5\overline{)40} = 8$

     4. $9\overline{)10.08} = 1.12$ **5.** $8\overline{)368} = 46$ **6.** $25\overline{)1.00} = 0.04$

     7. $24\overline{)26.4} = 1.1$ **8.** $125\overline{)5000} = 40$ **9.** $32\overline{)6.4} = 0.2$

     10. $3\overline{)2430} = 810$ **11.** $16\overline{)8.96} = 0.56$ **12.** $8\overline{)18.80} = 2.35$

     13. $6\overline{)10.00} = 1.66$ **14.** $4\overline{)27.00} = 6.75$ **15.** $5\overline{)63.00} = 12.60$

         $1 \div 0.6 \approx 1.7$ $2.7 \div 0.4 \approx 6.8$ $6.3 \div 0.5 = 12.6$

     16. $12\overline{)37.00} = 3.08$ **17.** $62\overline{)930.00} = 15.00$ **18.** $34\overline{)65.00} = 1.91$

         $3.7 \div 1.2 \approx 3.1$ $93 \div 6.2 = 15.0$ $0.65 \div 0.34 \approx 1.9$

     19. $7\overline{)4000.00} = 571.42$ **20.** $51\overline{)470.00} = 9.21$ **21.** $12\overline{)790.00} = 65.83$

         $40 \div 0.07 \approx 571.4$ $4.7 \div 0.51 \approx 9.2$ $7.9 \div 0.12 \approx 65.8$

     22. $23\overline{)600.00} = 26.08$ **23.** $36\overline{)92.00} = 2.55$ **24.** $18\overline{)82.50} = 4.58$

         $60 \div 2.3 \approx 26.1$ $0.92 \div 0.36 \approx 2.6$ $8.25 \div 1.8 \approx 4.6$

     25. $3\overline{)67.400} = 22.466$ **26.** $9\overline{)3.500} = 0.388$ **27.** $4\overline{)100.000} = 25.000$

         $6.74 \div 0.3 \approx 22.47$ $0.35 \div 0.9 \approx 0.39$ $10 \div 0.4 = 25.00$

     28. $17\overline{)620.000} = 36.470$ **29.** $8\overline{)260.000} = 32.500$ **30.** $34\overline{)310.000} = 9.117$

         $62 \div 1.7 \approx 36.47$ $2.6 \div 0.08 = 32.50$ $3.1 \div 0.34 \approx 9.12$

31.
$$\begin{array}{r} 238.461 \\ 26\overline{)6200.000} \end{array}$$
$62 \div 0.26 \approx 238.46$

32.
$$\begin{array}{r} 1680.000 \\ 5\overline{)8400.000} \end{array}$$
$84 \div 0.05 = 1680.00$

33.
$$\begin{array}{r} 53.333 \\ 75\overline{)4000.000} \end{array}$$
$40 \div 0.75 \approx 53.33$

34.
$$\begin{array}{r} 16.521 \\ 23\overline{)380.000} \end{array}$$
$38 \div 2.3 \approx 16.52$

35.
$$\begin{array}{r} 1266.666 \\ 6\overline{)7600.000} \end{array}$$
$76 \div 0.06 \approx 1266.67$

36.
$$\begin{array}{r} 296.296 \\ 54\overline{)16000.000} \end{array}$$
$160 \div 0.54 \approx 296.30$

Page 353 · WRITTEN EXERCISES

A

1.
$$\begin{array}{r} 0.125 \\ 8\overline{)1.000} \end{array}$$
$\dfrac{1}{8} = 0.125$

2.
$$\begin{array}{r} 0.625 \\ 8\overline{)5.000} \end{array}$$
$\dfrac{5}{8} = 0.625$

3.
$$\begin{array}{r} 0.6 \\ 5\overline{)3.0} \end{array}$$
$\dfrac{3}{5} = 0.6$

4.
$$\begin{array}{r} 0.555 \\ 9\overline{)5.000} \end{array}$$
$\dfrac{5}{9} = 0.55\ldots$

5. $\dfrac{3}{8} = 0.375$

6. $\dfrac{2}{3} = 0.66\ldots$

7. $\dfrac{2}{7} = 0.285714285714\ldots$

8. $\dfrac{1}{6} = 0.166\ldots$

9. $\dfrac{4}{7} = 0.571428571428\ldots$

10. $\dfrac{4}{9} = 0.44\ldots$

11. $\dfrac{1}{9} = 0.11\ldots$

12. $\dfrac{7}{9} = 0.77\ldots$

13. $\dfrac{6}{11} = 0.5454\ldots$

14. $\dfrac{5}{6} = 0.833\ldots$

15. $\dfrac{2}{5} = 0.4$

16. $\dfrac{15}{24} = 0.625$

17. $\dfrac{18}{25} = 0.720$

18. $\dfrac{9}{19} \approx 0.474$

19. $\dfrac{17}{26} \approx 0.654$

20. $\dfrac{20}{28} \approx 0.714$

Page 355 · WRITTEN EXERCISES

A

1. $0.2x + 0.3x = 25;\quad 10(0.2x + 0.3x) = 10(25);\quad 2x + 3x = 250;\quad 5x = 250;$
$x = 50$

2. $0.4a + 0.7 = 55;\quad 10(0.4a + 0.7) = 10(55);\quad 4a + 7 = 550;\quad 4a = 543;$
$a = 135\dfrac{3}{4}$, or 135.75

3. $0.6x = 2.4;\quad 10(0.6x) = 10(2.4);\quad 6x = 24;\quad x = 4$

4. $0.7y = 4.2;\quad 10(0.7y) = 10(4.2);\quad 7y = 42;\quad y = 6$

5. $0.10x + 0.12x = 1.9;\quad 100(0.10x + 0.12x) = 100(1.9);\quad 10x + 12x = 190;$
$22x = 190;\quad x = 8\dfrac{14}{22} = 8\dfrac{7}{11}$, or $8.6363\ldots$

6. $0.75x + 0.10x = 85;\quad 100(0.75x + 0.10x) = 100(85);\quad 75x + 10x = 8500;$
$85x = 8500;\quad x = 100$

7. $0.35x = 0.91;\quad 100(35x) = 100(0.91);\quad 35x = 91;\quad x = 2\dfrac{21}{35} = 2\dfrac{3}{5}$, or 2.6

8. $0.22y = 1.54;\quad 100(0.22y) = 100(1.54);\quad 22y = 154;\quad y = 7$

9. $0.2x + 0.04x = 1.2$; $100(0.2x + 0.04x) = 100(1.2)$; $20x + 4x = 120$; $24x = 120$; $x = 5$

10. $0.14a + 0.6a = 5.92$; $100(0.14a + 0.6a) = 100(5.92)$; $14a + 60a = 592$; $74a = 592$; $a = 8$

11. $0.3n - 0.24n = 0.8$; $100(0.3n - 0.24n) = 100(0.8)$; $30n - 24n = 80$; $6n = 80$; $n = 13\frac{2}{6} = 13\frac{1}{3}$, or $13.33\ldots$

12. $0.05x + 0.5x = 4.4$; $100(0.05x + 0.5x) = 100(4.4)$; $5x + 50x = 440$; $55x = 440$; $x = 8$

13. $0.12n = 0.3 - 0.03n$; $100(0.12n) = 100(0.3 - 0.03n)$; $12n = 30 - 3n$; $15n = 30$; $n = 2$

14. $0.05d = 2000$; $100(0.05d) = 100(2000)$; $5d = 200,000$; $d = 40,000$

15. $0.04x = 264$; $100(0.04x) = 100(264)$; $4x = 26,400$; $x = 6600$

16. $0.03x = 0.15(4 - x)$; $100(0.03x) = 100[0.15(4 - x)]$; $3x = 15(4 - x)$; $3x = 60 - 15x$; $18x = 60$; $x = 3\frac{6}{18} = 3\frac{1}{3}$, or $3.33\ldots$

17. $90 - x = 0.04(180 - x)$; $100(90 - x) = 100[0.04(180 - x)]$; $9000 - 100x = 4(180 - x)$; $9000 - 100x = 720 - 4x$; $9000 = 720 + 96x$; $8280 = 96x$; $x = 86\frac{24}{96} = 86\frac{1}{4}$, or 86.25

18. $0.2(x - 3) + 0.4x = 0$; $10[0.2(x - 3) + 0.4x] = 10(0)$; $2(x - 3) + 4x = 0$; $2x - 6 + 4x = 0$; $6x - 6 = 0$; $6x = 6$; $x = 1$

19. $0.2x + 0.3(x + 4) = 37$; $10[0.2x + 0.3(x + 4)] = 10(37)$; $2x + 3(x + 4) = 370$; $2x + 3x + 12 = 370$; $5x + 12 = 370$; $5x = 358$; $x = 71\frac{3}{5}$, or 71.6

20. $0.07x + 0.03(2x) = 390$; $100[0.07x + 0.03(2x)] = 100(390)$; $7x + 3(2x) = 39,000$; $7x + 6x = 39,000$; $13x = 39,000$; $x = 3000$

21. $0.05n + 0.03(n - 20) = 0$; $100[0.05n + 0.03(n - 20)] = 100(0)$; $5n + 3(n - 20) = 0$; $5n + 3n - 60 = 0$; $8n - 60 = 0$; $8n = 60$; $n = 7\frac{4}{8} = 7\frac{1}{2}$, or 7.5

22. $0.06x + 0.04(10 - x) = 70$; $100[0.06x + 0.04(10 - x)] = 100(70)$; $6x + 4(10 - x) = 7000$; $6x + 40 - 4x = 7000$; $2x + 40 = 7000$; $2x = 6960$; $x = 3480$

B 23. $0.2x + 0.25(9 - x) = 2$; $100[0.2x + 0.25(9 - x)] = 100(2)$; $20x + 25(9 - x) = 200$; $20x + 225 - 25x = 200$; $225 - 5x = 200$; $-5x = -25$; $x = 5$

24. $0.25(16 - n) + n = 0.4(16)$; $100[0.25(16 - n) + n] = 100(0.4)(16)$; $25(16 - n) + 100n = (40)(16)$; $400 - 25n + 100n = 640$; $400 + 75n = 640$; $75n = 240$; $n = 3\frac{15}{75} = 3\frac{1}{5}$, or 3.2

25. $0.6(15) + s = 0.7(15 + s)$; $10[0.6(15) + s] = 10(0.7)(15 + s)$; $6(15) + 10s =$
$7(15 + s)$; $90 + 10s = 105 + 7s$; $10s = 15 + 7s$; $3s = 15$; $s = 5$

26. $0.035x + 0.06(600 - x) = 27$; $1000[0.035x + 0.06(600 - x)] = 1000(27)$;
$35x + 60(600 - x) = 27{,}000$; $35x + 36{,}000 - 60x = 27{,}000$; $36{,}000 - 25x =$
$27{,}000$; $-25x = -9000$; $x = 360$

27. $0.4x + 0.24(x - 5) = 0.08$; $100[0.4x + 0.24(x - 5)] = 100(0.08)$; $40x +$
$24(x - 5) = 8$; $40x + 24x - 120 = 8$; $64x - 120 = 8$; $64x = 128$; $x = 2$

28. $0.025y - 0.05(20 - 2y) = 0.2$; $1000[0.025y - 0.05(20 - 2y)] = 1000(0.2)$;
$25y - 50(20 - 2y) = 200$; $25y - 1000 + 100y = 200$; $125y - 1000 = 200$;
$125y = 1200$; $y = 9\dfrac{75}{125} = 9\dfrac{3}{5}$, or 9.6

29. $0.36m - 0.045(m + 1) = 0.711$; $1000[0.36m - 0.045(m + 1)] = 1000(0.711)$;
$360m - 45(m + 1) = 711$; $360m - 45m - 45 = 711$; $315m - 45 = 711$;
$315m = 756$; $m = 2\dfrac{126}{315} = 2\dfrac{2}{5}$, or 2.4

30. $2.09z + 1.25(3 - z) = 4.8$; $100[2.09z + 1.25(3 - z)] = 100(4.8)$; $209z +$
$125(3 - z) = 480$; $209z + 375 - 125z = 480$; $84z + 375 = 480$; $84z = 105$;
$z = 1\dfrac{1}{4}$, or 1.25

C   **31.** $0.02(x - 3) + 0.41(x + 1) = 1.21$; $100[0.02(x - 3) + 0.41(x + 1)] = 100(1.21)$;
$2(x - 3) + 41(x + 1) = 121$; $2x - 6 + 41x + 41 = 121$; $43x + 35 = 121$;
$43x = 86$; $x = 2$

32. $0.55(2 - y) - 0.1(y - 3) = 0.75$; $100[0.55(2 - y) - 0.1(y - 3)] = 100(0.75)$;
$55(2 - y) - 10(y - 3) = 75$; $110 - 55y - 10y + 30 = 75$; $140 - 65y = 75$;
$-65y = -65$; $y = 1$

33. $3.2x + 0.8(x - 4) = 0.4(x + 10)$; $10[3.2x + 0.8(x - 4)] = 10[0.4(x + 10)]$;
$32x + 8(x - 4) = 4(x + 10)$; $32x + 8x - 32 = 4x + 40$; $40x - 32 = 4x + 40$;
$36x - 32 = 40$; $36x = 72$; $x = 2$

34. $0.5(z + 5) = 1.4(z - 1) + 0.15(10z + 2)$; $100[0.5(z + 5)] = 100[1.4(z - 1) +$
$0.15(10z + 2)]$; $50(z + 5) = 140(z - 1) + 15(10z + 2)$; $50z + 250 = 140z -$
$140 + 150z + 30$; $50z + 250 = 290z - 110$; $250 = 240z - 110$; $360 = 240z$;
$1\dfrac{1}{2} = z$, or $1.5 = z$

Page 357 • WRITTEN EXERCISES

A   **1.** Let x = smaller number. Then $x + 0.9$ = larger number. $x + x + 0.9 = 5.7$;
$2x + 0.9 = 5.7$; $10(2x + 0.9) = 10(5.7)$; $20x + 9 = 57$; $20x = 48$; $x = 2.4$ and
$x + 0.9 = 3.3$.   2.4 and 3.3

2. Let x = length (in km) of Forks Trail. Then $x + 2.4$ = length of Falls Trail.
$x + x + 2.4 = 10.2$; $2x + 2.4 = 10.2$; $10(2x + 2.4) = 10(10.2)$; $20x + 24 = 102$; $20x = 78$; $x = 3.9$ and $x + 2.4 = 6.3$. Forks Trail: 3.9 km; Falls Trail: 6.3 km

3. Let x = length (in cm) of shorter board. Then $2.1x$ = length of longer board.
$x + 2.1x = 133.3$; $10(x + 2.1x) = 10(133.3)$; $10x + 21x = 1333$; $31x = 1333$; $x = 43$ and $2.1x = 90.3$.   43 cm and 90.3 cm

4. Let x = kilograms of one suitcase. Then $x - 5.8$ = kilograms of other suitcase.
$x + x - 5.8 = 37.6$; $2x - 5.8 = 37.6$; $10(2x - 5.8) = 10(37.6)$; $20x - 58 = 376$; $20x = 434$; $x = 21.7$ and $x - 5.8 = 15.9$.   21.7 kg and 15.9 kg

5. Let x = km to office. $x + x + 9.8 = 32.5$; $2x + 9.8 = 32.5$; $10(2x + 9.8) = 10(32.5)$; $20x + 98 = 325$; $20x = 227$; $x = 11.35$.   11.35 km

6. Let x = length (in km) of St. Lawrence. Then $1.24x$ = length of Mississippi.
$1.24x = 3777.9$; $100(1.24x) = 1000(3777.9)$; $124x = 377{,}790$; $x = 3046.6935$. about 3046.7 km

Page 357 • CALCULATOR ACTIVITIES

1. $x = \dfrac{0.65}{0.26} = 2.5$

2. $a = \dfrac{12.7}{2} = 6.35$

3. $y = \dfrac{4.2}{5.6} = 0.75$

4. $x = \dfrac{0.0246}{0.123} = 0.2$

5. $b = (3.7)(2.9) = 10.73$

6. $z = (-7.1)(1.325) = -9.4075$

Pages 361–362 • WRITTEN EXERCISES

A

1. $0.40 = 40\%$

2. $0.23 = 23\%$

3. $0.65 = 65\%$

4. $0.89 = 89\%$

5. $0.50 = 50\%$

6. $0.14 = 14\%$

7. $0.07 = 7\%$

8. $0.03 = 3\%$

9. $0.09 = 9\%$

10. $0.06 = 6\%$

11. $0.045 = 4.5\%$

12. $0.036 = 3.6\%$

13. $0.214 = 21.4\%$

14. $0.525 = 52.5\%$

15. $0.125 = 12.5\%$

16. $\dfrac{2}{5} = 0.4 = 40\%$

17. $\dfrac{1}{5} = \dfrac{20}{100} = 20\%$

18. $\dfrac{4}{5} = \dfrac{80}{100} = 80\%$

19. $\dfrac{3}{4} = 0.75 = 75\%$

20. $\dfrac{5}{8} = 0.625 = 62.5\%$

21. $64\% = 0.64$

22. $86\% = 0.86$

23. $95\% = 0.95$

24. $30\% = 0.30$

25. $20\% = 0.20$

26. $56\% = 0.56$

27. $67\% = 0.67$

28. $66\dfrac{2}{3}\% = 0.66\dfrac{2}{3}$, or $0.66\ldots$

29. $47\% = 0.47$

30. $33\dfrac{1}{3}\% = 0.33\dfrac{1}{3}$, or $0.33\ldots$

31. $75\% = 0.75$

32. $6\% = 0.06$

33. $8\% = 0.08$

34. $2\% = 0.02$

35. $9\% = 0.09$

36. $3.5\% = 0.035$

37. $6.5\% = 0.065$

38. $7.5\% = 0.075$

39. $5.5\% = 0.055$

40. $5.25\% = 0.0525$

41. $33\frac{1}{3}\% = \frac{1}{3}$

42. $66\frac{2}{3}\% = \frac{2}{3}$

43. $25\% = \frac{25}{100} = \frac{1}{4}$

44. $75\% = \frac{75}{100} = \frac{3}{4}$

45. $50\% = \frac{50}{100} = \frac{1}{2}$

46. $20\% = \frac{20}{100} = \frac{1}{5}$

47. $10\% = \frac{10}{100} = \frac{1}{10}$

48. $40\% = \frac{40}{100} = \frac{2}{5}$

49. $60\% = \frac{60}{100} = \frac{3}{5}$

50. $80\% = \frac{80}{100} = \frac{4}{5}$

51. about 10%

52. about 25%

53. about 40%

54. about 25%

55. 55%; 45%; 35%

56. 25%; 30%; 35%; 40%

57. 10%; 15%; 20%; 25%

58. $100\% - (25\% + 35\%) = 100\% - 60\% = 40\%$

59. $100\% - (10\% + 25\% + 33\%) = 100\% - 68\% = 32\%$

60. $100\% - (27\% + 40\% + 10\%) = 100\% - 77\% = 23\%$

B   **61.** 4   **62.** $25\% + 20\% + 30\% = 75\%$

63.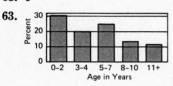

Page 364 • WRITTEN EXERCISES

A   **1.** $30\% \times 120 = 0.30 \times 120 = 36$

2. $60\% \times 46 = 0.60 \times 46 = 27.6$

3. $25\% \times 18 = 0.25 \times 18 = 4.5$

4. $75\% \times 50 = 0.75 \times 50 = 37.5$

5. $15\% \times 56 = 0.15 \times 56 = 8.4$

6. $40\% \times 65 = 0.40 \times 65 = 26$

7. $35\% \times 80 = 0.35 \times 80 = 28$

8. $45\% \times 25 = 0.45 \times 25 = 11.25$

9. $60\% \times 150 = 0.60 \times 150 = 90$

10. $80\% \times 210 = 0.80 \times 210 = 168$

11. $15\% \times 250 = 0.15 \times 250 = 37.5$

12. $45\% \times 500 = 0.45 \times 500 = 225$

13. $18\% \times 100 = 0.18 \times 100 = 18$

14. $55\% \times 200 = 0.55 \times 200 = 110$

15. $60\% \times 480 = 0.60 \times 480 = 288$

16. $2\% \times 80 = 0.02 \times 80 = 1.6$

17. $20\% \times \$27 = 0.20 \times \$27 = \$5.40$

18. $20\% \times \$18 = 0.20 \times \$18 = \$3.60$

19. $20\% \times \$25 = 0.20 \times \$25 = \$5.00$

20. $15\% \times \$7 = 0.15 \times \$7 = \$1.05$; $\$7 - \$1.05 = \$5.95$

21. $15\% \times \$5.60 = 0.15 \times \$5.60 = \$.84$; $\$5.60 - \$.84 = \$4.76$

22. $15\% \times \$11.90 = 0.15 \times \$11.90 = \$1.785 \approx \$1.79$; $\$11.90 - \$1.79 = \$10.11$

23. $33\frac{1}{3}\% \times \$75 = \frac{1}{3} \times \$75 = \$25$; $\$75 - \$25 = \$50$

24. $40\% \times \$12.40 = 0.40 \times \$12.40 = \$4.96$; $\$12.40 + \$4.96 = \$17.36$

Pages 367–368 • WRITTEN EXERCISES

A   **1.** $25\% \times 88 = n$; $0.25 \times 88 = n$; $n = 22$

2. $80\% \times 62 = n$; $0.80 \times 62 = n$; $n = 49.6$

3. $n \times 35 = 14$; $n = \dfrac{14}{35} = 0.4 = 40\%$ **4.** $n \times 56 = 14$; $n = \dfrac{14}{56} = 0.25 = 25\%$

5. $5\% \times 53 = n$; $0.05 \times 53 = n$; $n = 2.65$

6. $n \times 85 = 17$; $n = \dfrac{17}{85} = 0.2 = 20\%$

7. $40\% \times n = 22$; $0.40n = 22$; $100(0.40n) = 100(22)$; $40n = 2200$; $n = 55$

8. $65\% \times n = 84.5$; $0.65n = 84.5$; $100(0.65n) = 100(84.5)$; $65n = 8450$; $n = 130$

9. $95\% \times 1000 = n$; $0.95 \times 1000 = n$; $n = 950$

10. $n \times 280 = 42$; $n = \dfrac{42}{280} = 0.15 = 15\%$

11. $4\% \times 3.75 = x$; $0.04 \times 3.75 = x$; $x = 0.15$.   \$.15

12. $\dfrac{50}{60} = n$; $0.833\ldots = n$; $83\frac{1}{3}\% = n$

13. $60\% \times 53 = x$; $0.60 \times 53 = x$; $x = 31.8$.   31.8 kg

14. $n \times 700 = 1$; $n = \dfrac{1}{700} \approx 0.0014$.   0.1%

15. $68\% \times 1200 = n$; $0.68 \times 1200 = n$; $n = 816$.   816 people

16. $n \times 420 = 105$; $n = \dfrac{105}{420} = 0.25$.   25%

17. $n \times 420 = 84$; $n = \dfrac{84}{420} = 0.2$.   20%

B **18.** $6\% \times n = 87$; $0.06n = 87$; $100(0.06n) = 100(87)$; $6n = 8700$; $n = 1450$.
1450 students

19. $4\% \times 58 = n$; $0.04 \times 58 = n$; $n = 2.32$.   \$2.32

20. $15.6 = 26\% \times n$; $15.6 = 0.26n$; $n = 60$.   60 g

21. $n \times 5000 = 1200$; $n = \dfrac{1200}{5000} = 0.24$.   24%

22. $n \times 350 = 52.50$; $n = \dfrac{52.50}{350} = 0.15$.   15%

Page 368 • CALCULATOR ACTIVITIES

1. $15\% \times 30 = n$; $n = 4.5$ **2.** $17.5\% \times 200 = n$; $n = 35$

3. $n \times 20 = 15$; $n = 75\%$ **4.** $n \times 350 = 7$; $n = 2\%$

5. $20\% \times n = 2$; $n = 10$ **6.** $55.5\% \times n = 111$; $n = 200$

Page 370 • WRITTEN EXERCISES

A **1.** $I = prt$; $I = 500 \times 8\% \times 1 = 500 \times 0.08 \times 1 = 40$.   \$40

2. $I = prt$; $I = 450 \times 5\frac{1}{2}\% \times 1 = 450 \times 0.055 \times 1 = 24.75$.   \$24.75

3. $I = prt$; $I = 800 \times 6.25\% \times 1 = 800 \times 0.0625 \times 1 = 50$.   \$50

4. $I = prt$; $I = (785 \times 7\% \times 1) + \left(600 \times 7\% \times \dfrac{1}{2}\right) = (785 \times 0.07 \times 1) +$

$\left(600 \times 0.07 \times \dfrac{1}{2}\right) = 54.95 + 21 = 75.95.$ **\$75.95**

5. $I = prt$; $100 = 100 \times 6\% \times t$; $100 = 100 \times 0.06 \times t$; $100 = 6t$; $t \approx 16.7$; about 16.7 years

6. $I = prt$; $500 = 500 \times 6\% \times t$; $500 = 500 \times 0.06 \times t$; $500 = 30t$; $t \approx 16.7$; about 16.7 years

7. First bank: $I = prt$; $I = 2500 \times 5\% \times 1 = 2500 \times 0.05 \times 1 = 125.$ Second bank: $I = prt$; $I = 2500 \times 6\dfrac{1}{2}\% \times 1 = 2500 \times 0.065 \times 1 = 162.5.$ $162.5 - 125 = 37.5.$ **\$37.50**

8. $I = prt$; $I = 1000 \times 8.5\% \times 5 = 1000 \times 0.085 \times 5 = 425.$ The money will not earn enough interest. The interest earned will be \$500 − \$425 = \$75 less.

Page 372 • WRITTEN EXERCISES

A   1.

	p	r	t	I
Amount at 7%	1600	0.07	1	112
Amount at 8.5%	x	0.085	1	$0.085x$
				197

$112 + 0.085x = 197$; $112{,}000 + 85x = 197{,}000$; $85x = 85{,}000$; $x = 1000.$
\$1000 at 8.5%

2.

	p	r	t	I
Amount at 6%	x	0.06	1	$0.06x$
Amount at 8%	$1200 - x$	0.08	1	$0.08(1200 - x)$
				78

$0.06x + 0.08(1200 - x) = 78$; $6x + 8(1200 - x) = 7800$; $6x + 9600 - 8x = 7800$; $-2x + 9600 = 7800$; $-2x = -1800$; $x = 900$ and $1200 - x = 300.$
\$900 at 6% and \$300 at 8%

3.

	p	r	t	I
Amount at 7%	x	0.07	1	$0.07x$
Amount at 9%	$120{,}000 - x$	0.09	1	$0.09(120{,}000 - x)$
				9200

$0.07x + 0.09(120{,}000 - x) = 9200;$ $7x + 9(120{,}000 - x) = 920{,}000;$
$7x + 1{,}080{,}000 - 9x = 920{,}000;$ $-2x = -160{,}000;$ $x = 80{,}000$ and $120{,}000 -$
$x = 40{,}000.$ \$80,000 at 7% and \$40,000 at 9%

4.

	p	r	t	I
Amount at 7%	x	0.07	1	$0.07x$
Amount at 6%	$100{,}000 - x$	0.06	1	$0.06(100{,}000 - x)$
				6700

$0.07x + 0.06(100{,}000 - x) = 6700;$ $7x + 6(100{,}000 - x) = 670{,}000;$ $7x +$
$600{,}000 - 6x = 670{,}000;$ $x = 70{,}000$ and $100{,}000 - x = 30{,}000.$ \$70,000 at 7%,
\$30,000 at 6%

5.

	p	r	t	I
Stocks	x	0.06	1	$0.06x$
Bonds	$9000 - x$	0.07	1	$0.07(9000 - x)$
				575

$0.06x + 0.07(9000 - x) = 575;$ $6x + 7(9000 - x) = 57{,}500;$ $6x + 63{,}000 - 7x =$
$57{,}500;$ $-x = -5500;$ $x = 5500$ and $9000 - x = 3500.$ Stocks: \$5500;
Bonds: \$3500

6.

	p	r	t	I
Printing shop	x	0.08	1	$0.08x$
Grocery shop	$7000 - x$	0.05	1	$0.05(7000 - x)$
				431

$0.08x + 0.05(7000 - x) = 431;$ $8x + 5(7000 - x) = 43{,}100;$ $8x + 35{,}000 - 5x =$
$43{,}100;$ $3x = 8100;$ $x = 2700$ and $7000 - x = 4300.$ Printing shop: \$2700;
Grocery shop: \$4300

Pages 374–376 · WRITTEN EXERCISES

A   1. $70x + 95(100 - x) = 8500$; $70x + 9500 - 95x = 8500$; $-25x = -1000$; $x = 40$
and $100 - x = 60$.   40 kg of first kind, 60 kg of second kind

2.

	p	n	C
Keyrings	0.50	x	$0.50x$
Calendars	0.60	$5000 - x$	$0.60(5000 - x)$
			2650

$0.50x + 0.60(5000 - x) = 2650$; $50x + 60(5000 - x) = 265,000$; $50x +$
$300,000 - 60x = 265,000$; $-10x = -35,000$; $x = 3500$ and $5000 - x = 1500$.
3500 keyrings, 1500 calendars

3.

	p	n	C
Children	2.50	x	$2.50x$
Adults	4.25	$600 - x$	$4.25(600 - x)$
			1885

$2.50x + 4.25(600 - x) = 1885$; $250x + 425(600 - x) = 188,500$; $250x +$
$255,000 - 425x = 188,500$; $-175x = -66,500$; $x = 380$.   380 children's tickets

4.

	p	n	C
Safflower oil	5.25	x	$5.25x$
Corn oil	2.56	$100 - x$	$2.56(100 - x)$
			305

$5.25x + 2.56(100 - x) = 305$; $525x + 256(100 - x) = 30,500$; $525x +$
$25,600 - 256x = 30,500$; $269x = 4900$; $x \approx 18$.   about 18 liters of safflower oil

5.

	p	n	C
First kind	3.20	x	$3.20x$
Second kind	4.40	$60 - x$	$4.40(60 - x)$
Mixture	4.00	60	240

$3.20x + 4.40(60 - x) = 240$; $32x + 44(60 - x) = 2400$; $32x + 2640 - 44x =$
2400; $-12x = -240$; $x = 20$ and $60 - x = 40$.   20 pounds of the first kind,
40 pounds of the second kind

6.

	Amount of coolant	% of antifreeze	Amount of antifreeze
Old mixture	1000	30%	300
Water	x	0%	0
New mixture	1000 + x	25%	0.25(1000 + x)

$300 + 0 = 0.25(1000 + x)$; $30{,}000 = 25(1000 + x)$; $30{,}000 = 25{,}000 + 25x$;
$5000 = 25x$; $200 = x$. 200 liters of water

B **7.**

	Amount of repellent	% of active ingredients	Amount of active ingredients
Old repellent	1000	32%	320
Boiled-away ingredients	x	0%	0
New repellent	1000 − x	40%	0.40(1000 − x)

$320 - 0 = 0.40(1000 - x)$; $32{,}000 = 40(1000 - x)$; $32{,}000 = 40{,}000 - 40x$;
$-8000 = -40x$; $200 = x$. 200 kg of inactive ingredients

8.

	Amount of mixture	% of clay	Amount of clay
Old mixture	500	45%	225
Graphite added	x	0%	0
New mixture	500 + x	40%	0.40(500 + x)

$225 + 0 = 0.40(500 + x)$; $22{,}500 = 40(500 + x)$; $22{,}500 = 20{,}000 + 40x$;
$2500 = 40x$; $62.5 = x$. 62.5 liters of graphite

C **9.**

	Value of each coin	Number of coins	Value
Dimes	0.10	x	0.10x
Quarters	0.25	19 − x	0.25(19 − x)
			4.00

$0.10x + 0.25(19 - x) = 4.00$; $10x + 25(19 - x) = 400$; $10x + 475 - 25x = 400$; $-15x = -75$; $x = 5$ and $19 - x = 14$. 5 dimes, 14 quarters

Page 377 · CALCULATOR ACTIVITIES

1. 0.625 **2.** 0.0833... **3.** 0.533... **4.** 0.6363...

5. $\dfrac{1}{5} \cdot \dfrac{1}{2} = 0.1$ 6. $\dfrac{1}{5} + \dfrac{1}{2} = 0.7$ 7. $\dfrac{1}{5-2} = \dfrac{1}{3} = 0.33\ldots$

8. $\dfrac{1}{5^2 + 2} = \dfrac{1}{25 + 2} = \dfrac{1}{27} = 0.037037\ldots$

Page 379 · PROBLEM SOLVING STRATEGIES

1. A. Number of boys + Number of girls = 23
   B. 5 = Number of girls − Number of boys

2. A. 2000 = Amount invested at 8% + Amount invested at 5.5%
   B. Interest earned at 8% = $52 + Interest earned at 5.5%

3. Distance Marla travels + Distance Danny travels = 21

   Let t = time each travels; $8t + 6t = 21$; $14t = 21$; $t = 1\dfrac{1}{2}$.   They meet after

   $1\dfrac{1}{2}$ hours.

4. Part of the job done in one hour when they work together =
   Part Mr. Martinez does alone in one hour + Part daughter does alone in one
   hour. Let n = number of hours it would take the daughter alone;

   $\dfrac{1}{3} = \dfrac{1}{5} + \dfrac{1}{n}$; $\dfrac{1}{n} = \dfrac{2}{15}$; $n = 7.5$.   7.5 h

5. Cost of pear = 2 × cost of apple
   Cost of 3 apples + cost of 2 pears = 280
   Let p = cost of pear and a = cost of apple (in cents); $p = 2a$; $3a + 2p = 280$;
   $3a + 4a = 280$; $7a = 280$; $a = 40$; $p = 2(40) = 80$.   An apple costs 40¢ and a
   pear costs 80¢.

6. Mark's age + Joanne's age = 30
   Mark's age − Joanne's age = 6 (since Mark is older)
   Let m = Mark's age and j = Joanne's age; $m + j = 30$; $m - j = 6$; $2m = 36$;
   $m = 18$; $18 - j = 6$; $j = 12$.   Mark is 18 and Joanne is 12.

Page 380 · SKILLS REVIEW

1. increase: $3.50 − $2.98 = $.52   2. increase: $9.00 − $6.50 = $2.50

3. decrease: $4.25 − $3.50 = $.75   4. decrease: $25 − $16.80 = $8.20

5. increase: $12.59 − $10.35 = $2.24   6. increase: $255 − $230 = $25

7. increase = $1.00; $n \times \$2.00 = \$1.00$; $2n = 1$; $n = 0.5 = 50\%$

8. increase = $3.00; $n \times \$5.00 = \$3.00$; $5n = 3$; $n = 0.6 = 60\%$

9. increase = $2.00; $n \times \$4.00 = \$2.00$; $4n = 2$; $n = 0.5 = 50\%$

10. increase = $1.40; $n \times \$2.00 = \$1.40$; $2n = 1.4$; $20n = 14$; $n = 0.7 = 70\%$

11. increase = $1.60; $n \times \$8.00 = \$1.60$; $8n = 1.6$; $80n = 16$; $n = 0.2 = 20\%$

12. increase = $1.80; $n \times \$6.00 = \$1.80$; $6n = 1.8$; $60n = 18$; $n = 0.3 = 30\%$

13. increase = $5; $n \times \$20 = \$5$; $20n = 5$; $n = 0.25 = 25\%$

14. increase = $10; $n \times \$40 = \$10$; $40n = 10$; $n = 0.25 = 25\%$

15. increase = $24; $n \times \$60 = \$24$; $60n = 24$; $n = 0.4 = 40\%$

16. increase = $50; $n \times \$100 = \$50$; $100n = 50$; $n = 0.5 = 50\%$
17. decrease = $10; $n \times \$20 = \$10$; $20n = 10$; $n = 0.5 = 50\%$
18. decrease = $9; $n \times \$18 = \$9$; $18n = 9$; $n = 0.5 = 50\%$
19. decrease = $6; $n \times \$30 = \$6$; $30n = 6$; $n = 0.2 = 20\%$
20. decrease = $20; $n \times \$50 = \$20$; $50n = 20$; $n = 0.4 = 40\%$
21. decrease = $24; $n \times \$80 = \$24$; $80n = 24$; $n = 0.3 = 30\%$
22. decrease = $6; $n \times \$24 = \$6$; $24n = 6$; $n = 0.25 = 25\%$
23. decrease = $50; $n \times \$200 = \$50$; $200n = 50$; $n = 0.25 = 25\%$
24. decrease = $25; $n \times \$125 = \$25$; $125n = 25$; $n = 0.2 = 20\%$

Pages 381–382 · CHAPTER REVIEW EXERCISES

1. 5.70   **2.** 17.155   **3.** 51.15   **4.** 67.12   **5.** 4.9   **6.** 11.75

7. 78.08   **8.** 2.345   **9.** 12.24   **10.** 0.45   **11.** 0.812   **12.** 0.015

13.
$$\begin{array}{r} 52.40 \\ 5\overline{)262.00} \end{array}$$
$26.2 \div 0.5 = 52.4$

14.
$$\begin{array}{r} 1000.00 \\ 5\overline{)5000.00} \end{array}$$
$50 \div 0.05 = 1000.0$

15.
$$\begin{array}{r} 2142.50 \\ 2\overline{)4285.00} \end{array}$$
$42.85 \div 0.02 = 2142.5$

16.
$$\begin{array}{r} 1.65 \\ 19\overline{)31.39} \end{array}$$
$3.139 \div 1.9 \approx 1.7$

17. $\frac{3}{8} = 0.375$

18. $\frac{1}{8} = 0.125$

19. $\frac{5}{6} = 0.833\ldots$

20. $\frac{1}{11} = 0.0909\ldots$

21. $\frac{4}{9} = 0.44\ldots$

22. $0.3n + 0.9n = 4.8$; $10(0.3n + 0.9n) = 10(4.8)$; $3n + 9n = 48$; $12n = 48$; $n = 4$

23. $1.2x - 0.9x = 33$; $10(1.2x - 0.9x) = 10(33)$; $12x - 9x = 330$; $3x = 330$; $x = 110$

24. $0.7y = 4.9$; $10(0.7y) = 10(4.9)$; $7y = 49$; $y = 7$

25. $0.12x = 3.6$; $100(0.12x) = 100(3.6)$; $12x = 360$; $x = 30$

26. $0.12n = 45 - 0.03n$; $100(0.12n) = 100(45 - 0.03n)$; $12n = 4500 - 3n$; $15n = 4500$; $n = 300$

27. $5.5(2x - 3) = 10x$; $10[5.5(2x - 3)] = 10(10x)$; $55(2x - 3) = 100x$; $110x - 165 = 100x$; $-165 = -10x$; $16.5 = x$

28. $0.31x + 0.58 = 0.6x$; $100[0.31x + 0.58] = 100(0.6x)$; $31x + 58 = 60x$; $58 = 29x$; $2 = x$

29. $4.2z + 0.5(3z) = 57$; $10[4.2z + 0.5(3z)] = 10(57)$; $42z + 5(3z) = 570$; $42z + 15z = 570$; $57z = 570$; $z = 10$

30. $2.1(15 + n) = 63$; $10[2.1(15 + n)] = 10(63)$; $21(15 + n) = 630$; $315 + 21n = 630$; $21n = 315$; $n = 15$

31. Let x = cost of tire. Then $x + 21$ = cost of generator. $x + x + 21 = 109.50$; $2x + 21 = 109.50$; $100(2x + 21) = 100(109.50)$; $200x + 2100 = 10,950$; $200x = 8850$; $x = 44.25$. $44.25

32. Let x = cm of growth in first year. Then $x + 4.5$ = growth in second year. $x + x + 4.5 = 8.3$; $2x + 4.5 = 8.3$; $10(2x + 4.5) = 10(8.3)$; $20x + 45 = 83$; $20x = 38$; $x = 1.9$. 1.9 cm

33. $0.04 = 4\%$ **34.** $0.32 = 32\%$ **35.** $0.136 = 13.6\%$

36. $0.025 = 2.5\%$ **37.** $0.9 = 90\%$ **38.** $0.3 = 30\%$

39. $\dfrac{1}{4} = 0.25 = 25\%$ **40.** $\dfrac{1}{5} = \dfrac{20}{100} = 20\%$ **41.** $\dfrac{1}{8} = 0.125 = 12.5\%$

42. $\dfrac{1}{3} = 33\dfrac{1}{3}\%$ **43.** $36\% = 0.36$ **44.** $85\% = 0.85$

45. $2.5\% = 0.025$ **46.** $3.15\% = 0.0315$ **47.** $7\dfrac{1}{2}\% = 0.07\dfrac{1}{2}$, or 0.075

48. $25\% = \dfrac{25}{100} = \dfrac{1}{4}$ **49.** $10\% = \dfrac{10}{100} = \dfrac{1}{10}$ **50.** $66\dfrac{2}{3}\% = \dfrac{2}{3}$

51. $50\% = \dfrac{50}{100} = \dfrac{1}{2}$ **52.** $75\% = \dfrac{75}{100} = \dfrac{3}{4}$

53. $40\% \times 50 = n$; $0.40 \times 50 = n$; $n = 20$

54. $33\dfrac{1}{3}\% \times 15 = n$; $\dfrac{1}{3} \times 15 = n$; $n = 5$

55. $n \times 34 = 17$; $n = \dfrac{17}{34} = 0.5 = 50\%$

56. $n \times 16 = 2$; $n = \dfrac{2}{16} = 0.125 = 12.5\%$

57. $3\% \times n = 3$; $0.03n = 3$; $100(0.03n) = 100(3)$; $3n = 300$; $n = 100$

58. $32\% \times n = 48$; $0.32n = 48$; $100(0.32n) = 100(48)$; $32n = 4800$; $n = 150$

59.

	p	r	t	I
Amount at 6%	x	0.06	1	$0.06x$
Amount at 9%	$2000 - x$	0.09	1	$0.09(2000 - x)$
				141

$0.06x + 0.09(2000 - x) = 141$; $6x + 9(2000 - x) = 14{,}100$; $6x + 18{,}000 - 9x = 14{,}100$; $-3x = -3900$; $x = 1300$ and $2000 - x = 700$. $1300 at 6%, $700 at 9%

60.

	p	n	C
One dinner	15	x	$15x$
Other dinner	100	$550 - x$	$100(550 - x)$
			21,000

$15x + 100(550 - x) = 21,000$; $15x + 55,000 - 100x = 21,000$; $-85x = -34,000$; $x = 400$ and $550 - x = 150$.   400 people at the \$15 dinner, 150 people at the \$100 dinner

61.

	Amount of solution	% of salt	Amount of salt
Old solution	10	3%	0.3
Water	x	0%	0
New solution	$10 + x$	2%	$0.02(10 + x)$

$0.3 + 0 = 0.02(10 + x)$; $30 = 2(10 + x)$; $30 = 20 + 2x$; $10 = 2x$; $5 = x$.
5 liters of water

Page 383 · CHAPTER TEST

1. 15.168

2. 6.063

3. 0.8528

4. $27\overline{)83.7} \quad 3.1$

5. $3\overline{)80.00} \quad 26.66$

$8 \div 0.3 \approx 26.7$

6. $64\overline{)2400.0} \quad 37.5$

7. $\dfrac{2}{15} = 0.133\ldots$

8. $\dfrac{19}{20} = 0.95$

9. $\dfrac{17}{25} = 0.68$

10. $\dfrac{5}{18} = 0.277\ldots$

11. $0.3x = 1.36 - 0.02x$; $100(0.3x) = 100(1.36 - 0.02x)$; $30x = 136 - 2x$; $32x = 136$; $x = 4.25$

12. $0.4(a + 1) + 0.05a = 2.2$; $100[0.4(a + 1) + 0.05a] = 100(2.2)$; $40(a + 1) + 5a = 220$; $40a + 40 + 5a = 220$; $45a + 40 = 220$; $45a = 180$; $a = 4$

13. Let x = cost (in dollars) of hat. Then $x + 3.15$ = cost of shirt. $x + x + 3.15 = 13.35$; $2x + 3.15 = 13.35$; $100(2x + 3.15) = 100(13.35)$; $200x + 315 = 1335$; $200x = 1020$; $x = 5.1$ and $x + 3.15 = 8.25$.   hat: \$5.10;   shirt: \$8.25

14. a. $0.04 = \dfrac{4}{100} = 4\%$ **b.** $\dfrac{7}{10} = \dfrac{70}{100} = 70\%$ **15. a.** $29\% = 0.29$ **b.** $4.2\% = 0.042$

16. a. $33\dfrac{1}{3}\% = \dfrac{1}{3}$ **b.** $90\% = \dfrac{90}{100} = \dfrac{9}{10}$ **17.** $55\% \times 40 = 0.55 \times 40 = 22$

18. $90\% \times 240 = 0.90 \times 240 = 216$ **19.** $50\% \times 25 = 0.50 \times 25 = 12.5$

20. $20\% \times \$15 = 0.20 \times \$15 = \$3$; $\$15 - \$3 = \$12$

21. $n \times 68 = 51$; $n = \dfrac{51}{68} = 0.75 = 75\%$

22. $35\% \times n = 4.2$; $0.35 \times n = 4.2$; $n = \dfrac{4.2}{0.35} = 12$

23. $10\% \times \$28.50 = 0.10 \times \$28.50 = 2.85$; $\$28.50 + \$2.85 = \$31.35$

24. $I = prt$; $I = 300 \times 5.5\% \times 1 = 300 \times 0.055 \times 1 = 16.5$. $\$16.50$

25.

	p	r	t	I
Amount at 6%	x	0.06	1	$0.06x$
Amount at 9%	$6000 - x$	0.09	1	$0.09(6000 - x)$
				480

$0.06x + 0.09(6000 - x) = 480$; $6x + 9(6000 - x) = 48{,}000$; $6x + 54{,}000 - 9x = 48{,}000$; $-3x = -6000$; $x = 2000$ and $6000 - x = 4000$.   \$2000 at 6% and \$4000 at 9%

26.

	p	n	C
Children	2	x	$2x$
Adults	3	$438 - x$	$3(438 - x)$
			1018

$2x + 3(438 - x) = 1018$; $2x + 1314 - 3x = 1018$; $-x + 1314 = 1018$; $-x = -296$; $x = 296$.   296 children's tickets

Page 384 • CUMULATIVE REVIEW

1. $\dfrac{56xy^2}{24y} = \dfrac{7 \cdot 8 \cdot x \cdot y \cdot y}{3 \cdot 8 \cdot y} = \dfrac{7xy}{3}$

2. $\dfrac{68x^3y^2z}{24xyz^4} = \dfrac{17 \cdot 4 \cdot x \cdot x \cdot x \cdot y \cdot y \cdot z}{6 \cdot 4 \cdot x \cdot y \cdot z \cdot z \cdot z \cdot z} = \dfrac{17x^2y}{6z^3}$ **3.** $\dfrac{3z - 6}{4z - 8} = \dfrac{3(z - 2)}{4(z - 2)} = \dfrac{3}{4}$

4. $\dfrac{5 - b}{b^2 - 25} = \dfrac{-1(b - 5)}{(b + 5)(b - 5)} = \dfrac{-1}{b + 5} = -\dfrac{1}{b + 5}$

5. Let $5x$ and x = the lengths (in cm) of the pieces. $5x + x = 180$; $6x = 180$; $x = 30$, so $5x = 5 \cdot 30 = 150$.   150 cm

6. Let x = tax on an \$8000 car. $\dfrac{5400}{324} = \dfrac{8000}{x}$; $5400x = 2{,}592{,}000$; $x = 480$.   \$480

7. $\dfrac{8}{r^2 - 9} \cdot \dfrac{3 - r}{16} = \dfrac{8(-1)(r - 3)}{(r + 3)(r - 3)(16)} = \dfrac{-1}{(r + 3)(2)} = -\dfrac{1}{2(r + 3)}$

8. $\dfrac{x^2 - 4}{3x} \div \dfrac{x + 2}{3x - 6} = \dfrac{x^2 - 4}{3x} \cdot \dfrac{3x - 6}{x + 2} = \dfrac{(x + 2)(x - 2)(3)(x - 2)}{3x(x + 2)} = \dfrac{(x - 2)^2}{x}$

9. $\dfrac{5}{x} + \dfrac{7}{3x^2} = \dfrac{5 \cdot 3x}{3x^2} + \dfrac{7}{3x^2} = \dfrac{15x + 7}{3x^2}$

10. $\dfrac{x}{5} + \dfrac{x}{8} = 13$; $\dfrac{8x + 5x}{40} = 13$; $\dfrac{13x}{40} = \dfrac{13}{1}$; $13x = 520$; $x = 40$

11. $\dfrac{a - 1}{2} + \dfrac{a + 2}{3} = 1$; $\dfrac{3(a - 1) + 2(a + 2)}{6} = 1$; $\dfrac{3a - 3 + 2a + 4}{6} = 1$; $\dfrac{5a + 1}{6} = \dfrac{1}{1}$; $5a + 1 = 6$; $5a = 5$; $a = 1$

12.

	Pipe A	Pipe B	Together
Hours needed	16	8	x
Part done in one hour	$\dfrac{1}{16}$	$\dfrac{1}{8}$	$\dfrac{1}{x}$

$\dfrac{1}{16} + \dfrac{1}{8} = \dfrac{1}{x}$; $\dfrac{1}{16} + \dfrac{2}{16} = \dfrac{1}{x}$; $\dfrac{3}{16} = \dfrac{1}{x}$; $3x = 16$; $x = 5\dfrac{1}{3}$. $5\dfrac{1}{3}$ hours

13. 8.38 **14.** 13.16 **15.** 1.125 **16.** −5.31 **17.** 34.32

18. 13.363

19. $\dfrac{25}{34)\,850}$

20. $\dfrac{5.5}{71)\,390.5}$

21. a. $\dfrac{22}{25} = 0.88$ **b.** $\dfrac{8}{9} = 0.88\ldots$ **c.** $3\% = 0.03$

22. a. $0.65 = 65\%$ **b.** $0.732 = 73.2\%$ **c.** $\dfrac{7}{8} = 0.875 = 87.5\%$

23. a. $85\% = \dfrac{85}{100} = \dfrac{17}{20}$ **b.** $24\% = \dfrac{24}{100} = \dfrac{6}{25}$ **c.** $30\% = \dfrac{30}{100} = \dfrac{3}{10}$

24. $24\% \times 25 = n$; $0.24 \times 25 = n$; $6 = n$

25. $n \times 80 = 56$; $n = \dfrac{56}{80} = 0.7 = 70\%$

26.

	p	r	t	I
Amount at 5%	x	0.05	1	$0.05x$
Amount at 7%	$2000 - x$	0.07	1	$0.07(2000 - x)$
				125

$0.05x + 0.07(2000 - x) = 125$; $5x + 7(2000 - x) = 12{,}500$; $5x + 14{,}000 - 7x = 12{,}500$; $-2x = -1500$; $x = 750$ and $2000 - x = 1250$. \$750 at 5%, \$1250 at 7%

27.

	Amount of solution	% of acid	Amount of acid
Old solution	800	40%	320
Water added	x	0%	0
New solution	$800 + x$	25%	$0.25(800 + x)$

$320 + 0 = 0.25(800 + x)$; $32{,}000 = 25(800 + x)$; $32{,}000 = 20{,}000 + 25x$; $12{,}000 = 25x$; $480 = x$. 480 liters of water

CHAPTER 11 · Squares and Square Roots

Page 389 · WRITTEN EXERCISES

A   1. 2   2. −3   3. 5   4. −5   5. 4   6. −4   7. −2   8. 0
9. 1   10. 7   11. −7   12. −1   13. −6   14. 3   15. 8
16. 10   17. −8   18. 9   19. 6   20. −9   21. 8   22. 9
23. 3   24. −5   25. −10
26. $s = \sqrt{36} = 6$, so $P = 4s = 4 \cdot 6 = 24$.   24 cm
27. $s = \sqrt{25} = 5$, so $P = 4 \cdot 5 = 20$.   20 cm
28. $s = \sqrt{64} = 8$, so $P = 4 \cdot 8 = 32$.   32 cm
29. $s = \sqrt{16} = 4$, so $P = 4 \cdot 4 = 16$.   16 ft
30. $s = \sqrt{49} = 7$, so $P = 4 \cdot 7 = 28$.   28 in.
31. $s = \sqrt{100} = 10$, so $P = 4 \cdot 10 = 40$.   40 m

B   32. $\sqrt{3^2 + 4^2} = \sqrt{9 + 16} = \sqrt{25} = 5$ 33. $\sqrt{6^2 + 8^2} = \sqrt{36 + 64} = \sqrt{100} = 10$
34. $\sqrt{12^2 + 5^2} = \sqrt{144 + 25} = \sqrt{169} = 13$

Page 389 · PUZZLE PROBLEMS

Answers may vary. $5 = (\sqrt{4} \times \sqrt{4}) + (4 \div 4)$; $6 = \sqrt{4} + (4 \times 4) \div 4$; $7 = 4 + 4 - (4 \div 4)$; $8 = 4 + (4 \times 4) \div 4$; $9 = 4 + 4 + (4 \div 4)$; $10 = 4 + 4 + 4 - \sqrt{4}$;
$11 = 44 \div (\sqrt{4} + \sqrt{4})$; $12 = \sqrt{4} \times (4 + 4) - 4$; $13 = (44 \div 4) + \sqrt{4}$;
$14 = 4 + 4 + 4 + \sqrt{4}$; $15 = (4 \times 4) - (4 \div 4)$; $16 = 4 + 4 + 4 + 4$

Page 391 · WRITTEN EXERCISES

A   1. 2.828   2. 3.873   3. 6.325   4. 5.477   5. 4.243   6. 8.062
7. 4.472   8. 7.141   9. 8.888   10. 4.359   11. 9.434   12. 3.162
13. 5.916   14. 7.874   15. 8.660   16. −7.280   17. −6.164
18. −5.385   19. −7.071   20. −8.602   21. 7.550   22. ±9   23. ±4
24. ±5.916   25. ±8.185   26. 2.646 ≈ 2.6   27. 3.742 ≈ 3.7
28. 3.162 ≈ 3.2   29. 7.550 ≈ 7.6   30. 9.110 ≈ 9.1   31. 7.141 ≈ 7.1
32. 4.583 ≈ 4.6   33. 4.359 ≈ 4.4   34. 5.196 ≈ 5.2   35. 6.083 ≈ 6.1
36. 7.483 ≈ 7.5   37. −4.243 ≈ −4.2   38. −6.557 ≈ −6.6   39. −8.426 ≈ −8.4
40. −6.928 ≈ −6.9   41. 5.657 ≈ 5.7   42. 9.695 ≈ 9.7   43. −8.246 ≈ −8.2
44. ±4.472 ≈ ±4.5   45. 9.055 ≈ 9.1

B   46. $s = \sqrt{70} \approx 8.367 \approx 8.4$;   8.4 cm   47. $s = \sqrt{56} \approx 7.483 \approx 7.5$;   7.5 cm
48. $s = \sqrt{20} \approx 4.472 \approx 4.5$;   4.5 cm   49. $s = \sqrt{35} \approx 5.916 \approx 5.9$;   5.9 cm
50. $s = \sqrt{48} \approx 6.928$, so $P = 4s \approx 4(6.928) = 27.712$;   27.7 cm
51. $s = \sqrt{26} \approx 5.099$, so $P \approx 4(5.099) = 20.396$;   20.4 cm
52. $s = \sqrt{38} \approx 6.164$, so $P \approx 4(6.164) = 24.656$;   24.7 cm
53. $s = \sqrt{65} \approx 8.062$, so $P \approx 4(8.062) = 32.248$;   32.2 cm

54. $s = \sqrt{24} \approx 4.899$, so $P \approx 4(4.899) = 19.596$;   19.6 cm

55. $s = \sqrt{60} \approx 7.746$, so $P \approx 4(7.746) = 30.984$;   31.0 cm

56. $s = \sqrt{40} \approx 6.325$, so $P \approx 4(6.325) = 25.3$;   25.3 cm

57. $s = \sqrt{72} \approx 8.485$, so $P \approx 4(8.485) = 33.94$;   33.9 cm

Page 393 · WRITTEN EXERCISES

A   **1.** rational   **2.** rational   **3.** rational   **4.** rational   **5.** rational

6. irrational   **7.** rational   **8.** rational   **9.** irrational   **10.** rational

11. irrational   **12.** rational   **13.** rational   **14.** rational   **15.** rational

16. rational   **17.** rational   **18.** rational   **19.** rational   **20.** irrational

21. rational   **22.** irrational   **23.** rational   **24.** rational   **25.** rational

26. irrational   **27.** rational   **28.** rational   **29.** irrational   **30.** rational

31. $A = \pi r^2$; $A \approx (3.14)(7^2) = (3.14)(49) = 153.86$;   153.86 cm²

32. $A = \pi r^2$; $A \approx (3.14)(10^2) = (3.14)(100) = 314$;   314 cm²

33. $A = \pi r^2$; $A \approx (3.14)(3.5^2) = (3.14)(12.25) = 38.465$;   38.465 cm²

34. $A = \pi r^2$; $A \approx (3.14)(3^2) = (3.14)(9) = 28.26$;   28.26 cm²

35. perfect square

B   **36.** $\dfrac{2}{3} + \dfrac{5}{6} = \dfrac{4}{6} + \dfrac{5}{6} = \dfrac{9}{6} = \dfrac{3}{2}$ **37.** $\dfrac{4}{5} - \dfrac{3}{10} = \dfrac{8}{10} - \dfrac{3}{10} = \dfrac{5}{10} = \dfrac{1}{2}$

38. $\dfrac{4}{7} + \dfrac{3}{5} = \dfrac{20}{35} + \dfrac{21}{35} = \dfrac{41}{35}$ **39.** $\dfrac{2}{9} - \dfrac{5}{6} = \dfrac{4}{18} - \dfrac{15}{18} = -\dfrac{11}{18}$

40. Yes   **41.** Answers may vary. $\sqrt{2} - \sqrt{2} = 0$, which is rational.

Page 395 · WRITTEN EXERCISES

A   **1.** $\sqrt{144} = \sqrt{9 \cdot 16} = \sqrt{9} \cdot \sqrt{16} = 3 \cdot 4 = 12$

2. $\sqrt{243} = \sqrt{81 \cdot 3} = \sqrt{81} \cdot \sqrt{3} = 9\sqrt{3}$

3. $\sqrt{225} = \sqrt{9 \cdot 25} = \sqrt{9} \cdot \sqrt{25} = 3 \cdot 5 = 15$

4. $\sqrt{196} = \sqrt{4 \cdot 49} = \sqrt{4} \cdot \sqrt{49} = 2 \cdot 7 = 14$

5. $\sqrt{1000} = \sqrt{100 \cdot 10} = \sqrt{100} \cdot \sqrt{10} = 10\sqrt{10}$

6. $\sqrt{12} = \sqrt{4 \cdot 3} = \sqrt{4} \cdot \sqrt{3} = 2\sqrt{3}$

7. $\sqrt{2000} = \sqrt{400 \cdot 5} = \sqrt{400} \cdot \sqrt{5} = 20\sqrt{5}$

8. $\sqrt{324} = \sqrt{4 \cdot 81} = \sqrt{4} \cdot \sqrt{81} = 2 \cdot 9 = 18$

9. $\sqrt{128} = \sqrt{64 \cdot 2} = \sqrt{64} \cdot \sqrt{2} = 8\sqrt{2}$

10. $\sqrt{288} = \sqrt{144 \cdot 2} = \sqrt{144} \cdot \sqrt{2} = 12\sqrt{2}$

11. $\sqrt{180} = \sqrt{36 \cdot 5} = \sqrt{36} \cdot \sqrt{5} = 6\sqrt{5}$

12. $\sqrt{338} = \sqrt{169 \cdot 2} = \sqrt{169} \cdot \sqrt{2} = 13\sqrt{2}$

13. $\sqrt{108} = \sqrt{36 \cdot 3} = \sqrt{36} \cdot \sqrt{3} = 6\sqrt{3}$

14. $\sqrt{1764} = \sqrt{36 \cdot 49} = \sqrt{36} \cdot \sqrt{49} = 6 \cdot 7 = 42$

15. $\sqrt{192} = \sqrt{64 \cdot 3} = \sqrt{64} \cdot \sqrt{3} = 8\sqrt{3}$

B   16. $\sqrt{16x^2} = \sqrt{16 \cdot x^2} = \sqrt{16} \cdot \sqrt{x^2} = 4x$ 17. $\sqrt{12y^2} = \sqrt{3 \cdot 4 \cdot y^2} = 2y\sqrt{3}$

18. $\sqrt{64x^2y} = \sqrt{64 \cdot x^2 \cdot y} = 8x\sqrt{y}$ 19. $\sqrt{15x^2y^2} = \sqrt{15 \cdot x^2 \cdot y^2} = xy\sqrt{15}$

20. $\sqrt{75a^4} = \sqrt{3 \cdot 25 \cdot a^2 \cdot a^2} = 5a^2\sqrt{3}$ 21. $\sqrt{36xy^2} = \sqrt{36 \cdot x \cdot y^2} = 6y\sqrt{x}$

22. $\sqrt{125x^3} = \sqrt{5 \cdot 25 \cdot x \cdot x^2} = 5x\sqrt{5x}$ 23. $\sqrt{3a^2b^2} = \sqrt{3 \cdot a^2 \cdot b^2} = ab\sqrt{3}$

C   24. Let $x = -2$; $\sqrt{(-2)^2} = \sqrt{4} = 2 \neq -2$.

Page 397 · WRITTEN EXERCISES

A   1. $x^2 = 64$; $x = \pm\sqrt{64} = \pm 8$ 2. $x^2 = 81$; $x = \pm\sqrt{81} = \pm 9$

3. $m^2 = 16$; $m = \pm\sqrt{16} = \pm 4$ 4. $n^2 = 100$; $n = \pm\sqrt{100} = \pm 10$

5. $y^2 = 9$; $y = \pm 3$ 6. $n^2 = 49$; $n = \pm 7$

7. $n^2 = 121$; $n = \pm 11$ 8. $x^2 = 144$; $x = \pm 12$

9. $\dfrac{x}{3} = \dfrac{3}{x}$; $x^2 = 9$; $x = \pm 3$ 10. $\dfrac{x}{4} = \dfrac{25}{x}$; $x^2 = 100$; $x = \pm 10$

11. $\dfrac{4}{x} = \dfrac{x}{9}$; $x^2 = 36$; $x = \pm 6$ 12. $\dfrac{y}{2} = \dfrac{8}{y}$; $y^2 = 16$; $y = \pm 4$

13. $y^2 - 1 = 35$; $y^2 = 36$; $y = \pm 6$ 14. $y^2 - 5 = 20$; $y^2 = 25$; $y = \pm 5$

15. $x^2 + 6 = 31$; $x^2 = 25$; $x = \pm 5$ 16. $n^2 + 4 = 53$; $n^2 = 49$; $n = \pm 7$

17. $x^2 = 12$; $x = \pm\sqrt{12} = \pm\sqrt{4 \cdot 3} = \pm 2\sqrt{3}$

18. $n^2 = 10$; $n = \pm\sqrt{10}$

19. $y^2 = 6$; $y = \pm\sqrt{6}$

20. $x^2 = 27$; $x = \pm\sqrt{27} = \pm\sqrt{9 \cdot 3} = \pm 3\sqrt{3}$

21. $x^2 + 2 = 8$; $x^2 = 6$; $x = \pm\sqrt{6}$

22. $y^2 - 7 = 20$; $y^2 = \sqrt{27}$; $y = \pm\sqrt{27} = \pm\sqrt{9 \cdot 3} = \pm 3\sqrt{3}$

23. $5 + y^2 = 12$; $y^2 = 7$; $y = \pm\sqrt{7}$ 24. $8 + y^2 = 30$; $y^2 = 22$; $y = \pm\sqrt{22}$

25. $x^2 + 5 = 23$; $x^2 = 18$; $x = \pm\sqrt{18} = \pm\sqrt{9 \cdot 2} = \pm 3\sqrt{2}$

26. $n^2 - 2 = 40$; $n^2 = 42$; $n = \pm\sqrt{42}$ 27. $x^2 - 5 = 11$; $x^2 = 16$; $x = \pm 4$

28. $7 + a^2 = 9$; $a^2 = 2$; $a = \pm\sqrt{2}$ 29. $x^2 - 1 = 3$; $x^2 = 4$; $x = \pm 2$

30. $y^2 + 5 = 21$; $y^2 = 16$; $y = \pm 4$ 31. $x^2 - 5 = 31$; $x^2 = 36$; $x = \pm 6$

32. $a^2 + 1 = 82$; $a^2 = 81$; $a = \pm 9$

33. $n^2 = 40$; $n = \pm\sqrt{40} \approx \pm 6.325 \approx \pm 6.3$

34. $x^2 = 56$; $x = \pm\sqrt{56} \approx \pm 7.483 \approx \pm 7.5$

35. $y^2 = 38$; $y = \pm\sqrt{38} \approx \pm 6.164 \approx \pm 6.2$

36. $x^2 = 91$; $x = \pm\sqrt{91} \approx \pm 9.539 \approx \pm 9.5$

37. $x^2 - 1 = 75$; $x^2 = 76$; $x = \pm\sqrt{76} \approx \pm 8.7$

38. $y^2 + 3 = 80$; $y^2 = 77$; $y = \pm\sqrt{77} \approx \pm 8.8$

39. $x^2 + 7 = 52$; $x^2 = 45$; $x = \pm\sqrt{45} \approx \pm 6.7$

40. $n^2 - 4 = 75$; $n^2 = 79$; $n = \pm\sqrt{79} \approx \pm 8.9$

Pages 400–401 • WRITTEN EXERCISES

A   In Exs. 1–6, let x = the length of the third side.

1. $2^2 + 4^2 = x^2$; $4 + 16 = x^2$; $20 = x^2$; $x = \sqrt{20} = 2\sqrt{5}$

2. $4^2 + 5^2 = x^2$; $16 + 25 = x^2$; $41 = x^2$; $x = \sqrt{41}$

3. $6^2 + 2^2 = x^2$; $36 + 4 = x^2$; $40 = x^2$; $x = \sqrt{40} = 2\sqrt{10}$

4. $5^2 + 5^2 = x^2$; $25 + 25 = x^2$; $50 = x^2$; $x = \sqrt{50} = 5\sqrt{2}$

5. $x^2 + 2^2 = 8^2$; $x^2 + 4 = 64$; $x^2 = 60$; $x = \sqrt{60} = 2\sqrt{15}$

6. $5^2 + x^2 = 14^2$; $25 + x^2 = 196$; $x^2 = 171$; $x = \sqrt{171} = 3\sqrt{19}$

7. Let x = the height (in m). $x^2 + 1^2 = 6^2$; $x^2 + 1 = 36$; $x^2 = 35$;
   $x = \sqrt{35} \approx 5.9$.   5.9 m

8. Let x = the length (in cm). $12^2 + 8^2 = x^2$; $144 + 64 = x^2$; $208 = x^2$;
   $x = \sqrt{208} = 4\sqrt{13} \approx 4(3.606) = 14.424 = 14.4$.   14.4 cm

9. Let x = the distance (in km). $1^2 + 1^2 = x^2$; $1 + 1 = x^2$; $2 = x^2$;
   $x = \sqrt{2} \approx 1.4$.   1.4 km

10. Let x = the length (in cm) of the line (joining opposite corners). $20^2 + 30^2 = x^2$;
    $400 + 900 = x^2$; $1300 = x^2$; $x = \sqrt{1300} = 10\sqrt{13} \approx 10(3.606) = 36.06 \approx 36.1$.
    36.1 cm

11. Let x = length of umbrella (in cm). $45^2 + 18^2 = x^2$; $2025 + 324 = x^2$;
    $2349 = x^2$; $x = \sqrt{2349} = 9\sqrt{29} \approx 9(5.385) = 48.465 \approx 48.5$.   48.5 cm

12. $27^2 + 36^2 = c^2$; $729 + 1296 = c^2$; $2025 = c^2$; $45 = c$.   45 ft

B   13. $e^2 + e^2 = c^2$; $2e^2 = c^2$; $c = \sqrt{2e^2} = e\sqrt{2}$.   Thus, $d^2 = e^2 + c^2 = e^2 + 2e^2 = 3e^2$ and $d = \sqrt{3e^2} = e\sqrt{3}$.

Page 403 • WRITTEN EXERCISES

A   1. $\sqrt{\dfrac{4}{25}} = \dfrac{\sqrt{4}}{\sqrt{25}} = \dfrac{2}{5}$ 2. $\sqrt{\dfrac{16}{64}} = \dfrac{\sqrt{16}}{\sqrt{64}} = \dfrac{4}{8} = \dfrac{1}{2}$ 3. $\sqrt{\dfrac{121}{4}} = \dfrac{\sqrt{121}}{\sqrt{4}} = \dfrac{11}{2}$

4. $\sqrt{\dfrac{9}{144}} = \dfrac{\sqrt{9}}{\sqrt{144}} = \dfrac{3}{12} = \dfrac{1}{4}$ 5. $\sqrt{\dfrac{1}{16}} = \dfrac{\sqrt{1}}{\sqrt{16}} = \dfrac{1}{4}$

6. $\sqrt{\dfrac{16}{49}} = \dfrac{\sqrt{16}}{\sqrt{49}} = \dfrac{4}{7}$ 7. $\sqrt{\dfrac{36}{64}} = \dfrac{\sqrt{36}}{\sqrt{64}} = \dfrac{6}{8} = \dfrac{3}{4}$

8. $\sqrt{\dfrac{225}{625}} = \dfrac{\sqrt{225}}{\sqrt{625}} = \dfrac{15}{25} = \dfrac{3}{5}$ 9. $\sqrt{\dfrac{9}{196}} = \dfrac{\sqrt{9}}{\sqrt{196}} = \dfrac{3}{14}$

10. $\sqrt{\dfrac{169}{16}} = \dfrac{\sqrt{169}}{\sqrt{16}} = \dfrac{13}{4}$ 11. $\sqrt{\dfrac{1}{144}} = \dfrac{\sqrt{1}}{\sqrt{144}} = \dfrac{1}{12}$ 12. $\sqrt{\dfrac{1}{400}} = \dfrac{\sqrt{1}}{\sqrt{400}} = \dfrac{1}{20}$

13. $\sqrt{\dfrac{x^2}{36}} = \dfrac{\sqrt{x^2}}{\sqrt{36}} = \dfrac{x}{6}$ 14. $\sqrt{\dfrac{y^2}{81}} = \dfrac{\sqrt{y^2}}{\sqrt{81}} = \dfrac{y}{9}$ 15. $\sqrt{\dfrac{4y^2}{9}} = \dfrac{\sqrt{4y^2}}{\sqrt{9}} = \dfrac{2y}{3}$

16. $\sqrt{\dfrac{16x^2}{25}} = \dfrac{\sqrt{16x^2}}{\sqrt{25}} = \dfrac{4x}{5}$ 17. $\sqrt{\dfrac{3x^2}{4}} = \dfrac{\sqrt{3x^2}}{\sqrt{4}} = \dfrac{x\sqrt{3}}{2}$ 18. $\sqrt{\dfrac{x^2y^2}{100}} = \dfrac{\sqrt{x^2y^2}}{\sqrt{100}} = \dfrac{xy}{10}$

19. $\sqrt{\dfrac{7a^2}{64}} = \dfrac{\sqrt{7a^2}}{\sqrt{64}} = \dfrac{a\sqrt{7}}{8}$ 20. $\dfrac{1}{2}\sqrt{\dfrac{4n^2}{9}} = \dfrac{1}{2} \cdot \dfrac{\sqrt{4n^2}}{\sqrt{9}} = \dfrac{2n}{2 \cdot 3} = \dfrac{n}{3}$

21. $\sqrt{\dfrac{3}{4}} = \dfrac{\sqrt{3}}{2} \approx \dfrac{1.732}{2} = 0.866 \approx 0.87$ 22. $\sqrt{\dfrac{10}{16}} = \dfrac{\sqrt{10}}{4} \approx \dfrac{3.162}{4} = 0.7905 \approx 0.79$

23. $\sqrt{\dfrac{12}{9}} = \dfrac{\sqrt{12}}{3} \approx \dfrac{3.464}{3} \approx 1.154 \approx 1.15$ 24. $\sqrt{\dfrac{22}{49}} = \dfrac{\sqrt{22}}{7} \approx \dfrac{4.690}{7} = 0.67$

25. $\sqrt{\dfrac{20}{25}} = \dfrac{\sqrt{20}}{5} \approx \dfrac{4.472}{5} = 0.8944 \approx 0.89$ 26. $\sqrt{\dfrac{50}{4}} = \dfrac{\sqrt{50}}{2} \approx \dfrac{7.071}{2} = 3.5355 \approx 3.54$

27. $\sqrt{\dfrac{21}{9}} = \dfrac{\sqrt{21}}{3} \approx \dfrac{4.583}{3} \approx 1.527 \approx 1.53$ 28. $\sqrt{\dfrac{15}{36}} = \dfrac{\sqrt{15}}{6} \approx \dfrac{3.873}{6} = 0.6455 \approx 0.65$

Page 403 · PUZZLE PROBLEMS

By the law of Pythagoras, $(\text{diagonal})^2 = x^2 + x^2 = 2x^2$, so diagonal $=$
$\sqrt{2x^2} = x\sqrt{2}$. The length of the longer side of the rectangle and the length of the diagonal of the square are equal.

Page 405 · WRITTEN EXERCISES

A   1. $\sqrt{3} \cdot \sqrt{3} = \sqrt{3 \cdot 3} = \sqrt{9} = 3$ 2. $\sqrt{2} \cdot \sqrt{2} = \sqrt{2 \cdot 2} = \sqrt{4} = 2$

3. $\sqrt{7} \cdot \sqrt{7} = \sqrt{7 \cdot 7} = \sqrt{49} = 7$ 4. $\sqrt{5} \cdot \sqrt{5} = \sqrt{5 \cdot 5} = \sqrt{25} = 5$

5. $\sqrt{3} \cdot \sqrt{6} = \sqrt{3 \cdot 6} = \sqrt{18} = 3\sqrt{2}$ 6. $\sqrt{2} \cdot \sqrt{6} = \sqrt{2 \cdot 6} = \sqrt{12} = 2\sqrt{3}$

7. $\sqrt{2} \cdot \sqrt{8} = \sqrt{2 \cdot 8} = \sqrt{16} = 4$ 8. $\sqrt{10} \cdot \sqrt{2} = \sqrt{10 \cdot 2} = \sqrt{20} = 2\sqrt{5}$

9. $2 \cdot 2\sqrt{3} = 4\sqrt{3}$ 10. $2\sqrt{2} \cdot 2\sqrt{3} = 2 \cdot 2 \cdot \sqrt{2 \cdot 3} = 4\sqrt{6}$

11. $4\sqrt{3} \cdot \sqrt{3} = 4\sqrt{3 \cdot 3} = 4\sqrt{9} = 4 \cdot 3 = 12$

12. $4\sqrt{5} \cdot 2\sqrt{5} = 4 \cdot 2 \cdot \sqrt{5 \cdot 5} = 8\sqrt{25} = 8 \cdot 5 = 40$

13. $\sqrt{18} \cdot 5 = 3\sqrt{2} \cdot 5 = 15\sqrt{2}$ 14. $2\sqrt{18} \cdot 3 = 2 \cdot 3\sqrt{2} \cdot 3 = 18\sqrt{2}$

15. $\sqrt{2} \cdot \sqrt{18} = \sqrt{2 \cdot 18} = \sqrt{36} = 6$

16. $2\sqrt{6} \cdot \sqrt{12} = 2\sqrt{6 \cdot 12} = 2\sqrt{72} = 2 \cdot 6\sqrt{2} = 12\sqrt{2}$

17. $\sqrt{12} \cdot \sqrt{3} = \sqrt{12 \cdot 3} = \sqrt{36} = 6$

18. $\sqrt{24} \cdot \sqrt{2} = \sqrt{24 \cdot 2} = \sqrt{48} = 4\sqrt{3}$

19. $\sqrt{15} \cdot \sqrt{5} = \sqrt{15 \cdot 5} = \sqrt{75} = 5\sqrt{3}$

20. $2\sqrt{2} \cdot \sqrt{20} = 2\sqrt{2 \cdot 20} = 2\sqrt{40} = 2 \cdot 2\sqrt{10} = 4\sqrt{10}$

21. $\sqrt{x} \cdot \sqrt{x} = \sqrt{x \cdot x} = \sqrt{x^2} = x$ 22. $\sqrt{a} \cdot \sqrt{a} = \sqrt{a \cdot a} = \sqrt{a^2} = a$

23. $2\sqrt{a} \cdot 4\sqrt{a} = 2 \cdot 4 \cdot \sqrt{a \cdot a} = 8\sqrt{a^2} = 8a$

24. $3x\sqrt{2} \cdot x\sqrt{2} = 3x \cdot x \cdot \sqrt{2 \cdot 2} = 3x^2\sqrt{4} = 3x^2 \cdot 2 = 6x^2$

25. $4\sqrt{x} \cdot 2\sqrt{x} = 4 \cdot 2 \cdot \sqrt{x^2} = 8x$ 26. $2\sqrt{n} \cdot \sqrt{4n} = 2\sqrt{4n^2} = 2 \cdot 2n = 4n$

27. $\dfrac{\sqrt{10}}{\sqrt{2}} = \sqrt{\dfrac{10}{2}} = \sqrt{5}$ 28. $\dfrac{\sqrt{12}}{\sqrt{3}} = \sqrt{\dfrac{12}{3}} = \sqrt{4} = 2$

29. $\dfrac{\sqrt{14}}{\sqrt{2}} = \sqrt{\dfrac{14}{2}} = \sqrt{7}$ 30. $\dfrac{\sqrt{18}}{\sqrt{3}} = \sqrt{\dfrac{18}{3}} = \sqrt{6}$

31. $\dfrac{\sqrt{20}}{\sqrt{5}} = \sqrt{\dfrac{20}{5}} = \sqrt{4} = 2$ 32. $\dfrac{\sqrt{21}}{\sqrt{7}} = \sqrt{\dfrac{21}{7}} = \sqrt{3}$

33. $\dfrac{10\sqrt{6}}{\sqrt{3}} = 10\sqrt{\dfrac{6}{3}} = 10\sqrt{2}$ 34. $\dfrac{6\sqrt{10}}{\sqrt{2}} = 6\sqrt{\dfrac{10}{2}} = 6\sqrt{5}$

35. $\dfrac{8\sqrt{12}}{2\sqrt{3}} = 4\sqrt{\dfrac{12}{3}} = 4\sqrt{4} = 4 \cdot 2 = 8$ 36. $\dfrac{14\sqrt{14}}{7\sqrt{2}} = 2\sqrt{\dfrac{14}{2}} = 2\sqrt{7}$

37. $\dfrac{\sqrt{a^3}}{\sqrt{a}} = \sqrt{\dfrac{a^3}{a}} = \sqrt{a^2} = a$

38. $\dfrac{2\sqrt{x}}{\sqrt{x}} = 2\sqrt{\dfrac{x}{x}} = 2\sqrt{1} = 2$

39. $\dfrac{\sqrt{2n^2}}{\sqrt{n}} = \sqrt{\dfrac{2n^2}{n}} = \sqrt{2n}$

40. $\dfrac{\sqrt{8x}}{\sqrt{2x}} = \sqrt{\dfrac{8x}{2x}} = \sqrt{4} = 2$

41. $\dfrac{\sqrt{27a^2}}{\sqrt{3b^2}} = \sqrt{\dfrac{27a^2}{3b^2}} = \sqrt{\dfrac{9a^2}{b^2}} = \dfrac{3a}{b}$

42. $\dfrac{\sqrt{3xy^3}}{\sqrt{2xy}} = \sqrt{\dfrac{3xy^3}{2xy}} = \sqrt{\dfrac{3}{2}y^2} = y\sqrt{\dfrac{3}{2}}$

43. $\sqrt{\dfrac{1}{9}} \cdot \sqrt{\dfrac{1}{4}} = \sqrt{\dfrac{1}{36}} = \dfrac{1}{6}$

44. $\sqrt{\dfrac{2}{3}} \cdot \sqrt{\dfrac{2}{27}} = \sqrt{\dfrac{4}{81}} = \dfrac{2}{9}$

45. $\sqrt{\dfrac{2}{3}} \cdot \sqrt{\dfrac{3}{2}} = \sqrt{\dfrac{6}{6}} = \sqrt{1} = 1$

46. $\sqrt{\dfrac{4}{5}} \cdot \sqrt{\dfrac{5}{4}} = \sqrt{\dfrac{20}{20}} = \sqrt{1} = 1$

47. $\sqrt{\dfrac{10}{4}} \cdot \sqrt{\dfrac{4}{2}} = \sqrt{\dfrac{40}{8}} = \sqrt{5}$

48. $\sqrt{\dfrac{12}{5}} \cdot \sqrt{\dfrac{5}{3}} = \sqrt{\dfrac{60}{15}} = \sqrt{4} = 2$

B   **49.** $3\sqrt{x^3} \cdot 2\sqrt{x} = 6\sqrt{x^4} = 6x^2$

50. $5\sqrt{n} \cdot \dfrac{1}{5}\sqrt{\dfrac{1}{n}} = \dfrac{5}{5}\sqrt{\dfrac{n}{n}} = 1\sqrt{1} = 1$

51. $-3x\sqrt{2} \cdot x\sqrt{8} = -3x^2\sqrt{16} = -3x^2 \cdot 4 = -12x^2$

52. $\sqrt{\dfrac{1}{7}} \cdot -x\sqrt{14} = -x\sqrt{\dfrac{14}{7}} = -x\sqrt{2}$

53. $-x^3 \cdot -\sqrt{\dfrac{1}{4x^4}} = x^3 \cdot \dfrac{1}{2x^2} = \dfrac{x}{2}$

54. $\sqrt{2} \cdot \sqrt{3} \cdot \sqrt{6} = \sqrt{36} = 6$

55. $\sqrt{3x} \cdot \sqrt{2x} \cdot \sqrt{3} = \sqrt{18x^2} = 3x\sqrt{2}$

56. $3x\sqrt{9x^2} \cdot \sqrt{6x^4} = 3x \cdot 3x \cdot x^2\sqrt{6} = 9x^4\sqrt{6}$

57. $2a^2\sqrt{3a^2} \cdot 5\sqrt{5a^2} = 2a^2 \cdot a\sqrt{3} \cdot 5 \cdot a\sqrt{5} = 10a^4\sqrt{15}$

C   **58.** $\sqrt{n} \cdot \dfrac{\sqrt{n^4}}{\sqrt{n}} = \sqrt{n} \cdot \sqrt{n^3} = \sqrt{n^4} = n^2$

59. $x \cdot \dfrac{-2\sqrt{x^3}}{\sqrt{4x}} = x \cdot -2 \cdot \sqrt{\dfrac{x^2}{4}} = -2x \cdot \dfrac{x}{2} = -x^2$

60. $\dfrac{\sqrt{n^4}}{\sqrt{n^2}} \cdot 3n\sqrt{\dfrac{9n^3}{n}} = \sqrt{n^2} \cdot 3n\sqrt{9n^2} = n \cdot 3n \cdot 3n = 9n^3$

Page 407 · WRITTEN EXERCISES

A   **1.** $\sqrt{\dfrac{1}{2}} = \dfrac{\sqrt{1}}{\sqrt{2}} = \dfrac{\sqrt{1} \cdot \sqrt{2}}{\sqrt{2} \cdot \sqrt{2}} = \dfrac{\sqrt{2}}{2}$

2. $\sqrt{\dfrac{3}{2}} = \dfrac{\sqrt{3}}{\sqrt{2}} = \dfrac{\sqrt{3} \cdot \sqrt{2}}{\sqrt{2} \cdot \sqrt{2}} = \dfrac{\sqrt{6}}{2}$

3. $\sqrt{\dfrac{3}{4}} = \dfrac{\sqrt{3}}{\sqrt{4}} = \dfrac{\sqrt{3} \cdot \sqrt{4}}{\sqrt{4} \cdot \sqrt{4}} = \dfrac{2\sqrt{3}}{4} = \dfrac{\sqrt{3}}{2}$

4. $\sqrt{\dfrac{4}{5}} = \dfrac{\sqrt{4}}{\sqrt{5}} = \dfrac{\sqrt{4} \cdot \sqrt{5}}{\sqrt{5} \cdot \sqrt{5}} = \dfrac{2\sqrt{5}}{5}$

5. $\sqrt{\dfrac{1}{3}} = \dfrac{\sqrt{1} \cdot \sqrt{3}}{\sqrt{3} \cdot \sqrt{3}} = \dfrac{\sqrt{3}}{3}$

6. $\sqrt{\dfrac{3}{5}} = \dfrac{\sqrt{3} \cdot \sqrt{5}}{\sqrt{5} \cdot \sqrt{5}} = \dfrac{\sqrt{15}}{5}$

7. $\sqrt{\dfrac{5}{8}} = \dfrac{\sqrt{5} \cdot \sqrt{8}}{\sqrt{8} \cdot \sqrt{8}} = \dfrac{\sqrt{40}}{8} = \dfrac{2\sqrt{10}}{8} = \dfrac{\sqrt{10}}{4}$

8. $\sqrt{\dfrac{6}{7}} = \dfrac{\sqrt{6} \cdot \sqrt{7}}{\sqrt{7} \cdot \sqrt{7}} = \dfrac{\sqrt{42}}{7}$

9. $\sqrt{\dfrac{7}{10}} = \dfrac{\sqrt{7} \cdot \sqrt{10}}{\sqrt{10} \cdot \sqrt{10}} = \dfrac{\sqrt{70}}{10}$

10. $\sqrt{\dfrac{5}{6}} = \dfrac{\sqrt{5} \cdot \sqrt{6}}{\sqrt{6} \cdot \sqrt{6}} = \dfrac{\sqrt{30}}{6}$

11. $\sqrt{\dfrac{4}{7}} = \dfrac{\sqrt{4} \cdot \sqrt{7}}{\sqrt{7} \cdot \sqrt{7}} = \dfrac{2\sqrt{7}}{7}$

12. $\sqrt{\dfrac{5}{12}} = \dfrac{\sqrt{5} \cdot \sqrt{12}}{\sqrt{12} \cdot \sqrt{12}} = \dfrac{\sqrt{60}}{12} = \dfrac{2\sqrt{15}}{12} = \dfrac{\sqrt{15}}{6}$

13. $\dfrac{3x}{\sqrt{5}} = \dfrac{3x \cdot \sqrt{5}}{\sqrt{5} \cdot \sqrt{5}} = \dfrac{3x\sqrt{5}}{5}$

14. $\dfrac{a^2}{\sqrt{a}} = \dfrac{a^2 \cdot \sqrt{a}}{\sqrt{a} \cdot \sqrt{a}} = \dfrac{a^2\sqrt{a}}{a} = a\sqrt{a}$

15. $\sqrt{\dfrac{a^2}{2}} = \dfrac{\sqrt{a^2}}{\sqrt{2}} = \dfrac{a}{\sqrt{2}} = \dfrac{a \cdot \sqrt{2}}{\sqrt{2} \cdot \sqrt{2}} = \dfrac{a\sqrt{2}}{2}$

16. $\sqrt{\dfrac{x^2}{3}} = \dfrac{\sqrt{x^2}}{\sqrt{3}} = \dfrac{x}{\sqrt{3}} = \dfrac{x \cdot \sqrt{3}}{\sqrt{3} \cdot \sqrt{3}} = \dfrac{x\sqrt{3}}{3}$

17. $\dfrac{-x^2}{\sqrt{y}} = \dfrac{-x^2 \cdot \sqrt{y}}{\sqrt{y} \cdot \sqrt{y}} = \dfrac{-x^2\sqrt{y}}{y}$

18. $\dfrac{3\sqrt{18}}{\sqrt{2}} = \dfrac{3\sqrt{18} \cdot \sqrt{2}}{\sqrt{2} \cdot \sqrt{2}} = \dfrac{3\sqrt{36}}{2} = \dfrac{3 \cdot 6}{2} = 9$

19. $\dfrac{2\sqrt{8}}{\sqrt{2}} = \dfrac{2\sqrt{8} \cdot \sqrt{2}}{\sqrt{2} \cdot \sqrt{2}} = \dfrac{2\sqrt{16}}{2} = \sqrt{16} = 4$ 20. $3\sqrt{\dfrac{n^2}{5}} = \dfrac{3\sqrt{n^2}}{\sqrt{5}} = \dfrac{3n \cdot \sqrt{5}}{\sqrt{5} \cdot \sqrt{5}} = \dfrac{3n\sqrt{5}}{5}$

B   21. $5\sqrt{\dfrac{1}{2}} = \dfrac{5\sqrt{1} \cdot \sqrt{2}}{\sqrt{2} \cdot \sqrt{2}} = \dfrac{5\sqrt{2}}{2} \approx \dfrac{5(1.414)}{2} = 3.535 \approx 3.54$

22. $4\sqrt{\dfrac{1}{3}} = \dfrac{4\sqrt{1} \cdot \sqrt{3}}{\sqrt{3} \cdot \sqrt{3}} = \dfrac{4\sqrt{3}}{3} \approx \dfrac{4(1.732)}{3} \approx 2.309 \approx 2.31$

23. $6\sqrt{\dfrac{2}{5}} = \dfrac{6\sqrt{2} \cdot \sqrt{5}}{\sqrt{5} \cdot \sqrt{5}} = \dfrac{6\sqrt{10}}{5} \approx \dfrac{6(3.162)}{5} \approx 3.794 \approx 3.79$

24. $2\sqrt{\dfrac{3}{8}} = \dfrac{2\sqrt{3} \cdot \sqrt{8}}{\sqrt{8} \cdot \sqrt{8}} = \dfrac{2\sqrt{24}}{8} = \dfrac{2\sqrt{6}}{4} = \dfrac{\sqrt{6}}{2} \approx \dfrac{2.449}{2} \approx 1.224 \approx 1.22$

25. $\sqrt{\dfrac{1}{3}} \cdot \sqrt{\dfrac{1}{3}} = \sqrt{\dfrac{1}{9}} = \dfrac{1}{3} \approx 0.33$

26. $\sqrt{\dfrac{1}{3}} \cdot \sqrt{\dfrac{1}{2}} = \sqrt{\dfrac{1}{6}} = \dfrac{\sqrt{1} \cdot \sqrt{6}}{\sqrt{6} \cdot \sqrt{6}} = \dfrac{\sqrt{6}}{6} \approx \dfrac{2.449}{6} \approx 0.408 \approx 0.41$

27. $\sqrt{\dfrac{3}{4}} \cdot \sqrt{\dfrac{1}{2}} = \sqrt{\dfrac{3}{8}} = \dfrac{\sqrt{3} \cdot \sqrt{8}}{\sqrt{8} \cdot \sqrt{8}} = \dfrac{\sqrt{24}}{8} = \dfrac{2\sqrt{6}}{8} = \dfrac{\sqrt{6}}{4} \approx \dfrac{2.449}{4} \approx 0.612 \approx 0.61$

28. $\sqrt{\dfrac{5}{2}} \cdot \sqrt{\dfrac{2}{5}} = \sqrt{\dfrac{10}{10}} = \sqrt{1} = 1.00$

Page 409 · WRITTEN EXERCISES

A   1. $2\sqrt{2} + 6\sqrt{2} = 8\sqrt{2}$ 2. $6\sqrt{2} - 4\sqrt{2} = 2\sqrt{2}$

3. $8\sqrt{5} + 5\sqrt{5} = 13\sqrt{5}$ 4. $-\sqrt{3} + 2\sqrt{3} = \sqrt{3}$

5. $-\sqrt{5} + 5\sqrt{5} = 4\sqrt{5}$ 6. $10\sqrt{7} - \sqrt{7} = 9\sqrt{7}$

7. $6\sqrt{3} - 5\sqrt{3} = \sqrt{3}$ 8. $\sqrt{2} - \sqrt{2} = 0$

9. $\sqrt{18} + \sqrt{2} = 3\sqrt{2} + \sqrt{2} = 4\sqrt{2}$

10. $4\sqrt{12} - 2\sqrt{3} = 4 \cdot 2\sqrt{3} - 2\sqrt{3} = 8\sqrt{3} - 2\sqrt{3} = 6\sqrt{3}$

11. $2\sqrt{8} - \sqrt{2} = 2 \cdot 2\sqrt{2} - \sqrt{2} = 4\sqrt{2} - \sqrt{2} = 3\sqrt{2}$

12. $2\sqrt{3} + 3\sqrt{12} = 2\sqrt{3} + 3 \cdot 2\sqrt{3} = 2\sqrt{3} + 6\sqrt{3} = 8\sqrt{3}$

13. $6\sqrt{3} - 4\sqrt{27} = 6\sqrt{3} - 4 \cdot 3\sqrt{3} = 6\sqrt{3} - 12\sqrt{3} = -6\sqrt{3}$

14. $\sqrt{2} + \sqrt{3} + 2\sqrt{2} = 3\sqrt{2} + \sqrt{3}$ 15. $\sqrt{5} - \sqrt{3} + 3\sqrt{5} = 4\sqrt{5} - \sqrt{3}$

16. $\sqrt{6} + \sqrt{3} + 2\sqrt{6} = 3\sqrt{6} + \sqrt{3}$ 17. $4\sqrt{5} - 2\sqrt{2} + 6\sqrt{5} = 10\sqrt{5} - 2\sqrt{2}$

18. $\sqrt{2} - 6\sqrt{5} + 5\sqrt{2} = 6\sqrt{2} - 6\sqrt{5}$

19. $\sqrt{12} - 4\sqrt{3} + \sqrt{6} = 2\sqrt{3} - 4\sqrt{3} + \sqrt{6} = -2\sqrt{3} + \sqrt{6}$

20. $\sqrt{24} + 5\sqrt{6} - \sqrt{2} = 2\sqrt{6} + 5\sqrt{6} - \sqrt{2} = 7\sqrt{6} - \sqrt{2}$

21. $\sqrt{45} + 2\sqrt{5} - 4\sqrt{7} = 3\sqrt{5} + 2\sqrt{5} - 4\sqrt{7} = 5\sqrt{5} - 4\sqrt{7}$

B 22. $\sqrt{2a} + \sqrt{8a} = \sqrt{2a} + 2\sqrt{2a} = 3\sqrt{2a}$

23. $3\sqrt{b} + 3\sqrt{b^3} = 3\sqrt{b} + 3b\sqrt{b} = (3 + 3b)\sqrt{b}$

24. $\sqrt{49a^2} + 6\sqrt{b} - \sqrt{4b^3} = 7a + 6\sqrt{b} - 2b\sqrt{b} = 7a + (6 - 2b)\sqrt{b}$

25. $\sqrt{27a} - \sqrt{3a} - \sqrt{12a} = 3\sqrt{3a} - \sqrt{3a} - 2\sqrt{3a} = 0$

26. $\sqrt{5} - (2\sqrt{5} - \sqrt{125}) = \sqrt{5} - 2\sqrt{5} + 5\sqrt{5} = 4\sqrt{5}$

27. $\sqrt{6} - (\sqrt{24} + 6\sqrt{6}) = \sqrt{6} - 2\sqrt{6} - 6\sqrt{6} = -7\sqrt{6}$

28. $\sqrt{8} - 2\sqrt{\dfrac{1}{8}} + 6\sqrt{24} = 2\sqrt{2} - \dfrac{2\sqrt{8}}{8} + 6 \cdot 2\sqrt{6} = 2\sqrt{2} - \dfrac{2\sqrt{2}}{4} + 12\sqrt{6} =$
$\dfrac{3\sqrt{2}}{2} + 12\sqrt{6}$

29. $\sqrt{5} + 2\sqrt{\dfrac{1}{5}} + \sqrt{75} = \sqrt{5} + \dfrac{2\sqrt{5}}{5} + 5\sqrt{3} = \dfrac{7\sqrt{5}}{5} + 5\sqrt{3}$

Page 411 · WRITTEN EXERCISES

A 1. $(1 - \sqrt{2})(1 + \sqrt{2}) = 1^2 - (\sqrt{2})^2 = 1 - 2 = -1$

2. $(5 - \sqrt{5})(5 + \sqrt{5}) = 5^2 - (\sqrt{5})^2 = 25 - 5 = 20$

3. $(\sqrt{3} - 4)(\sqrt{3} + 4) = 3 - 16 = -13$ 4. $(2 - \sqrt{7})(2 + \sqrt{7}) = 4 - 7 = -3$

5. $(1 - \sqrt{6})(1 + \sqrt{6}) = 1 - 6 = -5$ 6. $(7 + \sqrt{5})(7 - \sqrt{5}) = 49 - 5 = 44$

7. $(\sqrt{6} - 2)(\sqrt{6} + 2) = 6 - 4 = 2$ 8. $(\sqrt{8} + 4)(\sqrt{8} - 4) = 8 - 16 = -8$

9. $(2\sqrt{2} + 1)(2\sqrt{2} - 1) = 4 \cdot 2 - 1 = 8 - 1 = 7$

10. $(4\sqrt{5} + 2)(4\sqrt{5} - 2) = 16 \cdot 5 - 4 = 80 - 4 = 76$

11. $(3\sqrt{5} - 1)(3\sqrt{5} + 1) = 9 \cdot 5 - 1 = 45 - 1 = 44$

12. $(3\sqrt{3} - 2)(3\sqrt{3} + 2) = 9 \cdot 3 - 4 = 27 - 4 = 23$

13. $\dfrac{1}{\sqrt{3} + 1} = \dfrac{\sqrt{3} - 1}{(\sqrt{3} + 1)(\sqrt{3} - 1)} = \dfrac{\sqrt{3} - 1}{3 - 1} = \dfrac{\sqrt{3} - 1}{2}$

14. $\dfrac{5}{2 + \sqrt{3}} = \dfrac{5(2 - \sqrt{3})}{(2 + \sqrt{3})(2 - \sqrt{3})} = \dfrac{5(2 - \sqrt{3})}{4 - 3} = \dfrac{5(2 - \sqrt{3})}{1} = 10 - 5\sqrt{3}$

15. $\dfrac{2}{\sqrt{2} + 3} = \dfrac{2(\sqrt{2} - 3)}{(\sqrt{2} + 3)(\sqrt{2} - 3)} = \dfrac{2(\sqrt{2} - 3)}{2 - 9} = \dfrac{2\sqrt{2} - 6}{-7}$

16. $\dfrac{\sqrt{5}}{2 + \sqrt{6}} = \dfrac{\sqrt{5}(2 - \sqrt{6})}{(2 + \sqrt{6})(2 - \sqrt{6})} = \dfrac{2\sqrt{5} - \sqrt{30}}{4 - 6} = \dfrac{2\sqrt{5} - \sqrt{30}}{-2}$

17. $\dfrac{\sqrt{6}}{1 - \sqrt{6}} = \dfrac{\sqrt{6}(1 + \sqrt{6})}{(1 - \sqrt{6})(1 + \sqrt{6})} = \dfrac{\sqrt{6} + \sqrt{36}}{1 - 6} = \dfrac{\sqrt{6} + 6}{-5}$

18. $\dfrac{\sqrt{8}}{4 - \sqrt{2}} = \dfrac{\sqrt{8}(4 + \sqrt{2})}{(4 - \sqrt{2})(4 + \sqrt{2})} = \dfrac{4\sqrt{8} + \sqrt{16}}{16 - 2} = \dfrac{8\sqrt{2} + 4}{14} = \dfrac{4\sqrt{2} + 2}{7}$

19. $\dfrac{\sqrt{10}}{3 + \sqrt{6}} = \dfrac{\sqrt{10}(3 - \sqrt{6})}{(3 + \sqrt{6})(3 - \sqrt{6})} = \dfrac{3\sqrt{10} - \sqrt{60}}{9 - 6} = \dfrac{3\sqrt{10} - 2\sqrt{15}}{3}$

20. $\dfrac{\sqrt{7}}{\sqrt{3} + 5} = \dfrac{\sqrt{7}(\sqrt{3} - 5)}{(\sqrt{3} + 5)(\sqrt{3} - 5)} = \dfrac{\sqrt{21} - 5\sqrt{7}}{3 - 25} = \dfrac{\sqrt{21} - 5\sqrt{7}}{-22}$

21. $\dfrac{2\sqrt{7}}{\sqrt{7} - 1} = \dfrac{2\sqrt{7}(\sqrt{7} + 1)}{(\sqrt{7} - 1)(\sqrt{7} + 1)} = \dfrac{2 \cdot 7 + 2\sqrt{7}}{7 - 1} = \dfrac{14 + 2\sqrt{7}}{6} = \dfrac{7 + \sqrt{7}}{3}$

22. $\dfrac{4\sqrt{3}}{5 - \sqrt{2}} = \dfrac{4\sqrt{3}\,(5 + \sqrt{2})}{(5 - \sqrt{2})\,(5 + \sqrt{2})} = \dfrac{20\sqrt{3} + 4\sqrt{6}}{25 - 2} = \dfrac{20\sqrt{3} + 4\sqrt{6}}{23}$

23. $\dfrac{6\sqrt{2}}{2 - \sqrt{5}} = \dfrac{6\sqrt{2}\,(2 + \sqrt{5})}{(2 - \sqrt{5})\,(2 + \sqrt{5})} = \dfrac{12\sqrt{2} + 6\sqrt{10}}{4 - 5} = \dfrac{12\sqrt{2} + 6\sqrt{10}}{-1} = -12\sqrt{2} - 6\sqrt{10}$

24. $\dfrac{4\sqrt{3}}{6 - \sqrt{6}} = \dfrac{4\sqrt{3}\,(6 + \sqrt{6})}{(6 - \sqrt{6})\,(6 + \sqrt{6})} = \dfrac{24\sqrt{3} + 4\sqrt{18}}{36 - 6} = \dfrac{24\sqrt{3} + 12\sqrt{2}}{30} = \dfrac{4\sqrt{3} + 2\sqrt{2}}{5}$

25. $\dfrac{\sqrt{3} - 1}{\sqrt{3} + 1} = \dfrac{(\sqrt{3} - 1)\,(\sqrt{3} - 1)}{(\sqrt{3} + 1)\,(\sqrt{3} - 1)} = \dfrac{3 - \sqrt{3} - \sqrt{3} + 1}{3 - 1} = \dfrac{4 - 2\sqrt{3}}{2} = 2 - \sqrt{3}$

26. $\dfrac{1 - \sqrt{3}}{\sqrt{3} + 1} = \dfrac{(1 - \sqrt{3})\,(\sqrt{3} - 1)}{(\sqrt{3} + 1)\,(\sqrt{3} - 1)} = \dfrac{\sqrt{3} - 1 - 3 + \sqrt{3}}{3 - 1} = \dfrac{2\sqrt{3} - 4}{2} = \sqrt{3} - 2$

27. $\dfrac{2 - \sqrt{7}}{\sqrt{7} - 4} = \dfrac{(2 - \sqrt{7})\,(\sqrt{7} + 4)}{(\sqrt{7} - 4)\,(\sqrt{7} + 4)} = \dfrac{2\sqrt{7} + 8 - 7 - 4\sqrt{7}}{7 - 16} = \dfrac{1 - 2\sqrt{7}}{-9}$

28. $\dfrac{2 - 3\sqrt{3}}{3\sqrt{2} + 2} = \dfrac{(2 - 3\sqrt{3})\,(3\sqrt{2} - 2)}{(3\sqrt{2} + 2)\,(3\sqrt{2} - 2)} = \dfrac{6\sqrt{2} - 4 - 9\sqrt{6} + 6\sqrt{3}}{9 \cdot 2 - 4} = \dfrac{6\sqrt{2} - 4 - 9\sqrt{6} + 6\sqrt{3}}{14}$

29. $(2 - \sqrt{3})^2 = 2^2 - 2 \cdot 2\sqrt{3} + (\sqrt{3})^2 = 4 - 4\sqrt{3} + 3 = 7 - 4\sqrt{3}$

30. $(5 + \sqrt{2})^2 = 5^2 + 2 \cdot 5\sqrt{2} + (\sqrt{2})^2 = 25 + 10\sqrt{2} + 2 = 27 + 10\sqrt{2}$

31. $(1 - \sqrt{5})^2 = 1^2 - 2 \cdot \sqrt{5} + (\sqrt{5})^2 = 1 - 2\sqrt{5} + 5 = 6 - 2\sqrt{5}$

32. $(3 + \sqrt{8})^2 = 3^2 + 2 \cdot 3\sqrt{8} + (\sqrt{8})^2 = 9 + 6\sqrt{8} + 8 = 17 + 6\sqrt{8}$

33. $(7 + \sqrt{5})^2 = 7^2 + 2 \cdot 7\sqrt{5} + (\sqrt{5})^2 = 49 + 14\sqrt{5} + 5 = 54 + 14\sqrt{5}$

34. $(6 - \sqrt{6})^2 = 6^2 - 2 \cdot 6\sqrt{6} + (\sqrt{6})^2 = 36 - 12\sqrt{6} + 6 = 42 - 12\sqrt{6}$

35. $(4 - \sqrt{10})^2 = 4^2 - 2 \cdot 4\sqrt{10} + (\sqrt{10})^2 = 16 - 8\sqrt{10} + 10 = 26 - 8\sqrt{10}$

36. $(2\sqrt{2} + 5)^2 = (2\sqrt{2})^2 + 2 \cdot 5 \cdot 2\sqrt{2} + 5^2 = 8 + 20\sqrt{2} + 25 = 33 + 20\sqrt{2}$

37. $(3\sqrt{5} - 2\sqrt{3})^2 = (3\sqrt{5})^2 - 2 \cdot 3\sqrt{5} \cdot 2\sqrt{3} + (2\sqrt{3})^2 = 45 - 12\sqrt{15} + 12 = $
$57 - 12\sqrt{15}$

38. $(4\sqrt{8} - 2\sqrt{7})^2 = (4\sqrt{8})^2 - 2 \cdot 4\sqrt{8} \cdot 2\sqrt{7} + (2\sqrt{7})^2 = 128 - 16\sqrt{56} + 28 = $
$128 - 32\sqrt{14} + 28 = 156 - 32\sqrt{14}$

39. $(2\sqrt{5} - 3\sqrt{8})^2 = (2\sqrt{5})^2 - 2 \cdot 2\sqrt{5} \cdot 3\sqrt{8} + (3\sqrt{8})^2 = 20 - 12\sqrt{40} + 72 = $
$20 - 24\sqrt{10} + 72 = 92 - 24\sqrt{10}$

40. $(4\sqrt{2} - 3\sqrt{3})^2 = (4\sqrt{2})^2 - 2 \cdot 4\sqrt{2} \cdot 3\sqrt{3} + (3\sqrt{3})^2 = 32 - 24\sqrt{6} + 27 = $
$59 - 24\sqrt{6}$

41. $A = (7 + 2\sqrt{3})^2 = 7^2 + 2 \cdot 7 \cdot 2\sqrt{3} + (2\sqrt{3})^2 = 49 + 28\sqrt{3} + 12 = 61 + 28\sqrt{3}$

Page 411 · CALCULATOR ACTIVITIES

1. $10.954451 \approx 10.954$ **2.** $13.416407 \approx 13.416$ **3.** -20

4. $30.854497 \approx 30.854$ **5.** ± 0.5 **6.** $1.5362291 \approx 1.536$

7. $92.325511 \approx 92.326$ **8.** $0.034641 \approx 0.035$

Page 412 · SKILLS REVIEW

1. $7x + 12y$ **2.** $11m^2 - 4m - 7$ **3.** -11

4.
$$\begin{array}{r} 2x - 3 \\ -5x + 1 \\ \hline -3x - 2 \end{array}$$

5.
$$\begin{array}{r} 15ab^2 - 3a + 7 \\ -9ab^2 - 3a - 4 \\ \hline 6ab^2 - 6a + 3 \end{array}$$

6.
$$\begin{array}{r} 2m^2n^2 - 3mn - 11 \\ -8m^2n^2 + 4mn - 2 \\ \hline -6m^2n^2 + mn - 13 \end{array}$$

7. $(y + 2)(y + 7) = y^2 + 9y + 14$

8. $(k + 5)(k + 4) = k^2 + 9k + 20$

9. $(b + 9)(b + 4) = b^2 + 13b + 36$

10. $(z - 3)(z - 1) = z^2 - 4z + 3$

11. $(t - 8)(t - 3) = t^2 - 11t + 24$

12. $(g - 7)(g + 1) = g^2 - 6g - 7$

13. $(c - 9)(c + 8) = c^2 - c - 72$

14. $(m - 11)(m + 3) = m^2 - 8m - 33$

15. $(k - 15)(k + 2) = k^2 - 13k - 30$

16. $(m + 7)(m + 7) = m^2 + 14m + 49$

17. $(y - 10z)(y + 10z) = y^2 - 100z^2$

18. $(2t - 1)(3t + 1) = 6t^2 - t - 1$

19. $6t + 12 = 6(t + 2)$

20. $2b^2 + 6b = 2b(b + 3)$

21. $3x^2 + 15y = 3(x^2 + 5y)$

22. $y^2 + 2y + 1 = (y + 1)^2$

23. $t^2 + 9t + 20 = (t + 4)(t + 5)$

24. $m^2 + 15m + 44 = (m + 4)(m + 11)$

25. $z^2 - 12z + 32 = (z - 4)(z - 8)$

26. $b^2 - 11b + 30 = (b - 5)(b - 6)$

27. $x^2 - 12x + 20 = (x - 2)(x - 10)$

28. $y^2 - 8y + 16 = (y - 4)^2$

29. $y^2 + 10yz + 25z^2 = (y + 5z)^2$

30. $m^2 - 18m + 81 = (m - 9)^2$

31. $a^2 - 4a - 21 = (a - 7)(a + 3)$

32. $n^2 - n - 42 = (n - 7)(n + 6)$

33. $t^2 + 7t - 60 = (t + 12)(t - 5)$

34. $9d^2 - 49 = (3d + 7)(3d - 7)$

35. $1 - 36c^2 = (1 + 6c)(1 - 6c)$

36. $25m^2 - 121n^2 = (5m + 11n)(5m - 11n)$

37. $2x^2 + 10x + 12 = 2(x^2 + 5x + 6) = 2(x + 2)(x + 3)$

38. $4z^2 - 8z - 32 = 4(z^2 - 2z - 8) = 4(z - 4)(z + 2)$

39. $5t^2 - 15t - 90 = 5(t^2 - 3t - 18) = 5(t - 6)(t + 3)$

40. $28y^2 - 7 = 7(4y^2 - 1) = 7(2y + 1)(2y - 1)$

41. $2d^3 - 50d = 2d(d^2 - 25) = 2d(d + 5)(d - 5)$

42. $p^2q^2 - 9pq^2 + 20q^2 = q^2(p^2 - 9p + 20) = q^2(p - 4)(p - 5)$

43. $-x^2 + 16 = 16 - x^2 = (4 + x)(4 - x)$

44. $3v^2 + 18v - 21 = 3(v^2 + 6v - 7) = 3(v + 7)(v - 1)$

45. $-m^2 + 16m - 64 = -(m^2 - 16m + 64) = -(m - 8)(m - 8) = -(m - 8)^2$

Pages 413–414 · CHAPTER REVIEW EXERCISES

1. 5   **2.** -6 **3.** 9   **4.** -10 **5.** 3   **6.** 8   **7.** -7 **8.** 8   **9.** 12   **10.** -15

11. $4.359 \approx 4.4$ **12.** $5.385 \approx 5.4$ **13.** $7.749 \approx 7.7$ **14.** $6.481 \approx 6.5$

15. $-2.646 \approx -2.6$ **16.** rational   **17.** rational   **18.** irrational   **19.** rational

20. rational   **21.** rational   **22.** rational   **23.** rational   **24.** rational   **25.** irrational

26. $\sqrt{18} = \sqrt{9 \cdot 2} = \sqrt{9} \cdot \sqrt{2} = 3\sqrt{2}$

27. $\sqrt{32} = \sqrt{16 \cdot 2} = \sqrt{16} \cdot \sqrt{2} = 4\sqrt{2}$

28. $\sqrt{28} = \sqrt{4 \cdot 7} = \sqrt{4} \cdot \sqrt{7} = 2\sqrt{7}$

29. $\sqrt{150} = \sqrt{25 \cdot 6} = \sqrt{25} \cdot \sqrt{6} = 5\sqrt{6}$

30. $\sqrt{45} = \sqrt{9 \cdot 5} = \sqrt{9} \cdot \sqrt{5} = 3\sqrt{5}$

31. $\sqrt{175} = \sqrt{25 \cdot 7} = \sqrt{25} \cdot \sqrt{7} = 5\sqrt{7}$

32. $\sqrt{140} = \sqrt{4 \cdot 35} = \sqrt{4} \cdot \sqrt{35} = 2\sqrt{35}$

33. $\sqrt{200} = \sqrt{100 \cdot 2} = \sqrt{100} \cdot \sqrt{2} = 10\sqrt{2}$

34. $n^2 = 9; \quad n = \pm\sqrt{9} = \pm 3$ 　　　35. $x^2 = 81; \quad x = \pm\sqrt{81} = \pm 9$

36. $y^2 = 48; \quad y = \pm\sqrt{48} = \pm\sqrt{16 \cdot 3} = \pm 4\sqrt{3}$

37. $a^2 - 6 = 43; \quad a^2 = 49; \quad a = \pm 7$

38. $5^2 + 7^2 = x^2; \quad 25 + 49 = x^2; \quad 74 = x^2; \quad x = \sqrt{74}$

39. $3^2 + 6^2 = x^2; \quad 9 + 36 = x^2; \quad 45 = x^2; \quad x = \sqrt{45} = 3\sqrt{5}$

40. $x^2 + 11^2 = 15^2; \quad x^2 + 121 = 225; \quad x^2 = 104; \quad x = \sqrt{104} = 2\sqrt{26}$

41. $\sqrt{\dfrac{4}{144}} = \dfrac{\sqrt{4}}{\sqrt{144}} = \dfrac{2}{12} = \dfrac{1}{6}$ 　　42. $\sqrt{\dfrac{121}{64}} = \dfrac{\sqrt{121}}{\sqrt{64}} = \dfrac{11}{8}$

43. $\sqrt{\dfrac{3}{81}} = \dfrac{\sqrt{3}}{\sqrt{81}} = \dfrac{\sqrt{3}}{9}$ 　　44. $\sqrt{\dfrac{14x^2}{64}} = \dfrac{\sqrt{14x^2}}{\sqrt{64}} = \dfrac{x\sqrt{14}}{8}$

45. $\sqrt{\dfrac{x^2y^2}{9}} = \dfrac{\sqrt{x^2y^2}}{\sqrt{9}} = \dfrac{xy}{3}$ 　　46. $\sqrt{10} \cdot \sqrt{10} = \sqrt{10 \cdot 10} = \sqrt{100} = 10$

47. $3\sqrt{3} \cdot 2\sqrt{3} = 3 \cdot 2 \cdot \sqrt{3 \cdot 3} = 6\sqrt{9} = 6 \cdot 3 = 18$

48. $2\sqrt{5} \cdot \sqrt{25} = 2\sqrt{5 \cdot 25} = 2\sqrt{125} = 2 \cdot 5\sqrt{5} = 10\sqrt{5}$

49. $3\sqrt{n} \cdot 6\sqrt{n} = 3 \cdot 6 \cdot \sqrt{n \cdot n} = 18\sqrt{n^2} = 18n$

50. $\dfrac{\sqrt{15}}{\sqrt{3}} = \sqrt{\dfrac{15}{3}} = \sqrt{5}$ 　　51. $\dfrac{\sqrt{21}}{\sqrt{7}} = \sqrt{\dfrac{21}{7}} = \sqrt{3}$

52. $\dfrac{\sqrt{2r^2}}{\sqrt{r}} = \sqrt{\dfrac{2r^2}{r}} = \sqrt{2r}$ 　　53. $\dfrac{\sqrt{24x}}{\sqrt{6x}} = \sqrt{\dfrac{24x}{6x}} = \sqrt{4} = 2$

54. $\sqrt{\dfrac{1}{5}} = \dfrac{\sqrt{1}}{\sqrt{5}} = \dfrac{\sqrt{1} \cdot \sqrt{5}}{\sqrt{5} \cdot \sqrt{5}} = \dfrac{\sqrt{5}}{5}$

55. $\sqrt{\dfrac{7}{8}} = \dfrac{\sqrt{7}}{\sqrt{8}} = \dfrac{\sqrt{7} \cdot \sqrt{8}}{\sqrt{8} \cdot \sqrt{8}} = \dfrac{\sqrt{56}}{8} = \dfrac{2\sqrt{14}}{8} = \dfrac{\sqrt{14}}{4}$

56. $\sqrt{\dfrac{a^2}{5}} = \dfrac{\sqrt{a^2}}{\sqrt{5}} = \dfrac{a \cdot \sqrt{5}}{\sqrt{5} \cdot \sqrt{5}} = \dfrac{a\sqrt{5}}{5}$ 　　57. $\sqrt{\dfrac{2x}{5}} = \dfrac{\sqrt{2x} \cdot \sqrt{5}}{\sqrt{5} \cdot \sqrt{5}} = \dfrac{\sqrt{10x}}{5}$

58. $4\sqrt{\dfrac{2}{\pi}} = \dfrac{4\sqrt{2} \cdot \sqrt{\pi}}{\sqrt{\pi} \cdot \sqrt{\pi}} = \dfrac{4\sqrt{2\pi}}{\pi}$ 　　59. $\dfrac{2n}{\sqrt{8}} = \dfrac{2n}{2\sqrt{2}} = \dfrac{n}{\sqrt{2}} = \dfrac{n \cdot \sqrt{2}}{\sqrt{2} \cdot \sqrt{2}} = \dfrac{n\sqrt{2}}{2}$

60. $3\sqrt{3} - \sqrt{3} = 2\sqrt{3}$

61. $\sqrt{2} + \sqrt{8} - \sqrt{12} = \sqrt{2} + 2\sqrt{2} - 2\sqrt{3} = 3\sqrt{2} - 2\sqrt{3}$

62. $2\sqrt{45} - \sqrt{5} + \sqrt{75} = 2 \cdot 3\sqrt{5} - \sqrt{5} + 5\sqrt{3} = 5\sqrt{5} + 5\sqrt{3}$

63. $5\sqrt{3} + \sqrt{48} - \sqrt{4} = 5\sqrt{3} + 4\sqrt{3} - 2 = 9\sqrt{3} - 2$

64. $3\sqrt{24} + \sqrt{8} - 6\sqrt{20} = 3 \cdot 2\sqrt{6} + 2\sqrt{2} - 6 \cdot 2\sqrt{5} = 6\sqrt{6} + 2\sqrt{2} - 12\sqrt{5}$

65. $\sqrt{7} - 3\sqrt{21} + \sqrt{49} = \sqrt{7} - 3\sqrt{21} + 7$

66. $(\sqrt{5} - 3)(\sqrt{5} + 3) = 5 - 9 = -4$

67. $(3\sqrt{2} - 1)(3\sqrt{2} + 1) = 9 \cdot 2 - 1 = 18 - 1 = 17$

68. $(4 - 2\sqrt{3})^2 = 4^2 - 2 \cdot 4 \cdot 2\sqrt{3} + (2\sqrt{3})^2 = 16 - 16\sqrt{3} + 12 = 28 - 16\sqrt{3}$

Page 415 • CHAPTER TEST

1. 5　 2. -10　 3. -6　 4. $\sqrt{12^2 + 9^2} = \sqrt{144 + 81} = \sqrt{225} = 15$

5. $5.292 \approx 5.3$　 6. $9.434 \approx 9.4$　 7. $-9.327 \approx -9.3$　 8. $9.849 \approx 9.8$

9. rational   **10.** rational   **11.** irrational   **12.** irrational

13. $A = \pi r^2$; $A \approx (3.14)(14)^2 = (3.14)(196) = 615.44$

14. $\sqrt{147} = \sqrt{49 \cdot 3} = \sqrt{49} \cdot \sqrt{3} = 7\sqrt{3}$

15. $\sqrt{126} = \sqrt{9 \cdot 14} = \sqrt{9} \cdot \sqrt{14} = 3\sqrt{14}$

16. $\sqrt{180} = \sqrt{36 \cdot 5} = \sqrt{36} \cdot \sqrt{5} = 6\sqrt{5}$

17. $\sqrt{27x^3 y} = \sqrt{3 \cdot 9 \cdot x^2 \cdot x \cdot y} = 3x\sqrt{3xy}$

18. $\dfrac{x}{2} = \dfrac{32}{x}$; $x^2 = 64$; $x = \pm\sqrt{64} = \pm 8$

19. $y^2 = 28$; $y = \pm\sqrt{28} = \pm\sqrt{4 \cdot 7} = \pm 2\sqrt{7}$

20. $1 + n^2 = 4$; $n^2 = 3$; $n = \pm\sqrt{3}$

21. $x^2 + 3^2 = 9^2$; $x^2 + 9 = 81$; $x^2 = 72$; $x = \sqrt{72} = 6\sqrt{2}$

22. $\sqrt{\dfrac{16}{100}} = \dfrac{\sqrt{16}}{\sqrt{100}} = \dfrac{4}{10} = \dfrac{2}{5}$ **23.** $\sqrt{\dfrac{144}{169}} = \dfrac{\sqrt{144}}{\sqrt{169}} = \dfrac{12}{13}$

24. $\sqrt{\dfrac{5x^2}{49}} = \dfrac{\sqrt{5x^2}}{\sqrt{49}} = \dfrac{x\sqrt{5}}{7}$

25. $4\sqrt{9x} \cdot 3\sqrt{x} = 4 \cdot 3 \cdot \sqrt{9x \cdot x} = 12\sqrt{9x^2} = 12 \cdot 3x = 36x$

26. $\dfrac{8\sqrt{18}}{4\sqrt{8}} = 2\sqrt{\dfrac{18}{8}} = 2\sqrt{\dfrac{9}{4}} = 2 \cdot \dfrac{3}{2} = 3$

27. $\sqrt{\dfrac{20}{11}} \cdot \sqrt{\dfrac{11}{5}} = \sqrt{\dfrac{20}{11} \cdot \dfrac{11}{5}} = \sqrt{4} = 2$

28. $\sqrt{\dfrac{25}{3}} = \dfrac{\sqrt{25} \cdot \sqrt{3}}{\sqrt{3} \cdot \sqrt{3}} = \dfrac{5\sqrt{3}}{3}$

29. $\sqrt{\dfrac{n^2}{8}} = \dfrac{\sqrt{n^2} \cdot \sqrt{8}}{\sqrt{8} \cdot \sqrt{8}} = \dfrac{n\sqrt{8}}{8} = \dfrac{n \cdot 2\sqrt{2}}{8} = \dfrac{n\sqrt{2}}{4}$

30. $4\sqrt{\dfrac{1}{5}} = \dfrac{4\sqrt{1} \cdot \sqrt{5}}{\sqrt{5} \cdot \sqrt{5}} = \dfrac{4\sqrt{5}}{5}$

31. $\sqrt{7} - 7\sqrt{2} + 2\sqrt{7} = 3\sqrt{7} - 7\sqrt{2}$

32. $\sqrt{50a} - \sqrt{32a} + \sqrt{8a} = 5\sqrt{2a} - 4\sqrt{2a} + 2\sqrt{2a} = 3\sqrt{2a}$

33. $\dfrac{3 - \sqrt{5}}{3 + \sqrt{5}} = \dfrac{(3 - \sqrt{5})(3 - \sqrt{5})}{(3 + \sqrt{5})(3 - \sqrt{5})} = \dfrac{9 - 6\sqrt{5} + 5}{9 - 5} = \dfrac{14 - 6\sqrt{5}}{4} = \dfrac{7 - 3\sqrt{5}}{2}$

Pages 416–417 • MIXED REVIEW

1. Let x = the length (in cm) of the third side. Then $1.6x$ = the length of each equal side. $18.9 = 2(1.6x) + x$; $18.9 = 3.2x + x$; $18.9 = 4.2x$; $4.5 = x$ and $1.6x = 1.6(4.5) = 7.2$.   4.5 cm, 7.2 cm, 7.2 cm

2. $(-1, 2)$

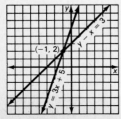

3. $\dfrac{\sqrt{24x^3}}{\sqrt{2x}} = \sqrt{\dfrac{24x^3}{2x}} = \sqrt{12x^2} = \sqrt{3 \cdot 4 \cdot x^2} = 2x\sqrt{3}$

4. $(3a - b)^2 - (8a^2 - 5ab) = 9a^2 - 6ab + b^2 - 8a^2 + 5ab = a^2 - ab + b^2$

5. $\dfrac{x^2 + 2x - 15}{x^2 - 25} \div (3 - x) = \dfrac{x^2 + 2x - 15}{x^2 - 25} \cdot \dfrac{1}{3 - x} = \dfrac{(x + 5)(x - 3) \cdot 1}{(x + 5)(x - 5)(-1)(x - 3)} =$

$\dfrac{1}{(-1)(x - 5)} = -\dfrac{1}{x - 5}$

6. $\dfrac{5}{6x} - \dfrac{3}{10x} = \dfrac{5 \cdot 5}{30x} - \dfrac{3 \cdot 3}{30x} = \dfrac{25 - 9}{30x} = \dfrac{16}{30x} = \dfrac{8}{15x}$

7. $\dfrac{2n}{n + 3} + \dfrac{6}{n + 3} = \dfrac{2n + 6}{n + 3} = \dfrac{2(n + 3)}{n + 3} = 2$

8. $\sqrt{9} - \sqrt{27} + \sqrt{18} = 3 - 3\sqrt{3} + 3\sqrt{2}$ **9.** $y = 2x + 4;$ 2

10. Let x = cost (in dollars) of a tire and y = cost of a battery. $4x + y = 167$ and
$5x + 2y = 241$; $y = -4x + 167$ and $5x + 2y = 241$; $5x + 2(-4x + 167) = 241$;
$5x - 8x + 334 = 241$; $-3x = -93$; $x = 31$; $y = -4x + 167 = -4(31) +$
$167 = -124 + 167 = 43.$ tire: \$31; battery: \$43

11. $\dfrac{7}{12} = 0.58333\ldots \approx 0.583$

12. Answers may vary. If $x = 0$, $y < 3 \cdot 0 - 1$; $y < -1.$ $(0, -2), (0, -3), (0, -4)$

13. $y^2 + 19y + 48 = (y + 3)(y + 16)$

14. $-12a^2 - 12a - 3 = -3(4a^2 + 4a + 1) = -3(2a + 1)^2$

15. $2x^3y + 4x^2y^2 + 14xy^3 = 2xy(x^2 + 2xy + 7y^2)$

16. Let $5x$ = kg of nuts and $2x$ = kg of raisins. $5x + 2x = 14$; $7x = 14$; $x = 2$, so
$5x = 5 \cdot 2 = 10.$ 10 kg of nuts

17. $\dfrac{r^2 + 3r}{6} \cdot \dfrac{4r^2}{r + 3} = \dfrac{r(r + 3) \cdot 4r^2}{6(r + 3)} = \dfrac{2r^3}{3}$

18. $(a^3 + 2b)(a^3 - 2b) = (a^3)^2 - (2b)^2 = a^6 - 4b^2$

19. $\begin{array}{r} 3m^2 - 2m - 5 \\ \underline{m - 4} \\ 3m^3 - 2m^2 - 5m \\ \underline{-12m^2 + 8m + 20} \\ 3m^3 - 14m^2 + 3m + 20 \end{array}$

20. $-7(4 - 3x) = 9x - 52$; $-28 +$
$21x = 9x - 52$; $21x = 9x - 24$;
$12x = -24$; $x = -2$

21. $0.04(120 - x) = 0.12x$; $4(120 - x) = 12x$; $480 - 4x = 12x$; $480 = 16x$;
$30 = x$

22. Let x = charge for 30 days of service. $\dfrac{49.60}{31} = \dfrac{x}{30}$; $31x = 1488$; $x = 48.$ \$48

23. $I = prt$; $800 = 800 \cdot 0.08 \cdot t$; $800 = 64t$; $12.5 = t.$ 12.5 years

24. **a.** the number of containers bought

b. $7 \div 3 = 2\dfrac{1}{3}$; 2 containers

25. $n \times 48 = 30$; $n = \dfrac{30}{48} = 0.625 = 62.5\%$

26. Let x = length of hypotenuse. $x^2 = 2^2 + 3^2$; $x^2 = 4 + 9$; $x^2 = 13$; $x = \sqrt{13}$

27.

	Carrier	Trainee	Together
Hours needed	6	9	x
Part done in one hour	$\frac{1}{6}$	$\frac{1}{9}$	$\frac{1}{x}$

$\frac{1}{6} + \frac{1}{9} = \frac{1}{x}$; $\frac{3}{18} + \frac{2}{18} = \frac{1}{x}$; $\frac{5}{18} = \frac{1}{x}$; $5x = 18$; $x = 3\frac{3}{5}$. $3\frac{3}{5}$ hours

28. $\frac{x+5}{3} = \frac{3x-1}{5}$; $5(x+5) = 3(3x-1)$; $5x + 25 = 9x - 3$; $25 = 4x - 3$;

$28 = 4x$; $7 = x$

29. $1 - 9y = -62$; $-9y = -63$; $y = 7$

30. $1.8s = 0.24 + 2.6s$; $180s = 24 + 260s$; $-80s = 24$; $s = -0.3$

31. $V = (x+2)(x-3)(4) = (x^2 - x - 6)(4) = 4x^2 - 4x - 24$

32. $(-7a^2b^3)(-2ab^2)^2 = (-7a^2b^3)(4a^2b^4) = (-7 \cdot 4)(a^2 \cdot a^2)(b^3 \cdot b^4) = -28a^4b^7$

33. $\frac{n^2 - 10n + 16}{3n^2 - 6n} = \frac{(n-2)(n-8)}{3n(n-2)} = \frac{n-8}{3n}$

34. $\frac{-8x^4 + 6x^2}{2x} + x^3 = \frac{-8x^4}{2x} + \frac{6x^2}{2x} + x^3 = -4x^3 + 3x + x^3 = -3x^3 + 3x$

35. If $x = -1$: $3(-1) - y = 6$; $-3 - y = 6$; $-y = 9$; $y = -9$.

If $y = 0$: $3x - 0 = 6$; $x = 2$. $(-1, -9), (2, 0)$

36. Let x = Shelley's age now. Then $x + 7$ = Haywood's age now, and $x + 4$ and $(x + 7) + 4$, or $x + 11$ are their ages in 4 years. $x + 11 = 2(x + 4)$; $x + 11 = 2x + 8$; $11 = x + 8$; $3 = x$ and $x + 7 = 10$.   Shelley: 3;   Haywood: 10

37.

	r	t	D
Going	10	2	20
Returning	0.80(10), or 8	x	$8x$

$20 = 8x$; $x = \frac{20}{8} = 2\frac{1}{2}$.

$2\frac{1}{2}$ hours

38.

	p	n	C
In advance	18	x	$18x$
At door	20	$573 - x$	$20(573 - x)$
			10,670

$18x + 20(573 - x) = 10{,}670$;

$18x + 11{,}460 - 20x = 10{,}670$;

$-2x + 11{,}460 = 10{,}670$; $-2x = -790$; $x = 395$.   395 tickets in advance

39. $-2 + x = -7$; $x = -5$

40. $(-2)^3 = (-2)(-2)(-2) = -8$ and $-(-3)^2 = -9$; $-8 > -9$; $>$

Page 421 · WRITTEN EXERCISES

A   **1.** $n(n + 3) = 0;$ $n = 0$ or $n + 3 = 0;$ $n = 0$ or $n = -3$

      2. $n(n - 3) = 0;$ $n = 0$ or $n - 3 = 0;$ $n = 0$ or $n = 3$

      3. $x(x + 9) = 0;$ $x = 0$ or $x + 9 = 0;$ $x = 0$ or $x = -9$

      4. $y(y + 6) = 0;$ $y = 0$ or $y + 6 = 0;$ $y = 0$ or $y = -6$

      5. $5m(m - 1) = 0;$ $5m = 0$ or $m - 1 = 0;$ $m = 0$ or $m = 1$

      6. $2m(m - 4) = 0;$ $2m = 0$ or $m - 4 = 0;$ $m = 0$ or $m = 4$

      7. $2n(n - 9) = 0;$ $2n = 0$ or $n - 9 = 0;$ $n = 0$ or $n = 9$

      8. $4x(x + 8) = 0;$ $4x = 0$ or $x + 8 = 0;$ $x = 0$ or $x = -8$

      9. $(y - 3)(y - 1) = 0;$ $y - 3 = 0$ or $y - 1 = 0;$ $y = 3$ or $y = 1$

    10. $(a + 8)(a + 1) = 0;$ $a + 8 = 0$ or $a + 1 = 0;$ $a = -8$ or $a = -1$

    11. $(x + 4)(x + 3) = 0;$ $x + 4 = 0$ or $x + 3 = 0;$ $x = -4$ or $x = -3$

    12. $(b + 1)(b + 6) = 0;$ $b + 1 = 0$ or $b + 6 = 0;$ $b = -1$ or $b = -6$

    13. $(y + 3)(y + 8) = 0;$ $y + 3 = 0$ or $y + 8 = 0;$ $y = -3$ or $y = -8$

    14. $(x - 10)(x - 20) = 0;$ $x - 10 = 0$ or $x - 20 = 0;$ $x = 10$ or $x = 20$

    15. $(n - 14)(n - 2) = 0;$ $n - 14 = 0$ or $n - 2 = 0;$ $n = 14$ or $n = 2$

    16. $(y - 10)(y - 3) = 0;$ $y - 10 = 0$ or $y - 3 = 0;$ $y = 10$ or $y = 3$

    17. $(x + 5)(x + 6) = 0;$ $x + 5 = 0$ or $x + 6 = 0;$ $x = -5$ or $x = -6$

    18. $(n + 20)(n + 40) = 0;$ $n + 20 = 0$ or $n + 40 = 0;$ $n = -20$ or $n = -40$

    19. $(a - 8)(a + 8) = 0;$ $a - 8 = 0$ or $a + 8 = 0;$ $a = 8$ or $a = -8$

    20. $(x - 25)(x + 25) = 0;$ $x - 25 = 0$ or $x + 25 = 0;$ $x = 25$ or $x = -25$

    21. $(y + 12)(y - 12) = 0;$ $y + 12 = 0$ or $y - 12 = 0;$ $y = -12$ or $y = 12$

    22. $(n - 6)(n + 4) = 0;$ $n - 6 = 0$ or $n + 4 = 0;$ $n = 6$ or $n = -4$

    23. $(x + 100)(x - 100) = 0;$ $x + 100 = 0$ or $x - 100 = 0;$ $x = -100$ or $x = 100$

    24. $(x - 4)(x - 75) = 0;$ $x - 4 = 0$ or $x - 75 = 0;$ $x = 4$ or $x = 75$

B   **25.** $(y - 12)(4y + 20) = 0;$ $y - 12 = 0$ or $4y + 20 = 0;$ $y = 12$ or $4y = -20;$
        $y = 12$ or $y = -5$

    26. $(m - 8)(3m - 12) = 0;$ $m - 8 = 0$ or $3m - 12 = 0;$ $m = 8$ or $3m = 12;$
        $m = 8$ or $m = 4$

    27. $(6r - 18)(7r + 49) = 0;$ $6r - 18 = 0$ or $7r + 49 = 0;$ $6r = 18$ or $7r = -49;$
        $r = 3$ or $r = -7$

    28. $(4n - 12)(6n + 72) = 0;$ $4n - 12 = 0$ or $6n + 72 = 0;$ $4n = 12$ or $6n = -72;$
        $n = 3$ or $n = -12$

    29. $(12x - 72)(5x + 20) = 0;$ $12x - 72 = 0$ or $5x + 20 = 0;$ $12x = 72$ or $5x = -20;$
        $x = 6$ or $x = -4$

    30. $x(x + 1)(x + 5) = 0;$ $x = 0,$ $x + 1 = 0,$ or $x + 5 = 0;$ $x = 0,$ $x = -1,$
        or $x = -5$

31. $n(n - 1)(n + 2) = 0$; $n = 0, n - 1 = 0$, or $n + 2 = 0$; $n = 0, n = 1$, or $n = -2$

32. $(18x - 54)(9x + 18) = 0$; $18x - 54 = 0$ or $9x + 18 = 0$; $18x = 54$ or $9x = -18$; $x = 3$ or $x = -2$

33. $(7x - 28)(15x - 90) = 0$; $7x - 28 = 0$ or $15x - 90 = 0$; $7x = 28$ or $15x = 90$; $x = 4$ or $x = 6$

34. $3y(y - 4)(y + 4) = 0$; $3y = 0, y - 4 = 0$, or $y + 4 = 0$; $y = 0, y = 4$, or $y = -4$

Page 423 • WRITTEN EXERCISES

A

1. $x^2 - 2x = x(x - 2)$

2. $y^2 - 3y = y(y - 3)$

3. $2x^2 - 4x = 2x(x - 2)$

4. $x^2 + 3x + 2 = (x + 1)(x + 2)$

5. $x^2 + 2x - 3 = (x - 1)(x + 3)$

6. $x^2 + x - 2 = (x - 1)(x + 2)$

7. $x^2 + 6x + 8 = (x + 2)(x + 4)$

8. $x^2 - 7x + 12 = (x - 3)(x - 4)$

9. $x^2 - 3x - 18 = (x + 3)(x - 6)$

10. $x^2 + x = 0$; $x(x + 1) = 0$; $x = 0$ or $x + 1 = 0$; $x = 0$ or $x = -1$

11. $x^2 - 3x = 0$; $x(x - 3) = 0$; $x = 0$ or $x - 3 = 0$; $x = 0$ or $x = 3$

12. $x^2 - 5x = 0$; $x(x - 5) = 0$; $x = 0$ or $x - 5 = 0$; $x = 0$ or $x = 5$

13. $6x + 2x^2 = 0$; $2x(3 + x) = 0$; $2x = 0$ or $3 + x = 0$; $x = 0$ or $x = -3$

14. $9x - 3x^2 = 0$; $3x(3 - x) = 0$; $3x = 0$ or $3 - x = 0$; $x = 0$ or $x = 3$

15. $16x - 4x^2 = 0$; $4x(4 - x) = 0$; $4x = 0$ or $4 - x = 0$; $x = 0$ or $x = 4$

16. $x^2 - 5x + 4 = 0$; $(x - 1)(x - 4) = 0$; $x - 1 = 0$ or $x - 4 = 0$; $x = 1$ or $x = 4$

17. $x^2 - 9x + 8 = 0$; $(x - 1)(x - 8) = 0$; $x - 1 = 0$ or $x - 8 = 0$; $x = 1$ or $x = 8$

18. $x^2 + 4x + 3 = 0$; $(x + 1)(x + 3) = 0$; $x + 1 = 0$ or $x + 3 = 0$; $x = -1$ or $x = -3$

19. $x^2 - 8x + 7 = 0$; $(x - 1)(x - 7) = 0$; $x - 1 = 0$ or $x - 7 = 0$; $x = 1$ or $x = 7$

20. $x^2 + 8x + 15 = 0$; $(x + 3)(x + 5) = 0$; $x + 3 = 0$ or $x + 5 = 0$; $x = -3$ or $x = -5$

21. $x^2 - 10x + 21 = 0$; $(x - 3)(x - 7) = 0$; $x - 3 = 0$ or $x - 7 = 0$; $x = 3$ or $x = 7$

22. $y^2 + 4y + 4 = 0$; $(y + 2)^2 = 0$; $y + 2 = 0$; $y = -2$

23. $y^2 + 4y - 21 = 0$; $(y - 3)(y + 7) = 0$; $y - 3 = 0$ or $y + 7 = 0$; $y = 3$ or $y = -7$

24. $y^2 - 7y + 6 = 0$; $(y - 1)(y - 6) = 0$; $y - 1 = 0$ or $y - 6 = 0$; $y = 1$ or $y = 6$

25. $x^2 - 2x - 8 = 0$; . $(x + 2)(x - 4) = 0$; $x + 2 = 0$ or $x - 4 = 0$; $x = -2$ or $x = 4$

26. $x^2 + x - 6 = 0$; $(x - 2)(x + 3) = 0$; $x - 2 = 0$ or $x + 3 = 0$; $x = 2$ or $x = -3$

27. $n^2 - 5n - 14 = 0$; $(n + 2)(n - 7) = 0$; $n + 2 = 0$ or $n - 7 = 0$; $n = -2$ or $n = 7$

28. $a^2 + 4a - 32 = 0$; $(a - 4)(a + 8) = 0$; $a - 4 = 0$ or $a + 8 = 0$; $a = 4$ or $a = -8$

29. $a^2 + 10a + 25 = 0$; $(a + 5)^2 = 0$; $a + 5 = 0$; $a = -5$

30. $a^2 - 12a + 35 = 0$; $(a - 5)(a - 7) = 0$; $a - 5 = 0$ or $a - 7 = 0$; $a = 5$ or $a = 7$

31. $x^2 - 11x + 28 = 0$; $(x - 4)(x - 7) = 0$; $x - 4 = 0$ or $x - 7 = 0$; $x = 4$ or $x = 7$

32. $x^2 - 16 = 0$; $(x + 4)(x - 4) = 0$; $x + 4 = 0$ or $x - 4 = 0$; $x = -4$ or $x = 4$

33. $x^2 - 9 = 0$; $(x + 3)(x - 3) = 0$; $x + 3 = 0$ or $x - 3 = 0$; $x = -3$ or $x = 3$

34. $x^2 + 12x + 36 = 0$; $(x + 6)^2 = 0$; $x + 6 = 0$; $x = -6$

35. $x^2 - 13x + 22 = 0$; $(x - 2)(x - 11) = 0$; $x - 2 = 0$ or $x - 11 = 0$; $x = 2$ or $x = 11$

36. $x^2 - 6x - 27 = 0$; $(x + 3)(x - 9) = 0$; $x + 3 = 0$ or $x - 9 = 0$; $x = -3$ or $x = 9$

37. $x^2 - 6x - 55 = 0$; $(x + 5)(x - 11) = 0$; $x + 5 = 0$ or $x - 11 = 0$; $x = -5$ or $x = 11$

38. $x^2 - 22x + 40 = 0$; $(x - 2)(x - 20) = 0$; $x - 2 = 0$ or $x - 20 = 0$; $x = 2$ or $x = 20$

39. $x^2 - 13x + 40 = 0$; $(x - 5)(x - 8) = 0$; $x - 5 = 0$ or $x - 8 = 0$; $x = 5$ or $x = 8$

40. $x^2 - 5x - 36 = 0$; $(x + 4)(x - 9) = 0$; $x + 4 = 0$ or $x - 9 = 0$; $x = -4$ or $x = 9$

41. $a^2 + 6a - 40 = 0$; $(a - 4)(a + 10) = 0$; $a - 4 = 0$ or $a + 10 = 0$; $a = 4$ or $a = -10$

42. $y^2 + 5y - 24 = 0$; $(y - 3)(y + 8) = 0$; $y - 3 = 0$ or $y + 8 = 0$; $y = 3$ or $y = -8$

43. $n^2 - 6n - 7 = 0$; $(n + 1)(n - 7) = 0$; $n + 1 = 0$ or $n - 7 = 0$; $n = -1$ or $n = 7$

44. $m^2 + 20m + 64 = 0$; $(m + 4)(m + 16) = 0$; $m + 4 = 0$ or $m + 16 = 0$; $m = -4$ or $m = -16$

45. $x^2 + 11x - 80 = 0$; $(x - 5)(x + 16) = 0$; $x - 5 = 0$ or $x + 16 = 0$; $x = 5$ or $x = -16$

Page 425 • **WRITTEN EXERCISES**

A   1. $x^2 = 3x$; $x^2 - 3x = 0$; $x(x - 3) = 0$; $x = 0$ or $x - 3 = 0$; $x = 0$ or $x = 3$

2. $x^2 = -7x$; $x^2 + 7x = 0$; $x(x + 7) = 0$; $x = 0$ or $x + 7 = 0$; $x = 0$ or $x = -7$

3. $x^2 = -4x$; $x^2 + 4x = 0$; $x(x + 4) = 0$; $x = 0$ or $x + 4 = 0$; $x = 0$ or $x = -4$

4. $x^2 + 5x = 6x$; $x^2 - x = 0$; $x(x - 1) = 0$; $x = 0$ or $x - 1 = 0$; $x = 0$ or $x = 1$

5. $y^2 + 4y = 32$; $y^2 + 4y - 32 = 0$; $(y - 4)(y + 8) = 0$; $y - 4 = 0$ or $y + 8 = 0$; $y = 4$ or $y = -8$

6. $x^2 + 5x = 6$; $x^2 + 5x - 6 = 0$; $(x - 1)(x + 6) = 0$; $x - 1 = 0$ or $x + 6 = 0$; $x = 1$ or $x = -6$

7. $y^2 - 2y = 8$; $y^2 - 2y - 8 = 0$; $(y + 2)(y - 4) = 0$; $y + 2 = 0$ or $y - 4 = 0$; $y = -2$ or $y = 4$

8. $6 - x = x^2$; $0 = x^2 + x - 6$; $0 = (x - 2)(x + 3)$; $x - 2 = 0$ or $x + 3 = 0$; $x = 2$ or $x = -3$

9. $6y - 9 = y^2$; $0 = y^2 - 6y + 9$; $0 = (y - 3)^2$; $y - 3 = 0$; $y = 3$

10. $x^2 + 22 = 13x$; $x^2 - 13x + 22 = 0$; $(x - 2)(x - 11) = 0$; $x - 2 = 0$ or $x - 11 = 0$; $x = 2$ or $x = 11$

11. $x^2 + 20 = 12x$; $x^2 + 12x + 20 = 0$; $(x + 2)(x + 10) = 0$; $x + 2 = 0$ or $x + 10 = 0$; $x = -2$ or $x = -10$

12. $x^2 - 6 = 5x$; $x^2 - 5x - 6 = 0$; $(x + 1)(x - 6) = 0$; $x + 1 = 0$ or $x - 6 = 0$; $x = -1$ or $x = 6$

13. $y^2 + y = 42$; $y^2 + y - 42 = 0$; $(y - 6)(y + 7) = 0$; $y - 6 = 0$ or $y + 7 = 0$; $y = 6$ or $y = -7$

14. $y^2 = 9y - 20$; $y^2 - 9y + 20 = 0$; $(y - 4)(y - 5) = 0$; $y - 4 = 0$ or $y - 5 = 0$; $y = 4$ or $y = 5$

15. $x^2 = 7x + 18$; $x^2 - 7x - 18 = 0$; $(x + 2)(x - 9) = 0$; $x + 2 = 0$ or $x - 9 = 0$; $x = -2$ or $x = 9$

16. $a^2 - 3a = 18$; $a^2 - 3a - 18 = 0$; $(a + 3)(a - 6) = 0$; $a + 3 = 0$ or $a - 6 = 0$; $a = -3$ or $a = 6$

17. $a^2 = 12a - 35$; $a^2 - 12a + 35 = 0$; $(a - 5)(a - 7) = 0$; $a - 5 = 0$ or $a - 7 = 0$; $a = 5$ or $a = 7$

18. $x^2 = 5x + 24$; $x^2 - 5x - 24 = 0$; $(x + 3)(x - 8) = 0$; $x + 3 = 0$ or $x - 8 = 0$; $x = -3$ or $x = 8$

19. $x(x + 4) = 12$; $x^2 + 4x = 12$; $x^2 + 4x - 12 = 0$; $(x - 2)(x + 6) = 0$; $x - 2 = 0$ or $x + 6 = 0$; $x = 2$ or $x = -6$

20. $y(y + 3) = 40$; $y^2 + 3y = 40$; $y^2 + 3y - 40 = 0$; $(y - 5)(y + 8) = 0$; $y - 5 = 0$ or $y + 8 = 0$; $y = 5$ or $y = -8$

21. $(y-1)(y+1) = 48$; $y^2 - 1 = 48$; $y^2 - 49 = 0$; $(y+7)(y-7) = 0$;

   $y + 7 = 0$ or $y - 7 = 0$; $y = -7$ or $y = 7$

22. $(x+2)(x-3) = 6$; $x^2 - x - 6 = 6$; $x^2 - x - 12 = 0$; $(x+3)(x-4) = 0$;

   $x + 3 = 0$ or $x - 4 = 0$; $x = -3$ or $x = 4$

23. $(n-2)(n+5) = 18$; $n^2 + 3n - 10 = 18$; $n^2 + 3n - 28 = 0$;

   $(n-4)(n+7) = 0$; $n - 4 = 0$ or $n + 7 = 0$; $n = 4$ or $n = -7$

24. $(a-1)(a+1) = 15$; $a^2 - 1 = 15$; $a^2 - 16 = 0$; $(a+4)(a-4) = 0$;

   $a + 4 = 0$ or $a - 4 = 0$; $a = -4$ or $a = 4$

B 25. $\dfrac{x}{2} = \dfrac{6}{x+1}$; $x(x+1) = 12$; $x^2 + x = 12$; $x^2 + x - 12 = 0$;

   $(x-3)(x+4) = 0$; $x - 3 = 0$ or $x + 4 = 0$; $x = 3$ or $x = -4$

26. $\dfrac{x}{4} = \dfrac{7}{x+3}$; $x(x+3) = 28$; $x^2 + 3x = 28$; $x^2 + 3x - 28 = 0$;

   $(x-4)(x+7) = 0$; $x - 4 = 0$ or $x + 7 = 0$; $x = 4$ or $x = -7$

27. $\dfrac{x-1}{3} = \dfrac{4}{x}$; $x(x-1) = 12$; $x^2 - x = 12$; $x^2 - x - 12 = 0$;

   $(x+3)(x-4) = 0$; $x + 3 = 0$ or $x - 4 = 0$; $x = -3$ or $x = 4$

28. $\dfrac{x-2}{x} = \dfrac{x}{8}$; $8(x-2) = x^2$; $8x - 16 = x^2$; $0 = x^2 - 8x + 16$; $0 = (x-4)^2$;

   $x - 4 = 0$; $x = 4$

29. $\dfrac{y-3}{5} = \dfrac{8}{y+3}$; $(y-3)(y+3) = 40$; $y^2 - 9 = 40$; $y^2 - 49 = 0$;

   $(y+7)(y-7) = 0$; $y + 7 = 0$ or $y - 7 = 0$; $y = -7$ or $y = 7$

30. $\dfrac{4}{x+3} = \dfrac{x-3}{4}$; $16 = (x+3)(x-3)$; $16 = x^2 - 9$; $0 = x^2 - 25$;

   $0 = (x+5)(x-5)$; $x + 5 = 0$ or $x - 5 = 0$; $x = -5$ or $x = 5$

C 31. $2x^2 + x = 3$; $2x^2 + x - 3 = 0$; $(2x+3)(x-1) = 0$; $2x + 3 = 0$ or

   $x - 1 = 0$; $x = -\dfrac{3}{2}$ or $x = 1$

32. $2x^2 - x = 1$; $2x^2 - x - 1 = 0$; $(2x+1)(x-1) = 0$; $2x + 1 = 0$ or

   $x - 1 = 0$; $x = -\dfrac{1}{2}$ or $x = 1$

33. $x + 10 = 2x^2$; $0 = 2x^2 - x - 10$; $0 = (2x-5)(x+2)$; $2x - 5 = 0$ or

   $x + 2 = 0$; $x = \dfrac{5}{2}$ or $x = -2$

34. $(2x+1)(2x-1) = 0$; $2x + 1 = 0$ or $2x - 1 = 0$; $x = -\dfrac{1}{2}$ or $x = \dfrac{1}{2}$

35. $(2n+1)(n+1) = 28$; $2n^2 + 3n + 1 = 28$; $2n^2 + 3n - 27 = 0$;

   $(2n+9)(n-3) = 0$; $2n + 9 = 0$ or $n - 3 = 0$; $n = -\dfrac{9}{2}$ or $n = 3$

36. $\dfrac{1 + n}{3} = 2n^2;\quad 1 + n = 6n^2;\quad 0 = 6n^2 - n - 1;\quad 0 = (3n + 1)(2n - 1);$

$3n + 1 = 0$ or $2n - 1 = 0;\quad n = -\dfrac{1}{3}$ or $n = \dfrac{1}{2}$

Page 427 · WRITTEN EXERCISES

A

1. $x^2 = 4;\quad x = \pm 2;\quad x = 2$ or $x = -2$

2. $x^2 = 9;\quad x = \pm 3;\quad x = 3$ or $x = -3$

3. $x^2 = 16;\quad x = \pm 4;\quad x = 4$ or $x = -4$

4. $y^2 = 49;\quad y = \pm 7;\quad y = 7$ or $y = -7$

5. $x^2 = 64;\quad x = \pm 8;\quad x = 8$ or $x = -8$

6. $a^2 = 81;\quad a = \pm 9;\quad a = 9$ or $a = -9$

7. $y^2 = 121;\quad y = \pm 11;\quad y = 11$ or $y = -11$

8. $x^2 = 100;\quad x = \pm 10;\quad x = 10$ or $x = -10$

9. $x^2 = 400;\quad x = \pm 20;\quad x = 20$ or $x = -20$

10. $x^2 = 3;\quad x = \pm\sqrt{3};\quad x = \sqrt{3}$ or $x = -\sqrt{3}$

11. $y^2 = 5;\quad y = \pm\sqrt{5};\quad y = \sqrt{5}$ or $y = -\sqrt{5}$

12. $n^2 = 7;\quad n = \pm\sqrt{7};\quad n = \sqrt{7}$ or $n = -\sqrt{7}$

13. $a^2 = 8;\quad a = \pm\sqrt{8} = \pm 2\sqrt{2};\quad a = 2\sqrt{2}$ or $a = -2\sqrt{2}$

14. $y^2 = 32;\quad y = \pm\sqrt{32} = \pm 4\sqrt{2};\quad y = 4\sqrt{2}$ or $y = -4\sqrt{2}$

15. $n^2 = 1;\quad n = \pm 1;\quad n = 1$ or $n = -1$ **16.** $n^2 = 0;\quad n = 0$

17. $a^2 = 12;\quad a = \pm\sqrt{12} = \pm 2\sqrt{3};\quad a = 2\sqrt{3}$ or $a = -2\sqrt{3}$

18. $2x^2 = 18;\quad x^2 = 9;\quad x = \pm 3;\quad x = 3$ or $x = -3$

19. $3x^2 = 12;\quad x^2 = 4;\quad x = \pm 2;\quad x = 2$ or $x = -2$

20. $2x^2 = 50;\quad x^2 = 25;\quad x = \pm 5;\quad x = 5$ or $x = -5$

21. $3y^2 = -6;\quad y^2 = -2;\quad$ no solution

22. $2n^2 = -200;\quad n^2 = -100;\quad$ no solution

23. $x^2 - 1 = 3;\quad x^2 = 4;\quad x = \pm 2;\quad x = 2$ or $x = -2$

24. $x^2 - 5 = 4;\quad x^2 = 9;\quad x = \pm 3;\quad x = 3$ or $x = -3$

25. $x^2 + 1 = 50;\quad x^2 = 49;\quad x = \pm 7;\quad x = 7$ or $x = -7$

26. $y^2 + 2 = 30;\quad y^2 = 28;\quad y = \pm\sqrt{28} = \pm 2\sqrt{7};\quad y = 2\sqrt{7}$ or $y = -2\sqrt{7}$

27. $y^2 - 4 = 20;\quad y^2 = 24;\quad y = \pm\sqrt{24} = \pm 2\sqrt{6};\quad y = 2\sqrt{6}$ or $y = -2\sqrt{6}$

28. $n^2 - 2 = 5;\quad n^2 = 7;\quad n = \pm\sqrt{7};\quad n = \sqrt{7}$ or $n = -\sqrt{7}$

29. $a^2 + 7 = 4;\quad a^2 = -3;\quad$ no solution   **30.** $x^2 + 3 = 2;\quad x^2 = -1;\quad$ no solution

31. $n^2 - 5 = 10;\quad n^2 = 15;\quad n = \pm\sqrt{15};\quad n = \sqrt{15}$ or $n = -\sqrt{15}$

32. $y^2 + 3 = 21;\quad y^2 = 18;\quad y = \pm\sqrt{18} = \pm 3\sqrt{2};\quad y = 3\sqrt{2}$ or $y = -3\sqrt{2}$

33. $m^2 - 11 = 49;\quad m^2 = 60;\quad m = \pm\sqrt{60};\quad m = \pm 2\sqrt{15};\quad m = 2\sqrt{15}$ or $m = -2\sqrt{15}$

34. $x^2 + 8 = 7;\quad x^2 = -1;\quad$ no solution

35. $b^2 - 15 = 12$; $b^2 = 27$; $b = \pm\sqrt{27}$; $b = \pm 3\sqrt{3}$; $b = 3\sqrt{3}$ or $b = -3\sqrt{3}$

36. $z^2 - 12 = -4$; $z^2 = 8$; $z = \pm\sqrt{8}$; $z = \pm 2\sqrt{2}$; $z = 2\sqrt{2}$ or $z = -2\sqrt{2}$

B **37.** $2x^2 - 1 = 17$; $2x^2 = 18$; $x^2 = 9$; $x = \pm 3$; $x = 3$ or $x = -3$

38. $3y^2 + 5 = 50$; $3y^2 = 45$; $y^2 = 15$; $y = \pm\sqrt{15}$; $y = \sqrt{15}$ or $y = -\sqrt{15}$

39. $2n^2 - 7 = 13$; $2n^2 = 20$; $n^2 = 10$; $n = \pm\sqrt{10}$; $n = \sqrt{10}$ or $n = -\sqrt{10}$

40. $3n^2 + 9 = 42$; $3n^2 = 33$; $n^2 = 11$; $n = \pm\sqrt{11}$; $n = \sqrt{11}$ or $n = -\sqrt{11}$

41. $x^2 = \dfrac{1}{9}$; $x = \pm\sqrt{\dfrac{1}{9}} = \pm\dfrac{1}{3}$; $x = \dfrac{1}{3}$ or $x = -\dfrac{1}{3}$

42. $x^2 = \dfrac{4}{9}$; $x = \pm\sqrt{\dfrac{4}{9}} = \pm\dfrac{2}{3}$; $x = \dfrac{2}{3}$ or $x = -\dfrac{2}{3}$

43. $\dfrac{x}{4} = \dfrac{9}{x}$; $x^2 = 36$; $x = \pm 6$; $x = 6$ or $x = -6$

44. $\dfrac{y}{4} = \dfrac{16}{y}$; $y^2 = 64$; $y = \pm 8$; $y = 8$ or $y = -8$

45.

time	1	2	4	5	10
distance	5	20	80	125	500

Page 429 · WRITTEN EXERCISES

A **1.** $(x - 3)^2 = 36$; $x - 3 = \pm 6$; $x - 3 = 6$ or $x - 3 = -6$; $x = 9$ or $x = -3$

2. $(x + 5)^2 = 49$; $x + 5 = \pm 7$; $x + 5 = 7$ or $x + 5 = -7$; $x = 2$ or $x = -12$

3. $(x - 4)^2 = 25$; $x - 4 = \pm 5$; $x - 4 = 5$ or $x - 4 = -5$; $x = 9$ or $x = -1$

4. $(x + 1)^2 = 16$; $x + 1 = \pm 4$; $x + 1 = 4$ or $x + 1 = -4$; $x = 3$ or $x = -5$

5. $(y + 8)^2 = 100$; $y + 8 = \pm 10$; $y + 8 = 10$ or $y + 8 = -10$; $y = 2$ or $y = -18$

6. $(y - 2)^2 = 81$; $y - 2 = \pm 9$; $y - 2 = 9$ or $y - 2 = -9$; $y = 11$ or $y = -7$

7. $(y - 3)^2 = 64$; $y - 3 = \pm 8$; $y - 3 = 8$ or $y - 3 = -8$; $y = 11$ or $y = -5$

8. $(n + 2)^2 = 9$; $n + 2 = \pm 3$; $n + 2 = 3$ or $n + 2 = -3$; $n = 1$ or $n = -5$

9. $(a - 7)^2 = 121$; $a - 7 = \pm 11$; $a - 7 = 11$ or $a - 7 = -11$; $a = 18$ or $a = -4$

10. $(2x - 1)^2 = 25$; $2x - 1 = \pm 5$; $2x - 1 = 5$ or $2x - 1 = -5$; $2x = 6$ or $2x = -4$; $x = 3$ or $x = -2$

11. $(4n - 2)^2 = 36$; $4n - 2 = \pm 6$; $4n - 2 = 6$ or $4n - 2 = -6$; $4n = 8$ or $4n = -4$; $n = 2$ or $n = -1$

12. $(2y - 2)^2 = 64$; $2y - 2 = \pm 8$; $2y - 2 = 8$ or $2y - 2 = -8$; $2y = 10$ or $2y = -6$; $y = 5$ or $y = -3$

13. $(2x - 3)^2 = 49$; $2x - 3 = 7$ or $2x - 3 = -7$; $2x = 10$ or $2x = -4$; $x = 5$ or $x = -2$

14. $(2x - 1)^2 = 81$; $2x - 1 = \pm 9$; $2x - 1 = 9$ or $2x - 1 = -9$; $2x = 10$ or $2x = -8$; $x = 5$ or $x = -4$

15. $(2x + 3)^2 = 25$; $2x + 3 = \pm5$; $2x + 3 = 5$ or $2x + 3 = -5$; $2x = 2$ or $2x = -8$; $x = 1$ or $x = -4$

16. $(5p + 5)^2 = 100$; $5p + 5 = \pm10$; $5p + 5 = 10$ or $5p + 5 = -10$; $5p = 5$ or $5p = -15$; $p = 1$ or $p = -3$

17. $(2p + 3)^2 = 81$; $2p + 3 = \pm9$; $2p + 3 = 9$ or $2p + 3 = -9$; $2p = 6$ or $2p = -12$; $p = 3$ or $p = -6$

18. $(3x - 6)^2 = 9$; $3x - 6 = \pm3$; $3x - 6 = 3$ or $3x - 6 = -3$; $3x = 9$ or $3x = 3$; $x = 3$ or $x = 1$

19. $(4a - 8)^2 = 0$; $4a - 8 = 0$; $4a = 8$; $a = 2$

20. $(3x - 3)^2 = 1$; $3x - 3 = \pm1$; $3x - 3 = 1$ or $3x - 3 = -1$; $3x = 4$ or $3x = 2$; $x = \dfrac{4}{3}$ or $x = \dfrac{2}{3}$

21. $(6n - 6)^2 = 36$; $6n - 6 = \pm6$; $6n - 6 = 6$ or $6n - 6 = -6$; $6n = 12$ or $6n = 0$; $n = 2$ or $n = 0$

22. $(2m + 5)^2 = 49$; $2m + 5 = \pm7$; $2m + 5 = 7$ or $2m + 5 = -7$; $2m = 2$ or $2m = -12$; $m = 1$ or $m = -6$

B 23. $(x - 2)^2 = 7$; $x - 2 = \pm\sqrt{7}$; $x - 2 = \sqrt{7}$ or $x - 2 = -\sqrt{7}$; $x = 2 + \sqrt{7}$ or $x = 2 - \sqrt{7}$

24. $(x - 4)^2 = 3$; $x - 4 = \pm\sqrt{3}$; $x - 4 = \sqrt{3}$ or $x - 4 = -\sqrt{3}$; $x = 4 + \sqrt{3}$ or $x = 4 - \sqrt{3}$

25. $(y + 1)^2 = 2$; $y + 1 = \pm\sqrt{2}$; $y + 1 = \sqrt{2}$ or $y + 1 = -\sqrt{2}$; $y = -1 + \sqrt{2}$ or $y = -1 - \sqrt{2}$

26. $(y + 8)^2 = 6$; $y + 8 = \pm\sqrt{6}$; $y + 8 = \sqrt{6}$ or $y + 8 = -\sqrt{6}$; $y = -8 + \sqrt{6}$ or $y = -8 - \sqrt{6}$

27. $(a - 5)^2 = 8$; $a - 5 = \pm\sqrt{8} = \pm2\sqrt{2}$; $a - 5 = 2\sqrt{2}$ or $a - 5 = -2\sqrt{2}$; $a = 5 + 2\sqrt{2}$ or $a = 5 - 2\sqrt{2}$

28. $3(x - 1)^2 = 12$; $(x - 1)^2 = 4$; $x - 1 = \pm2$; $x - 1 = 2$ or $x - 1 = -2$; $x = 3$ or $x = -1$

29. $2(x - 7)^2 = 18$; $(x - 7)^2 = 9$; $x - 7 = \pm3$; $x - 7 = 3$ or $x - 7 = -3$; $x = 10$ or $x = 4$

30. $6(a + 1)^2 = 24$; $(a + 1)^2 = 4$; $a + 1 = \pm2$; $a + 1 = 2$ or $a + 1 = -2$; $a = 1$ or $a = -3$

31. $5(n + 2)^2 = 45$; $(n + 2)^2 = 9$; $n + 2 = \pm3$; $n + 2 = 3$ or $n + 2 = -3$; $n = 1$ or $n = -5$

32. $3(a - 1)^2 = 75$; $(a - 1)^2 = 25$; $a - 1 = \pm5$; $a - 1 = 5$ or $a - 1 = -5$; $a = 6$ or $a = -4$

33. $7(x + 3)^2 = 28$; $(x + 3)^2 = 4$; $x + 3 = \pm2$; $x + 3 = 2$ or $x + 3 = -2$; $x = -1$ or $x = -5$

34. $3(y - 6)^2 = 27$; $(y - 6)^2 = 9$; $y - 6 = \pm 3$; $y - 6 = 3$ or $y - 6 = -3$; $y = 9$ or $y = 3$

35. $3(2x + 2)^2 = 48$; $(2x + 2)^2 = 16$; $2x + 2 = \pm 4$; $2x + 2 = 4$ or $2x + 2 = -4$; $2x = 2$ or $2x = -6$; $x = 1$ or $x = -3$

36. $2(2x - 1)^2 = 50$; $(2x - 1)^2 = 25$; $2x - 1 = \pm 5$; $2x - 1 = 5$ or $2x - 1 = -5$; $2x = 6$ or $2x = -4$; $x = 3$ or $x = -2$

37. $2(2y + 4)^2 = 72$; $(2y + 4)^2 = 36$; $2y + 4 = \pm 6$; $2y + 4 = 6$ or $2y + 4 = -6$; $2y = 2$ or $2y = -10$; $y = 1$ or $y = -5$

38. $4(a - 7)^2 = 20$; $(a - 7)^2 = 5$; $a - 7 = \pm\sqrt{5}$; $a - 7 = \sqrt{5}$ or $a - 7 = -\sqrt{5}$; $a = 7 + \sqrt{5}$ or $a = 7 - \sqrt{5}$

Page 432 · WRITTEN EXERCISES

A

1. $2x^2 + 3x + 1 = 0$; $x = \dfrac{-3 \pm \sqrt{3^2 - (4 \cdot 2 \cdot 1)}}{2 \cdot 2} = \dfrac{-3 \pm \sqrt{9 - 8}}{4} = \dfrac{-3 \pm \sqrt{1}}{4} = \dfrac{-3 \pm 1}{4}$;

$x = \dfrac{-3 + 1}{4}$ or $x = \dfrac{-3 - 1}{4}$; $x = \dfrac{-2}{4} = -\dfrac{1}{2}$ or $x = \dfrac{-4}{4} = -1$

2. $4x^2 + 5x + 1 = 0$; $x = \dfrac{-5 \pm \sqrt{5^2 - (4 \cdot 4 \cdot 1)}}{2 \cdot 4} = \dfrac{-5 \pm \sqrt{25 - 16}}{8} = \dfrac{-5 \pm \sqrt{9}}{8} = $

$\dfrac{-5 \pm 3}{8}$; $x = \dfrac{-5 + 3}{8}$ or $x = \dfrac{-5 - 3}{8}$; $x = \dfrac{-2}{8} = -\dfrac{1}{4}$ or $x = \dfrac{-8}{8} = -1$

3. $3x^2 - 7x + 2 = 0$; $x = \dfrac{-(-7) \pm \sqrt{(-7)^2 - (4 \cdot 3 \cdot 2)}}{2 \cdot 3} = \dfrac{7 \pm \sqrt{49 - 24}}{6} = \dfrac{7 \pm \sqrt{25}}{6} = $

$\dfrac{7 \pm 5}{6}$; $x = \dfrac{7 + 5}{6}$ or $x = \dfrac{7 - 5}{6}$; $x = \dfrac{12}{6} = 2$ or $x = \dfrac{2}{6} = \dfrac{1}{3}$

4. $2x^2 - 5x - 3 = 0$; $x = \dfrac{-(-5) \pm \sqrt{(-5)^2 - (4 \cdot 2 \cdot -3)}}{2 \cdot 2} = \dfrac{5 \pm \sqrt{25 + 24}}{4} = \dfrac{5 \pm \sqrt{49}}{4} = $

$\dfrac{5 \pm 7}{4}$; $x = \dfrac{5 + 7}{4}$ or $x = \dfrac{5 - 7}{4}$; $x = \dfrac{12}{4} = 3$ or $x = \dfrac{-2}{4} = -\dfrac{1}{2}$

5. $4x^2 + 7x - 2 = 0$; $x = \dfrac{-7 \pm \sqrt{7^2 - (4 \cdot 4 \cdot -2)}}{2 \cdot 4} = \dfrac{-7 \pm \sqrt{49 + 32}}{8} = \dfrac{-7 \pm \sqrt{81}}{8} = $

$\dfrac{-7 \pm 9}{8}$; $x = \dfrac{-7 + 9}{8}$ or $x = \dfrac{-7 - 9}{8}$; $x = \dfrac{2}{8} = \dfrac{1}{4}$ or $x = \dfrac{-16}{8} = -2$

6. $x^2 - 5x - 14 = 0$; $x = \dfrac{-(-5) \pm \sqrt{(-5)^2 - (4 \cdot 1 \cdot -14)}}{2 \cdot 1} = \dfrac{5 \pm \sqrt{25 + 56}}{2} = $

$\dfrac{5 \pm \sqrt{81}}{2} = \dfrac{5 \pm 9}{2}$; $x = \dfrac{5 + 9}{2}$ or $x = \dfrac{5 - 9}{2}$; $x = \dfrac{14}{2} = 7$ or $x = \dfrac{-4}{2} = -2$

7. $2x^2 + 7x - 4 = 0$; $x = \dfrac{-7 \pm \sqrt{7^2 - (4 \cdot 2 \cdot -4)}}{2 \cdot 2} = \dfrac{-7 \pm \sqrt{49 + 32}}{4} = \dfrac{-7 \pm \sqrt{81}}{4} = $

$\dfrac{-7 \pm 9}{4}$; $x = \dfrac{-7 + 9}{4}$ or $x = \dfrac{-7 - 9}{4}$; $x = \dfrac{2}{4} = \dfrac{1}{2}$ or $x = \dfrac{-16}{4} = -4$

8. $x^2 + 4x - 21 = 0$; $x = \dfrac{-4 \pm \sqrt{4^2 - (4 \cdot 1 \cdot -21)}}{2 \cdot 1} = \dfrac{-4 \pm \sqrt{16 + 84}}{2} = \dfrac{-4 \pm \sqrt{100}}{2} = $

$\dfrac{-4 \pm 10}{2} = -2 \pm 5$; $x = -2 + 5 = 3$ or $x = -2 - 5 = -7$

9. $3y^2 - 7y + 2 = 0$; $y = \dfrac{-(-7) \pm \sqrt{(-7)^2 - (4 \cdot 3 \cdot 2)}}{2 \cdot 3} = \dfrac{7 \pm \sqrt{49 - 24}}{6} = \dfrac{7 \pm \sqrt{25}}{6} =$

$\dfrac{7 \pm 5}{6}$; $y = \dfrac{7 + 5}{6}$ or $y = \dfrac{7 - 5}{6}$; $y = \dfrac{12}{6} = 2$ or $y = \dfrac{2}{6} = \dfrac{1}{3}$

10. $5x^2 + 3x - 2 = 0$; $x = \dfrac{-3 \pm \sqrt{3^2 - (4 \cdot 5 \cdot -2)}}{2 \cdot 5} = \dfrac{-3 \pm \sqrt{9 + 40}}{10} = \dfrac{-3 \pm \sqrt{49}}{10} =$

$\dfrac{-3 \pm 7}{10}$; $x = \dfrac{-3 + 7}{10}$ or $x = \dfrac{-3 - 7}{10}$; $x = \dfrac{4}{10} = \dfrac{2}{5}$ or $x = \dfrac{-10}{10} = -1$

11. $3x^2 - x - 2 = 0$; $x = \dfrac{-(-1) \pm \sqrt{(-1)^2 - (4 \cdot 3 \cdot -2)}}{2 \cdot 3} = \dfrac{1 \pm \sqrt{1 + 24}}{6} = \dfrac{1 \pm \sqrt{25}}{6} =$

$\dfrac{1 \pm 5}{6}$; $x = \dfrac{1 + 5}{6}$ or $x = \dfrac{1 - 5}{6}$; $x = \dfrac{6}{6} = 1$ or $x = \dfrac{-4}{6} = -\dfrac{2}{3}$

12. $2x^2 - 2x - 3 = 0$; $x = \dfrac{-(-2) \pm \sqrt{(-2)^2 - (4 \cdot 2 \cdot -3)}}{2 \cdot 2} = \dfrac{2 \pm \sqrt{4 + 24}}{4} = \dfrac{2 \pm \sqrt{28}}{4} =$

$\dfrac{2 \pm 2\sqrt{7}}{4} = \dfrac{1 \pm \sqrt{7}}{2}$; $x = \dfrac{1 + \sqrt{7}}{2}$ or $x = \dfrac{1 - \sqrt{7}}{2}$

13. $4x^2 - 4 = 0$; $x = \dfrac{-0 \pm \sqrt{0^2 - (4 \cdot 4 \cdot -4)}}{2 \cdot 4} = \dfrac{0 \pm \sqrt{64}}{8} = \dfrac{\pm 8}{8} = \pm 1$; $x = 1$ or

$x = -1$

14. $x^2 - 9 = 0$; $x = \dfrac{-0 \pm \sqrt{0^2 - (4 \cdot 1 \cdot -9)}}{2 \cdot 1} = \dfrac{0 \pm \sqrt{36}}{2} = \dfrac{\pm 6}{2} = \pm 3$; $x = 3$ or $x = -3$

15. $4x^2 - 1 = 0$; $x = \dfrac{-0 \pm \sqrt{0^2 - (4 \cdot 4 \cdot -1)}}{2 \cdot 4} = \dfrac{0 \pm \sqrt{16}}{8} = \dfrac{\pm 4}{8} = \pm \dfrac{1}{2}$; $x = \dfrac{1}{2}$ or

$x = -\dfrac{1}{2}$

16. $y^2 - y - 1 = 0$; $y = \dfrac{-(-1) \pm \sqrt{(-1)^2 - (4 \cdot 1 \cdot -1)}}{2 \cdot 1} = \dfrac{1 \pm \sqrt{1 + 4}}{2} = \dfrac{1 \pm \sqrt{5}}{2}$;

$y = \dfrac{1 + \sqrt{5}}{2}$ or $y = \dfrac{1 - \sqrt{5}}{2}$

17. $2x^2 + x - 2 = 0$; $x = \dfrac{-1 \pm \sqrt{1^2 - (4 \cdot 2 \cdot -2)}}{2 \cdot 2} = \dfrac{-1 \pm \sqrt{1 + 16}}{4} = \dfrac{-1 \pm \sqrt{17}}{4}$;

$x = \dfrac{-1 + \sqrt{17}}{4}$ or $x = \dfrac{-1 - \sqrt{17}}{4}$

18. $x^2 - 3x + 1 = 0$; $x = \dfrac{-(-3) \pm \sqrt{(-3)^2 - (4 \cdot 1 \cdot 1)}}{2 \cdot 1} = \dfrac{3 \pm \sqrt{9 - 4}}{2} = \dfrac{3 \pm \sqrt{5}}{2}$;

$x = \dfrac{3 + \sqrt{5}}{2}$ or $x = \dfrac{3 - \sqrt{5}}{2}$

19. $5y^2 + 7y + 2 = 0$; $y = \dfrac{-7 \pm \sqrt{7^2 - (4 \cdot 5 \cdot 2)}}{2 \cdot 5} = \dfrac{-7 \pm \sqrt{49 - 40}}{10} = \dfrac{-7 \pm \sqrt{9}}{10} =$

$\dfrac{-7 \pm 3}{10}$; $y = \dfrac{-7 + 3}{10}$ or $y = \dfrac{-7 - 3}{10}$; $y = \dfrac{-4}{10} = -\dfrac{2}{5}$ or $y = \dfrac{-10}{10} = -1$

20. $2y^2 + 4y + 1 = 0$; $y = \dfrac{-4 \pm \sqrt{4^2 - (4 \cdot 2 \cdot 1)}}{2 \cdot 2} = \dfrac{-4 \pm \sqrt{16 - 8}}{4} = \dfrac{-4 \pm \sqrt{8}}{4} =$

$\dfrac{-4 \pm 2\sqrt{2}}{4} = \dfrac{-2 \pm \sqrt{2}}{4}$; $y = \dfrac{-2 + \sqrt{2}}{2}$ or $y = \dfrac{-2 - \sqrt{2}}{2}$

21. $x^2 - 2x - 5 = 0;\quad x = \dfrac{-(-2) \pm \sqrt{(-2)^2 - (4 \cdot 1 \cdot -5)}}{2 \cdot 1} = \dfrac{2 \pm \sqrt{4 + 20}}{2} = \dfrac{2 \pm \sqrt{24}}{2} =$

$\dfrac{2 \pm 2\sqrt{6}}{2} = 1 \pm \sqrt{6};\quad x = 1 + \sqrt{6}$ or $x = 1 - \sqrt{6}$

22. $2x^2 - 3x = 5;\quad 2x^2 - 3x - 5 = 0;\quad x = \dfrac{-(-3) \pm \sqrt{(-3)^2 - (4 \cdot 2 \cdot -5)}}{2 \cdot 2} =$

$\dfrac{3 \pm \sqrt{9 + 40}}{4} = \dfrac{3 \pm \sqrt{49}}{4} = \dfrac{3 \pm 7}{4};\quad x = \dfrac{3 + 7}{4}$ or $x = \dfrac{3 - 7}{4};\quad x = \dfrac{10}{4} = \dfrac{5}{2}$ or

$x = \dfrac{-4}{4} = -1$

23. $2x^2 + x = 7;\quad 2x^2 + x - 7 = 0;\quad x = \dfrac{-1 \pm \sqrt{1^2 - (4 \cdot 2 \cdot -7)}}{2 \cdot 2} = \dfrac{-1 \pm \sqrt{1 + 56}}{4} =$

$\dfrac{-1 \pm \sqrt{57}}{4};\quad x = \dfrac{-1 + \sqrt{57}}{4}$ or $x = \dfrac{-1 - \sqrt{57}}{4}$

24. $3x^2 = x + 1;\quad 3x^2 - x - 1 = 0;\quad x = \dfrac{-(-1) \pm \sqrt{(-1)^2 - (4 \cdot 3 \cdot -1)}}{2 \cdot 3} =$

$\dfrac{1 \pm \sqrt{1 + 12}}{6} = \dfrac{1 \pm \sqrt{13}}{6};\quad x = \dfrac{1 + \sqrt{13}}{6}$ or $x = \dfrac{1 - \sqrt{13}}{6}$

25. $2x^2 = 5 + x;\quad 2x^2 - x - 5 = 0;\quad x = \dfrac{-(-1) \pm \sqrt{(-1)^2 - (4 \cdot 2 \cdot -5)}}{2 \cdot 2} =$

$\dfrac{1 \pm \sqrt{1 + 40}}{4} = \dfrac{1 \pm \sqrt{41}}{4};\quad x = \dfrac{1 + \sqrt{41}}{4}$ or $x = \dfrac{1 - \sqrt{41}}{4}$

26. $3x^2 = 8 - x;\quad 3x^2 + x - 8 = 0;\quad x = \dfrac{-1 \pm \sqrt{1^2 - (4 \cdot 3 \cdot -8)}}{2 \cdot 3} = \dfrac{-1 \pm \sqrt{1 + 96}}{6} =$

$\dfrac{-1 \pm \sqrt{97}}{6};\quad x = \dfrac{-1 + \sqrt{97}}{6}$ or $x = \dfrac{-1 - \sqrt{97}}{6}$

27. $x^2 + 3 = 6x;\quad x^2 - 6x + 3 = 0;\quad x = \dfrac{-(-6) \pm \sqrt{(-6)^2 - (4 \cdot 1 \cdot 3)}}{2 \cdot 1} =$

$\dfrac{6 \pm \sqrt{36 - 12}}{2} = \dfrac{6 \pm \sqrt{24}}{2} = \dfrac{6 \pm 2\sqrt{6}}{2} = 3 \pm \sqrt{6};\quad x = 3 + \sqrt{6}$ or $x = 3 - \sqrt{6}$

28. $2x^2 = 5x - 3;\quad 2x^2 - 5x + 3 = 0;\quad x = \dfrac{-(-5) \pm \sqrt{(-5)^2 - (4 \cdot 2 \cdot 3)}}{2 \cdot 2} =$

$\dfrac{5 \pm \sqrt{25 - 24}}{4} = \dfrac{5 \pm \sqrt{1}}{4} = \dfrac{5 \pm 1}{4};\quad x = \dfrac{5 + 1}{4}$ or $x = \dfrac{5 - 1}{4};\quad x = \dfrac{6}{4} = \dfrac{3}{2}$ or

$x = \dfrac{4}{4} = 1$

29. $2x^2 = 7x - 2;\quad 2x^2 - 7x + 2 = 0;\quad x = \dfrac{-(-7) \pm \sqrt{(-7)^2 - (4 \cdot 2 \cdot 2)}}{2 \cdot 2} =$

$\dfrac{7 \pm \sqrt{49 - 16}}{4} = \dfrac{7 \pm \sqrt{33}}{4};\quad x = \dfrac{7 + \sqrt{33}}{4}$ or $x = \dfrac{7 - \sqrt{33}}{4}$

30. $4y^2 = 6y + 2;\quad 4y^2 - 6y - 2 = 0;\quad y = \dfrac{-(-6) \pm \sqrt{(-6)^2 - (4 \cdot 4 \cdot -2)}}{2 \cdot 4} =$

$\dfrac{6 \pm \sqrt{36 + 32}}{8} = \dfrac{6 \pm \sqrt{68}}{8} = \dfrac{6 \pm 2\sqrt{17}}{8} = \dfrac{3 \pm \sqrt{17}}{4};\quad y = \dfrac{3 + \sqrt{17}}{4}$ or $y = \dfrac{3 - \sqrt{17}}{4}$

B **31. a.** $x^2 - 5x = 0$; $x(x - 5) = 0$; $x = 0$ or $x - 5 = 0$; $x = 0$ or $x = 5$

  b. $x = \dfrac{-(-5) \pm \sqrt{(-5)^2 - (4 \cdot 1 \cdot 0)}}{2 \cdot 1} = \dfrac{5 \pm \sqrt{25}}{2} = \dfrac{5 \pm 5}{2}$; $x = \dfrac{5 + 5}{2} = \dfrac{10}{2} = 5$ or

  $x = \dfrac{5 - 5}{2} = \dfrac{0}{2} = 0$

32. a. $x^2 + 8x + 7 = 0$; $(x + 1)(x + 7) = 0$; $x + 1 = 0$ or $x + 7 = 0$; $x = -1$ or

  $x = -7$

  b. $x = \dfrac{-8 \pm \sqrt{8^2 - (4 \cdot 1 \cdot 7)}}{2 \cdot 1} = \dfrac{-8 \pm \sqrt{64 - 28}}{2} = \dfrac{-8 \pm \sqrt{36}}{2} = \dfrac{-8 \pm 6}{2}$;

  $x = \dfrac{-8 + 6}{2} = \dfrac{-2}{2} = -1$ or $x = \dfrac{-8 - 6}{2} = \dfrac{-14}{2} = -7$

33. a. $x^2 - 4x - 5 = 0$; $(x + 1)(x - 5) = 0$; $x + 1 = 0$ or $x - 5 = 0$; $x = -1$ or

  $x = 5$

  b. $x = \dfrac{-(-4) \pm \sqrt{(-4)^2 - (4 \cdot 1 \cdot -5)}}{2 \cdot 1} = \dfrac{4 \pm \sqrt{16 + 20}}{2} = \dfrac{4 \pm \sqrt{36}}{2} = \dfrac{4 \pm 6}{2}$;

  $x = \dfrac{4 + 6}{2} = \dfrac{10}{2} = 5$ or $x = \dfrac{4 - 6}{2} = \dfrac{-2}{2} = -1$

34. a. $y^2 + y = 0$; $y(y + 1) = 0$; $y = 0$ or $y + 1 = 0$; $y = 0$ or $y = -1$

  b. $y = \dfrac{-1 \pm \sqrt{1^2 - (4 \cdot 1 \cdot 0)}}{2 \cdot 1} = \dfrac{-1 \pm \sqrt{1}}{2} = \dfrac{-1 \pm 1}{2}$; $y = \dfrac{-1 + 1}{2} = \dfrac{0}{2} = 0$ or

  $y = \dfrac{-1 - 1}{2} = \dfrac{-2}{2} = -1$

35. a. $y^2 - 7y + 10 = 0$; $(y - 2)(y - 5) = 0$; $y - 2 = 0$ or $y - 5 = 0$; $y = 2$ or

  $y = 5$

  b. $y = \dfrac{-(-7) \pm \sqrt{(-7)^2 - (4 \cdot 1 \cdot 10)}}{2 \cdot 1} = \dfrac{7 \pm \sqrt{49 - 40}}{2} = \dfrac{7 \pm \sqrt{9}}{2} = \dfrac{7 \pm 3}{2}$;

  $y = \dfrac{7 + 3}{2} = \dfrac{10}{2} = 5$ or $y = \dfrac{7 - 3}{2} = \dfrac{4}{2} = 2$

36. a. $s^2 - 6s + 9 = 0$; $(s - 3)^2 = 0$; $s - 3 = 0$; $s = 3$

  b. $s = \dfrac{-(-6) \pm \sqrt{(-6)^2 - (4 \cdot 1 \cdot 9)}}{2 \cdot 1} = \dfrac{6 \pm \sqrt{36 - 36}}{2} = \dfrac{6 \pm \sqrt{0}}{2} = \dfrac{6 \pm 0}{2} = \dfrac{6}{2} = 3$

37. a. $x^2 - 6x + 8 = 0$; $(x - 2)(x - 4) = 0$; $x - 2 = 0$ or $x - 4 = 0$; $x = 2$ or

  $x = 4$

  b. $x = \dfrac{-(-6) \pm \sqrt{(-6)^2 - (4 \cdot 1 \cdot 8)}}{2 \cdot 1} = \dfrac{6 \pm \sqrt{36 - 32}}{2} = \dfrac{6 \pm \sqrt{4}}{2} = \dfrac{6 \pm 2}{2}$;

  $x = \dfrac{6 + 2}{2} = \dfrac{8}{2} = 4$ or $x = \dfrac{6 - 2}{2} = \dfrac{4}{2} = 2$

38. a. $x^2 - 6x - 7 = 0$; $(x + 1)(x - 7) = 0$; $x + 1 = 0$ or $x - 7 = 0$; $x = -1$ or

  $x = 7$

  b. $x = \dfrac{-(-6) \pm \sqrt{(-6)^2 - (4 \cdot 1 \cdot -7)}}{2 \cdot 1} = \dfrac{6 \pm \sqrt{36 + 28}}{2} = \dfrac{6 \pm \sqrt{64}}{2} = \dfrac{6 \pm 8}{2}$;

  $x = \dfrac{6 + 8}{2} = \dfrac{14}{2} = 7$ or $x = \dfrac{6 - 8}{2} = \dfrac{-2}{2} = -1$

39. a. $x^2 - x - 12 = 0$; $(x + 3)(x - 4) = 0$; $x + 3 = 0$ or $x - 4 = 0$; $x = -3$ or $x = 4$

b. $x = \dfrac{-(-1) \pm \sqrt{(-1)^2 - (4 \cdot 1 \cdot -12)}}{2 \cdot 1} = \dfrac{1 \pm \sqrt{1 + 48}}{2} = \dfrac{1 \pm \sqrt{49}}{2} = \dfrac{1 \pm 7}{2};$

$x = \dfrac{1 + 7}{2} = \dfrac{8}{2} = 4$ or $x = \dfrac{1 - 7}{2} = \dfrac{-6}{2} = -3$

40. a. $x^2 - 4x - 12 = 0$; $(x + 2)(x - 6) = 0$; $x + 2 = 0$ or $x - 6 = 0$; $x = -2$ or $x = 6$

b. $x = \dfrac{-(-4) \pm \sqrt{(-4)^2 - (4 \cdot 1 \cdot -12)}}{2 \cdot 1} = \dfrac{4 \pm \sqrt{16 + 48}}{2} = \dfrac{4 \pm \sqrt{64}}{2} = \dfrac{4 \pm 8}{2};$

$x = \dfrac{4 + 8}{2} = \dfrac{12}{2} = 6$ or $x = \dfrac{4 - 8}{2} = \dfrac{-4}{2} = -2$

41. a. $m^2 + 3m - 10 = 0$; $(m - 2)(m + 5) = 0$; $m - 2 = 0$ or $m + 5 = 0$; $m = 2$ or $m = -5$

b. $m = \dfrac{-3 \pm \sqrt{3^2 - (4 \cdot 1 \cdot -10)}}{2 \cdot 1} = \dfrac{-3 \pm \sqrt{9 + 40}}{2} = \dfrac{-3 \pm \sqrt{49}}{2} = \dfrac{-3 \pm 7}{2};$

$m = \dfrac{-3 + 7}{2} = \dfrac{4}{2} = 2$ or $m = \dfrac{-3 - 7}{2} = \dfrac{-10}{2} = -5$

42. a. $y^2 + y - 20 = 0$; $(y - 4)(y + 5) = 0$; $y - 4 = 0$ or $y + 5 = 0$; $y = 4$ or $y = -5$

b. $y = \dfrac{-1 \pm \sqrt{1^2 - (4 \cdot 1 \cdot -20)}}{2 \cdot 1} = \dfrac{-1 \pm \sqrt{1 + 80}}{2} = \dfrac{-1 \pm \sqrt{81}}{2} = \dfrac{-1 \pm 9}{2};$

$y = \dfrac{-1 + 9}{2} = \dfrac{8}{2} = 4$ or $y = \dfrac{-1 - 9}{2} = \dfrac{-10}{2} = -5$

43. a. $x^2 + 5x = 6$; $x^2 + 5x - 6 = 0$; $(x - 1)(x + 6) = 0$; $x - 1 = 0$ or $x + 6 = 0$; $x = 1$ or $x = -6$

b. $x = \dfrac{-5 \pm \sqrt{5^2 - (4 \cdot 1 \cdot -6)}}{2 \cdot 1} = \dfrac{-5 \pm \sqrt{25 + 24}}{2} = \dfrac{-5 \pm \sqrt{49}}{2} = \dfrac{-5 \pm 7}{2};$

$x = \dfrac{-5 + 7}{2} = \dfrac{2}{2} = 1$ or $x = \dfrac{-5 - 7}{2} = \dfrac{-12}{2} = -6$

44. a. $z^2 - 2z = 15$; $z^2 - 2z - 15 = 0$; $(z + 3)(z - 5) = 0$; $z + 3 = 0$ or $z - 5 = 0$; $z = -3$ or $z = 5$

b. $z = \dfrac{-(-2) \pm \sqrt{(-2)^2 - (4 \cdot 1 \cdot -15)}}{2 \cdot 1} = \dfrac{2 \pm \sqrt{4 + 60}}{2} = \dfrac{2 \pm \sqrt{64}}{2} = \dfrac{2 \pm 8}{2};$

$z = \dfrac{2 + 8}{2} = \dfrac{10}{2} = 5$ or $z = \dfrac{2 - 8}{2} = \dfrac{-6}{2} = -3$

45. a. $x^2 - 8 = 2x$; $x^2 - 2x - 8 = 0$; $(x + 2)(x - 4) = 0$; $x + 2 = 0$ or $x - 4 = 0$; $x = -2$ or $x = 4$

b. $x = \dfrac{-(-2) \pm \sqrt{(-2)^2 - (4 \cdot 1 \cdot -8)}}{2 \cdot 1} = \dfrac{2 \pm \sqrt{4 + 32}}{2} = \dfrac{2 \pm \sqrt{36}}{2} = \dfrac{2 \pm 6}{2};$

$x = \dfrac{2 + 6}{2} = \dfrac{8}{2} = 4$ or $x = \dfrac{2 - 6}{2} = \dfrac{-4}{2} = -2$

46. a. $n^2 - 12 = 11n$; $n^2 - 11n - 12 = 0$; $(n + 1)(n - 12) = 0$; $n + 1 = 0$ or
$n - 12 = 0$; $n = -1$ or $n = 12$

b. $n = \dfrac{-(-11) \pm \sqrt{(-11)^2 - (4 \cdot 1 \cdot -12)}}{2 \cdot 1} = \dfrac{11 \pm \sqrt{121 + 48}}{2} = \dfrac{11 \pm \sqrt{169}}{2} = \dfrac{11 \pm 13}{2}$;

$n = \dfrac{11 + 13}{2} = \dfrac{24}{2} = 12$ or $n = \dfrac{11 - 13}{2} = \dfrac{-2}{2} = -1$

47. a. $b^2 = -b + 30$; $b^2 + b - 30 = 0$; $(b - 5)(b + 6) = 0$; $b - 5 = 0$ or
$b + 6 = 0$; $b = 5$ or $b = -6$

b. $b = \dfrac{-1 \pm \sqrt{1^2 - (4 \cdot 1 \cdot -30)}}{2 \cdot 1} = \dfrac{-1 \pm \sqrt{1 + 120}}{2} = \dfrac{-1 \pm \sqrt{121}}{2} = \dfrac{-1 \pm 11}{2}$;

$b = \dfrac{-1 + 11}{2} = \dfrac{10}{2} = 5$ or $b = \dfrac{-1 - 11}{2} = \dfrac{-12}{2} = -6$

48. a. $a^2 = 10a - 24$; $a^2 - 10a + 24 = 0$; $(a - 4)(a - 6) = 0$; $a - 4 = 0$ or
$a - 6 = 0$; $a = 4$ or $a = 6$

b. $a = \dfrac{-(-10) \pm \sqrt{(-10)^2 - (4 \cdot 1 \cdot 24)}}{2 \cdot 1} = \dfrac{10 \pm \sqrt{100 - 96}}{2} = \dfrac{10 \pm \sqrt{4}}{2} = \dfrac{10 \pm 2}{2}$;

$a = \dfrac{10 + 2}{2} = \dfrac{12}{2} = 6$ or $a = \dfrac{10 - 2}{2} = \dfrac{8}{2} = 4$

Pages 432–433 • MIXED PRACTICE EXERCISES

1. $x^2 - 2x = 0$; $x(x - 2) = 0$; $x = 0$ or $x - 2 = 0$; $x = 0$ or $x = 2$

2. $y^2 + 3y = 0$; $y(y + 3) = 0$; $y = 0$ or $y + 3 = 0$; $y = 0$ or $y = -3$

3. $x^2 - 4x = 0$; $x(x - 4) = 0$; $x = 0$ or $x - 4 = 0$; $x = 0$ or $x = 4$

4. $2x^2 - 4x = 0$; $2x(x - 2) = 0$; $2x = 0$ or $x - 2 = 0$; $x = 0$ or $x = 2$

5. $5x^2 - 10x = 0$; $5x(x - 2) = 0$; $5x = 0$ or $x - 2 = 0$; $x = 0$ or $x = 2$

6. $2x^2 + x = 0$; $x(2x + 1) = 0$; $x = 0$ or $2x + 1 = 0$; $x = 0$ or $2x = -1$;
$x = 0$ or $x = -\dfrac{1}{2}$

7. $x^2 + 3x + 2 = 0$; $(x + 1)(x + 2) = 0$; $x + 1 = 0$ or $x + 2 = 0$; $x = -1$ or
$x = -2$

8. $x^2 - 2x - 3 = 0$; $(x + 1)(x - 3) = 0$; $x + 1 = 0$ or $x - 3 = 0$; $x = -1$ or
$x = 3$

9. $x^2 + 2x - 8 = 0$; $(x - 2)(x + 4) = 0$; $x - 2 = 0$ or $x + 4 = 0$; $x = 2$ or
$x = -4$

10. $x^2 + 10x + 21 = 0$; $(x + 3)(x + 7) = 0$; $x + 3 = 0$ or $x + 7 = 0$; $x = -3$ or
$x = -7$

11. $x^2 - x = 30$; $x^2 - x - 30 = 0$; $(x + 5)(x - 6) = 0$; $x + 5 = 0$ or $x - 6 = 0$;
$x = -5$ or $x = 6$

12. $x^2 + 2x = 35$; $x^2 + 2x - 35 = 0$; $(x - 5)(x + 7) = 0$; $x - 5 = 0$ or
$x + 7 = 0$; $x = 5$ or $x = -7$

13. $x^2 = 9$; $x = \pm 3$; $x = 3$ or $x = -3$

14. $x^2 = 36$; $x = \pm 6$; $x = 6$ or $x = -6$

15. $y^2 = 35$; $y = \pm\sqrt{35}$; $y = \sqrt{35}$ or $y = -\sqrt{35}$

16. $x^2 = 42$; $x = \pm\sqrt{42}$; $x = \sqrt{42}$ or $x = -\sqrt{42}$

17. $y^2 = 19$; $y = \pm\sqrt{19}$; $y = \sqrt{19}$ or $y = -\sqrt{19}$

18. $x^2 - 40 = 0$; $x^2 = 40$; $x = \pm\sqrt{40} = \pm2\sqrt{10}$; $x = 2\sqrt{10}$ or $x = -2\sqrt{10}$

19. $x^2 - 18 = 0$; $x^2 = 18$; $x = \pm\sqrt{18} = \pm3\sqrt{2}$; $x = 3\sqrt{2}$ or $x = -3\sqrt{2}$

20. $(x + 2)^2 = 25$; $x + 2 = \pm5$; $x + 2 = 5$ or $x + 2 = -5$; $x = 3$ or $x = -7$

21. $(x - 1)^2 = 16$; $x - 1 = \pm4$; $x - 1 = 4$ or $x - 1 = -4$; $x = 5$ or $x = -3$

22. $(x + 4)^2 = 49$; $x + 4 = \pm7$; $x + 4 = 7$ or $x + 4 = -7$; $x = 3$ or $x = -11$

23. $(y - 3)^2 = 17$; $y - 3 = \pm\sqrt{17}$; $y - 3 = \sqrt{17}$ or $y - 3 = -\sqrt{17}$; $y = 3 + \sqrt{17}$ or $y = 3 - \sqrt{17}$

24. $(y + 2)^2 = 28$; $y + 2 = \pm\sqrt{28} = \pm2\sqrt{7}$; $y + 2 = 2\sqrt{7}$ or $y + 2 = -2\sqrt{7}$; $y = -2 + 2\sqrt{7}$ or $y = -2 - 2\sqrt{7}$

25. $x^2 - 2x - 3 = 0$; $x = \dfrac{-(-2) \pm \sqrt{(-2)^2 - (4 \cdot 1 \cdot -3)}}{2 \cdot 1} = \dfrac{2 \pm \sqrt{4 + 12}}{2} = \dfrac{2 \pm \sqrt{16}}{2} = \dfrac{2 \pm 4}{2} = 1 \pm 2$; $x = 1 + 2 = 3$ or $x = 1 - 2 = -1$

26. $x^2 + 4x - 5 = 0$; $x = \dfrac{-4 \pm \sqrt{4^2 - (4 \cdot 1 \cdot -5)}}{2 \cdot 1} = \dfrac{-4 \pm \sqrt{16 + 20}}{2} = \dfrac{-4 \pm \sqrt{36}}{2} = \dfrac{-4 \pm 6}{2} = -2 \pm 3$; $x = -2 + 3 = 1$ or $x = -2 - 3 = -5$

27. $x^2 + 2x - 8 = 0$; $x = \dfrac{-2 \pm \sqrt{2^2 - (4 \cdot 1 \cdot -8)}}{2 \cdot 1} = \dfrac{-2 \pm \sqrt{4 + 32}}{2} = \dfrac{-2 \pm \sqrt{36}}{2} = \dfrac{-2 \pm 6}{2} = -1 \pm 3$; $x = -1 + 3 = 2$ or $x = -1 - 3 = -4$

28. $2x^2 - 3x + 1 = 0$; $x = \dfrac{-(-3) \pm \sqrt{(-3)^2 - (4 \cdot 2 \cdot 1)}}{2 \cdot 2} = \dfrac{3 \pm \sqrt{9 - 8}}{4} = \dfrac{3 \pm \sqrt{1}}{4} = \dfrac{3 \pm 1}{4}$; $x = \dfrac{3 + 1}{4} = \dfrac{4}{4} = 1$ or $x = \dfrac{3 - 1}{4} = \dfrac{2}{4} = \dfrac{1}{2}$

29. $3x^2 + 4x + 1 = 0$; $x = \dfrac{-4 \pm \sqrt{4^2 - (4 \cdot 3 \cdot 1)}}{2 \cdot 3} = \dfrac{-4 \pm \sqrt{16 - 12}}{6} = \dfrac{-4 \pm \sqrt{4}}{6} = \dfrac{-4 \pm 2}{6}$; $x = \dfrac{-4 + 2}{6} = \dfrac{-2}{6} = -\dfrac{1}{3}$ or $x = \dfrac{-4 - 2}{6} = \dfrac{-6}{6} = -1$

30. $x^2 - 5x + 3 = 0$; $x = \dfrac{-(-5) \pm \sqrt{(-5)^2 - (4 \cdot 1 \cdot 3)}}{2 \cdot 1} = \dfrac{5 \pm \sqrt{25 - 12}}{2} = \dfrac{5 \pm \sqrt{13}}{2}$; $x = \dfrac{5 + \sqrt{13}}{2}$ or $x = \dfrac{5 - \sqrt{13}}{2}$

31. $3x^2 - 2x - 5 = 0$; $x = \dfrac{-(-2) \pm \sqrt{(-2)^2 - (4 \cdot 3 \cdot -5)}}{2 \cdot 3} = \dfrac{2 \pm \sqrt{4 + 60}}{6} = \dfrac{2 \pm \sqrt{64}}{6} = \dfrac{2 \pm 8}{6}$; $x = \dfrac{2 + 8}{6} = \dfrac{10}{6} = \dfrac{5}{3}$ or $x = \dfrac{2 - 8}{6} = \dfrac{-6}{6} = -1$

32. $2x^2 + 5x - 3 = 0$; $x = \dfrac{-5 \pm \sqrt{5^2 - (4 \cdot 2 \cdot -3)}}{2 \cdot 2} = \dfrac{-5 \pm \sqrt{25 + 24}}{4} = \dfrac{-5 \pm \sqrt{49}}{4} =$

$\dfrac{-5 \pm 7}{4}$; $x = \dfrac{-5 + 7}{4} = \dfrac{2}{4} = \dfrac{1}{2}$ or $x = \dfrac{-5 - 7}{4} = \dfrac{-12}{4} = -3$

33. $x^2 - 3x + 2 = 0$; $x = \dfrac{-(-3) \pm \sqrt{(-3)^2 - (4 \cdot 1 \cdot 2)}}{2 \cdot 1} = \dfrac{3 \pm \sqrt{9 - 8}}{2} = \dfrac{3 \pm \sqrt{1}}{2} =$

$\dfrac{3 \pm 1}{2}$; $x = \dfrac{3 + 1}{2} = \dfrac{4}{2} = 2$ or $x = \dfrac{3 - 1}{2} = \dfrac{2}{2} = 1$

34. $x^2 + 4x - 3 = 0$; $x = \dfrac{-4 \pm \sqrt{4^2 - (4 \cdot 1 \cdot -3)}}{2 \cdot 1} = \dfrac{-4 \pm \sqrt{16 + 12}}{2} = \dfrac{-4 \pm \sqrt{28}}{2} =$

$\dfrac{-4 \pm 2\sqrt{7}}{2} = -2 \pm \sqrt{7}$; $x = -2 + \sqrt{7}$ or $x = -2 - \sqrt{7}$

35. $2x^2 - x = 4$; $2x^2 - x - 4 = 0$; $x = \dfrac{-(-1) \pm \sqrt{(-1)^2 - (4 \cdot 2 \cdot -4)}}{2 \cdot 2} =$

$\dfrac{1 \pm \sqrt{1 + 32}}{4} = \dfrac{1 \pm \sqrt{33}}{4}$; $x = \dfrac{1 + \sqrt{33}}{4}$ or $x = \dfrac{1 - \sqrt{33}}{4}$

36. $2x^2 - 2x = 3$; $2x^2 - 2x - 3 = 0$; $x = \dfrac{-(-2) \pm \sqrt{(-2)^2 - (4 \cdot 2 \cdot -3)}}{2 \cdot 2} =$

$\dfrac{2 \pm \sqrt{4 + 24}}{4} = \dfrac{2 \pm \sqrt{28}}{4} = \dfrac{2 \pm 2\sqrt{7}}{4} = \dfrac{1 \pm \sqrt{7}}{2}$; $x = \dfrac{1 + \sqrt{7}}{2}$ or $x = \dfrac{1 - \sqrt{7}}{2}$

Page 435 · WRITTEN EXERCISES

A

1. Let x = width (in cm). Then $x + 3$ = length. $x(x + 3) = 40$; $x^2 + 3x = 40$; $x^2 + 3x - 40 = 0$; $(x - 5)(x + 8) = 0$; $x - 5 = 0$ or $x + 8 = 0$; $x = 5$ or $x = -8$ (reject); $x = 5$ and $x + 3 = 8$. width: 5 cm; length: 8 cm

2. Let x = width (in cm). Then $2x$ = length. $x(2x) = 72$; $2x^2 = 72$; $x^2 = 36$; $x = \pm 6$; $x = 6$ or $x = -6$ (reject); $x = 6$ and $2x = 12$. width: 6 cm; length: 12 cm

3. Let x and $x + 2$ = the numbers. $x(x + 2) = 35$; $x^2 + 2x = 35$; $x^2 + 2x - 35 = 0$; $(x - 5)(x + 7) = 0$; $x - 5 = 0$ or $x + 7 = 0$; $x = 5$ or $x = -7$, so $x + 2 = 7$ or $x + 2 = -5$. 5 and 7, or -7 and -5

4. Let x = the number. Then $x + 30$ = square. $x^2 = x + 30$; $x^2 - x - 30 = 0$; $(x + 5)(x - 6) = 0$; $x + 5 = 0$ or $x - 6 = 0$; $x = -5$ (reject) or $x = 6$. 6

5. $(3x \cdot 4x) - (x \cdot 2x) = 40$; $12x^2 - 2x^2 = 40$; $10x^2 = 40$; $x^2 = 4$; $x = \pm 2$; $x = 2$ or $x = -2$ (reject). $x = 2$

6. $(3x \cdot 6x) - (x \cdot 2x) = 400$; $18x^2 - 2x^2 = 400$; $16x^2 = 400$; $x^2 = 25$; $x = \pm 5$; $x = 5$ or $x = -5$ (reject). $x = 5$

7. $2 \cdot x \cdot (x + 1) = 60$; $2x^2 + 2x = 60$; $x^2 + x = 30$; $x^2 + x - 30 = 0$; $(x - 5)(x + 6) = 0$; $x - 5 = 0$ or $x + 6 = 0$; $x = 5$ or $x = -6$ (reject). $x = 5$

8. $x^2 + (x + 2)^2 = 10^2$; $x^2 + x^2 + 4x + 4 = 100$; $2x^2 + 4x - 96 = 0$; $x^2 + 2x - 48 = 0$; $(x - 6)(x + 8) = 0$; $x - 6 = 0$ or $x + 8 = 0$; $x = 6$ or $x = -8$ (reject). $x = 6$

9. $x^2 + 5^2 = (x + 1)^2$; $x^2 + 25 = x^2 + 2x + 1$; $25 = 2x + 1$; $24 = 2x$; $12 = x$.

10. $x^2 + 4^2 = (3x)^2$; $x^2 + 16 = 9x^2$; $16 = 8x^2$; $2 = x^2$; $x = \pm\sqrt{2}$; $x = \sqrt{2}$ or $x = -\sqrt{2}$ (reject). $x = \sqrt{2}$

B **11.** $2l + 2w = 32$; $l + w = 16$; $l = 16 - w$. $A = lw$, so $48 = (16 - w)w$; $48 = 16w - w^2$; $w^2 - 16w + 48 = 0$; $(w - 4)(w - 12) = 0$; $w - 4 = 0$ or $w - 12 = 0$; $w = 4$ or $w = 12$, so $l = 12$ or $l = 4$. 4 cm by 12 cm

12. Let x = one number. Then $13 - x$ = the other number. $42 = x(13 - x)$; $42 = 13x - x^2$; $x^2 - 13x + 42 = 0$; $(x - 6)(x - 7) = 0$; $x - 6 = 0$ or $x - 7 = 0$; $x = 6$ or $x = 7$, so $13 - x = 7$ or 6.   6 and 7

C **13.**

	D	r	t
Al	36	x	$\dfrac{36}{x}$
Vince	36	$x + 3$	$\dfrac{36}{x + 3}$

$\dfrac{36}{x} = \dfrac{36}{x + 3} + 1$; $\dfrac{36}{x} = \dfrac{36 + (x + 3)}{x + 3}$; $\dfrac{36}{x} = \dfrac{x + 39}{x + 3}$; $36(x + 3) = x(x + 39)$; $36x + 108 = x^2 + 39x$; $x^2 + 3x - 108 = 0$;

$(x - 9)(x + 12) = 0$; $x - 9 = 0$ or $x + 12 = 0$; $x = 9$ or $x = -12$ (reject); $x = 9$ and $x + 3 = 12$. Al: 9 km/h; Vince: 12 km/h

14.

	D	t	r
Andrea	30	x	$\dfrac{30}{x}$
Liane	30	$x + 1$	$\dfrac{30}{x + 1}$

$\dfrac{30}{x} = \dfrac{30}{x + 1} + 1$; $\dfrac{30}{x} = \dfrac{30 + (x + 1)}{x + 1}$; $\dfrac{30}{x} = \dfrac{x + 31}{x + 1}$; $30(x + 1) = x(x + 31)$; $30x + 30 = x^2 + 31x$; $x^2 + x - 30 = 0$;

$(x + 6)(x - 5) = 0$; $x + 6 = 0$ or $x - 5 = 0$; $x = -6$ (reject) or $x = 5$; $x = 5$ and $x + 1 = 6$. Andrea: 5 hours; Liane: 6 hours

Pages 438–439 • WRITTEN EXERCISES

A **1. a.**

$y = x^2$	
x	y
-2	4
-1	1
0	0
1	1
2	4

b.

c. $(0, 0)$

2. a. $y = x^2 + 1$

x	y
-2	5
-1	2
0	1
1	2
2	5

b.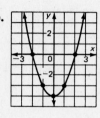

c. $(0, 1)$

3. a. $y = x^2 - 4$

x	y
-2	0
-1	-3
0	-4
1	-3
2	0

b.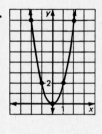

c. $(0, -4)$

4. a. $y = 2x^2$

x	y
-2	8
-1	2
0	0
1	2
2	8

b.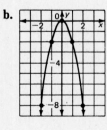

c. $(0, 0)$

5. a. $y = -2x^2$

x	y
-2	-8
-1	-2
0	0
1	-2
2	-8

b. (graph)

c. $(0, 0)$

6. a.

x	y
$y = \frac{1}{2}x^2$	
-3	$\frac{9}{2}$
-2	2
-1	$\frac{1}{2}$
0	0
1	$\frac{1}{2}$
2	2
3	$\frac{9}{2}$

b.

c. $(0, 0)$

7. a.

x	y
$y = x^2 + 2x$	
-3	3
-2	0
-1	-1
0	0
1	3

b.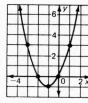

c. $(-1, -1)$

8. a.

x	y
$y = x^2 - 2x$	
-2	8
-1	3
0	0
1	-1
2	0
3	3
4	8

b.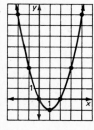

c. $(1, -1)$

9. a.

$y = 2x - x^2$	
x	y
-1	-3
0	0
1	1
2	0
3	-3

b.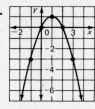

c. $(1, 1)$

10. a.

$y = x^2 - 2x - 1$	
x	y
-2	7
-1	2
0	-1
1	-2
2	-1
3	2
4	7

b.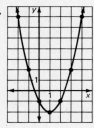

c. $(1, -2)$

11. a.

$y = x^2 - 4x + 3$	
x	y
0	3
1	0
2	-1
3	0
4	3

b.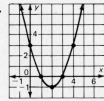

c. $(2, -1)$

12. a.

$y = x^2 + 4x - 2$	
x	y
-5	3
-4	-2
-3	-5
-2	-6
-1	-5
0	-2
1	3

b.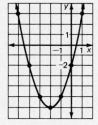

c. $(-2, -6)$

13. a.

$y = x^2 - 2x - 2$	
x	y
-1	1
0	-2
1	-3
2	-2
3	1

b.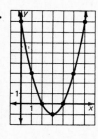

c. $(1, -3)$

14. a.

$y = x^2 - 6x + 8$	
x	y
0	8
1	3
2	0
3	-1
4	0
5	3
6	8

b.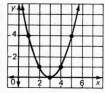

c. $(3, -1)$

15. a.

$y = x^2 - 6x + 9$	
x	y
1	4
2	1
3	0
4	1
5	4

b.

c. $(3, 0)$

16. a.

$y = x^2 + 6x - 6$	
x	y
-6	-6
-5	-11
-4	-14
-3	-15
-2	-14
-1	-11
0	-6

b.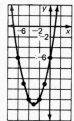

c. $(-3, -15)$

17. a.

$y = (3 - x)(3 + x)$	
x	y
-3	0
-2	5
-1	8
0	9
1	8
2	5
3	0

b.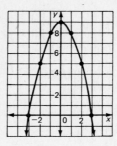

c. $(0, 9)$

18. a.

$y = (2 - x)(2 - x)$	
x	y
-1	9
0	4
1	1
2	0
3	1
4	4
5	9

b.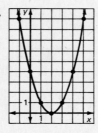

c. $(2, 0)$

19.

$y = x^2 + 2$		$y = -(x^2 + 2)$	
x	y	x	y
-2	6	-2	-6
-1	3	-1	-3
0	2	0	-2
1	3	1	-3
2	6	2	-6

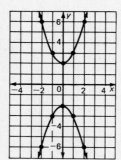

20. $(2, -2)$ **21.** 1 and 3   **22.** 2   **23.** 2   **24.** 1

B   **25.**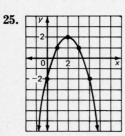

26. $(4, -1)$

27. $y = x^2 - 2x$: $(1, -1)$;
$y = x^2 - 2x + 2$: $(1, 1)$;
$y = x^2 - 2x + 4$: $(1, 3)$

28. $(1, 5)$

C   **29.** $y = x^2 - 5x + 2$; $b^2 - 4ac = (-5)^2 - (4 \cdot 1 \cdot 2) = 25 - 8 = 17 > 0$;   2 times

30. $y = x^2 - 4x + 6$; $b^2 - 4ac = (-4)^2 - (4 \cdot 1 \cdot 6) = 16 - 24 = -8 < 0$;   0 times

31. $y = x^2 - 10x + 25$; $b^2 - 4ac = (-10)^2 - (4 \cdot 1 \cdot 25) = 100 - 100 = 0$;   1 time

Page 441 • PROBLEM SOLVING STRATEGIES

1. Let n be the first integer; if $n + (n + 2) + (n + 4) = 18$, then $3n + 6 = 18$ and $n = 4$, an even integer; the problem has no solution.

2. Let $x =$ amount invested at each rate; $0.06x + 0.08x = 280$; $0.14x = 280$; $x = 2000$; $2000 + 2000 = 4000$. \$4000

3. Not enough information is given. Information about the amount of either the old solution or the new solution is missing.

4. The problem contains contradictory information. Let n and q be the numbers of nickels and quarters, respectively; the two equations suggested by the problem are $n = 5q - 1$ and $5n = 25q - 15$, or $n = 5q - 3$; it is not possible for both equations to be true.

5. Not enough information is given. Information about both the actual cost of a pen or pencil and the total amount Seth spent is missing.

6. Let $n =$ total number of stamps; $\dfrac{2}{5}x + \dfrac{1}{3}x + 16 = x$; $\dfrac{11}{15}x + 16 = x$; $\dfrac{4}{15}x = 16$; $x = 60$. 60 stamps

7. Let $x =$ number of minutes it would take them working together; $\dfrac{1}{30} + \dfrac{1}{45} = \dfrac{1}{x}$; $45x + 30x = 1350$; $75x = 1350$; $x = 18$. 18 min

8. Not enough information is given. Information about the actual rates of either Elaine or Judy is missing.

9. Let $r =$ Rose's age and $r + 2 =$ Arthur's age; $r + 3 + [(r + 2) + 3] = 8$; $2r + 8 = 8$; $r = 0$; the problem has no solution.

10. Let a and s be the numbers of adult and student tickets, respectively; $s = 3a$ and $2s + 4a = 324$; $2(3a) + 4a = 324$; $6a + 4a = 324$; $10a = 324$; $a = 32.4$; since the number of tickets must be a whole number, the problem has no solution.

Page 443 • COMPUTER ACTIVITIES

1. **a.** 0.625 **b.** $0.\overline{5}$ **c.** $0.\overline{692307}$ **d.** $0.5\overline{71428}$ **e.** 0.65 **f.** 0.64 **g.** $0.\overline{5294117647058823}$
h. $0.6\overline{1}$ **i.** $0.6\overline{956521739130434782608}$ **j.** 0.384 **k.** $0.11\overline{6}$ **l.** $0.\overline{358974}$

2. Answers may vary. Example: $\dfrac{5}{17} = 0.\overline{2941176470588235}$

3. Answers may vary. Example: $\dfrac{12}{29} = 0.\overline{4137931034482758620689655172}$

4. $\dfrac{5}{16}$ 5. $\dfrac{13}{48}$ 6. Answers may vary. **a.** $\dfrac{8}{15} = 0.53$; $\dfrac{9}{16} = 0.5623$; $\dfrac{5}{9} = 0.5\overline{5}$;

$\dfrac{8}{15} < \dfrac{5}{9} < \dfrac{9}{16}$ **b.** $\dfrac{6}{13} = 0.\overline{461538}$; $\dfrac{16}{31} = 0.\overline{516129032258064}$; $\dfrac{1}{2} = 0.5$; $\dfrac{6}{13} < \dfrac{1}{2} < \dfrac{16}{31}$

c. $\frac{10}{11} = 0.\overline{90}; \quad \frac{11}{12} = 0.91\overline{6}; \quad \frac{91}{100} = 0.91; \quad \frac{10}{11} < \frac{91}{100} < \frac{11}{12}$ **d.** $\frac{6}{7} = 0.\overline{857142};$

$\frac{16}{17} = 0.9411764705882352; \quad \frac{9}{10} = 0.9; \quad \frac{6}{7} < \frac{9}{10} < \frac{16}{17}$

Page 444 • SKILLS REVIEW

1. $P = 2l + 2w = 2 \cdot 9 + 2 \cdot 6 = 18 + 12 = 30.$ 30 cm
2. $P = 6s = 6 \cdot 4 = 24$
3. $P = 3 \cdot 5x + 11x = 15x + 11x = 26x$
4. $A = (5y \cdot 3x) - x^2 = 15xy - x^2$
5. $A = bh = 5 \cdot 4 = 20$
6. $A = \frac{1}{2}h(a + b) = \frac{1}{2} \cdot 4 \cdot (4 + 8) = 2 \cdot 12 = 24$
7. $A = \pi r^2 = \pi \cdot 5^2 = 25\pi$
8. $A = \pi r^2 - s^2 = \pi(\sqrt{2})^2 - 2^2 = 2\pi - 4$
9. $A = \pi(\sqrt{10})^2 - \pi(\sqrt{6})^2 = 10\pi - 6\pi = 4\pi$
10. $V = Bh = (15 \cdot 5) \cdot 12 = 900$
11. Base Area $= (5x \cdot 3x) - (x \cdot x) = 15x^2 - x^2 = 14x^2; \quad V = Bh = 14x^2 \cdot 2x = 28x^3$
12. Base Area $= 6^2 - 1^2 = 36 - 1 = 35; \quad V = Bh = 35 \cdot 6 = 210.$ 210 m³

Pages 445–446 • CHAPTER REVIEW EXERCISES

1. $x(x - 2) = 0; \quad x = 0$ or $x - 2 = 0; \quad x = 0$ or $x = 2$
2. $y(y + 4) = 0; \quad y = 0$ or $y + 4 = 0; \quad y = 0$ or $y = -4$
3. $x(x - 7) = 0; \quad x = 0$ or $x - 7 = 0; \quad x = 0$ or $x = 7$
4. $2a(a - 9) = 0; \quad 2a = 0$ or $a - 9 = 0; \quad a = 0$ or $a = 9$
5. $4x(x + 6) = 0; \quad 4x = 0$ or $x + 6 = 0; \quad x = 0$ or $x = -6$
6. $5n(n + 8) = 0; \quad 5n = 0$ or $n + 8 = 0; \quad n = 0$ or $n = -8$
7. $(x + 1)(x + 5) = 0; \quad x + 1 = 0$ or $x + 5 = 0; \quad x = -1$ or $x = -5$
8. $(x - 3)(x + 6) = 0; \quad x - 3 = 0$ or $x + 6 = 0; \quad x = 3$ or $x = -6$
9. $(y + 5)(y - 3) = 0; \quad y + 5 = 0$ or $y - 3 = 0; \quad y = -5$ or $y = 3$
10. $x^2 - 3x = 0; \quad x(x - 3) = 0; \quad x = 0$ or $x - 3 = 0; \quad x = 0$ or $x = 3$
11. $n^2 + 4n = 0; \quad n(n + 4) = 0; \quad n = 0$ or $n + 4 = 0; \quad n = 0$ or $n = -4$
12. $5x + x^2 = 0; \quad x(5 + x) = 0; \quad x = 0$ or $5 + x = 0; \quad x = 0$ or $x = -5$
13. $2y^2 - 4y = 0; \quad 2y(y - 2) = 0; \quad 2y = 0$ or $y - 2 = 0; \quad y = 0$ or $y = 2$
14. $x^2 - 8x + 15 = 0; \quad (x - 3)(x - 5) = 0; \quad x - 3 = 0$ or $x - 5 = 0; \quad x = 3$ or $x = 5$
15. $x^2 - 6x + 8 = 0; \quad (x - 2)(x - 4) = 0; \quad x - 2 = 0$ or $x - 4 = 0; \quad x = 2$ or $x = 4$
16. $x^2 - x - 20 = 0; \quad (x + 4)(x - 5) = 0; \quad x + 4 = 0$ or $x - 5 = 0; \quad x = -4$ or $x = 5$

17. $x^2 + 3x - 18 = 0$; $(x - 3)(x + 6) = 0$; $x - 3 = 0$ or $x + 6 = 0$; $x = 3$ or $x = -6$

18. $x^2 + 2x - 35 = 0$; $(x - 5)(x + 7) = 0$; $x - 5 = 0$ or $x + 7 = 0$; $x = 5$ or $x = -7$

19. $x^2 = 2x$; $x^2 - 2x = 0$; $x(x - 2) = 0$; $x = 0$ or $x - 2 = 0$; $x = 0$ or $x = 2$

20. $x^2 = -3x$; $x^2 + 3x = 0$; $x(x + 3) = 0$; $x = 0$ or $x + 3 = 0$; $x = 0$ or $x = -3$

21. $y^2 = -y$; $y^2 + y = 0$; $y(y + 1) = 0$; $y = 0$ or $y + 1 = 0$; $y = 0$ or $y = -1$

22. $n^2 = -4n$; $n^2 + 4n = 0$; $n(n + 4) = 0$; $n = 0$ or $n + 4 = 0$; $n = 0$ or $n = -4$

23. $x^2 + 7x = -10$; $x^2 + 7x + 10 = 0$; $(x + 2)(x + 5) = 0$; $x + 2 = 0$ or $x + 5 = 0$; $x = -2$ or $x = -5$

24. $x^2 + 6x = -9$; $x^2 + 6x + 9 = 0$; $(x + 3)^2 = 0$; $x + 3 = 0$; $x = -3$

25. $x^2 = -11x - 30$; $x^2 + 11x + 30 = 0$; $(x + 5)(x + 6) = 0$; $x + 5 = 0$ or $x + 6 = 0$; $x = -5$ or $x = -6$

26. $x^2 = 9x - 14$; $x^2 - 9x + 14 = 0$; $(x - 2)(x - 7) = 0$; $x - 2 = 0$ or $x - 7 = 0$; $x = 2$ or $x = 7$

27. $x^2 = x + 42$; $x^2 - x - 42 = 0$; $(x + 6)(x - 7) = 0$; $x + 6 = 0$ or $x - 7 = 0$; $x = -6$ or $x = 7$

28. $x^2 = 9$; $x = \pm 3$; $x = 3$ or $x = -3$

29. $y^2 = 81$; $y = \pm 9$; $y = 9$ or $y = -9$

30. $n^2 = 20$; $n = \pm\sqrt{20} = \pm 2\sqrt{5}$; $n = 2\sqrt{5}$ or $n = -2\sqrt{5}$

31. $y^2 = 27$; $y = \pm\sqrt{27} = \pm 3\sqrt{3}$; $y = 3\sqrt{3}$ or $y = -3\sqrt{3}$

32. $2x^2 = 50$; $x^2 = 25$; $x = \pm 5$; $x = 5$ or $x = -5$

33. $3y^2 = 18$; $y^2 = 6$; $y = \pm\sqrt{6}$; $y = \sqrt{6}$ or $y = -\sqrt{6}$

34. $n^2 - 3 = 13$; $n^2 = 16$; $n = \pm 4$; $n = 4$ or $n = -4$

35. $x^2 + 2 = 38$; $x^2 = 36$; $x = \pm 6$; $x = 6$ or $x = -6$

36. $y^2 - 6 = 43$; $y^2 = 49$; $y = \pm 7$; $y = 7$ or $y = -7$

37. $(x - 1)^2 = 16$; $x - 1 = \pm 4$; $x - 1 = 4$ or $x - 1 = -4$; $x = 5$ or $x = -3$

38. $(x + 3)^2 = 81$; $x + 3 = \pm 9$; $x + 3 = 9$ or $x + 3 = -9$; $x = 6$ or $x = -12$

39. $(x + 4)^2 = 64$; $x + 4 = \pm 8$; $x + 4 = 8$ or $x + 4 = -8$; $x = 4$ or $x = -12$

40. $(y - 2)^2 = 36$; $y - 2 = \pm 6$; $y - 2 = 6$ or $y - 2 = -6$; $y = 8$ or $y = -4$

41. $(y + 5)^2 = 49$; $y + 5 = \pm 7$; $y + 5 = 7$ or $y + 5 = -7$; $y = 2$ or $y = -12$

42. $(a - 7)^2 = 100$; $a - 7 = \pm 10$; $a - 7 = 10$ or $a - 7 = -10$; $a = 17$ or $a = -3$

43. $(4x + 4)^2 = 16$; $4x + 4 = \pm 4$; $4x + 4 = 4$ or $4x + 4 = -4$; $4x = 0$ or $4x = -8$; $x = 0$ or $x = -2$

44. $(3x - 9)^2 = 36$; $3x - 9 = \pm 6$; $3x - 9 = 6$ or $3x - 9 = -6$; $3x = 15$ or $3x = 3$; $x = 5$ or $x = 1$

45. $(5x - 15)^2 = 100$; $5x - 15 = \pm 10$; $5x - 15 = 10$ or $5x - 15 = -10$; $5x = 25$ or $5x = 5$; $x = 5$ or $x = 1$

46. $x^2 + 3x - 10 = 0$; $x = \dfrac{-3 \pm \sqrt{3^2 - (4 \cdot 1 \cdot -10)}}{2 \cdot 1} = \dfrac{-3 \pm \sqrt{9 + 40}}{2} = \dfrac{-3 \pm \sqrt{49}}{2} =$

$\dfrac{-3 \pm 7}{2}$; $x = \dfrac{-3 + 7}{2} = \dfrac{4}{2} = 2$ or $x = \dfrac{-3 - 7}{2} = \dfrac{-10}{2} = -5$

47. $2x^2 + 5x - 3 = 0$; $x = \dfrac{-5 \pm \sqrt{5^2 - (4 \cdot 2 \cdot -3)}}{2 \cdot 2} = \dfrac{-5 \pm \sqrt{25 + 24}}{4} = \dfrac{-5 \pm \sqrt{49}}{4} =$

$\dfrac{-5 \pm 7}{4}$; $x = \dfrac{-5 + 7}{4} = \dfrac{2}{4} = \dfrac{1}{2}$ or $x = \dfrac{-5 - 7}{4} = \dfrac{-12}{4} = -3$

48. $3x^2 + 5x + 2 = 0$; $x = \dfrac{-5 \pm \sqrt{5^2 - (4 \cdot 3 \cdot 2)}}{2 \cdot 3} = \dfrac{-5 \pm \sqrt{25 - 24}}{6} = \dfrac{-5 \pm \sqrt{1}}{6} =$

$\dfrac{-5 \pm 1}{6}$; $x = \dfrac{-5 + 1}{6} = \dfrac{-4}{6} = -\dfrac{2}{3}$ or $x = \dfrac{-5 - 1}{6} = \dfrac{-6}{6} = -1$

49. $4x^2 - 2x - 2 = 0$; $x = \dfrac{-(-2) \pm \sqrt{(-2)^2 - (4 \cdot 4 \cdot -2)}}{2 \cdot 4} = \dfrac{2 \pm \sqrt{4 + 32}}{8} = \dfrac{2 \pm \sqrt{36}}{8} =$

$\dfrac{2 \pm 6}{8}$; $x = \dfrac{2 + 6}{8} = \dfrac{8}{8} = 1$ or $x = \dfrac{2 - 6}{8} = \dfrac{-4}{8} = -\dfrac{1}{2}$

50. $2x^2 - x - 3 = 0$; $x = \dfrac{-(-1) \pm \sqrt{(-1)^2 - (4 \cdot 2 \cdot -3)}}{2 \cdot 2} = \dfrac{1 \pm \sqrt{1 + 24}}{4} = \dfrac{1 \pm \sqrt{25}}{4} =$

$\dfrac{1 \pm 5}{4}$; $x = \dfrac{1 + 5}{4} = \dfrac{6}{4} = \dfrac{3}{2}$ or $x = \dfrac{1 - 5}{4} = \dfrac{-4}{4} = -1$

51. $2x^2 + 5x + 3 = 0$; $x = \dfrac{-5 \pm \sqrt{5^2 - (4 \cdot 2 \cdot 3)}}{2 \cdot 2} = \dfrac{-5 \pm \sqrt{25 - 24}}{4} = \dfrac{-5 \pm \sqrt{1}}{4} =$

$\dfrac{-5 \pm 1}{4}$; $x = \dfrac{-5 + 1}{4} = \dfrac{-4}{4} = -1$ or $x = \dfrac{-5 - 1}{4} = \dfrac{-6}{4} = -\dfrac{3}{2}$

52. $5x^2 + 2x - 2 = 0$; $x = \dfrac{-2 \pm \sqrt{2^2 - (4 \cdot 5 \cdot -2)}}{2 \cdot 5} = \dfrac{-2 \pm \sqrt{4 + 40}}{10} = \dfrac{-2 \pm \sqrt{44}}{10} =$

$\dfrac{-2 \pm 2\sqrt{11}}{10} = \dfrac{-1 \pm \sqrt{11}}{5}$; $x = \dfrac{-1 + \sqrt{11}}{5}$ or $x = \dfrac{-1 - \sqrt{11}}{5}$

53. $4x^2 + 7x + 2 = 0$; $x = \dfrac{-7 \pm \sqrt{7^2 - (4 \cdot 4 \cdot 2)}}{2 \cdot 4} = \dfrac{-7 \pm \sqrt{49 - 32}}{8} = \dfrac{-7 \pm \sqrt{17}}{8}$;

$x = \dfrac{-7 + \sqrt{17}}{8}$ or $x = \dfrac{-7 - \sqrt{17}}{8}$

54. $3x^2 - 2x - 4 = 0$; $x = \dfrac{-(-2) \pm \sqrt{(-2)^2 - (4 \cdot 3 \cdot -4)}}{2 \cdot 3} = \dfrac{2 \pm \sqrt{4 + 48}}{6} = \dfrac{2 \pm \sqrt{52}}{6} =$

$\dfrac{2 \pm 2\sqrt{13}}{6} = \dfrac{1 \pm \sqrt{13}}{3}$; $x = \dfrac{1 + \sqrt{13}}{3}$ or $x = \dfrac{1 - \sqrt{13}}{3}$

55. Let $x =$ width (in cm). Then $x + 5 =$ length. $x(x + 5) = 36$; $x^2 + 5x = 36$; $x^2 + 5x - 36 = 0$; $(x - 4)(x + 9) = 0$; $x - 4 = 0$ or $x + 9 = 0$; $x = 4$ or $x = -9$ (reject); $x = 4$ and $x + 5 = 9$. width: 4 cm; length: 9 cm

56. Let x and $x + 6 =$ the numbers. $x^2 + (x + 6)^2 = 90$; $x^2 + x^2 + 12x + 36 = 90$; $2x^2 + 12x - 54 = 0$; $x^2 + 6x - 27 = 0$; $(x - 3)(x + 9) = 0$; $x - 3 = 0$ or $x + 9 = 0$; $x = 3$ or $x = -9$, so $x + 6 = 9$ or $x + 6 = -3$. 3 and 9, or -9 and -3

57.

$y = x^2 + 2$	
x	y
-3	11
-2	6
-1	3
0	2
1	3
2	6
3	11

If $y = 6$, $x = -2$ or $x = 2$.

58.

$y = x^2 + 3x - 5$	
x	y
-3	-5
-2	-7
-1	-7
0	-5
1	-1
2	5
3	13

If $y = -7$, $x = -2$ or $x = -1$.

59.

$y = x^2 - 3$	
x	y
-2	1
-1	-2
0	-3
1	-2
2	1

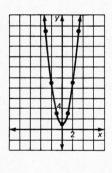

60.

$y = 2x^2 + 1$	
x	y
-3	19
-2	9
-1	3
0	1
1	3
2	9
3	19

61.

$y = x^2 + 2x + 1$	
x	y
-3	4
-2	1
-1	0
0	1
1	4

Page 447 · CHAPTER TEST

1. $3x(x + 2) = 0$; $3x = 0$ or $x + 2 = 0$; $x = 0$ or $x = -2$

2. $(y - 3)(y + 9) = 0$; $y - 3 = 0$ or $y + 9 = 0$; $y = 3$ or $y = -9$

3. $(2a - 10)(7a - 42) = 0$; $2a - 10 = 0$ or $7a - 42 = 0$; $2a = 10$ or
   $7a = 42$; $a = 5$ or $a = 6$

4. $r(r + 8)(r - 6) = 0$; $r = 0$, $r + 8 = 0$, or $r - 6 = 0$; $r = 0$, $r = -8$, or $r = 6$

5. $n^2 + 10n + 9 = 0$; $(n + 1)(n + 9) = 0$; $n + 1 = 0$ or $n + 9 = 0$; $n = -1$ or
   $n = -9$

6. $x^2 - 12x + 36 = 0$; $(x - 6)^2 = 0$; $x - 6 = 0$; $x = 6$

7. $y^2 - y - 20 = 0$; $(y - 5)(y + 4) = 0$; $y - 5 = 0$ or $y + 4 = 0$; $y = 5$ or
   $y = -4$

8. $a^2 - 1 = 0$; $(a + 1)(a - 1) = 0$; $a + 1 = 0$ or $a - 1 = 0$; $a = -1$ or $a = 1$

9. $x^2 = 14x - 48$; $x^2 - 14x + 48 = 0$; $(x - 6)(x - 8) = 0$; $x - 6 = 0$ or
   $x - 8 = 0$; $x = 6$ or $x = 8$

10. $y^2 + 9y = 22$; $y^2 + 9y - 22 = 0$; $(y + 11)(y - 2) = 0$; $y + 11 = 0$ or
    $y - 2 = 0$; $y = -11$ or $y = 2$

11. $(n + 4)(n - 2) = -8$; $n^2 + 2n - 8 = -8$; $n^2 + 2n = 0$; $n(n + 2) = 0$; $n = 0$
    or $n + 2 = 0$; $n = 0$ or $n = -2$

12. $\dfrac{x + 1}{4} = \dfrac{3}{x - 3}$; $(x + 1)(x - 3) = 12$; $x^2 - 2x - 3 = 12$; $x^2 - 2x - 15 = 0$;

    $(x - 5)(x + 3) = 0$; $x - 5 = 0$ or $x + 3 = 0$; $x = 5$ or $x = -3$

13. $3y^2 = 27$; $y^2 = 9$; $y = \pm 3$; $y = 3$ or $y = -3$

14. $x^2 = 56$; $x = \pm\sqrt{56} = \pm 2\sqrt{14}$; $x = 2\sqrt{14}$ or $x = -2\sqrt{14}$

15. $n^2 - 8 = 2$; $n^2 = 10$; $n = \pm\sqrt{10}$; $n = \sqrt{10}$ or $n = -\sqrt{10}$

16. $n^2 + 8 = 2$; $n^2 = -6$;   no solution

17. $5y^2 + 1 = 81$; $5y^2 = 80$; $y^2 = 16$; $y = \pm 4$; $y = 4$ or $y = -4$

18. $\dfrac{8x}{9} = \dfrac{1}{2x}$; $16x^2 = 9$; $x^2 = \dfrac{9}{16}$; $x = \pm\sqrt{\dfrac{9}{16}} = \pm\dfrac{3}{4}$; $x = \dfrac{3}{4}$ or $x = -\dfrac{3}{4}$

19. $(2x + 5)^2 = 1$; $2x + 5 = \pm 1$; $2x + 5 = 1$ or $2x + 5 = -1$; $2x = -4$ or
    $2x = -6$; $x = -2$ or $x = -3$

20. $(y - 9)^2 = 4$; $y - 9 = \pm 2$; $y - 9 = 2$ or $y - 9 = -2$; $y = 11$ or $y = 7$

21. $(3x - 7)^2 = 0$; $3x - 7 = 0$; $3x = 7$; $x = \dfrac{7}{3}$

22. $(n - 3)^2 = 5$; $n - 3 = \pm\sqrt{5}$; $n - 3 = \sqrt{5}$ or $n - 3 = -\sqrt{5}$; $n = 3 + \sqrt{5}$ or
    $n = 3 - \sqrt{5}$

23. $4(a + 5)^2 = 64$; $(a + 5)^2 = 16$; $a + 5 = \pm 4$; $a + 5 = 4$ or $a + 5 = -4$;
    $a = -1$ or $a = -9$

24. $2(3y + 18)^2 = 72$; $(3y + 18)^2 = 36$; $3y + 18 = \pm 6$; $3y + 18 = 6$ or
    $3y + 18 = -6$; $3y = -12$ or $3y = -24$; $y = -4$ or $y = -8$

25. $x^2 + 6x + 4 = 0$; $x = \dfrac{-6 \pm \sqrt{6^2 - (4 \cdot 1 \cdot 4)}}{2 \cdot 1} = \dfrac{-6 \pm \sqrt{36 - 16}}{2} = \dfrac{-6 \pm \sqrt{20}}{2} =$

$\dfrac{-6 \pm 2\sqrt{5}}{2} = -3 \pm \sqrt{5}$; $x = -3 + \sqrt{5}$ or $x = -3 - \sqrt{5}$

26. $5x^2 + 3x - 4 = 0$; $x = \dfrac{-3 \pm \sqrt{3^2 - (4 \cdot 5 \cdot -4)}}{2 \cdot 5} = \dfrac{-3 \pm \sqrt{9 + 80}}{10} = \dfrac{-3 \pm \sqrt{89}}{10}$;

$x = \dfrac{-3 + \sqrt{89}}{10}$ or $x = \dfrac{-3 - \sqrt{89}}{10}$

27. $3x^2 + 4x - 4 = 0$; $x = \dfrac{-4 \pm \sqrt{4^2 - (4 \cdot 3 \cdot -4)}}{2 \cdot 3} = \dfrac{-4 \pm \sqrt{16 + 48}}{6} = \dfrac{-4 \pm \sqrt{64}}{6} =$

$\dfrac{-4 \pm 8}{6}$; $x = \dfrac{-4 + 8}{6}$ or $x = \dfrac{-4 - 8}{6}$; $x = \dfrac{4}{6} = \dfrac{2}{3}$ or $x = \dfrac{-12}{6} = -2$

28. $2y^2 - 4y + 1 = 0$; $y = \dfrac{-(-4) \pm \sqrt{(-4)^2 - (4 \cdot 2 \cdot 1)}}{2 \cdot 2} = \dfrac{4 \pm \sqrt{16 - 8}}{4} = \dfrac{4 \pm \sqrt{8}}{4} =$

$\dfrac{4 \pm 2\sqrt{2}}{4} = \dfrac{2 \pm \sqrt{2}}{2}$; $y = \dfrac{2 + \sqrt{2}}{2}$ or $y = \dfrac{2 - \sqrt{2}}{2}$

29. **a.** $s^2 + 2s + 1 = 0$; $(s + 1)^2 = 0$; $s + 1 = 0$; $s = -1$

b. $s = \dfrac{-2 \pm \sqrt{2^2 - (4 \cdot 1 \cdot 1)}}{2 \cdot 1} = \dfrac{-2 \pm \sqrt{4 - 4}}{2} = \dfrac{-2 \pm \sqrt{0}}{2} = \dfrac{-2}{2} = -1$

30. $(x + 3)(x + 1)(4) = 192$; $(x + 3)(x + 1) = 48$; $x^2 + 4x + 3 = 48$;

$x^2 + 4x - 45 = 0$; $(x + 9)(x - 5) = 0$; $x + 9 = 0$ or $x - 5 = 0$; $x = -9$

(reject) or $x = 5$. $x = 5$

31. Let x and $x - 4 =$ the numbers. $x^2 + (x - 4)^2 = 40$; $x^2 + x^2 - 8x + 16 = 40$;
$2x^2 - 8x - 24 = 0$; $x^2 - 4x - 12 = 0$; $(x - 6)(x + 2) = 0$; $x - 6 = 0$ or
$x + 2 = 0$; $x = 6$ or $x = -2$, so $x - 4 = 2$ or $x - 4 = -6$. 6 and 2 or -2
and -6

32.

$y = x^2 + 3x - 6$	
x	y
-4	-2
-3	-6
-2	-8
-1	-8
0	-6
1	-2
2	4

If $y = -2$, $x = -4$ or $x = 1$.

33. $y = x^2 - 4x + 1$

x	y
-1	6
0	1
1	-2
2	-3
3	-2
4	1

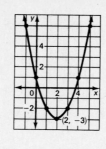

Vertex: $(2, -3)$

34. $y = 6x - x^2$

x	y
5	5
4	8
3	9
2	8
1	5

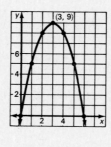

Vertex: $(3, 9)$

35. 9

Page 448 · CUMULATIVE REVIEW

1. 20   2. -10 3. 9   4. 11   5. -23 6. 7.937   7. -4.123 8. 6.403

9. -4.690 10. ±1.732 11. $\sqrt{54} = \sqrt{9 \cdot 6} = \sqrt{9} \cdot \sqrt{6} = 3\sqrt{6}$

12. $\sqrt{75} = \sqrt{25 \cdot 3} = \sqrt{25} \cdot \sqrt{3} = 5\sqrt{3}$

13. $\sqrt{432} = \sqrt{144 \cdot 3} = \sqrt{144} \cdot \sqrt{3} = 12\sqrt{3}$

14. $\sqrt{48a^3} = \sqrt{16 \cdot 3 \cdot a^2 \cdot a} = 4a\sqrt{3a}$

15. $\sqrt{72x^2y} = \sqrt{36 \cdot 2 \cdot x^2 \cdot y} = 6x\sqrt{2y}$

16. $x^2 = 1$; $x = \pm1$; $x = 1$ or $x = -1$

17. $y^2 - 3 = 13$; $y^2 = 16$; $y = \pm4$; $y = 4$ or $y = -4$

18. $x^2 = 18$; $x = \pm\sqrt{18} = \pm3\sqrt{2}$; $x = 3\sqrt{2}$ or $x = -3\sqrt{2}$

19. $b^2 + 17 = 66$; $b^2 = 49$; $b = \pm7$; $b = 7$ or $b = -7$

20. Let x = length of hypotenuse. $x^2 = 4^2 + 7^2$; $x^2 = 16 + 49$; $x^2 = 65$;
$x = \sqrt{65}$. $\sqrt{65}$

21. $\sqrt{\dfrac{9}{16}} = \dfrac{\sqrt{9}}{\sqrt{16}} = \dfrac{3}{4}$

22. $\sqrt{\dfrac{4y^2}{121}} = \dfrac{\sqrt{4y^2}}{\sqrt{121}} = \dfrac{2y}{11}$

23. $\sqrt{3a} \cdot \sqrt{12a^3} = \sqrt{3a \cdot 12a^3} = \sqrt{36a^4} = 6a^2$

24. $\sqrt{\dfrac{5}{8}} \cdot \sqrt{\dfrac{1}{10}} = \sqrt{\dfrac{5}{80}} = \sqrt{\dfrac{1}{16}} = \dfrac{1}{4}$

25. $\sqrt{\dfrac{3}{8}} = \dfrac{\sqrt{3} \cdot \sqrt{8}}{\sqrt{8} \cdot \sqrt{8}} = \dfrac{\sqrt{24}}{8} = \dfrac{2\sqrt{6}}{8} = \dfrac{\sqrt{6}}{4}$

26. $2\sqrt{\dfrac{1}{5}} = \dfrac{2\sqrt{1}\cdot\sqrt{5}}{\sqrt{5}\cdot\sqrt{5}} = \dfrac{2\sqrt{5}}{5}$ **27.** $5\sqrt{3} - 3\sqrt{3} = 2\sqrt{3}$

28. $2\sqrt{18} - 4\sqrt{2} = 2\cdot 3\sqrt{2} - 4\sqrt{2} = 6\sqrt{2} - 4\sqrt{2} = 2\sqrt{2}$

29. $5y(y + 4) = 0$; $5y = 0$ or $y + 4 = 0$; $y = 0$ or $y = -4$

30. $(x - 3)(x + 6) = 0$; $x - 3 = 0$ or $x + 6 = 0$; $x = 3$ or $x = -6$

31. $(3n - 6)(2n + 8) = 0$; $3n - 6 = 0$ or $2n + 8 = 0$; $3n = 6$ or $2n = -8$; $n = 2$ or $n = -4$

32. $t^2 - 7t = 0$; $t(t - 7) = 0$; $t = 0$ or $t - 7 = 0$; $t = 0$ or $t = 7$

33. $y^2 - 8y + 7 = 0$; $(y - 1)(y - 7) = 0$; $y - 1 = 0$ or $y - 7 = 0$; $y = 1$ or $y = 7$

34. $x^2 - x - 6 = 0$; $(x - 3)(x + 2) = 0$; $x - 3 = 0$ or $x + 2 = 0$; $x = 3$ or $x = -2$

35. $x^2 = -5x$; $x^2 + 5x = 0$; $x(x + 5) = 0$; $x = 0$ or $x + 5 = 0$; $x = 0$ or $x = -5$

36. $t^2 + 2t = 15$; $t^2 + 2t - 15 = 0$; $(t + 5)(t - 3) = 0$; $t + 5 = 0$ or $t - 3 = 0$; $t = -5$ or $t = 3$

37. $n^2 = 8n - 16$; $n^2 - 8n + 16 = 0$; $(n - 4)^2 = 0$; $n - 4 = 0$; $n = 4$

38. $z^2 = 11$; $z = \pm\sqrt{11}$; $z = \sqrt{11}$ or $z = -\sqrt{11}$

39. $m^2 - 3 = 17$; $m^2 = 20$; $m = \pm\sqrt{20} = \pm 2\sqrt{5}$; $m = 2\sqrt{5}$ or $m = -2\sqrt{5}$

40. $2y^2 + 7 = 19$; $2y^2 = 12$; $y^2 = 6$; $y = \pm\sqrt{6}$; $y = \sqrt{6}$ or $y = -\sqrt{6}$

41. $(n - 1)^2 = 16$; $n - 1 = \pm 4$; $n - 1 = 4$ or $n - 1 = -4$; $n = 5$ or $n = -3$

42. $(2t + 2)^2 = 36$; $2t + 2 = \pm 6$; $2t + 2 = 6$ or $2t + 2 = -6$; $2t = 4$ or $2t = -8$; $t = 2$ or $t = -4$

43. $(a - 4)^2 = 64$; $a - 4 = \pm 8$; $a - 4 = 8$ or $a - 4 = -8$; $a = 12$ or $a = -4$

44. $6x^2 + 5x + 1 = 0$; $x = \dfrac{-5 \pm \sqrt{5^2 - (4\cdot 6\cdot 1)}}{2\cdot 6} = \dfrac{-5 \pm \sqrt{25 - 24}}{12} = \dfrac{-5 \pm \sqrt{1}}{12} =$

$\dfrac{-5 \pm 1}{12}$; $x = \dfrac{-5 + 1}{12} = \dfrac{-4}{12} = -\dfrac{1}{3}$ or $x = \dfrac{-5 - 1}{12} = \dfrac{-6}{12} = -\dfrac{1}{2}$

45. $4x^2 + 4x = 3$; $4x^2 + 4x - 3 = 0$; $x = \dfrac{-4 \pm \sqrt{4^2 - (4\cdot 4\cdot -3)}}{2\cdot 4} = \dfrac{-4 \pm \sqrt{16 + 48}}{8} =$

$\dfrac{-4 \pm \sqrt{64}}{8} = \dfrac{-4 \pm 8}{8}$; $x = \dfrac{-4 + 8}{8} = \dfrac{4}{8} = \dfrac{1}{2}$ or $x = \dfrac{-4 - 8}{8} = \dfrac{-12}{8} = -\dfrac{3}{2}$

46. $x^2 + x - 1 = 0$; $x = \dfrac{-1 \pm \sqrt{1^2 - (4\cdot 1\cdot -1)}}{2\cdot 1} = \dfrac{-1 \pm \sqrt{1 + 4}}{2} = \dfrac{-1 \pm \sqrt{5}}{2}$;

$x = \dfrac{-1 + \sqrt{5}}{2}$ or $x = \dfrac{-1 - \sqrt{5}}{2}$

47.

$y = x^2 - 4x + 5$	
x	y
4	5
3	2
2	1
1	2
0	5

(2, 1)

48.

$y = 1 - x^2$	
x	y
2	−3
1	0
0	1
−1	0
−2	−3

(0, 1)

49.

$y = x^2 - 2x$	
x	y
3	3
2	0
1	−1
0	0
−1	3

(1, −1)

EXTRA PRACTICE EXERCISES

CHAPTER 1

Page 451 · FOR PAGES 4–9

A   1. $7 \cdot 6 = 42$ 2. $5 \cdot 8 = 40$ 3. $9 \cdot 7 = 63$ 4. $6 \cdot 4 = 24$

     5. $2 \cdot 3 + 4 = 6 + 4 = 10$ 6. $5 \cdot 2 - 6 = 10 - 6 = 4$

     7. $(6 - 1) \cdot 3 = 5 \cdot 3 = 15$ 8. $(4 + 2) \div 3 = 6 \div 3 = 2$

     9. $2(3 + 2) = 2 \cdot 5 = 10$ 10. $4(5 + 1) = 4 \cdot 6 = 24$

     11. $8 - (2 \cdot 3) = 8 - 6 = 2$ 12. $10 + (6 \cdot 1) = 10 + 6 = 16$

     13. $9n = 9 \cdot 3 = 27$ 14. $9 + n = 9 + 3 = 12$

     15. $9 - n = 9 - 3 = 6$ 16. $9 \div n = 9 \div 3 = 3$

     17. $x + y = 10 + 5 = 15$ 18. $x - y = 10 - 5 = 5$

     19. $x + x + y = 10 + 10 + 5 = 25$ 20. $x + 5 - y = 10 + 5 - 5 = 10$

     21. $6(y + 2) = 6(2 + 2) = 6 \cdot 4 = 24$ 22. $6y + 2 = 6 \cdot 2 + 2 = 12 + 2 = 14$

     23. $(y - 1) \cdot 4 = (2 - 1) \cdot 4 = 1 \cdot 4 = 4$ 24. $2y - 3 = 2 \cdot 2 - 3 = 4 - 3 = 1$

Page 451 · FOR PAGES 10–13

A   1. $n + n = 2n$ 2. $n + n + n = 3n$ 3. $2n + n = 3n$

     4. $n + 2n = 3n$ 5. $3x - x = 2x$ 6. $5y - 2y = 3y$

     7. $7m + 3m = 10m$ 8. $10x - x = 9x$ 9. $4a + 5a = 9a$

     10. $6x + 3x = 9x$ 11. $4a + 2a - 3a = 3a$ 12. $5x + 5x - 3x = 7x$

     13. $3x + 8x - x = 10x$ 14. $n + 6n - 7 = 7n - 7$ 15. $a + 2a - b = 3a - b$

     16. $3n - n + m = 2n + m$

     17. $2x - 5 + 3x = (2x + 3x) - 5 = 5x - 5$

     18. $6n - 6 + 2n = (6n + 2n) - 6 = 8n - 6$

     19. $5m + 2 + 6m - 1 = (5m + 6m) + (2 - 1) = 11m + 1$

     20. $6 + 7n - 6n + 7 = (6 + 7) + (7n - 6n) = 13 + n$

     21. $8 + 6y + 5y - 7 = (8 - 7) + (6y + 5y) = 1 + 11y$

     22. $7 + 3x + 9x + 9 = (7 + 9) + (3x + 9x) = 16 + 12x$

     23. $5x + 3y - y + 11x = (5x + 11x) + (3y - y) = 16x + 2y$

     24. $2x + 5y - y + 9x = (2x + 9x) + (5y - y) = 11x + 4y$

     25. $4n + r + r - 3n = (4n - 3n) + (r + r) = n + 2r$

Page 452 · FOR PAGES 14–17

A   1. $3^2 = 3 \cdot 3 = 9$ 2. $4^2 = 4 \cdot 4 = 16$ 3. $2^2 = 2 \cdot 2 = 4$

     4. $6^2 = 6 \cdot 6 = 36$ 5. $4^1 = 4$ 6. $2^3 = 2 \cdot 2 \cdot 2 = 8$

     7. $3^3 = 3 \cdot 3 \cdot 3 = 27$ 8. $5^2 = 5 \cdot 5 = 25$ 9. $y \cdot y \cdot y = y^3$

     10. $x \cdot x = x^2$ 11. $a \cdot a \cdot a = a^3$ 12. $c \cdot c \cdot c \cdot c = c^4$

13. $r \cdot s \cdot s = r \cdot (s \cdot s) = rs^2$ **14.** $a \cdot a \cdot b \cdot b = (a \cdot a) \cdot (b \cdot b) = a^2b^2$

15. $x \cdot y \cdot y \cdot y = x \cdot (y \cdot y \cdot y) = xy^3$

16. $a \cdot b \cdot b \cdot c \cdot c = a \cdot (b \cdot b) \cdot (c \cdot c) = ab^2c^2$

17. $3 \cdot a \cdot a = 3a^2$ **18.** $2 \cdot c \cdot c = 2c^2$

19. $5 \cdot 2 \cdot n \cdot n = 10n^2$ **20.** $6 \cdot a \cdot a \cdot b = 6a^2b$

21. $2 \cdot (4r) = (2 \cdot 4) \cdot r = 8r$ **22.** $6 \cdot 3n = (6 \cdot 3) \cdot n = 18n$

23. $(4n) \cdot 4 = (4 \cdot 4) \cdot n = 16n$ **24.** $6n \cdot 5 = (6 \cdot 5) \cdot n = 30n$

25. $(2r)(4r) = (2 \cdot 4)(r \cdot r) = 8r^2$ **26.** $(10x)(5x) = (10 \cdot 5)(x \cdot x) = 50x^2$

27. $7y \cdot 8y = (7 \cdot 8)(y \cdot y) = 56y^2$ **28.** $6s \cdot 5s = (6 \cdot 5)(s \cdot s) = 30s^2$

29. $(2a)(4b) = (2 \cdot 4) \cdot a \cdot b = 8ab$ **30.** $(3x)(4y) = (3 \cdot 4) \cdot x \cdot y = 12xy$

31. $2 \cdot a \cdot 5 \cdot a = (2 \cdot 5)(a \cdot a) = 10a^2$ **32.** $x \cdot x \cdot 4 \cdot x = 4 \cdot (x \cdot x \cdot x) = 4x^3$

Page 452 · FOR PAGES 18–19

A   **1.** $3(x + 3) = 3 \cdot x + 3 \cdot 3 = 3x + 9$ **2.** $7(y + 1) = 7 \cdot y + 7 \cdot 1 = 7y + 7$

3. $5(a - 1) = 5 \cdot a - 5 \cdot 1 = 5a - 5$ **4.** $3(n + 2) = 3 \cdot n + 3 \cdot 2 = 3n + 6$

5. $8(n + 4) = 8 \cdot n + 8 \cdot 4 = 8n + 32$ **6.** $6(c - 2) = 6 \cdot c - 6 \cdot 2 = 6c - 12$

7. $4(x + 3) = 4 \cdot x + 4 \cdot 3 = 4x + 12$ **8.** $a(a + 1) = a \cdot a + a \cdot 1 = a^2 + a$

9. $x(x - 2) = x \cdot x - x \cdot 2 = x^2 - 2x$ **10.** $y(y + 7) = y \cdot y + y \cdot 7 = y^2 + 7y$

11. $n(n + 2) = n \cdot n + n \cdot 2 = n^2 + 2n$ **12.** $c(c + 9) = c \cdot c + c \cdot 9 = c^2 + 9c$

13. $6(2n - 8) = 6 \cdot 2n - 6 \cdot 8 = 12n - 48$

14. $4(5x - 4) = 4 \cdot 5x - 4 \cdot 4 = 20x - 16$ **15.** $3(2a - 1) = 3 \cdot 2a - 3 \cdot 1 = 6a - 3$

16. $5(3x + 2) = 5 \cdot 3x + 5 \cdot 2 = 15x + 10$

17. $3(2a - 3b) = 3 \cdot 2a - 3 \cdot 3b = 6a - 9b$

18. $2(4x - 2y) = 2 \cdot 4x - 2 \cdot 2y = 8x - 4y$

19. $a(6a + b) = a \cdot 6a + a \cdot b = 6a^2 + ab$

20. $x(4x - 2y) = x \cdot 4x - x \cdot 2y = 4x^2 - 2xy$

21. $2(n + 3) + 1 = 2n + 6 + 1 = 2n + 7$

22. $4(y + 1) + 3 = 4y + 4 + 3 = 4y + 7$

23. $3(n + 1) + 8 = 3n + 3 + 8 = 3n + 11$

24. $4(x + 3) + 2x = 4x + 12 + 2x = 6x + 12$

25. $2(x + 4) - 2x = 2x + 8 - 2x = 8$

26. $8(3n + 1) - 2 = 24n + 8 - 2 = 24n + 6$

B   **27.** $4(2a + 3) - 7a + 1 = 8a + 12 - 7a + 1 = a + 13$

28. $3(5x + 7) + 2x - 20 = 15x + 21 + 2x - 20 = 17x + 1$

29. $2(3y + 5) - 5y + 4 = 6y + 10 - 5y + 4 = y + 14$

30. $5(2 + 3x) - 4x - 7 = 10 + 15x - 4x - 7 = 3 + 11x$

31. $6(a + 1) + 4(a + 2) = 6a + 6 + 4a + 8 = 10a + 14$

32. $7(n + 3) + 2(n - 6) = 7n + 21 + 2n - 12 = 9n + 9$

Page 453 • FOR PAGES 20–23

A   **1.** $6 \cdot 0 = 0$ **2.** $5 \cdot 1 = 5$ **3.** $8 \cdot 0 = 0$ **4.** $\dfrac{0}{4} = 0$ **5.** $\dfrac{0}{7} = 0$

6. $\dfrac{9}{0}$; impossible   **7.** $6 \div 1 = 6$ **8.** $\dfrac{8}{0}$; impossible   **9.** $\dfrac{5 \cdot 0}{3} = \dfrac{0}{3} = 0$

10. $\dfrac{7}{1} = 7$ **11.** $\dfrac{7 + 1}{0} = \dfrac{8}{0}$; impossible   **12.** $12 \div 0$; impossible

13. $7(x - 1) = 7(1 - 1) = 7 \cdot 0 = 0$ **14.** $x \div x = 1 \div 1 = 1$

15. $3x \div x = 3 \cdot 1 \div 1 = 3 \div 1 = 3$ **16.** $\dfrac{4}{4} x = \dfrac{4}{4} \cdot 1 = 1 \cdot 1 = 1$

17. $(a - 1) \div 5 = (6 - 1) \div 5 = 5 \div 5 = 1$

18. $(2a - 2) \div 10 = (2 \cdot 6 - 2) \div 10 = (12 - 2) \div 10 = 10 \div 10 = 1$

19. $\dfrac{a + 2}{8} = \dfrac{6 + 2}{8} = \dfrac{8}{8} = 1$ **20.** $\dfrac{a}{6} \cdot 1 = \dfrac{6}{6} \cdot 1 = 1 \cdot 1 = 1$

21. $3 + 2x - 3 = 2x + (3 - 3) = 2x + 0 = 2x$

22. $7 + 4y - 7 = 4y + (7 - 7) = 4y + 0 = 4y$

23. $4a + 5 + 2a - 5 = (4a + 2a) + (5 - 5) = 6a + 0 = 6a$

24. $2(x + 4) - 8 = 2x + 8 - 8 = 2x + 0 = 2x$

25. $\dfrac{6}{6} \cdot n = 1 \cdot n = n$ **26.** $\dfrac{9}{9} \cdot c = 1 \cdot c = c$

27. $\dfrac{x}{3} \cdot 3 = x \cdot \dfrac{3}{3} = x \cdot 1 = x$ **28.** $\dfrac{2n}{2} = \dfrac{2}{2} \cdot n = 1 \cdot n = n$

Page 453 • FOR PAGES 25–28

A   **1.** If $n = 2$: $n + 3 = 2 + 3 = 5$; if $n = 1$: $1 + 3 = 4$; if $n = 0$: $0 + 3 = 3$.   1

2. If $x = 7$: $x - 2 = 7 - 2 = 5$; if $x = 8$: $8 - 2 = 6$; if $x = 9$: $9 - 2 = 7$.   8

3. If $y = 3$: $5 + y = 5 + 3 = 8$; if $y = 2$: $5 + 2 = 7$; if $y = 1$: $5 + 1 = 6$.   3

4. If $a = 14$: $a - 4 = 14 - 4 = 10$; if $a = 13$: $13 - 4 = 9$; if $a = 12$: $12 -$
4 = 8.   12

5. If $a = 4$: $6 - a = 6 - 4 = 2$; if $a = 3$: $6 - 3 = 3$; if $a = 2$: $6 - 2 = 4$.   2

6. If $n = 14$: $n - 5 = 14 - 5 = 9$; if $n = 15$: $15 - 5 = 10$; if $n = 16$: $16 -$
5 = 11.   15

7. $n < 4$; $1 < 4, 2 < 4, 4 = 4, 5 > 4$.   1, 2

8. $x > 6$; $6 = 6, 7 > 6, 9 > 6, 10 > 6$.   7, 9, 10

9. $y < 17$; $15 < 17, 16 < 17, 19 > 17, 20 > 17$.   15, 16

10. $a > 9$; $6 < 9, 8 < 9, 10 > 9, 12 > 9$.   10, 12

11. $x + 1 > 3$; $1 + 1 = 2 < 3, 3 + 1 = 4 > 3, 4 + 1 = 5 > 3, 10 + 1 = 11 > 3$.
3, 4, 10

12. $2c > 3$; $2 \cdot 4 = 8 > 3, 2 \cdot 2 = 4 > 3, 2 \cdot 1 = 2 < 3, 2 \cdot 9 = 18 > 3$.   4, 2, 9

CHAPTER 2

Page 454 · FOR PAGES 38–41

A   1. $x - 2 = 4$; $x - 2 + 2 = 4 + 2$; $x = 6$

     2. $x - 4 = 2$; $x - 4 + 4 = 2 + 4$; $x = 6$

     3. $n - 3 = 3$; $n - 3 + 3 = 3 + 3$; $n = 6$

     4. $t - 4 = 4$; $t - 4 + 4 = 4 + 4$; $t = 8$

     5. $y - 4 = 7$; $y - 4 + 4 = 7 + 4$; $y = 11$

     6. $x - 3 = 6$; $x - 3 + 3 = 6 + 3$; $x = 9$

     7. $n - 7 = 6$; $n - 7 + 7 = 6 + 7$; $n = 13$

     8. $m - 3 = 8$; $m - 3 + 3 = 8 + 3$; $m = 11$

     9. $n + 2 = 6$; $n + 2 - 2 = 6 - 2$; $n = 4$

   10. $x + 1 = 7$; $x + 1 - 1 = 7 - 1$; $x = 6$

   11. $x + 7 = 8$; $x + 7 - 7 = 8 - 7$; $x = 1$

   12. $y + 6 = 10$; $y + 6 - 6 = 10 - 6$; $y = 4$

   13. $x + 5 = 20$; $x + 5 - 5 = 20 - 5$; $x = 15$

   14. $y + 6 = 14$; $y + 6 - 6 = 14 - 6$; $y = 8$

   15. $n + 9 = 15$; $n + 9 - 9 = 15 - 9$; $n = 6$

   16. $m + 12 = 18$; $m + 12 - 12 = 18 - 12$; $m = 6$

   17. $y - 5 = 20$; $y = 25$ 18. $x - 6 = 13$; $x = 19$

   19. $x + 6 = 19$; $x = 13$ 20. $n + 7 = 14$; $n = 7$

   21. $n + 9 = 17$; $n = 8$ 22. $n - 5 = 16$; $n = 21$

   23. $x + 14 = 21$; $x = 7$ 24. $r - 11 = 22$; $r = 33$

   25. $t + 16 = 22$; $t = 6$ 26. $n + 18 = 25$; $n = 7$

   27. $x - 6 = 18$; $x = 24$ 28. $30 = n - 5$; $35 = n$

   29. $15 = y + 8$; $7 = y$ 30. $x - 4 = 0$; $x = 4$

   31. $y + 3 = 27$; $y = 24$ 32. $16 = x - 14$; $30 = x$

Page 454 · FOR PAGES 42–45

A   1. $2x = 12$; $\dfrac{2x}{2} = \dfrac{12}{2}$; $x = 6$ 2. $3y = 12$; $\dfrac{3y}{3} = \dfrac{12}{3}$; $y = 4$

     3. $4n = 12$; $n = 3$ 4. $5n = 25$; $n = 5$ 5. $3x = 15$; $x = 5$

     6. $6x = 18$; $x = 3$ 7. $7y = 28$; $y = 4$ 8. $9x = 63$; $x = 7$

     9. $\dfrac{x}{3} = 4$; $3 \cdot \dfrac{x}{3} = 3 \cdot 4$; $x = 12$ 10. $\dfrac{n}{2} = 9$; $2 \cdot \dfrac{n}{2} = 2 \cdot 9$; $n = 18$

   11. $\dfrac{a}{4} = 3$; $a = 12$ 12. $\dfrac{x}{2} = 6$; $x = 12$ 13. $9y = 54$; $y = 6$

   14. $7x = 21$; $x = 3$ 15. $\dfrac{n}{3} = 8$; $n = 24$ 16. $\dfrac{a}{6} = 7$; $a = 42$

   17. $4x = 36$; $x = 9$ 18. $8x = 72$; $x = 9$ 19. $7y = 42$; $y = 6$

20. $9a = 81;$ $a = 9$ **21.** $\dfrac{x}{7} = 5;$ $x = 35$ **22.** $9 = \dfrac{a}{4};$ $36 = a$

23. $\dfrac{n}{8} = 8;$ $n = 64$ **24.** $9x = 54;$ $x = 6$ **25.** $15x = 150;$ $x = 10$

26. $\dfrac{m}{6} = 5;$ $m = 30$ **27.** $7a = 196;$ $a = 28$ **28.** $\dfrac{x}{12} = 10;$ $x = 120$

Page 455 • FOR PAGES 46–49

A

1. $2x + 4 = 6;$ $2x + 4 - 4 = 6 - 4;$ $2x = 2;$ $x = 1$

2. $4y - 3 = 9;$ $4y - 3 + 3 = 9 + 3;$ $4y = 12;$ $y = 3$

3. $3n - 7 = 11;$ $3n - 7 + 7 = 11 + 7;$ $3n = 18;$ $n = 6$

4. $5n - 3 = 22;$ $5n - 3 + 3 = 22 + 3;$ $5n = 25;$ $n = 5$

5. $5y + 5 = 15;$ $5y + 5 - 5 = 15 - 5;$ $5y = 10;$ $y = 2$

6. $6x + 18 = 36;$ $6x + 18 - 18 = 36 - 18;$ $6x = 18;$ $x = 3$

7. $4n + 5 = 45;$ $4n + 5 - 5 = 45 - 5;$ $4n = 40;$ $n = 10$

8. $2a - 6 = 20;$ $2a - 6 + 6 = 20 + 6;$ $2a = 26;$ $a = 13$

9. $3a + 6 = 36;$ $3a = 30;$ $a = 10$ **10.** $7n - 8 = 41;$ $7n = 49;$ $n = 7$

11. $3x - 8 = 13;$ $3x = 21;$ $x = 7$ **12.** $8x - 5 = 35;$ $8x = 40;$ $x = 5$

13. $7a + 14 = 21;$ $7a = 7;$ $a = 1$ **14.** $2m + 1 = 13;$ $2m = 12;$ $m = 6$

15. $4n + 3 = 51;$ $4n = 48;$ $n = 12$ **16.** $9n - 40 = 32;$ $9n = 72;$ $n = 8$

17. $6x - 12 = 30;$ $6x = 42;$ $x = 7$ **18.** $4a + 2 = 46;$ $4a = 44;$ $a = 11$

19. $7x + 12 = 75;$ $7x = 63;$ $x = 9$ **20.** $5b - 13 = 17;$ $5b = 30;$ $b = 6$

Page 455 • FOR PAGES 52–53

A

1. $2x + 3x = 15;$ $5x = 15;$ $x = 3$ **2.** $4a - 2a = 10;$ $2a = 10;$ $a = 5$

3. $6y + 3y = 18;$ $9y = 18;$ $y = 2$

4. $2x + 3x - 5 = 30;$ $5x - 5 = 30;$ $5x = 35;$ $x = 7$

5. $5c - 2c + 8 = 20;$ $3c + 8 = 20;$ $3c = 12;$ $c = 4$

6. $9a + a - 2 = 28;$ $10a - 2 = 28;$ $10a = 30;$ $a = 3$

7. $2n - 5 + 8n = 45;$ $10n - 5 = 45;$ $10n = 50;$ $n = 5$

8. $7x + 3 - 2x = 18;$ $5x + 3 = 18;$ $5x = 15;$ $x = 3$

9. $4n + 7 + 3n = 21;$ $7n + 7 = 21;$ $7n = 14;$ $n = 2$

10. $15n - 6 - 7n = 34;$ $8n - 6 = 34;$ $8n = 40;$ $n = 5$

11. $8y - 3 + y = 24;$ $9y - 3 = 24;$ $9y = 27;$ $y = 3$

12. $10a + 6 - 7a = 30;$ $3a + 6 = 30;$ $3a = 24;$ $a = 8$

13. $28 = y - 4 + 3y;$ $28 = 4y - 4;$ $32 = 4y;$ $8 = y$

14. $40 = 13x - 6x - 2;$ $40 = 7x - 2;$ $42 = 7x;$ $6 = x$

15. $56 = 7a - 16 + 2a;$ $56 = 9a - 16;$ $72 = 9a;$ $8 = a$

16. $46 = 2x + 5x - 3;$ $46 = 7x - 3;$ $49 = 7x;$ $7 = x$

17. $19 = 8a + 7 - 6a;$ $19 = 2a + 7;$ $12 = 2a;$ $6 = a$

18. $21 = 6x + 3x - 4 - 4x;$ $21 = 5x - 4;$ $25 = 5x;$ $5 = x$

Page 455 • FOR PAGES 58–61

A   **1.** Let n = the number. $3n + 5n = 64$; $8n = 64$; $n = 8$.   8

2. Let x = Cynthia's age. Then $x + 26$ = her father's age. $x + (x + 26) = 42$; $2x + 26 = 42$; $2x = 16$; $x = 8$.   8 years old

3. Let x = amount Brian has. Then $2x$ = amount Amy has. $x + 2x = 15$; $3x = 15$; $x = 5$, so $2x = 10$.   \$10

4. Let x = number of boys. Then $x - 4$ = number of girls. $x + (x - 4) = 90$; $2x - 4 = 90$; $2x = 94$; $x = 47$.   47 boys

Page 456 • FOR PAGES 62–63

A   **1.** $2n = 10 + n$; $2n - n = 10 + n - n$; $n = 10$

2. $4x = x + 15$; $4x - x = x + 15 - x$; $3x = 15$; $x = 5$

3. $8x = 2x + 6$; $8x - 2x = 2x + 6 - 2x$; $6x = 6$; $x = 1$

4. $16 - 2y = 6y$; $16 - 2y + 2y = 6y + 2y$; $16 = 8y$; $2 = y$

5. $5x - 20 = 3x$; $5x - 20 - 3x = 3x - 3x$; $2x - 20 = 0$; $2x = 20$; $x = 10$

6. $10x - 14 = 3x$; $10x - 14 - 3x = 3x - 3x$; $7x - 14 = 0$; $7x = 14$; $x = 2$

7. $7x - 12 = 4x$; $7x - 12 - 4x = 4x - 4x$; $3x - 12 = 0$; $3x = 12$; $x = 4$

8. $3y + 9 = 4y$; $3y + 9 - 3y = 4y - 3y$; $9 = y$

9. $14 - 2x = 5x$; $14 - 2x + 2x = 5x + 2x$; $14 = 7x$; $2 = x$

10. $6y - 2 = 4y + 8$; $6y - 2 - 4y = 4y + 8 - 4y$; $2y - 2 = 8$; $2y = 10$; $y = 5$

11. $3x + 1 = x + 7$; $3x + 1 - x = x + 7 - x$; $2x + 1 = 7$; $2x = 6$; $x = 3$

12. $5a - 4 = 8 - a$; $5a - 4 + a = 8 - a + a$; $6a - 4 = 8$; $6a = 12$; $a = 2$

13. $2n + 11 = 7 + 6n$; $2n + 11 - 2n = 7 + 6n - 2n$; $11 = 7 + 4n$; $4 = 4n$; $1 = n$

14. $5x - 8 = 10 + 2x$; $5x - 8 - 2x = 10 + 2x - 2x$; $3x - 8 = 10$; $3x = 18$; $x = 6$

15. $7c - 4 = 2c + 6$; $7c - 4 - 2c = 2c + 6 - 2c$; $5c - 4 = 6$; $5c = 10$; $c = 2$

Page 456 • FOR PAGES 64–65

A   **1.** $2(n - 4) = 16$; $2n - 8 = 16$; $2n = 24$; $n = 12$

2. $7(a - 3) = 4a$; $7a - 21 = 4a$; $3a - 21 = 0$; $3a = 21$; $a = 7$

3. $5(z + 2) = 10z$; $5z + 10 = 10z$; $10 = 5z$; $2 = z$

4. $6(y - 3) = 0$; $6y - 18 = 0$; $6y = 18$; $y = 3$

5. $8(x - 2) = 0$; $8x - 16 = 0$; $8x = 16$; $x = 2$

6. $5(m + 2) = 25$; $5m + 10 = 25$; $5m = 15$; $m = 3$

7. $4(x + 3) = 6x$; $4x + 12 = 6x$; $12 = 2x$; $6 = x$

8. $6(y + 3) = 18$; $6y + 18 = 18$; $6y = 0$; $y = 0$

9. $3(2 - x) = 3x$; $6 - 3x = 3x$; $6 = 6x$; $1 = x$

10. $4(x - 2) = 3x + 2$; $4x - 8 = 3x + 2$; $4x = 3x + 10$; $x = 10$

11. $3(x + 1) = 2x + 10$; $3x + 3 = 2x + 10$; $3x = 2x + 7$; $x = 7$

12. $2(5 - x) = 3x$; $10 - 2x = 3x$; $10 = 5x$; $2 = x$

B   **13.** $2(c + 3) = 3(c - 1)$; $2c + 6 = 3c - 3$; $2c + 9 = 3c$; $9 = c$

14. $4(r - 2) = 2(r + 1)$; $4r - 8 = 2r + 2$; $4r = 2r + 10$; $2r = 10$; $r = 5$

15. $4(x - 4) = 2(x - 1)$; $4x - 16 = 2x - 2$; $4x = 2x + 14$; $2x = 14$; $x = 7$

Page 456 · FOR PAGES 66–67

A   **1.** Let x = amount Bob has. Then $3x$ = amount Al has, and $x - 10$ = amount Ed has. $x + 3x + (x - 10) = 65$; $5x - 10 = 65$; $5x = 75$; $x = 15, 3x = 45$, and $x - 10 = 5$. Bob: \$15; Al: \$45; Ed: \$5

2. Let x = number of dimes May has. Then $2x$ = number of dimes Jo has, and $x + 4$ = number of dimes Sara has. $x + 2x + (x + 4) = 100$; $4x + 4 = 100$; $4x = 96$; $x = 24, 2x = 48$, and $x + 4 = 28$. May: 24 dimes; Jo: 48 dimes; Sara: 28 dimes

B   **3.** Let x = Nadine's age now. Then $x - 11$ = Rob's age now. $x + 1 = 2(x - 11 + 1)$; $x + 1 = 2(x - 10)$; $x + 1 = 2x - 20$; $x + 21 = 2x$; $21 = x$. Nadine is 21.

CHAPTER 3

Page 457 · FOR PAGES 78–82

A   **1.** >   **2.** >   **3.** >   **4.** <   **5.** >   **6.** >   **7.** >   **8.** <

9. $n + 1 = 4$; $n = 3$

10. $y - 3 = 7$; $y = 10$

11. $4 + x = 6$; $x = 2$

12. $a - 4 = 3$; $a = 7$

13. $x < 3$

14. $x > -1$

15. $y < -2$

16. $x > 3$

Page 457 · FOR PAGES 83–87

A   **1.** $2 + (-6) = -4$ **2.** $6 + (-2) = 4$ **3.** $-8 + 8 = 0$

4. $-3 + 5 = 2$ **5.** $-4 + (-1) = -5$ **6.** $-2 + (-8) = -10$

7. $3 + (-1) + (-8) = 2 + (-8) = -6$ **8.** $-4 + 8 + (-4) = 4 + (-4) = 0$

9. $5 + 0 + (-7) = 5 + (-7) = -2$ **10.** $4 - (-8) = 4 + (8) = 12$

11. $7 - 11 = 7 + (-11) = -4$ **12.** $-9 - (-2) = -9 + (2) = -7$

13. $-6 - 4 = -6 + (-4) = -10$ **14.** $18 - 4 = 18 + (-4) = 14$

15. $3 - (-9) = 3 + (9) = 12$ **16.** $-4 - 7 = -4 + (-7) = -11$

17. $8 - (-5) = 8 + (5) = 13$

18. $-3 - 10 = -3 + (-10) = -13$

19. $-2 - (-5) = -2 + (5) = 3$

20. $10 - 15 = 10 + (-15) = -5$

21. $6 - (-3) = 6 + (3) = 9$

Page 457 • FOR PAGES 89–90

A

1. $2n - 5n = (2 - 5)n = -3n$

2. $-a + 6a = (-1 + 6)a = 5a$

3. $-4b - 6b = (-4 - 6)b = -10b$

4. $x + 9x = 10x$

5. $-2y - 4y = -6y$

6. $x^2 - 3x^2 = -2x^2$

7. $-5a + 6a + 1 = a + 1$

8. $7x - (-x) + y = 7x + x + y = 8x + y$

9. $-4y + 3 + 4y = 3 + 0 = 3$

10. $15 - (-3x) + x = 15 + 3x + x = 15 + 4x$

11. $-7x - (-8x) + 2 = -7x + 8x + 2 = x + 2$

12. $6y + 2x - (-6y) = 6y + 2x + 6y = 12y + 2x$

13. $x^2 - y + 2x^2 = 3x^2 - y$

14. $8a - (-2a) + a^2 = 8a + 2a + a^2 = 10a + a^2$

15. $2n - n^2 - (-3n) = 2n - n^2 + 3n = 5n - n^2$

Page 458 • FOR PAGES 91–94

A

1. $12 \cdot 4 = 48$

2. $-3(6) = -18$

3. $-8(-7) = 56$

4. $3(-9) = -27$

5. $5^2 = 5 \cdot 5 = 25$

6. $(-5)^2 = (-5)(-5) = 25$

7. $-(5)^2 = -(5 \cdot 5) = -25$

8. $-(-4)^2 = -(-4 \cdot -4) = -16$

9. $3n(-7) = -21n$

10. $-2y(-8x) = (-2)(-8)(y \cdot x) = 16yx$

11. $2(2a - b) = (2 \cdot 2a) - (2 \cdot b) = 4a - 2b$

12. $-5(x + y) = (-5 \cdot x) + (-5 \cdot y) = -5x - 5y$

13. $-(m - 4) = -1(m - 4) = (-1 \cdot m) + (-1 \cdot -4) = -m + 4$

14. $-(8 - a) = -1(8 - a) = (-1 \cdot 8) + (-1 \cdot -a) = -8 + a$

15. $x - (x + 2) = x + (-1)(x + 2) = x + (-1 \cdot x) + (-1 \cdot 2) = x - x - 2 = -2$

16. $2y - (y - 3) = 2y + (-1)(y - 3) = 2y + (-1 \cdot y) + (-1 \cdot -3) = 2y - y + 3 = y + 3$

Page 458 • FOR PAGES 95–96

A

1. $72 \div -8 = -9$

2. $14 \div -14 = -1$

3. $-63 \div -9 = 7$

4. $-54 \div 6 = -9$

5. $56 \div -8 = -7$

6. $-24 \div -6 = 4$

7. $15 \div -5 = -3$

8. $-30 \div -3 = 10$

9. $\dfrac{40}{-5} = -8$

10. $\dfrac{-18}{-2} = 9$

11. $\dfrac{-16}{4} = -4$

12. $\dfrac{81}{9} = 9$

Page 458 • FOR PAGES 97–98

A   1. $n + 3 = 6$; $n + 3 - 3 = 6 - 3$; $n = 3$.   Check: $3 + 3 = 6$

     2. $n - 3 = 7$; $n - 3 + 3 = 7 + 3$; $n = 10$.   Check: $10 - 3 = 7$

     3. $x + 9 = 6$; $x + 9 - 9 = 6 - 9$; $x = -3$.   Check: $-3 + 9 = 6$

     4. $a + 4 = -4$; $a + 4 - 4 = -4 - 4$; $a = -8$.   Check: $-8 + 4 = -4$

     5. $b - 2 = -2$; $b - 2 + 2 = -2 + 2$; $b = 0$.   Check: $0 - 2 = -2$

     6. $15 = y - 12$; $15 + 12 = y - 12 + 12$; $27 = y$.   Check: $15 = 27 - 12$

     7. $3x = 15$; $x = 5$.   Check: $3 \cdot 5 = 15$

     8. $-3x = -15$; $x = 5$.   Check: $-3 \cdot 5 = -15$

     9. $6x = -24$; $x = -4$.   Check: $6(-4) = -24$

    10. $33 = -11y$; $-3 = y$.   Check: $33 = (-11)(-3)$

    11. $5n - n = -16$; $4n = -16$; $n = -4$.   Check: $5(-4) - (-4) = -20 + 4 = -16$

    12. $9y - 11y = 20$; $-2y = 20$; $y = -10$.   Check: $9(-10) - 11(-10) = -90 + 110 = 20$

    13. $5n = 50 - 45$; $5n = 5$; $n = 1$.   Check: $5 \cdot 1 = 5 = 50 - 45$

    14. $3n + 60 = 2n$; $60 = -n$; $-60 = n$.   Check: $3(-60) + 60 = 2(-60)$? $-180 + 60 = -120$? $-120 = -120$

    15. $-3n + 2 = -4n$; $2 = -n$; $-2 = n$.   Check: $-3(-2) + 2 = -4(-2)$? $6 + 2 = 8$? $8 = 8$

    16. $3x - 9 = 12x$; $-9 = 9x$; $-1 = x$.   Check: $3(-1) - 9 = 12(-1)$? $-3 - 9 = -12$? $-12 = -12$

    17. $5n - 4 = 3n + 8$; $5n = 3n + 12$; $2n = 12$; $n = 6$.   Check: $5 \cdot 6 - 4 = 3 \cdot 6 + 8$? $30 - 4 = 18 + 8$? $26 = 26$

    18. $14x - 4 = 13x$; $-4 = -x$; $4 = x$.   Check: $14 \cdot 4 - 4 = 13 \cdot 4$? $56 - 4 = 52$? $52 = 52$

    19. $-4x + 1 = -x + 19$; $-4x = -x + 18$; $-3x = 18$; $x = -6$.   Check: $-4(-6) + 1 = -(-6) + 19$? $24 + 1 = 6 + 19$? $25 = 25$

    20. $6x - 14 = 2x - 2$; $6x = 2x + 12$; $4x = 12$; $x = 3$.   Check: $6 \cdot 3 - 14 = 2 \cdot 3 - 2$? $18 - 14 = 6 - 2$? $4 = 4$

    21. $-7y - 3 = y + 29$; $-7y = y + 32$; $-8y = 32$; $y = -4$.   Check: $-7(-4) - 3 = -4 + 29$? $28 - 3 = 25$; $25 = 25$

B   22. $x - (4 - x) = 20$; $x - 4 + x = 20$; $2x - 4 = 20$; $2x = 24$; $x = 12$.   Check: $12 - (4 - 12) = 12 - (-8) = 12 + 8 = 20$

    23. $5n - (n + 3) = 25$; $5n - n - 3 = 25$; $4n - 3 = 25$; $4n = 28$; $n = 7$.   Check: $5 \cdot 7 - (7 + 3) = 35 - 10 = 25$

    24. $11x - (x - 20) = 100$; $11x - x + 20 = 100$; $10x + 20 = 100$; $10x = 80$; $x = 8$.   Check: $11 \cdot 8 - (8 - 20) = 88 - (-12) = 88 + 12 = 100$

25. $4 - (10 - x) = 34$; $4 - 10 + x = 34$; $-6 + x = 34$; $x = 40$. Check: $4 -$
$(10 - 40) = 4 - (-30) = 4 + 30 = 34$

26. $(w - 4) - 3 = 11$; $w - 7 = 11$; $w = 18$. Check: $(18 - 4) - 3 = 14 - 3 = 11$

27. $-10(v + 4) - 20 = 10v$; $-10v - 40 - 20 = 10v$; $-10v - 60 = 10v$; $-60 =$
$20v$; $-3 = v$. Check: $-10(-3 + 4) - 20 = 10(-3)$? $-10 \cdot 1 - 20 = -30$?
$-30 = -30$

CHAPTER 4

Page 459 • FOR PAGES 112–114

A   **1.** $P = 4s$; $P = 4 \cdot 7 = 28$; 28 cm

2. $P = 2l + 2w$; $P = 2 \cdot 9 + 2 \cdot 6 = 18 + 12 = 30$; 30 m

3. $P = 2l + 2w$; $P = 2 \cdot 2 + 2 \cdot 8 = 4 + 16 = 20$; 20 m

4. $P = 2l + 2w$; $P = 2 \cdot 9 + 2 \cdot 7 = 18 + 14 = 32$; 32 m

5. Length of each unlabeled side $= a - b$; $P = 2 \cdot a + 2(a - b) + 2 \cdot b =$
$2a + 2a - 2b + 2b = 4a$

6. $P = 3y + 2x + x + 3y + x + 2x = 6x + 6y$

7. $P = 5x + 2 \cdot 3x + 4 \cdot 2x + x = 5x + 6x + 8x + x = 20x$

8. Let x = width (in in.). Then $x + 4$ = length. $P = 2l + 2w$; $20 = 2(x + 4) +$
$2x$; $20 = 2x + 8 + 2x$; $20 = 4x + 8$; $12 = 4x$; $3 = x$ and $x + 4 = 7$.
width: 3 in.; length: 7 in.

9. $2x + x + 2x + 3 = 23$; $5x + 3 = 23$; $5x = 20$; $x = 4$

10. $C = 3.14d$; $C = 3.14 \cdot 4$; $C = 12.56$. 12.56 in.

Page 459 • FOR PAGES 115–118

A   **1.** $A = lw = 7 \cdot 4 = 28$ **2.** $A = \dfrac{1}{2}bh$

3. $A = (3x \cdot x) + (x \cdot x) = 3x^2 + x^2 = 4x^2$

4. Area $= 30 \cdot 35 = 1050$ (m²), so value $= \$12.50 \times 1050 = \$13,125$

5. Area $= 3 \cdot 4 = 12$ (m²), so cost $= \$3.95 \times 12 = \$47.40$.

Page 460 • FOR PAGES 119–122

A   **1.** $V = Bh$ (Base area $\times$ height)   **2.** $V = Bh$; $V = 20 \cdot 3 = 60$; 60 cm³

3. $B = 6 \cdot 6 = 36$, so $V = Bh = 36 \cdot 6 = 216$; 216 cm³

4. $V = 6x \cdot 4x \cdot 2x = 48x^3$; $48x^3$ cubic units

Page 460 • FOR PAGES 124–133

A   **1.** hK **2.** $7x$ dollars

3.

	rate	time	Distance
Bike	20	$x + 3$	$20(x + 3)$
Ride	80	x	$80x$

$20(x + 3) = 80x$; $20x + 60 = 80x$;

$60 = 60x$; $1 = x$, so $80x = 80$. 80 km

4.

	pay/hour	number of hours	pay/week
Di	8	$x + 3$	$8(x + 3)$
Jo	8	x	$8x$
			$664

$8(x + 3) + 8x = 664$; $8x + 24 + 8x = 664$; $16x + 24 = 664$; $16x = 640$; $x = 40$. 40 hours

Page 460 • FOR PAGES 134–137

A 1. $y + x = n$; $y + x - y = n - y$; $x = n - y$

2. $2x - y = 4$; $2x = 4 + y$; $x = \dfrac{4 + y}{2}$

3. $4y - x = 12$; $-x = 12 - 4y$; $(-1)(-x) = (-1)(12 - 4y)$; $x = -12 + 4y$

4. $4x = y$; $x = \dfrac{y}{4}$ 5. $nx = y$; $x = \dfrac{y}{n}$

6. $kx = y$; $x = \dfrac{y}{k}$ 7. $10y + x = 15$; $x = 15 - 10y$

8. $3y - x = 9$; $-x = 9 - 3y$; $x = -9 + 3y$

9. $3x - y = 0$; $3x = y$; $x = \dfrac{y}{3}$

10. $5x + 5 = 15y$; $5x = 15y - 5$; $x = \dfrac{15y - 5}{5}$, or $3y - 1$

11. $x - y = k$; $x = k + y$

12. $-x + y = -2$; $-x = -2 - y$; $x = 2 + y$

CHAPTER 5

Page 461 • FOR PAGES 152–155

A 1. $n + 3$
$\underline{3n + 1}$
$4n + 4$

2. $7y^2 + 3$
$\underline{8y^2 + 9}$
$15y^2 + 12$

3. $a^2 + b^2$
$\underline{a^2 - b^2}$
$2a^2$

4. $3x + 7y$
$\underline{-5x + 8y}$
$-2x + 15y$

5. $-4xy - z$
$\underline{4xy + 6z}$
$5z$

6. $a^2 + ab$
$\underline{3a^2 + ab}$
$4a^2 + 2ab$

7. $4x - 6y$
$\underline{-7x + 4y}$
$-3x - 2y$

8. $9x^2 + 4y^2$
$\underline{6x^2 - 8y^2}$
$15x^2 - 4y^2$

9. $5a - b$
$\underline{8a + 4b}$
$13a + 3b$

10.
$$\begin{array}{r} x + 2y + 9 \\ -3x - y + 7 \\ \hline -2x + y + 16 \end{array}$$

11.
$$\begin{array}{r} x^2 - 3x + 4 \\ 2x^2 + 3x - 4 \\ \hline 3x^2 \end{array}$$

12.
$$\begin{array}{r} k^2 \quad\;\; + 1 \\ -3k^2 + 2k - 1 \\ \hline -2k^2 + 2k \end{array}$$

13.
$$\begin{array}{r} 3a - b \\ -2a + b \\ \hline a \end{array}$$

14.
$$\begin{array}{r} -6n + 5 \\ 7n - 2 \\ \hline n + 3 \end{array}$$

15.
$$\begin{array}{r} y^3 + 4 \\ -y^3 + 4 \\ \hline 8 \end{array}$$

16.
$$\begin{array}{r} 3x + y \\ -7x + y \\ \hline -4x + 2y \end{array}$$

17.
$$\begin{array}{r} 2n^2 + 2n \\ -4n^2 - 2n \\ \hline -2n^2 \end{array}$$

18.
$$\begin{array}{r} 5x - 3 \\ -3x - 4 \\ \hline 2x - 7 \end{array}$$

19.
$$\begin{array}{r} 3a - 7 \\ -7a - 1 \\ \hline -4a - 8 \end{array}$$

20.
$$\begin{array}{r} 6x^2y^2 + 3 \\ -x^2y^2 + 3 \\ \hline 5x^2y^2 + 6 \end{array}$$

21.
$$\begin{array}{r} 2x + 1 \\ -x + 4 \\ \hline x + 5 \end{array}$$

22.
$$\begin{array}{r} 3x - 6y \\ -2x + 5y \\ \hline x - y \end{array}$$

23.
$$\begin{array}{r} 6a + 2b \\ -a + b \\ \hline 5a + 3b \end{array}$$

24.
$$\begin{array}{r} 7m - 2n \\ -m - 3n \\ \hline 6m - 5n \end{array}$$

B 25. $2n + (n + 1) + (2n + 4) = 5n + 5$; $(5n + 5)$ km

26. $(5n + 4) - (4n - 1) = 5n + 4 - 4n + 1 = n + 5$; $(n + 5)$ km

Pages 461–462 • FOR PAGES 156–159

A 1. $y \cdot y^3 = y^{1+3} = y^4$ 2. $(3x^2)(7x^2) = (3 \cdot 7)x^{2+2} = 21x^4$

3. $(-2n)(-n^2) = (-2 \cdot -1)n^{1+2} = 2n^3$ 4. $(-4x^2)(7x^3) = (-4 \cdot 7)x^{2+3} = -28x^5$

5. $x(-xy) = -x^{1+1}y = -x^2y$

6. $(4x^2y)(3xy^2) = (4 \cdot 3)x^{2+1}y^{1+2} = 12x^3y^3$ 7. $n^5 \cdot n^2 = n^{5+2} = n^7$

8. $x^4 \cdot x^3 = x^{4+3} = x^7$ 9. $(n^3)^4 = n^{3 \cdot 4} = n^{12}$

10. $(x^2)^5 = x^{2 \cdot 5} = x^{10}$ 11. $(y^5)^2 = y^{5 \cdot 2} = y^{10}$

12. $(x^4)^4 = x^{4 \cdot 4} = x^{16}$ 13. $(ab)^2 = a^2b^2$

14. $(xy)^2 = x^2y^2$ 15. $(mn)^4 = m^4n^4$

16. $(2a)^2 = 2^2a^2 = 4a^2$ 17. $(-xy)^3 = (-1)^3x^3y^3 = -x^3y^3$

18. $(3a^2)^2 = 3^2(a^2)^2 = 9a^4$ 19. $(-3x^2)^3 = (-3)^3(x^2)^3 = -27x^6$

20. $5(-ab^2)^5 = 5(-1)^5(a)^5(b^2)^5 = -5a^5b^{10}$

B 21. $-(ab^2)(ab)^2 = -(ab^2)(a^2b^2) = -a^{1+2}b^{2+2} = -a^3b^4$

22. $(-xy)^2(-xy) = (-1)^2(x^2y^2)(-xy) = -x^{2+1}y^{2+1} = -x^3y^3$

23. $(4rs)^3(2s)^3 = (64r^3s^3)(8s^3) = 512r^3s^6$

24. $(3mn)^3(-3mn) = (27m^3n^3)(-3mn) = -81m^4n^4$

25. $(2x^2y^2)(2x^3y^3)^2 = (2x^2y^2)(4x^6y^6) = 8x^8y^8$

26. $-(2ab)^4(4a^4b^4) = -(16a^4b^4)(4a^4b^4) = -64a^8b^8$

Page 462 • FOR PAGES 160–163

A 1. $3(n + 5) = (3 \cdot n) + (3 \cdot 5) = 3n + 15$

2. $-4(x - y) = (-4 \cdot x) - (-4 \cdot y) = -4x + 4y$

3. $6(a^2 + b^2) = (6 \cdot a^2) + (6 \cdot b^2) = 6a^2 + 6b^2$

4. $-2(a - 3b) = (-2 \cdot a) - (-2 \cdot 3b) = -2a + 6b$

5. $a(x - y) = ax - ay$

6. $-3x(2x + 3y) = (-3x \cdot 2x) + (-3x \cdot 3y) = -6x^2 - 9xy$

7. $-3(2a + 4b + 6c) = (-3 \cdot 2a) + (-3 \cdot 4b) + (-3 \cdot 6c) = -6a - 12b - 18c$

8. $-1(5x + 2y + z) = (-1 \cdot 5x) + (-1 \cdot 2y) + (-1 \cdot z) = -5x - 2y - z$

9. $4x(x^2y + xy^2 + 3y^3) = (4x \cdot x^2y) + (4x \cdot xy^2) + (4x \cdot 3y^3) = 4x^3y + 4x^2y^2 + 12xy^3$

10. $5a(a^3 + a^2 - a + 1) = 5a^4 + 5a^3 - 5a^2 + 5a$

11. $-x(3x^3 - 5x^2 + x - 1) = -3x^4 + 5x^3 - x^2 + x$

12. $m^2(m^2 + 2mn + n) = m^4 + 2m^3n + m^2n$

13.
$$\begin{array}{r} 4 - a \\ 5 - a \\ \hline 20 - 5a \\ -4a + a^2 \\ \hline 20 - 9a + a^2 \end{array}$$

14.
$$\begin{array}{r} n - 3 \\ n + 5 \\ \hline n^2 - 3n \\ + 5n - 15 \\ \hline n^2 + 2n - 15 \end{array}$$

15.
$$\begin{array}{r} x + 8 \\ x + 4 \\ \hline x^2 + 8x \\ + 4x + 32 \\ \hline x^2 + 12x + 32 \end{array}$$

16.
$$\begin{array}{r} a + 1 \\ 3a - 1 \\ \hline 3a^2 + 3a \\ - a - 1 \\ \hline 3a^2 + 2a - 1 \end{array}$$

17.
$$\begin{array}{r} 5x - 4 \\ 5x - 2 \\ \hline 25x^2 - 20x \\ - 10x + 8 \\ \hline 25x^2 - 30x + 8 \end{array}$$

18.
$$\begin{array}{r} n - 9 \\ n + 6 \\ \hline n^2 - 9n \\ + 6n - 54 \\ \hline n^2 - 3n - 54 \end{array}$$

B

19.
$$\begin{array}{r} a^2 + 2a + 1 \\ a - 1 \\ \hline a^3 + 2a^2 + a \\ - a^2 - 2a - 1 \\ \hline a^3 + a^2 - a - 1 \end{array}$$

20.
$$\begin{array}{r} x^2 - 2xy - y^2 \\ x - y \\ \hline x^3 - 2x^2y - xy^2 \\ - x^2y + 2xy^2 + y^3 \\ \hline x^3 - 3x^2y + xy^2 + y^3 \end{array}$$

21.
$$\begin{array}{r} m^2 - n^2 \\ m - n \\ \hline m^3 - mn^2 \\ - m^2n + n^3 \\ \hline m^3 - mn^2 - m^2n + n^3 \end{array}$$

22.
$$\begin{array}{r} x^2 + 2xy + y^2 \\ x + y \\ \hline x^3 + 2x^2y + xy^2 \\ + x^2y + 2xy^2 + y^3 \\ \hline x^3 + 3x^2y + 3xy^2 + y^3 \end{array}$$

23.
$$\begin{array}{r} a^2 + 1 \\ a^2 - 1 \\ \hline a^4 + a^2 \\ - a^2 - 1 \\ \hline a^4 - 1 \end{array}$$

24.
$$\begin{array}{r} n^2 + 4n + 4 \\ n + 2 \\ \hline n^3 + 4n^2 + 4n \\ + 2n^2 + 8n + 8 \\ \hline n^3 + 6n^2 + 12n + 8 \end{array}$$

Pages 462–463 · FOR PAGES 164–169

A

1. $(n + 3)(n + 4) = n^2 + 4n + 3n + 12 = n^2 + 7n + 12$

2. $(x + 6)(x - 3) = x^2 - 3x + 6x - 18 = x^2 + 3x - 18$

3. $(m - 5)(m - 2) = m^2 - 2m - 5m + 10 = m^2 - 7m + 10$

4. $(3x - 2)(2x + 3) = 6x^2 + 9x - 4x - 6 = 6x^2 + 5x - 6$

5. $(5y - 1)(5y + 1) = 25y^2 + 5y - 5y - 1 = 25y^2 - 1$

6. $(7a + 3)(3a - 1) = 21a^2 - 7a + 9a - 3 = 21a^2 + 2a - 3$

7. $(2x + y)(4x - y) = 8x^2 - 2xy + 4xy - y^2 = 8x^2 + 2xy - y^2$

8. $(2a + b)(a - 2b) = 2a^2 - 4ab + ab - 2b^2 = 2a^2 - 3ab - 2b^2$

9. $(5x - y)(3x - y) = 15x^2 - 5xy - 3xy + y^2 = 15x^2 - 8xy + y^2$

10. $A = (3n + 1)(2n + 5) = (6n^2 + 17n + 5) \text{ cm}^2$

11. $(n + 2)^2 = n^2 + 2 \cdot 2n + 2^2 = n^2 + 4n + 4$

12. $(y - 5)^2 = y^2 - 2 \cdot 5y + (-5)^2 = y^2 - 10y + 25$

13. $(n + 7)^2 = n^2 + 2 \cdot 7n + 7^2 = n^2 + 14n + 49$

14. $(x - 10)^2 = x^2 - 2 \cdot 10x + (-10)^2 = x^2 - 20x + 100$

15. $(x + 3y)^2 = x^2 + 6xy + 9y^2$ **16.** $(a - 2b)^2 = a^2 - 4ab + 4b^2$

17. $(a - 10b)^2 = a^2 - 20ab + 100b^2$ **18.** $(3m + 2n)^2 = 9m^2 + 12mn + 4n^2$

19. $-1(n + 2)^2 = -1(n^2 + 4n + 4) = -n^2 - 4n - 4$

20. $-2(4x + y)^2 = -2(16x^2 + 8xy + y^2) = -32x^2 - 16xy - 2y^2$

21. $4(2r + s)^2 = 4(4r^2 + 4rs + s^2) = 16r^2 + 16rs + 4s^2$

22. $-3(2x - 3y)^2 = -3(4x^2 - 12xy + 9y^2) = -12x^2 + 36xy - 27y^2$

B   **23.** $A = \frac{1}{2}bh = \frac{1}{2}(8n + 1)(4n) = \frac{1}{2}(32n^2 + 4n) = 16n^2 + 2n$

      24. $A = lw = (3n + 1)(6n) = 18n^2 + 6n$

Page 463 • FOR PAGES 170–171, 174–175

A   **1.** $\frac{3x}{x} = \frac{3 \cdot x}{x} = 3$ **2.** $\frac{a^8}{a^2} = \frac{a \cdot a \cdot a \cdot a \cdot a \cdot a \cdot a \cdot a}{a \cdot a} = a^6$

     3. $\frac{5x}{5} = \frac{5 \cdot x}{5} = x$ **4.** $\frac{4ab}{4b} = \frac{4 \cdot a \cdot b}{4 \cdot b} = a$

     5. $\frac{6x^3}{2x} = \frac{3 \cdot 2 \cdot x \cdot x \cdot x}{2 \cdot x} = 3x^2$

     6. $\frac{-12a^5}{2a^3} = \frac{-6 \cdot 2 \cdot a \cdot a \cdot a \cdot a \cdot a}{2 \cdot a \cdot a \cdot a} = -6a^2$

     7. $\frac{63n^4}{-7n^2} = \frac{9 \cdot 7 \cdot n \cdot n \cdot n \cdot n}{-1 \cdot 7 \cdot n \cdot n} = \frac{9n^2}{-1} = -9n^2$

     8. $\frac{-49a^2b^2}{-7ab} = \frac{-7 \cdot 7 \cdot a \cdot a \cdot b \cdot b}{-7 \cdot a \cdot b} = 7ab$ **9.** $\frac{4x - 8y}{2} = \frac{4x}{2} - \frac{8y}{2} = 2x - 4y$

     10. $-\frac{6a - 6b}{6} = -\left(\frac{6a}{6} - \frac{6b}{6}\right) = -(a - b) = -a + b$

     11. $\frac{x^2y^2 - 3xy}{xy} = \frac{x^2y^2}{xy} - \frac{3xy}{xy} = xy - 3$

     12. $\frac{15m + 30mn}{5m} = \frac{15m}{5m} + \frac{30mn}{5m} = 3 + 6n$

     13. $\frac{20n^3 - 10n^2 - 5n}{5n} = \frac{20n^3}{5n} - \frac{10n^2}{5n} - \frac{5n}{5n} = 4n^2 - 2n - 1$

     14. $\frac{x^3 + 3x^4 - 5x^5}{-x^2} = \frac{x^3}{-x^2} + \frac{3x^4}{-x^2} - \frac{5x^5}{-x^2} = -x - 3x^2 + 5x^3$

15. $\dfrac{18a^3b^3 - 12a^2b^2 + 6ab}{6ab} = \dfrac{18a^3b^3}{6ab} - \dfrac{12a^2b^2}{6ab} + \dfrac{6ab}{6ab} = 3a^2b^2 - 2ab + 1$

16. $\dfrac{8x^2y + 12xy^2 - 16y^3}{4y} = \dfrac{8x^2y}{4y} + \dfrac{12xy^2}{4y} - \dfrac{16y^3}{4y} = 2x^2 + 3xy - 4y^2$

17. $\dfrac{a^3 - 6a^5 + 8a^7}{-a^3} = \dfrac{a^3}{-a^3} - \dfrac{6a^5}{-a^3} + \dfrac{8a^7}{-a^3} = -1 + 6a^2 - 8a^4$

18. $\dfrac{11x^4y^5 + 9x^3y^4 + 3x^3y^3}{x^2y^2} = \dfrac{11x^4y^5}{x^2y^2} + \dfrac{9x^3y^4}{x^2y^2} + \dfrac{3x^3y^3}{x^2y^2} = 11x^2y^3 + 9xy^2 + 3xy$

B **19.** $\dfrac{12xy}{4x} - \dfrac{10y^3}{5y^2} = 3y - 2y = y$ **20.** $\dfrac{-21ab}{7b} + \dfrac{4a^2}{a} = -3a + 4a = a$

21. $\dfrac{3x^3y^5}{xy^3} - \dfrac{x^4y^5}{x^2y^3} = 3x^2y^2 - x^2y^2 = 2x^2y^2$

22. $\dfrac{9x^4y^4 - 12x^3y^3 + 21x^2y^2 - 3xy}{-3xy} = -3x^3y^3 + 4x^2y^2 - 7xy + 1$

23. $\dfrac{8a^5b^5 + 6a^4b^6 - a^3b^7 + 3a^2b^8}{a^2b^3} = 8a^3b^2 + 6a^2b^3 - ab^4 + 3b^5$

CHAPTER 6

Page 464 · FOR PAGES 186–189

A **1.** $20 = 2 \cdot 2 \cdot 5$ **2.** $40 = 2 \cdot 2 \cdot 2 \cdot 5$ **3.** $250 = 2 \cdot 5 \cdot 5 \cdot 5$

4. $110 = 2 \cdot 5 \cdot 11$ **5.** $140 = 2 \cdot 2 \cdot 5 \cdot 7$ **6.** $60 = 2 \cdot 2 \cdot 3 \cdot 5$

7. $150 = 2 \cdot 3 \cdot 5 \cdot 5$ **8.** $200 = 2 \cdot 2 \cdot 2 \cdot 5 \cdot 5$ **9.** $180 = 2 \cdot 2 \cdot 3 \cdot 3 \cdot 5$

10. $630 = 2 \cdot 3 \cdot 3 \cdot 5 \cdot 7$

11. $10 = 2 \cdot 5$ and $25 = 5 \cdot 5$; g.c.f. $= 5$

12. $32 = 2 \cdot 2 \cdot 2 \cdot 2 \cdot 2$ and $28 = 2 \cdot 2 \cdot 7$; g.c.f. $= 2 \cdot 2 = 4$

13. $26 = 2 \cdot 13$ and $39 = 3 \cdot 13$; g.c.f. $= 13$

14. $63 = 3 \cdot 3 \cdot 7$ and $14 = 2 \cdot 7$; g.c.f. $= 7$

15. $15 = 3 \cdot 5$ and $90 = 2 \cdot 3 \cdot 3 \cdot 5$; g.c.f. $= 3 \cdot 5 = 15$

16. $40 = 2 \cdot 2 \cdot 2 \cdot 5$ and $48 = 2 \cdot 2 \cdot 2 \cdot 2 \cdot 3$; g.c.f. $= 2 \cdot 2 \cdot 2 = 8$

17. $70 = 2 \cdot 5 \cdot 7$ and $105 = 3 \cdot 5 \cdot 7$; g.c.f. $= 5 \cdot 7 = 35$

18. $55 = 5 \cdot 11$ and $242 = 2 \cdot 11 \cdot 11$; g.c.f. $= 11$

19. $14 - 2x = (2 \cdot 7) - (2 \cdot x) = 2(7 - x)$

20. $20n - 4 = (4 \cdot 5n) - (4 \cdot 1) = 4(5n - 1)$

21. $12 - 36x = (12 \cdot 1) - (12 \cdot 3x) = 12(1 - 3x)$

22. $2x^2 + 6x = 2x(x + 3)$ **23.** $y^2 - 3y = y(y - 3)$

24. $14x^2 + 7x = 7x(2x + 1)$ **25.** $10xy - y^2 = y(10x - y)$

26. $11n^2 - 44n = 11n(n - 4)$ **27.** $6ab^2 + 24a^2b = 6ab(b + 4a)$

28. $3n^2 - 9n + 12 = 3(n^2 - 3n + 4)$

29. $4x^2 + 12xy + 24y^2 = 4(x^2 + 3xy + 6y^2)$

30. $7a + 14b + 70c = 7(a + 2b + 10c)$

B **31.** $5 - 25n + 75n^2 + 125n^3 = 5(1 - 5n + 15n^2 + 25n^3)$

32. $6xy + 42x^2y^2 - 66x^3y^3 + 72x^4y^4 = 6xy(1 + 7xy - 11x^2y^2 + 12x^3y^3)$

33. $a^4b - a^3b^2 + a^2b^3 - ab^4 = ab(a^3 - a^2b + ab^2 - b^3)$

34. $12a^2b - 36ab^2 + 48b^2 = 12b(a^2 - 3ab + 4b)$

35. $100x^4y^4 - 75x^3y^3 + 50x^2y^2 + 25xy = 25xy(4x^3y^3 - 3x^2y^2 + 2xy + 1)$

36. $15m^2n^4 + 45m^3n^5 - 60m^4n^6 = 15m^2n^4(1 + 3mn - 4m^2n^2)$

Pages 464–465 · FOR PAGES 192–197

A

1. $y^2 + 7y + 12 = (y + 3)(y + 4)$

2. $n^2 + 10n + 24 = (n + 4)(n + 6)$

3. $n^2 + 16n + 60 = (n + 6)(n + 10)$

4. $x^2 + 14x + 33 = (x + 3)(x + 11)$

5. $x^2 + 13x + 36 = (x + 4)(x + 9)$

6. $n^2 + 16n + 55 = (n + 5)(n + 11)$

7. $b^2 + 19b + 90 = (b + 9)(b + 10)$

8. $m^2 + 15m + 36 = (m + 3)(m + 12)$

9. $n^2 + 13n + 12 = (n + 1)(n + 12)$

10. $a^2 + 14a + 24 = (a + 2)(a + 12)$

11. $n^2 + 19n + 34 = (n + 2)(n + 17)$

12. $m^2 + 17m + 16 = (m + 1)(m + 16)$

13. $x^2 + 14x + 13 = (x + 1)(x + 13)$

14. $y^2 + 10y + 16 = (y + 2)(y + 8)$

15. $x^2 + 16x + 28 = (x + 2)(x + 14)$

16. $b^2 - 10b + 16 = (b - 2)(b - 8)$

17. $m^2 - 9m + 20 = (m - 4)(m - 5)$

18. $n^2 - 8n + 15 = (n - 3)(n - 5)$

19. $y^2 - 14y + 45 = (y - 5)(y - 9)$

20. $n^2 - 10n + 9 = (n - 1)(n - 9)$

21. $y^2 - 17y + 60 = (y - 5)(y - 12)$

22. $x^2 - 19x + 90 = (x - 9)(x - 10)$

23. $n^2 - 18n + 80 = (n - 8)(n - 10)$

24. $y^2 - 15y + 44 = (y - 4)(y - 11)$

B

25. $x^2 + 42x + 80 = (x + 2)(x + 40)$

26. $x^2 + 29x + 100 = (x + 4)(x + 25)$

27. $x^2 - 101x + 100 = (x - 1)(x - 100)$

28. $y^2 - 34y + 93 = (y - 3)(y - 31)$

29. $m^2 - 32m + 60 = (m - 2)(m - 30)$

30. $y^2 - 22y + 72 = (y - 4)(y - 18)$

31. $x^2 + 20x + 64 = (x + 4)(x + 16)$

32. $n^2 - 40n + 144 = (n - 4)(n - 36)$

33. $n^2 + 21n + 68 = (n + 4)(n + 17)$

Page 465 · FOR PAGES 198–203

A

1. $y^2 - 6y + 9 = (y - 3)^2$

2. $x^2 + 16x + 64 = (x + 8)^2$

3. $n^2 - 14n + 49 = (n - 7)^2$

4. $36 - 12x + x^2 = (6 - x)^2$

5. $n^2 + 20n + 100 = (n + 10)^2$

6. $81 - 18x + x^2 = (9 - x)^2$

7. $n^2 + 5n + 4 = (n + 1)(n + 4)$

8. $x^2 - 13x + 42 = (x - 6)(x - 7)$

9. $y^2 - 22y + 121 = (y - 11)^2$

10. $25 - 10x + x^2 = (5 - x)^2$

11. $m^2n^2 - 4mn + 4 = (mn - 2)^2$

12. $b^2 - 10b + 25 = (b - 5)^2$

13. $y^2 - 7y - 8 = (y + 1)(y - 8)$

14. $n^2 - 2n - 15 = (n + 3)(n - 5)$

15. $y^2 + 6y - 40 = (y - 4)(y + 10)$

16. $n^2 - 8n - 20 = (n + 2)(n - 10)$

17. $m^2 - 5m - 50 = (m + 5)(m - 10)$

18. $x^2 - 7x - 60 = (x + 5)(x - 12)$

19. $n^2 - 8n - 9 = (n + 1)(n - 9)$

20. $n^2 + 13n - 14 = (n - 1)(n + 14)$

21. $n^2 - 3n - 4 = (n + 1)(n - 4)$

Pages 465–466 · FOR PAGES 204–207

A
1. $(x + 4)(x - 4) = x^2 - 16$
2. $(m - n)(m + n) = m^2 - n^2$
3. $(1 - 5x)(1 + 5x) = 1 - 25x^2$
4. $(3y + 1)(3y - 1) = 9y^2 - 1$
5. $(2x - y)(2x + y) = 4x^2 - y^2$
6. $(a + 2b)(a - 2b) = a^2 - 4b^2$
7. $(5x + 2y)(5x - 2y) = 25x^2 - 4y^2$
8. $(8m - n)(8m + n) = 64m^2 - n^2$
9. $(4n + 3)(4n - 3) = 16n^2 - 9$
10. $(1 - 6a)(1 + 6a) = 1 - 36a^2$
11. $(9 - 2b)(9 + 2b) = 81 - 4b^2$
12. $(8x - 3)(8x + 3) = 64x^2 - 9$
13. $x^2 - 25 = (x + 5)(x - 5)$
14. $n^2 - 36 = (n + 6)(n - 6)$
15. $x^2 - 100 = (x + 10)(x - 10)$
16. $4a^2 - 9b^2 = (2a + 3b)(2a - 3b)$
17. $9x^2 - 36y^2 = 9(x^2 - 4y^2) = 9(x + 2y)(x - 2y)$
18. $m^2 - 49 = (m + 7)(m - 7)$
19. $16a^2 - 1 = (4a + 1)(4a - 1)$
20. $64 - 4a^2 = 4(16 - a^2) = 4(4 + a)(4 - a)$
21. $16s^2 - 4t^2 = 4(4s^2 - t^2) = 4(2s + t)(2s - t)$
22. $25a^2 - b^2 = (5a + b)(5a - b)$
23. $121 - 9x^2 = (11 + 3x)(11 - 3x)$
24. $9b^2 - 4c^2 = (3b + 2c)(3b - 2c)$

B
25. $(n^2 + 1)(n^2 - 1) = n^4 - 1$
26. $(x^2 - 5)(x^2 + 5) = x^4 - 25$
27. $(y^2 + 8)(y^2 - 8) = y^4 - 64$
28. $-1 + x^2 = x^2 - 1 = (x + 1)(x - 1)$
29. $-9 + 16y^2 = 16y^2 - 9 = (4y + 3)(4y - 3)$
30. $-b^2 + a^2 = a^2 - b^2 = (a + b)(a - b)$
31. $-64 + a^2 = a^2 - 64 = (a + 8)(a - 8)$
32. $-4m^2 + k^2 = k^2 - 4m^2 = (k + 2m)(k - 2m)$
33. $-16b^2 + 81a^2 = 81a^2 - 16b^2 = (9a + 4b)(9a - 4b)$
34. $m^2 - 144 = (m + 12)(m - 12)$
35. $n^2 - 625 = (n + 25)(n - 25)$
36. $a^2 - 169 = (a + 13)(a - 13)$
37. $400 - r^2 = (20 + r)(20 - r)$
38. $x^2y^2 - 16 = (xy + 4)(xy - 4)$
39. $m^2n^2 - 25p^2 = (mn + 5p)(mn - 5p)$

Page 466 · FOR PAGES 208–209

A
1. $2x^2 - 50 = 2(x^2 - 25) = 2(x + 5)(x - 5)$
2. $3y^2 - 12y + 12 = 3(y^2 - 4y + 4) = 3(y - 2)^2$
3. $4n^2 + 20n + 16 = 4(n^2 + 5n + 4) = 4(n + 1)(n + 4)$
4. $8 + 4n - 4n^2 = 4(2 + n - n^2) = 4(2 - n)(1 + n)$
5. $4n^2 + 12n - 16 = 4(n^2 + 3n - 4) = 4(n - 1)(n + 4)$
6. $4x^2 - 64 = 4(x^2 - 16) = 4(x + 4)(x - 4)$
7. $10x^2 - 40 = 10(x^2 - 4) = 10(x + 2)(x - 2)$
8. $a^3 - 16a = a(a^2 - 16) = a(a + 4)(a - 4)$
9. $3n^2 - 9n - 12 = 3(n^2 - 3n - 4) = 3(n + 1)(n - 4)$
10. $5n^2 - 40n - 100 = 5(n^2 - 8n - 20) = 5(n + 2)(n - 10)$

11. $12x^2 + 36x + 24 = 12(x^2 + 3x + 2) = 12(x + 1)(x + 2)$

12. $3x^2 - 24x + 45 = 3(x^2 - 8x + 15) = 3(x - 3)(x - 5)$

13. $3n^2 + 12n + 12 = 3(n^2 + 4n + 4) = 3(n + 2)^2$

14. $4n^2 + 28n + 40 = 4(n^2 + 7n + 10) = 4(n + 2)(n + 5)$

15. $5a^2 - 5b^2 = 5(a^2 - b^2) = 5(a + b)(a - b)$

16. $12x^2 + 36x + 24 = 12(x^2 + 3x + 2) = 12(x + 1)(x + 2)$

17. $4x^2 - 24x + 36 = 4(x^2 - 6x + 9) = 4(x - 3)^2$

18. $7x^2 + 42x + 63 = 7(x^2 + 6x + 9) = 7(x + 3)^2$

19. $4n^2 - 8n - 96 = 4(n^2 - 2n - 24) = 4(n + 4)(n - 6)$

20. $4x^2 - 36 = 4(x^2 - 9) = 4(x + 3)(x - 3)$

21. $x - 3x^2 = x(1 - 3x)$

22. $7x^2 - 56x + 49 = 7(x^2 - 8x + 7) = 7(x - 7)(x - 1)$

23. $x^2y^2 - y^3 = y^2(x^2 - y)$ 24. $4x^2 + 48 = 4(x^2 + 12)$

B   25. $10a^2 - 10ab - 60b^2 = 10(a^2 - ab - 6b^2) = 10(a + 2b)(a - 3b)$

26. $3x^2 - 6x - 189 = 3(x^2 - 2x - 63) = 3(x + 7)(x - 9)$

27. $-4b^2 + 9a^2 = 9a^2 - 4b^2 = (3a + 2b)(3a - 2b)$

28. $-6 - 6x^4 = -6(1 + x^4)$

29. $4r^2 - 24rs + 32s^2 = 4(r^2 - 6rs + 8s^2) = 4(r - 2s)(r - 4s)$

30. $9b^2 + 36b - 45 = 9(b^2 + 4b - 5) = 9(b - 1)(b + 5)$

31. $5x^2 - 55x - 300 = 5(x^2 - 11x - 60) = 5(x + 4)(x - 15)$

32. $24n^2 + 216n - 864 = 24(n^2 + 9n - 36) = 24(n - 3)(n + 12)$

33. $x^4 - y^4 = (x^2 + y^2)(x^2 - y^2) = (x^2 + y^2)(x + y)(x - y)$

CHAPTER 7

Page 467 · FOR PAGES 224–227

A   1.

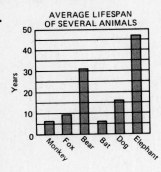

AVERAGE LIFESPAN OF SEVERAL ANIMALS

2. **a.** 23 years

   b. about 7 times

Page 467 · FOR PAGES 233–236

A 1.–4.

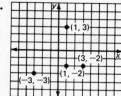

B 5.

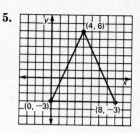

Pages 467–468 · FOR PAGES 237–240

A **1.** $y = x + 4$ **2.** $y = -x - 1$ **3.** $y = 3x + 2$ **4.** $y = -2x + 3$

x	y
0	4
1	5
−1	3
2	6

x	y
0	−1
−1	0
2	−3
3	−4

x	y
0	2
1	5
3	11
4	14

x	y
0	3
1	1
−1	5
2	−1

5. $x + y = 7$; $x + y - x = 7 - x$; $y = 7 - x$

6. $x + y = -7$; $x + y - x = -7 - x$; $y = -7 - x$

7. $y - x = 3$; $y = x + 3$ **8.** $4x + y = 8$; $y = 8 - 4x$

9. $x - y = 5$; $-y = 5 - x$; $y = x - 5$

10. $2x - y = 9$; $-y = 9 - 2x$; $y = 2x - 9$

11. $10x + y = -2$; $y = -2 - 10x$ **12.** $5x - y = 0$; $5x = y$

Answers to Exs. 13–20 may vary.

13. $y = x - 3$ **14.** $y = 5 - x$ **15.** $y = 1 - 2x$

x	y
0	−3
1	−2
2	−1

x	y
0	5
1	4
5	0

x	y
0	1
1	−1
−1	3

$(0, -3), (1, -2), (2, -1)$ $(0, 5), (1, 4), (5, 0)$ $(0, 1), (1, -1), (-1, 3)$

16. $y = 7 - 4x$

x	y
0	7
1	3
2	-1

$(0, 7), (1, 3), (2, -1)$

17. $y = -2x$

x	y
0	0
1	-2
-1	2

$(0, 0), (1, -2), (-1, 2)$

18. $y = 12 - 4x$

x	y
0	12
1	8
2	4

$(0, 12), (1, 8), (2, 4)$

19. $y = 2x$

x	y
0	0
1	2
2	4

$(0, 0), (1, 2), (2, 4)$

20. $y = 10 + 5x$

x	y
0	10
1	15
-2	0

$(0, 10), (1, 15), (-2, 0)$

B   **21.** $y = x - 5$ **22.** $y = 2x$ **23.** $y = 3x - 1$ **24.** $y = 4x + 3$

Page 468 • FOR PAGES 241–244

A   **1.**
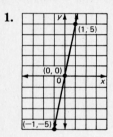

2.
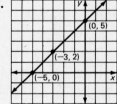

3.
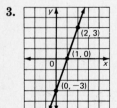

4.
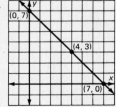

5.
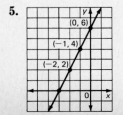

6.
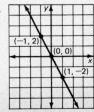

7.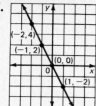

8.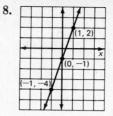

9.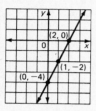

10. 2

11. −3

12. −1

13. $\frac{1}{2}$

Page 468 · FOR PAGES 245–248

A   **1.** $y = 3x$

2. $y = x + 2$

3.

Rate (km/h)	Distance (km)
10	30
30	90
40	120

4.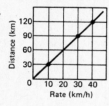

CHAPTER 8

Page 469 · FOR PAGES 262–265

A   **1.** $(-1, 1)$

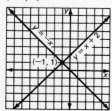

2. $(3, 6)$

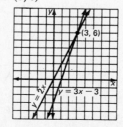

3. $(2, 1)$

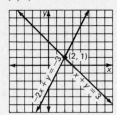

4. $(-2, -5)$

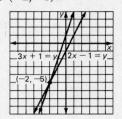

B   **5.** about $\left(-2\dfrac{1}{2}, -5\dfrac{1}{2}\right)$

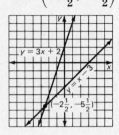

6. about $\left(3\dfrac{1}{3}, -3\dfrac{2}{3}\right)$

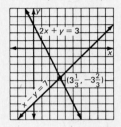

7. about $\left(\dfrac{3}{4}, -1\dfrac{3}{4}\right)$

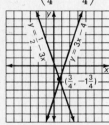

8. about $\left(1\dfrac{1}{2}, -1\dfrac{1}{2}\right)$

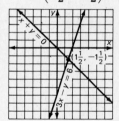

Page 469 · FOR PAGES 266–269

A   **1.** 3   **2.** 2   **3.** -3 **4.** $\dfrac{1}{2}$

5. Slope of each equation = 3;   no pair

6. $y = -x + 6$ and $y = x - 2$;   slopes = -1 and 1;   one pair

7. $y = -x$ and $3y = -3x$, or $y = -x$;   same equation;   all pairs

8. $y = -x + 3$ and $y = -x + 12$;   slope of each equation = -1;   no pair

Pages 469–470 · FOR PAGES 270–275

A   **1.** $y = 3x$ and $x + y = 12$; $x + (3x) = 12$; $4x = 12$; $x = 3$; $y = 3x = 3 \cdot 3 = 9$.
$(3, 9)$

2. $y = 4x$ and $y - x = 6$; $(4x) - x = 6$; $3x = 6$; $x = 2$; $y = 4x = 4 \cdot 2 = 8$. $(2, 8)$

3. $x - 2y = -4$ and $3x + y = 2$; $x = 2y - 4$ and $3x + y = 2$; $3(2y - 4) + y = 2$; $6y - 12 + y = 2$; $7y - 12 = 2$; $7y = 14$; $y = 2$; $x = 2y - 4 = 2 \cdot 2 - 4 = 4 - 4 = 0$. $(0, 2)$

4. $3x - y = 3$ and $x + y = 5$; $3x - y = 3$ and $y = 5 - x$; $3x - (5 - x) = 3$; $3x - 5 + x = 3$; $4x - 5 = 3$; $4x = 8$; $x = 2$; $y = 5 - x = 5 - 2 = 3$. $(2, 3)$

5.
$$x + y = 4$$
$$\underline{x - y = 6} \quad \text{Add}$$
$$2x \quad\;\; = 10; \quad x = 5$$
$5 + y = 4$; $y = -1$. $(5, -1)$

6.
$$2x + 2y = 5$$
$$\underline{2x + 3y = 4} \quad \text{Subt.}$$
$$-y = 1; \quad y = -1$$
$2x + 2(-1) = 5$; $2x - 2 = 5$;
$2x = 7$; $x = \dfrac{7}{2}$. $\left(\dfrac{7}{2}, -1\right)$

7.
$$3x - 2y = 0$$
$$\underline{x - 2y = 4} \quad \text{Subt.}$$
$$2x \quad\;\; = -4; \quad x = -2$$
$-2 - 2y = 4$; $-2y = 6$; $y = -3$.
$(-2, -3)$

8.
$$x - 2y = 18$$
$$\underline{5x + 2y = 0} \quad \text{Add}$$
$$6x \quad\;\; = 18; \quad x = 3$$
$3 - 2y = 18$; $-2y = 15$;
$y = -\dfrac{15}{2}$. $\left(3, -\dfrac{15}{2}\right)$

9. $3x - 7 = y$ and $4x + 2 = y$; $3x - 7 = 4x + 2$; $3x - 9 = 4x$; $-9 = x$; $y = 3x - 7 = 3(-9) - 7 = -27 - 7 = -34$. $(-9, -34)$

10. $y = 4x$ and $3x - y = 0$; $3x - (4x) = 0$; $-x = 0$; $x = 0$; $y = 4x = 4 \cdot 0 = 0$. $(0, 0)$

11. $x = y$ and $3x + 5y = 8$; $3y + 5y = 8$; $8y = 8$; $y = 1$; $x = y = 1$. $(1, 1)$

12.
$$x - 4y = 3 \qquad\qquad x - 4 \cdot 1 = 3; \quad x - 4 = 3; \quad x = 7. \quad (7, 1)$$
$$\underline{x - 6y = 1} \quad \text{Subt.}$$
$$2y = 2; \quad y = 1$$

B

13.
$$x - 5y = 0$$
$$\underline{x - 4y = 9} \quad \text{Subt.}$$
$$-y = -9; \quad y = 9$$
$x - 5 \cdot 9 = 0$; $x - 45 = 0$; $x = 45$.
$(45, 9)$

14.
$$3x + 4y = 6$$
$$\underline{x + 4y = 7} \quad \text{Subt.}$$
$$2x \quad\;\; = -1; \quad x = -\dfrac{1}{2}$$
$-\dfrac{1}{2} + 4y = 7$; $4y = 7\dfrac{1}{2}$;
$y = \dfrac{15}{2} \div 4 = \dfrac{15}{8}$. $\left(-\dfrac{1}{2}, \dfrac{15}{8}\right)$

15.
$$5x - y = 5$$
$$\underline{15x + y = 10} \quad \text{Add}$$
$$20x \quad\;\; = 15; \quad x = \dfrac{15}{20} = \dfrac{3}{4}$$
$5 \cdot \dfrac{3}{4} - y = 5$; $\dfrac{15}{4} - y = 5$; $-y = \dfrac{5}{4}$;
$y = -\dfrac{5}{4}$. $\left(\dfrac{3}{4}, -\dfrac{5}{4}\right)$

16.
$$4x - 12y = 3$$
$$\underline{4x + 10y = -2} \quad \text{Subt.}$$
$$-22y = 5; \quad y = -\dfrac{5}{22}$$
$4x - 12\left(-\dfrac{5}{22}\right) = 3$; $4x + \dfrac{60}{22} = 3$;
$4x = \dfrac{6}{22}$; $x = \dfrac{3}{44}$. $\left(\dfrac{3}{44}, -\dfrac{5}{22}\right)$

Page 470 · FOR PAGES 276–283

A

1. $x + 2y = 5 \longrightarrow 2x + 4y = 10$
$2x + 3y = 4 \longrightarrow \underline{2x + 3y = 4}$ Subt.
$y = 6$

$x + 2 \cdot 6 = 5; \quad x + 12 = 5;$
$x = -7. \quad (-7, 6)$

2. $2x - y = 4 \longrightarrow 4x - 2y = 8$
$3x + 2y = 6 \longrightarrow \underline{3x + 2y = 6}$ Add
$7x = 14; \quad x = 2$

$2 \cdot 2 - y = 4; \quad 4 - y = 4;$
$-y = 0; \quad y = 0. \quad (2, 0)$

3. $x - 4y = 2 \longrightarrow 3x - 12y = 6$
$3x - 10y = 4 \longrightarrow \underline{3x - 10y = 4}$ Subt.
$ -2y = 2; \quad y = -1$

$x - 4(-1) = 2; \quad x + 4 = 2;$
$x = -2. \quad (-2, -1)$

4. $4x - 5y = 9 \longrightarrow 4x - 5y = 9$
$x + y = 0 \longrightarrow \underline{5x + 5y = 0}$ Add
$9x = 9; \quad x = 1$

$1 + y = 0; \quad y = -1. \quad (1, -1)$

5. $3x - y = 6 \longrightarrow 6x - 2y = 12$
$x + 2y = 2 \longrightarrow \underline{x + 2y = 2}$ Add
$7x = 14; \quad x = 2$

$2 + 2y = 2; \quad 2y = 0;$
$y = 0. \quad (2, 0)$

6. $2x + 3y = 1 \longrightarrow 2x + 3y = 1$
$x + y = 4 \longrightarrow \underline{2x + 2y = 8}$ Subt.
$y = -7$

$x + (-7) = 4; \quad x = 11. \quad (11, -7)$

7. $x - 3y = 4 \longrightarrow 3x - 9y = 12$
$3x - y = 4 \longrightarrow \underline{3x - y = 4}$ Subt.
$ -8y = 8; \quad y = -1$

$x - 3(-1) = 4; \quad x + 3 = 4;$
$x = 1. \quad (1, -1)$

8. $3x - 5y = 2 \longrightarrow 3x - 5y = 2$
$x - 2y = 5 \longrightarrow \underline{3x - 6y = 15}$ Subt.
$y = -13$

$x - 2(-13) = 5; \quad x + 26 = 5;$
$x = -21. \quad (-21, -13)$

9. Let x and y = the numbers. $x + y = 15$ and $x = 2y$; $(2y) + y = 15$; $3y = 15$; $y = 5$; $x = 2y = 10$.   5 and 10

10. Let h = cost of a hat (in cents) and t = cost of a tie.

$2h + 3t = 8850 \longrightarrow 2h + 3t = 8850$
$h + 2t = 5100 \longrightarrow \underline{2h + 4t = 10200}$ Subt.
$ -t = -1350; \quad t = 1350$

$h + 2(1350) = 5100; \quad h + 2700 = 5100; \quad h = 2400.$ hat: \$24.00;   tie: \$13.50

11. Let b = cost of a bat (in dollars) and x = cost of a ball.

$2b + x = 29 \longrightarrow 4b + 2x = 58$
$b + 2x = 22 \longrightarrow \underline{b + 2x = 22}$ Subt.
$3b = 36; \quad b = 12$

bat: \$12

12. Let x and y = the numbers.

$x + y = 15 \longrightarrow 2x + 2y = 30$
$x + 2y = 25 \longrightarrow \underline{x + 2y = 25}$ Subt.
$x = 5$

$5 + y = 15; \quad y = 10.$ 5 and 10

13. Let x = Al's score and y = Maya's score. $x = y + 6$ and $x + y = 16$;
$(y + 6) + y = 16$; $2y + 6 = 16$; $2y = 10$; $y = 5$; $x = y + 6 = 11$.
Al: 11;   Maya: 5

14. Let e = cost of an egg (in cents) and j = cost of a glass of juice.

$\begin{aligned} e + 2j &= 230 \longrightarrow 2e + 4j = 460 \\ 2e + j &= 340 \longrightarrow 2e + j = 340 \quad \text{Subt.} \\ & \underline{} \\ & 3j = 120; \quad j = 40 \end{aligned}$

$2e + 40 = 340$; $2e = 300$;
$e = 150$.   Thus, $e + j =$
$150 + 40 = 190$.   \$1.90

B

15. $\begin{aligned} 2x + 3y &= 0 \longrightarrow 6x + 9y = 0 \\ 3x + 4y &= 6 \longrightarrow \underline{6x + 8y = 12} \quad \text{Subt.} \\ & y = -12 \end{aligned}$

$2x + 3(-12) = 0$; $2x - 36 =$
0; $2x = 36$; $x = 18$.
$(18, -12)$

16. $\begin{aligned} 4x - 3y &= 1 \longrightarrow 12x - 9y = 3 \\ 3x - 2y &= 0 \longrightarrow \underline{12x - 8y = 0} \quad \text{Subt.} \\ & -y = 3; \quad y = -3 \end{aligned}$

$3x - 2(-3) = 0$; $3x + 6 = 0$;
$3x = -6$; $x = -2$. $(-2, -3)$

17. $\begin{aligned} 2x + 5y &= 0 \longrightarrow 4x + 10y = 0 \\ 5x - 2y &= 0 \longrightarrow \underline{25x - 10y = 0} \quad \text{Add} \\ & 29x = 0; \quad x = 0 \end{aligned}$

$2 \cdot 0 + 5y = 0$; $5y = 0$;
$y = 0$. $(0, 0)$

18. $\begin{aligned} 3x - 4y &= 5 \longrightarrow 6x - 8y = 10 \\ 2x - 5y &= 1 \longrightarrow \underline{6x - 15y = 3} \quad \text{Subt.} \\ & 7y = 7; \quad y = 1 \end{aligned}$

$2x - 5 \cdot 1 = 1$; $2x - 5 = 1$;
$2x = 6$; $x = 3$. $(3, 1)$

19. $\begin{aligned} 2x - 4y &= 4 \longrightarrow 6x - 12y = 12 \\ 3x - 5y &= 1 \longrightarrow \underline{6x - 10y = 2} \quad \text{Subt.} \\ & -2y = 10; \quad y = -5 \end{aligned}$

$3x - 5(-5) = 1$; $3x + 25 = 1$;
$3x = -24$; $x = -8$. $(-8, -5)$

20. $\begin{aligned} 4x - 3y &= -2 \longrightarrow 16x - 12y = -8 \\ 3x + 4y &= 1 \longrightarrow \underline{9x + 12y = 3} \quad \text{Add} \\ & 25x = -5; \quad x = -\frac{1}{5} \end{aligned}$

$3\left(-\frac{1}{5}\right) + 4y = 1$; $-\frac{3}{5} + 4y = 1$; $4y = \frac{8}{5}$; $y = \frac{2}{5}$. $\left(-\frac{1}{5}, \frac{2}{5}\right)$

21.

	price	number	Cost
Pencils	25	x	$25x$
Pens	75	y	$75y$
		25	675

$\begin{aligned} x + y &= 25 \longrightarrow 75x + 75y = 1875 \\ 25x + 75y &= 675 \longrightarrow \underline{25x + 75y = 675} \quad \text{Subt.} \\ & 50x = 1200; \quad x = 24 \end{aligned}$

$24 + y = 25$; $y = 1$.   24 pencils, 1 pen

CHAPTER 9

Page 471 · FOR PAGES 298–303

A 1. $\dfrac{4}{16} = \dfrac{4 \cdot 1}{4 \cdot 4} = \dfrac{1}{4}$

2. $\dfrac{9}{15} = \dfrac{3 \cdot 3}{3 \cdot 5} = \dfrac{3}{5}$

3. $\dfrac{3}{12} = \dfrac{3 \cdot 1}{3 \cdot 4} = \dfrac{1}{4}$

4. $\dfrac{4}{20} = \dfrac{4 \cdot 1}{4 \cdot 5} = \dfrac{1}{5}$

5. $\dfrac{6}{18} = \dfrac{6 \cdot 1}{6 \cdot 3} = \dfrac{1}{3}$

6. $\dfrac{5}{25a} = \dfrac{5}{5 \cdot 5 \cdot a} = \dfrac{1}{5a}$

7. $-\dfrac{2x}{14x} = -\dfrac{2 \cdot x}{7 \cdot 2 \cdot x} = -\dfrac{1}{7}$

8. $\dfrac{21r}{7r^2} = \dfrac{3 \cdot 7 \cdot r}{7 \cdot r \cdot r} = \dfrac{3}{r}$

9. $\dfrac{5y}{-20xy^2} = \dfrac{5 \cdot y}{-4 \cdot 5 \cdot x \cdot y \cdot y} = \dfrac{1}{-4xy} = -\dfrac{1}{4xy}$

10. $\dfrac{2mn^2}{10m^2n} = \dfrac{2 \cdot m \cdot n \cdot n}{5 \cdot 2 \cdot m \cdot m \cdot n} = \dfrac{n}{5m}$

11. $\dfrac{-4xy^2}{12x^2} = \dfrac{-4 \cdot x \cdot y \cdot y}{4 \cdot 3 \cdot x \cdot x} = \dfrac{-y^2}{3x} = -\dfrac{y^2}{3x}$

12. $\dfrac{5x^3y^3}{30xy} = \dfrac{5 \cdot x \cdot x \cdot x \cdot y \cdot y \cdot y}{6 \cdot 5 \cdot x \cdot y} = \dfrac{x^2y^2}{6}$

13. $-\dfrac{40n^2s}{5n^3s^2} = -\dfrac{5 \cdot 8 \cdot n \cdot n \cdot s}{5 \cdot n \cdot n \cdot n \cdot s \cdot s} = -\dfrac{8}{ns}$

14. $\dfrac{6d^2e^2}{24d^4e^4} = \dfrac{6 \cdot d \cdot d \cdot e \cdot e}{4 \cdot 6 \cdot d \cdot d \cdot d \cdot d \cdot e \cdot e \cdot e \cdot e} = \dfrac{1}{4d^2e^2}$

15. $\dfrac{24x^4y^3}{12x^3y^2} = \dfrac{2 \cdot 12 \cdot x \cdot x \cdot x \cdot x \cdot y \cdot y \cdot y}{12 \cdot x \cdot x \cdot x \cdot y \cdot y} = 2xy$

16. $\dfrac{n+1}{2n+2} = \dfrac{n+1}{2(n+1)} = \dfrac{1}{2}$

17. $\dfrac{3x+3y}{10x+10y} = \dfrac{3(x+y)}{10(x+y)} = \dfrac{3}{10}$

18. $\dfrac{x-y}{x^2-y^2} = \dfrac{x-y}{(x+y)(x-y)} = \dfrac{1}{x+y}$

19. $\dfrac{4a-4b}{12a-12b} = \dfrac{4(a-b)}{3 \cdot 4(a-b)} = \dfrac{1}{3}$

20. $1 - n = -1(-1+n)$, or $-1(n-1)$

21. $x = -1(-x)$

22. $6 - 4y = -1(-6+4y)$, or $-1(4y-6)$

23. $-n^2 + z^2 = -1(n^2 - z^2)$

24. $\dfrac{m-n}{n-m} = \dfrac{m-n}{-1(m-n)} = \dfrac{1}{-1} = -1$

25. $\dfrac{d-z}{z-d} = \dfrac{-1(z-d)}{z-d} = \dfrac{-1}{1} = -1$

26. $\dfrac{3x-3y}{y-x} = \dfrac{3(x-y)}{-1(x-y)} = \dfrac{3}{-1} = -3$

27. $\dfrac{x^2-y^2}{y-x} = \dfrac{(x+y)(x-y)}{-1(x-y)} = \dfrac{x+y}{-1} = -x - y$

28. $\dfrac{4n^2-4}{1-n} = \dfrac{4(n^2-1)}{1-n} = \dfrac{4(n+1)(n-1)}{-1(n-1)} = \dfrac{4(n+1)}{-1} = -4n - 4$

29. $\dfrac{x-3}{9-x^2} = \dfrac{-1(3-x)}{(3+x)(3-x)} = \dfrac{-1}{3+x} = -\dfrac{1}{3+x}$

30. $\dfrac{4n-8}{48-12n^2} = \dfrac{4(n-2)}{-12(n^2-4)} = \dfrac{4(n-2)}{-12(n+2)(n-2)} = \dfrac{1}{-3(n+2)} = -\dfrac{1}{3n+6}$

31. $\dfrac{1-n}{3n^2-3} = \dfrac{1-n}{3(n^2-1)} = \dfrac{-1(n-1)}{3(n+1)(n-1)} = \dfrac{-1}{3(n+1)} = -\dfrac{1}{3n+3}$

B 32. $\dfrac{x^2+2x+1}{x^2+6x+5} = \dfrac{(x+1)(x+1)}{(x+1)(x+5)} = \dfrac{x+1}{x+5}$

33. $\dfrac{x^2 - 7x + 12}{x^2 - 2x - 8} = \dfrac{(x-3)(x-4)}{(x+2)(x-4)} = \dfrac{x-3}{x+2}$

34. $\dfrac{x^2 - 3x - 10}{x^2 - 6x - 16} = \dfrac{(x+2)(x-5)}{(x+2)(x-8)} = \dfrac{x-5}{x-8}$

Pages 471–472 • FOR PAGES 304–313

A   **1.** $\dfrac{18}{12} = \dfrac{3}{2}$ **2.** $\dfrac{4}{40} = \dfrac{1}{10}$ **3.** $\dfrac{6}{24} = \dfrac{1}{4}$ **4.** $\dfrac{9}{36} = \dfrac{1}{4}$

5. Let $3x$ = number of girls and $4x$ = number of boys. $3x + 4x = 21$; $7x = 21$; $x = 3$, so $3x = 9$ and $4x = 12$.   12 boys, 9 girls

6. $\dfrac{2}{5} = \dfrac{x}{20}$; $2 \cdot 20 = 5x$; $40 = 5x$; $8 = x$

7. $\dfrac{3}{9} = \dfrac{4}{x}$; $3x = 9 \cdot 4$; $3x = 36$; $x = 12$

8. $\dfrac{5}{a} = \dfrac{10}{100}$; $5 \cdot 100 = a \cdot 10$; $500 = 10a$; $50 = a$

9. $\dfrac{3}{1} = \dfrac{n}{8}$; $3 \cdot 8 = n$; $24 = n$

10. $\dfrac{x}{3+x} = \dfrac{2}{3}$; $x \cdot 3 = (3+x)2$; $3x = 6 + 2x$; $x = 6$

11. $\dfrac{2n-3}{n+1} = \dfrac{3}{4}$; $(2n-3)4 = (n+1)3$; $8n - 12 = 3n + 3$; $8n = 3n - 15$; $5n = 15$; $n = 3$

12. $\dfrac{x+2}{x-2} = \dfrac{1}{5}$; $(x+2)5 = x - 2$; $5x + 10 = x - 2$; $5x = x - 12$; $4x = -12$; $x = -3$

13. $\dfrac{2y+6}{y} = \dfrac{4}{5}$; $(2y+6)5 = y \cdot 4$; $10y + 30 = 4y$; $30 = -6y$; $-5 = y$

14.

Number	5	15
Cost	70¢	x

$\dfrac{5}{70} = \dfrac{15}{x}$; $5x = 1050$; $x = 210$. $2.10

15.

Distance	4 km	x
Time	2	3

$\dfrac{4}{2} = \dfrac{x}{3}$; $12 = 2x$; $6 = x$.   6 km

B   **16.** $\dfrac{2 \text{ weeks}}{21 \text{ days}} = \dfrac{14 \text{ days}}{21 \text{ days}} = \dfrac{2 \text{ days}}{3 \text{ days}}$ or $\dfrac{2}{3}$ **17.** $\dfrac{50¢}{\$2.50} = \dfrac{50¢}{250¢} = \dfrac{1¢}{5¢}$ or $\dfrac{1}{5}$

18. $\dfrac{25 \text{ cm}}{6 \text{ m}} = \dfrac{25 \text{ cm}}{600 \text{ cm}} = \dfrac{1 \text{ cm}}{24 \text{ cm}}$ or $\dfrac{1}{24}$ **19.** $\dfrac{15 \text{ s}}{3 \text{ min}} = \dfrac{15 \text{ s}}{180 \text{ s}} = \dfrac{1 \text{ s}}{12 \text{ s}}$ or $\dfrac{1}{12}$

20. Let $6x$, $5x$, and $2x$ = number of people watching each show. $6x + 5x + 2x = 260{,}000$; $13x = 260{,}000$; $x = 20{,}000$, so $6x = 120{,}000$, $5x = 100{,}000$ and $2x = 40{,}000$.   120,000 people, 100,000 people, 40,000 people

Page 472 • FOR PAGES 314–324

A   1. $\dfrac{1}{2} \cdot \dfrac{1}{8} = \dfrac{1 \cdot 1}{2 \cdot 8} = \dfrac{1}{16}$ 2. $\dfrac{4}{7} \cdot \dfrac{21}{2} = \dfrac{4 \cdot 21}{7 \cdot 2} = 6$ 3. $\dfrac{a}{b} \cdot \dfrac{2a}{b} = \dfrac{a \cdot 2 \cdot a}{b \cdot b} = \dfrac{2a^2}{b^2}$

4. $\dfrac{x^2 - y^2}{2x + 2y} \cdot \dfrac{6}{5} = \dfrac{(x + y)(x - y) \cdot 6}{2(x + y) \cdot 5} = \dfrac{3(x - y)}{5} = \dfrac{3x - 3y}{5}$

5. $\dfrac{9}{10} \div \dfrac{3}{20} = \dfrac{9}{10} \cdot \dfrac{20}{3} = 6$

6. $\dfrac{3x - 3}{2x} \div \dfrac{3}{4x} = \dfrac{3x - 3}{2x} \cdot \dfrac{4x}{3} = \dfrac{3(x - 1) \cdot 4x}{2x \cdot 3} = 2(x - 1) = 2x - 2$

7. $\dfrac{x^2 - 9}{3} \div (x + 3) = \dfrac{(x + 3)(x - 3)}{3} \cdot \dfrac{1}{x + 3} = \dfrac{x - 3}{3}$

8. $\dfrac{14}{x - 1} \div \dfrac{7}{x^2 - 1} = \dfrac{14}{x - 1} \cdot \dfrac{(x + 1)(x - 1)}{7} = 2(x + 1) = 2x + 2$

9. $\dfrac{5n}{3} - \dfrac{9n}{3} = \dfrac{5n - 9n}{3} = \dfrac{-4n}{3} = -\dfrac{4n}{3}$ 10. $\dfrac{12}{10x} - \dfrac{4}{10x} = \dfrac{12 - 4}{10x} = \dfrac{8}{10x} = \dfrac{4}{5x}$

11. $\dfrac{3n}{n - 1} - \dfrac{1}{n - 1} = \dfrac{3n - 1}{n - 1}$

12. $\dfrac{a - 2}{a + b} - \dfrac{a - 4}{a + b} = \dfrac{a - 2 - (a - 4)}{a + b} = \dfrac{a - 2 - a + 4}{a + b} = \dfrac{2}{a + b}$

13. $\dfrac{3}{4} + \dfrac{1}{2} = \dfrac{3}{4} + \dfrac{2}{4} = \dfrac{5}{4}$ 14. $\dfrac{6}{5} + \dfrac{1}{4} = \dfrac{24}{20} + \dfrac{5}{20} = \dfrac{29}{20}$

15. $\dfrac{x}{2} - \dfrac{x}{5} = \dfrac{5x}{10} - \dfrac{2x}{10} = \dfrac{3x}{10}$

16. $\dfrac{2n}{3} - \dfrac{n}{2} = \dfrac{2 \cdot 2n}{6} - \dfrac{3n}{6} = \dfrac{4n}{6} - \dfrac{3n}{6} = \dfrac{n}{6}$

17. $\dfrac{1}{x} + \dfrac{3}{y} = \dfrac{y}{xy} + \dfrac{3x}{xy} = \dfrac{y + 3x}{xy}$ 18. $\dfrac{2}{a} - \dfrac{3}{a^2} = \dfrac{2a}{a^2} - \dfrac{3}{a^2} = \dfrac{2a - 3}{a^2}$

19. $\dfrac{9}{ab^2} - \dfrac{1}{a^2b} = \dfrac{9a}{a^2b^2} - \dfrac{b}{a^2b^2} = \dfrac{9a - b}{a^2b^2}$

20. $\dfrac{3}{2xy} - \dfrac{2}{xy} = \dfrac{3}{2xy} - \dfrac{4}{2xy} = \dfrac{-1}{2xy} = -\dfrac{1}{2xy}$

B 21. $\dfrac{x^2 - y^2}{x^2 - 64} \cdot \dfrac{x - 8}{x + y} = \dfrac{(x + y)(x - y)(x - 8)}{(x + 8)(x - 8)(x + y)} = \dfrac{x - y}{x + 8}$

22. $\dfrac{x - 3}{1 - x} \cdot \dfrac{x - 1}{x^2 - 9} = \dfrac{(x - 3)(x - 1)}{-1(x - 1)(x + 3)(x - 3)} = \dfrac{1}{-(x + 3)} = -\dfrac{1}{x + 3}$

23. $\dfrac{3a - 6}{5a} \cdot \dfrac{5a^3 - 15a^2}{a - 2} = \dfrac{3(a - 2) \cdot 5a^2(a - 3)}{5a(a - 2)} = 3a(a - 3) = 3a^2 - 9a$

Page 473 • FOR PAGES 327–332

A   1. $\dfrac{x}{5} - \dfrac{x}{2} = 30;\quad \dfrac{2x}{10} - \dfrac{5x}{10} = 30;\quad \dfrac{-3x}{10} = \dfrac{30}{1};\quad -3x = 300;\quad x = -100$

2. $\dfrac{x}{6} - \dfrac{x}{5} = 3;\quad \dfrac{5x}{30} - \dfrac{6x}{30} = 3;\quad \dfrac{-x}{30} = \dfrac{3}{1};\quad -x = 90;\quad x = -90$

3. $\dfrac{x}{3} = \dfrac{1}{6} + \dfrac{1}{2};\quad \dfrac{x}{3} = \dfrac{1}{6} + \dfrac{3}{6};\quad \dfrac{x}{3} = \dfrac{4}{6};\quad 6x = 12;\quad x = 2$

4. $\dfrac{1}{4} - \dfrac{1}{6} = \dfrac{3}{n}$; $\dfrac{3}{12} - \dfrac{2}{12} = \dfrac{3}{n}$; $\dfrac{1}{12} = \dfrac{3}{n}$; $n = 36$

5. $\dfrac{2x}{3} - \dfrac{x}{5} = \dfrac{14}{15}$; $\dfrac{10x}{15} - \dfrac{3x}{15} = \dfrac{14}{15}$; $\dfrac{7x}{15} = \dfrac{14}{15}$; $7x = 14$; $x = 2$

6. $\dfrac{2}{n} + \dfrac{3}{4n} = \dfrac{1}{8}$; $\dfrac{8}{4n} + \dfrac{3}{4n} = \dfrac{1}{8}$; $\dfrac{11}{4n} = \dfrac{1}{8}$; $4n = 88$; $n = 22$

7. $\dfrac{n}{3} + 10 = \dfrac{4n}{2}$; $\dfrac{n}{3} + \dfrac{30}{3} = \dfrac{4n}{2}$; $\dfrac{n + 30}{3} = \dfrac{4n}{2}$; $2(n + 30) = 12n$; $2n + 60 = 12n$;
   $60 = 10n$; $6 = n$

8. $\dfrac{x}{2} + \dfrac{3x}{8} = 7$; $\dfrac{4x}{8} + \dfrac{3x}{8} = 7$; $\dfrac{7x}{8} = \dfrac{7}{1}$; $7x = 56$; $x = 8$

9. $\dfrac{3n}{4} + 5 = \dfrac{n}{3}$; $\dfrac{3n}{4} + \dfrac{20}{4} = \dfrac{n}{3}$; $\dfrac{3n + 20}{4} = \dfrac{n}{3}$; $3(3n + 20) = 4n$; $9n + 60 = 4n$;
   $60 = -5n$; $-12 = n$

10.

	Frank	Mark	Together
Hours needed	2	3	h
Part done in one hour	$\dfrac{1}{2}$	$\dfrac{1}{3}$	$\dfrac{1}{h}$

$\dfrac{1}{2} + \dfrac{1}{3} = \dfrac{1}{h}$; $\dfrac{3}{6} + \dfrac{2}{6} = \dfrac{1}{h}$; $\dfrac{5}{6} = \dfrac{1}{h}$; $5h = 6$; $h = 1\dfrac{1}{5}$. $1\dfrac{1}{5}$ hours

11.

	Mary	Brenda	Together
Hours needed	4	6	n
Part done in one hour	$\dfrac{1}{4}$	$\dfrac{1}{6}$	$\dfrac{1}{n}$

$\dfrac{1}{4} + \dfrac{1}{6} = \dfrac{1}{n}$; $\dfrac{3}{12} + \dfrac{2}{12} = \dfrac{1}{n}$; $\dfrac{5}{12} = \dfrac{1}{n}$; $5n = 12$; $n = 2\dfrac{2}{5}$. $2\dfrac{2}{5}$ hours

B 12. $\dfrac{x - 1}{2} + \dfrac{x + 2}{4} = 9$; $\dfrac{2(x - 1)}{4} + \dfrac{x + 2}{4} = 9$; $\dfrac{2x - 2 + x + 2}{4} = 9$; $\dfrac{3x}{4} = \dfrac{9}{1}$; $3x =$
   36; $x = 12$

13. $\dfrac{12 - x}{3} + \dfrac{3 + x}{2} = \dfrac{1}{2}$; $\dfrac{2(12 - x)}{6} + \dfrac{3(3 + x)}{6} = \dfrac{1}{2}$; $\dfrac{24 - 2x + 9 + 3x}{6} = \dfrac{1}{2}$; $\dfrac{x + 33}{6} =$
   $\dfrac{1}{2}$; $2(x + 33) = 6$; $2x + 66 = 6$; $2x = -60$; $x = -30$

14. $\dfrac{n + 5}{2} - \dfrac{n - 5}{6} = \dfrac{5}{8}$; $\dfrac{3(n + 5)}{6} - \dfrac{n - 5}{6} = \dfrac{5}{8}$; $\dfrac{3n + 15 - n + 5}{6} = \dfrac{5}{8}$; $\dfrac{2n + 20}{6} =$
   $\dfrac{5}{8}$; $8(2n + 20) = 30$; $16n + 160 = 30$; $16n = -130$; $n = -8\dfrac{1}{8}$

15.

	Gardener	Helper	Together
Hours needed	5	x	3
Part done in one hour	$\dfrac{1}{5}$	$\dfrac{1}{x}$	$\dfrac{1}{3}$

$\dfrac{1}{5} + \dfrac{1}{x} = \dfrac{1}{3}; \quad \dfrac{x}{5x} + \dfrac{5}{5x} = \dfrac{1}{3}; \quad \dfrac{x+5}{5x} = \dfrac{1}{3}; \quad 5x = 3(x+5); \quad 5x = 3x + 15;$

$2x = 15; \quad x = 7\dfrac{1}{2}. \quad 7\dfrac{1}{2}$ hours

CHAPTER 10

Page 474 · FOR PAGES 348–353

A **1.** 0.61 **2.** 9.17 **3.** 5.30 **4.** 0.75 **5.** 0.505 **6.** 1.88 **7.** 1.64 **8.** 4.405

9. 47.5 **10.** 0.93 **11.** 62.5 **12.** 450 **13.** 3.5577 **14.** 1.0530 **15.** 4.275

16. 4.6921

17.
$$\begin{array}{r} 1.23 \\ 3\overline{)3.70} \end{array}$$
$3.7 \div 3 \approx 1.2$

18.
$$\begin{array}{r} 124.66 \\ 3\overline{)374.00} \end{array}$$
$3.74 \div 0.03 \approx 124.7$

19.
$$\begin{array}{r} 6.85 \\ 9\overline{)61.70} \end{array}$$
$6.17 \div 0.9 \approx 6.9$

20.
$$\begin{array}{r} 6.70 \\ 15\overline{)100.60} \end{array}$$
$10.06 \div 1.5 \approx 6.7$

21.
$$\begin{array}{r} 0.49 \\ 62\overline{)30.60} \end{array}$$
$3.06 \div 6.2 \approx 0.5$

22.
$$\begin{array}{r} 1.19 \\ 78\overline{)93.10} \end{array}$$
$9.31 \div 7.8 \approx 1.2$

23.
$$\begin{array}{r} 52.81 \\ 16\overline{)845.00} \end{array}$$
$8.45 \div 0.16 \approx 52.8$

24.
$$\begin{array}{r} 12.63 \\ 55\overline{)695.00} \end{array}$$
$6.95 \div 0.55 \approx 12.6$

25. $\dfrac{2}{5} = 0.4$

26. $\dfrac{1}{6} = 0.166\ldots$

27. $\dfrac{1}{7} = 0.142857142857\ldots$

28. $\dfrac{5}{6} = 0.833\ldots$

29. $\dfrac{5}{7} = 0.714285714285\ldots$

30. $\dfrac{7}{8} = 0.875$ **31.** $\dfrac{1}{12} = 0.0833\ldots$ **32.** $\dfrac{1}{3} = 0.33\ldots$

33. $\dfrac{2}{3} = 0.66\ldots$ **34.** $\dfrac{1}{4} = 0.25$

Page 474 · FOR PAGES 354–357

A **1.** $0.20x = 0.1x + 1; \quad 100(0.20x) = 100(0.1x + 1); \quad 20x = 10x + 100; \quad 10x = 100;$
$x = 10$

2. $0.5 - 0.5n = 0.2n - 3; \quad 10(0.5 - 0.5n) = 10(0.2n - 3); \quad 5 - 5n = 2n - 30;$
$5 = 7n - 30; \quad 35 = 7n; \quad 5 = n$

3. $0.06x = 360; \quad 100(0.06x) = 100(360); \quad 6x = 36,000; \quad x = 6000$

4. $0.05y - 5 = 3y$; $100(0.05y - 5) = 100(3y)$; $5y - 500 = 300y$; $-500 = 295y$;
$y = -1.69 \approx -1.7$

5. $0.07x + 25 = 0.124$; $1000(0.07x + 25) = 1000(0.124)$; $70x + 25,000 = 124$;
$70x = -24,876$; $x = -355.37 \approx -355.4$

6. $0.01(5 - n) = 0.04(6)$; $100(0.01)(5 - n) = 100(0.04)(6)$; $5 - n = 24$;
$-n = 19$; $n = -19$

7. Let n = smaller number. Then $50 - n$ = greater number. $0.75n = 0.5(50 - n)$;
$100(0.75n) = 100(0.5)(50 - n)$; $75n = 50(50 - n)$; $75n = 2500 - 50n$; $125n = 2500$; $n = 20$, and $50 - n = 30$. 20 and 30

B   **8.** $0.83n + 0.63(45 - n) = 0.71 - 45$; $100[0.83n + 0.63(45 - n)] = 100(0.71 - 45)$;
$83n + 63(45 - n) = 71 - 4500$; $83n + 2835 - 63n = -4429$; $20n = -7264$;
$n = -363.2$

9. $0.07(8000 + n) = 600 + 0.05n$; $100[0.07(8000 + n)] = 100(600 + 0.05n)$;
$7(8000 + n) = 60,000 + 5n$; $56,000 + 7n = 60,000 + 5n$; $2n = 4000$; $n = 2000$

10. Let x = width (in m). Then $1.4x$ = length. $2x + 2(1.4x) = 19.6$; $2x + 2.8x = 19.6$; $10(2x + 2.8x) = 10(19.6)$; $20x + 28x = 196$; $48x = 196$; $x \approx 4.08 \approx 4.1$, so $1.4x = 5.74 \approx 5.7$. about 5.7 m

Page 475 · FOR PAGES 359–364

A

	Percent	Decimal	Fraction
1.	5%	0.05	$\frac{1}{20}$
2.	10%	0.10	$\frac{1}{10}$
3.	80%	0.80	$\frac{4}{5}$
4.	12.5%	0.125	$\frac{1}{8}$
5.	9%	0.09	$\frac{9}{100}$
6.	50%	0.50	$\frac{1}{2}$
7.	6%	0.06	$\frac{3}{50}$

8. $20\% \times \$5.80 = 0.20 \times \$5.80 = \$1.16$

9. $10\% \times \$6.30 = 0.10 \times \$6.30 = \$.63$

10. $25\% \times \$5 = 0.25 \times \$5 = \$1.25$

11. $33\% \times \$9.30 = 0.33 \times \$9.30 = \$3.069 \approx \$3.07$

12. $50\% \times \$45 = 0.50 \times \$45 = \$22.50$

13. $50\% \times \$6.20 = 0.50 \times \$6.20 = \$3.10$

14. $50\% \times \$22 = 0.50 \times \$22 = \$11$

15. $33\frac{1}{3}\% \times \$27 = \frac{1}{3} \times \$27 = \$9$
$\$27 - \$9 = \$18$

16. $33\frac{1}{3}\% \times \$63 = \frac{1}{3} \times \$63 = \$21$; $\$63 - \$21 = \$42$

17. $33\frac{1}{3}\% \times \$360 = \frac{1}{3} \times \$360 = \$120$; $\$360 - \$120 = \$240$

Pages 475–476 • FOR PAGES 365–368

A   1. $n \times 20 = 4$; $n = \dfrac{4}{20} = 0.20$.   20%

      2. $10\% \times \$50 = n$; $0.10 \times 50 = n$; $5 = n$.   \$5

      3. $35\% \times 1000 = n$; $0.35 \times 1000 = n$; $350 = n$.   350

      4. $5\% \times n = 45$; $0.05n = 45$; $5n = 4500$; $n = 900$.   900

      5. $60\% \times n = 36$; $0.60n = 36$; $60n = 3600$; $n = 60$.   60

      6. $n \times 300 = 60$; $n = \dfrac{60}{300} = 0.20$.   20%

      7. Let t = tax. $8\% \times \$12.25 = t$; $0.08(12.25) = t$; $0.98 = t$.   \$.98

      8. $26\% \times 1000 = 260$

B   9. $180 = n \times 7200$; $n = \dfrac{180}{7200} = 0.025 = 2.5\%$

Page 476 • FOR PAGES 369–372

A   1. $I = prt$; $I = 6500 \times 6\% \times 1 = 6500 \times 0.06 \times 1 = 390$.   \$390

      2. $I = prt$; $2000 = 2000 \times 5\% \times t$; $2000 = 2000 \times 0.05 \times t$; $2000 = 100t$; $20 = t$.   20 years

      3. $I = prt$; $300 = p \times 0.05 \times 1$; $300 = 0.05p$; $6000 = p$.   \$6000

      4.

	p	r	t	I
Amount at 5%	x	0.05	1	$0.05x$
Amount at 7%	$10{,}000 - x$	0.07	1	$0.07(10{,}000 - x)$
				620

$0.05x + 0.07(10{,}000 - x) = 620$; $5x + 7(10{,}000 - x) = 62{,}000$; $5x + 70{,}000 - 7x = 62{,}000$; $-2x = -8000$; $x = 4000$ and $10{,}000 - x = 6000$.   \$6000 at 7%, \$4000 at 5%

B   5.

	p	r	t	I
Amount at 9%	x	0.09	1	$0.09x$
Amount at 18%	$200 - x$	0.18	1	$0.18(200 - x)$
				22.50

$0.09x + 0.18(200 - x) = 22.50$; $9x + 18(200 - x) = 2250$; $9x + 3600 - 18x = 2250$; $-9x = -1350$; $x = 150$, and $200 - x = 50$.   \$150 at 9%, \$50 at 18%

Page 476 · FOR PAGES 373–376

A   **1.**

	Number	Unit price	Receipts
Student	n	1.50	$1.50n$
Others	$2000 - n$	2.25	$2.25(2000 - n)$
			4125

$1.50n + 2.25(2000 - n) = 4125$; $150n + 225(2000 - n) = 412{,}500$; $150n +$
$450{,}000 - 225n = 412{,}500$; $-75n = -37{,}500$; $n = 500$.   500 student tickets

2.

	Amount of solution	% salt	Amount salt
Original solution	15	4%	0.60
Water added	x	0	0
New solution	$15 + x$	2%	$0.02(15 + x)$

$0.60 + 0 = 0.02(15 + x)$; $60 = 2(15 + x)$; $60 = 30 + 2x$; $30 = 2x$; $15 = x$.
15 liters

CHAPTER 11

Page 477 · FOR PAGES 388–393

A   **1.** $\sqrt{49} = 7$ **2.** $\sqrt{100} = 10$ **3.** $\sqrt{25} = 5$
    4. $-\sqrt{16} = -4$ **5.** $\sqrt{1} = 1$ **6.** $-\sqrt{64} = -8$
    7. $\sqrt{0} = 0$ **8.** $\sqrt{4^2} = 4$ **9.** $\sqrt{6^2} = 6$
   10. $\sqrt{8^2} = 8$ **11.** $-\sqrt{12} \approx -3.464$ **12.** $-\sqrt{7} \approx -2.646$
   13. $-\sqrt{3} \approx -1.732$ **14.** $\pm\sqrt{66} \approx \pm8.124$ **15.** $-\sqrt{76} \approx -8.718$
   16. $\sqrt{13} \approx 3.606 \approx 3.6$ **17.** $\sqrt{98} \approx 9.899 \approx 9.9$ **18.** $-\sqrt{52} \approx -7.211 \approx -7.2$
   19. $\sqrt{63} \approx 7.937 \approx 7.9$ **20.** $\sqrt{70} \approx 8.367 \approx 8.4$ **21.** rational
   22. rational   **23.** irrational   **24.** rational   **25.** irrational
B   **26.** $A = s^2$, so $s = \sqrt{A} = \sqrt{60} \approx 7.746 \approx 7.75$; 7.75 cm

Page 477 · FOR PAGES 394–395

A   **1.** $\sqrt{500} = \sqrt{100 \cdot 5} = \sqrt{100} \cdot \sqrt{5} = 10\sqrt{5}$ **2.** $\sqrt{320} = \sqrt{64 \cdot 5} = \sqrt{64} \cdot \sqrt{5} = 8\sqrt{5}$
   3. $\sqrt{6400} = \sqrt{64 \cdot 100} = \sqrt{64} \cdot \sqrt{100} = 8 \cdot 10 = 80$
   4. $\sqrt{405} = \sqrt{81 \cdot 5} = \sqrt{81} \cdot \sqrt{5} = 9\sqrt{5}$ **5.** $\sqrt{147} = \sqrt{49 \cdot 3} = \sqrt{49} \cdot \sqrt{3} = 7\sqrt{3}$
   6. $\sqrt{112} = \sqrt{16 \cdot 7} = \sqrt{16} \cdot \sqrt{7} = 4\sqrt{7}$
   7. $\sqrt{729} = \sqrt{81 \cdot 9} = \sqrt{81} \cdot \sqrt{9} = 9 \cdot 3 = 27$
   8. $\sqrt{120} = \sqrt{4 \cdot 30} = \sqrt{4} \cdot \sqrt{30} = 2\sqrt{30}$

B   9. $\sqrt{81x^2} = \sqrt{81 \cdot x^2} = \sqrt{81} \cdot \sqrt{x^2} = 9x$

10. $\sqrt{49x^4} = \sqrt{49 \cdot x^4} = \sqrt{49} \cdot \sqrt{x^4} = 7x^2$

11. $\sqrt{5x^4y^2} = \sqrt{5 \cdot x^4 \cdot y^2} = \sqrt{5} \cdot \sqrt{x^4} \cdot \sqrt{y^2} = x^2y\sqrt{5}$

12. $\sqrt{10a^2b^3} = \sqrt{10 \cdot a^2 \cdot b^2 \cdot b} = \sqrt{10} \cdot \sqrt{a^2} \cdot \sqrt{b^2} \cdot \sqrt{b} = ab\sqrt{10b}$

Page 477 · FOR PAGES 396–397

A   1. $x^2 = 49;\quad x = \pm7$ 2. $n^2 = 25;\quad n = \pm5$

3. $y^2 = 36;\quad y = \pm6$ 4. $\dfrac{n}{4} = \dfrac{4}{n};\quad n^2 = 16;\quad n = \pm4$

5. $x^2 = 28;\quad x = \pm\sqrt{28} = \pm\sqrt{4 \cdot 7} = \pm2\sqrt{7}$

6. $x^2 + 1 = 65;\quad x^2 = 64;\quad x = \pm8$

B   7. $x^2 = 18;\quad x = \pm\sqrt{18} \approx \pm4.243 \approx \pm4.2$

8. $y^2 = 24;\quad y = \pm\sqrt{24} \approx \pm4.899 \approx \pm4.9$

9. $x^2 - 5 = 15;\quad x^2 = 20;\quad x = \pm\sqrt{20} \approx \pm4.472 \approx \pm4.5$

Page 478 · FOR PAGES 398–401

A   1. $c^2 = a^2 + b^2;\quad c^2 = 4^2 + 6^2 = 16 + 36 = 52;\quad c = \sqrt{52} \approx 7.211 \approx 7.2.\quad$ 7.2 cm

2. $a^2 + b^2 = c^2;\quad a^2 + 10^2 = 15^2;\quad a^2 + 100 = 225;\quad a^2 = 125;\quad a = \sqrt{125} =$ $\sqrt{25 \cdot 5} = 5\sqrt{5} \approx 5(2.236) = 11.18 \approx 11.2.\quad$ 11.2 m

3. $a^2 + b^2 = c^2;\quad a^2 + 5^2 = 9^2;\quad a^2 + 25 = 81;\quad a^2 = 56;\quad a = \sqrt{56} \approx 7.483 \approx 7.5;$ 7.5 cm

B   4. Let x = length of wire (in m). $x^2 = 5^2 + 10^2;\quad x^2 = 25 + 100 = 125;\quad x = \sqrt{125} = 5\sqrt{5} \approx 5(2.236) = 11.18 \approx 11.2.\quad$ 11.2 m

Page 478 · FOR PAGES 402–407

A   1. $\sqrt{\dfrac{1}{9}} = \dfrac{\sqrt{1}}{\sqrt{9}} = \dfrac{1}{3}$ 2. $\sqrt{\dfrac{4}{49}} = \dfrac{\sqrt{4}}{\sqrt{49}} = \dfrac{2}{7}$ 3. $\sqrt{\dfrac{9}{100}} = \dfrac{\sqrt{9}}{\sqrt{100}} = \dfrac{3}{10}$

4. $\sqrt{\dfrac{n^2}{4}} = \dfrac{\sqrt{n^2}}{\sqrt{4}} = \dfrac{n}{2}$ 5. $\sqrt{\dfrac{x^2}{25}} = \dfrac{\sqrt{x^2}}{\sqrt{25}} = \dfrac{x}{5}$ 6. $\sqrt{\dfrac{4n^2}{36}} = \dfrac{\sqrt{4n^2}}{\sqrt{36}} = \dfrac{2n}{6} = \dfrac{n}{3}$

7. $\sqrt{6} \cdot \sqrt{6} = \sqrt{6 \cdot 6} = \sqrt{36} = 6$ 8. $\sqrt{8} \cdot \sqrt{2} = \sqrt{8 \cdot 2} = \sqrt{16} = 4$

9. $\sqrt{4} \cdot 2\sqrt{9} = 2\sqrt{4 \cdot 9} = 2\sqrt{36} = 2 \cdot 6 = 12$

10. $\sqrt{n} \cdot \sqrt{4n} = \sqrt{n \cdot 4n} = \sqrt{4n^2} = 2n$

11. $5\sqrt{3x} \cdot 3\sqrt{12x} = 5 \cdot 3 \cdot \sqrt{3x \cdot 12x} = 15\sqrt{36x^2} = 15 \cdot 6x = 90x$

12. $y\sqrt{6} \cdot 8\sqrt{2y^2} = y \cdot 8 \cdot \sqrt{6 \cdot 2y^2} = 8y\sqrt{12y^2} = 8y\sqrt{3 \cdot 4y^2} = 8y \cdot 2y\sqrt{3} = 16y^2\sqrt{3}$

B   13. $\sqrt{3} \cdot \sqrt{5} \cdot \sqrt{15} = \sqrt{3 \cdot 5 \cdot 15} = \sqrt{15 \cdot 15} = 15$

14. $-x\sqrt{5} \cdot x\sqrt{75} = -x \cdot x \cdot \sqrt{5 \cdot 75} = -x^2\sqrt{375} = -x^2\sqrt{25 \cdot 15} = -5x^2\sqrt{15}$

15. $\sqrt{\dfrac{1}{2}} \cdot \sqrt{8x^4} = \sqrt{\dfrac{1}{2} \cdot 8x^4} = \sqrt{4x^4} = 2x^2$

16. $5\sqrt{\dfrac{2}{3}} = 5 \cdot \dfrac{\sqrt{2}}{\sqrt{3}} = 5 \cdot \dfrac{\sqrt{2} \cdot \sqrt{3}}{\sqrt{3} \cdot \sqrt{3}} = \dfrac{5\sqrt{6}}{3} \approx \dfrac{5(2.449)}{3} = \dfrac{12.245}{3} \approx 4.081 \approx 4.08$

17. $\sqrt{\dfrac{5}{8}} = \dfrac{\sqrt{5}}{\sqrt{8}} = \dfrac{\sqrt{5} \cdot \sqrt{8}}{\sqrt{8} \cdot \sqrt{8}} = \dfrac{\sqrt{40}}{8} \approx \dfrac{6.325}{8} \approx 0.790 \approx 0.79$

18. $\sqrt{\dfrac{1}{2}} \cdot \sqrt{\dfrac{1}{3}} = \sqrt{\dfrac{1}{6}} = \dfrac{\sqrt{1}}{\sqrt{6}} = \dfrac{1 \cdot \sqrt{6}}{\sqrt{6} \cdot \sqrt{6}} = \dfrac{\sqrt{6}}{6} \approx \dfrac{2.449}{6} \approx 0.408 \approx 0.41$

19. $\sqrt{\dfrac{1}{6}} \cdot \sqrt{\dfrac{5}{12}} = \sqrt{\dfrac{5}{72}} = \dfrac{\sqrt{5}}{\sqrt{72}} = \dfrac{\sqrt{5} \cdot \sqrt{2}}{6\sqrt{2} \cdot \sqrt{2}} = \dfrac{\sqrt{10}}{12} \approx \dfrac{3.162}{12} \approx 0.263 \approx 0.26$

Page 478 · FOR PAGES 408–409

A
  1. $3\sqrt{3} - 9\sqrt{3} = -6\sqrt{3}$ **2.** $10\sqrt{2} - \sqrt{2} = 9\sqrt{2}$

  3. $\sqrt{8} + \sqrt{2} = 2\sqrt{2} + \sqrt{2} = 3\sqrt{2}$ **4.** $-\sqrt{6} + 3\sqrt{6} = 2\sqrt{6}$

  5. $4\sqrt{18} + \sqrt{50} = 4 \cdot 3\sqrt{2} + 5\sqrt{2} = 12\sqrt{2} + 5\sqrt{2} = 17\sqrt{2}$

  6. $6\sqrt{5} - \sqrt{3} + \sqrt{5} = 7\sqrt{5} - \sqrt{3}$

B
  7. $\sqrt{3n} - \sqrt{27n} = \sqrt{3n} - 3\sqrt{3n} = -2\sqrt{3n}$

  8. $\sqrt{n} + 4\sqrt{n^3} = \sqrt{n} + 4n\sqrt{n} = (1 + 4n)\sqrt{n}$

  9. $\sqrt{18} - \sqrt{32} = 3\sqrt{2} - 4\sqrt{2} = -\sqrt{2}$

 10. $\sqrt{3} - (\sqrt{12} + \sqrt{45}) = \sqrt{3} - (2\sqrt{3} + 3\sqrt{5}) = \sqrt{3} - 2\sqrt{3} - 3\sqrt{5} = -\sqrt{3} - 3\sqrt{5}$

CHAPTER 12

Page 479 · FOR PAGES 420–425

A
  1. $n(n + 5) = 0$; $n = 0$ or $n + 5 = 0$; $n = 0$ or $n = -5$

  2. $x(x - 10) = 0$; $x = 0$ or $x - 10 = 0$; $x = 0$ or $x = 10$

  3. $m(m - 2) = 0$; $m = 0$ or $m - 2 = 0$; $m = 0$ or $m = 2$

  4. $(y + 2)(y + 3) = 0$; $y + 2 = 0$ or $y + 3 = 0$; $y = -2$ or $y = -3$

  5. $(n - 5)(n - 2) = 0$; $n - 5 = 0$ or $n - 2 = 0$; $n = 5$ or $n = 2$

  6. $(x + 3)(x - 4) = 0$; $x + 3 = 0$ or $x - 4 = 0$; $x = -3$ or $x = 4$

  7. $n^2 + n = 0$; $n(n + 1) = 0$; $n = 0$ or $n + 1 = 0$; $n = 0$ or $n = -1$

  8. $x^2 - 4x = 0$; $x(x - 4) = 0$; $x = 0$ or $x - 4 = 0$; $x = 0$ or $x = 4$

  9. $y^2 - 9y = 0$; $y(y - 9) = 0$; $y = 0$ or $y - 9 = 0$; $y = 0$ or $y = 9$

 10. $x^2 + 2x - 15 = 0$; $(x - 3)(x + 5) = 0$; $x - 3 = 0$ or $x + 5 = 0$; $x = 3$ or $x = -5$

 11. $n^2 + 11n + 24 = 0$; $(n + 3)(n + 8) = 0$; $n + 3 = 0$ or $n + 8 = 0$; $n = -3$ or $n = -8$

 12. $y^2 + 3y - 18 = 0$; $(y - 3)(y + 6) = 0$; $y - 3 = 0$ or $y + 6 = 0$; $y = 3$ or $y = -6$

 13. $n^2 = 3n - 2$; $n^2 - 3n + 2 = 0$; $(n - 1)(n - 2) = 0$; $n - 1 = 0$ or $n - 2 = 0$; $n = 1$ or $n = 2$

 14. $x^2 + x = 12$; $x^2 + x - 12 = 0$; $(x - 3)(x + 4) = 0$; $x - 3 = 0$ or $x + 4 = 0$; $x = 3$ or $x = -4$

15. $y^2 - 10 = -3y$; $y^2 + 3y - 10 = 0$; $(y - 2)(y + 5) = 0$; $y - 2 = 0$ or
$y + 5 = 0$; $y = 2$ or $y = -5$

B 16. $(4x - 5)(x + 1) = 0$; $4x - 5 = 0$ or $x + 1 = 0$; $4x = 5$ or $x = -1$; $x = \dfrac{5}{4}$ or
$x = -1$

17. $(3n - 18)(4n - 4) = 0$; $3n - 18 = 0$ or $4n - 4 = 0$; $3n = 18$ or $4n = 4$;
$n = 6$ or $n = 1$

18. $\dfrac{x}{3} = \dfrac{5}{x - 2}$; $x(x - 2) = 15$; $x^2 - 2x = 15$; $x^2 - 2x - 15 = 0$;
$(x + 3)(x - 5) = 0$; $x + 3 = 0$ or $x - 5 = 0$; $x = -3$ or $x = 5$

Page 479 • FOR PAGES 426–429

A 1. $x^2 = 0$; $x = 0$ 2. $2x^2 = 72$; $x^2 = 36$; $x = \pm6$; $x = 6$ or $x = -6$
3. $4y^2 = -8$; $y^2 = -2$; no solution
4. $x^2 + 64 = 100$; $x^2 = 36$; $x = \pm6$; $x = 6$ or $x = -6$
5. $(x - 2)^2 = 36$; $x - 2 = \pm6$; $x - 2 = 6$ or $x - 2 = -6$; $x = 8$ or $x = -4$
6. $(2x - 1)^2 = 49$; $2x - 1 = \pm7$; $2x - 1 = 7$ or $2x - 1 = -7$; $2x = 8$ or
$2x = -6$: $x = 4$ or $x = -3$
7. $(x - 4)^2 = 2$; $x - 4 = \pm\sqrt{2}$; $x - 4 = \sqrt{2}$ or $x - 4 = -\sqrt{2}$; $x = 4 + \sqrt{2}$ or
$x = 4 - \sqrt{2}$
8. $n^2 = 1$; $n = \pm1$; $n = 1$ or $n = -1$
9. $(3n + 1)^2 = 8$; $3n + 1 = \pm\sqrt{8} = \pm2\sqrt{2}$; $3n + 1 = 2\sqrt{2}$ or $3n + 1 = -2\sqrt{2}$;
$3n = -1 + 2\sqrt{2}$ or $3n = -1 - 2\sqrt{2}$; $n = \dfrac{-1 + 2\sqrt{2}}{3}$ or $n = \dfrac{-1 - 2\sqrt{2}}{3}$

B 10. $2y^2 + 3 = 19$; $2y^2 = 16$; $y^2 = 8$; $y = \pm\sqrt{8} = \pm2\sqrt{2}$; $y = 2\sqrt{2}$ or $y = -2\sqrt{2}$
11. $3n^2 - 9 = 15$; $3n^2 = 24$; $n^2 = 8$; $n = \pm\sqrt{8} = \pm2\sqrt{2}$; $n = 2\sqrt{2}$ or $n = -2\sqrt{2}$
12. $3n^2 + 11 = 38$; $3n^2 = 27$; $n^2 = 9$; $n = \pm3$; $n = 3$ or $n = -3$

Pages 479–480 • FOR PAGES 430–433

A 1. $a^2 - 5a + 6 = 0$; $a = \dfrac{-(-5) \pm \sqrt{(-5)^2 - (4 \cdot 1 \cdot 6)}}{2 \cdot 1} = \dfrac{5 \pm \sqrt{25 - 24}}{2} = \dfrac{5 \pm \sqrt{1}}{2} =$
$\dfrac{5 \pm 1}{2}$; $a = \dfrac{5 + 1}{2} = \dfrac{6}{2} = 3$ or $a = \dfrac{5 - 1}{2} = \dfrac{4}{2} = 2$
2. $n^2 + 5n - 3 = 0$; $n = \dfrac{-5 \pm \sqrt{5^2 - (4 \cdot 1 \cdot -3)}}{2 \cdot 1} = \dfrac{-5 \pm \sqrt{25 + 12}}{2} = \dfrac{-5 \pm \sqrt{37}}{2}$;
$n = \dfrac{-5 + \sqrt{37}}{2}$ or $n = \dfrac{-5 - \sqrt{37}}{2}$
3. $x^2 - 5x + 1 = 0$; $x = \dfrac{-(-5) \pm \sqrt{(-5)^2 - (4 \cdot 1 \cdot 1)}}{2 \cdot 1} = \dfrac{5 \pm \sqrt{25 - 4}}{2} = \dfrac{5 \pm \sqrt{21}}{2}$;
$x = \dfrac{5 + \sqrt{21}}{2}$ or $x = \dfrac{5 - \sqrt{21}}{2}$

4. $x^2 - 2x - 3 = 0;$ $x = \dfrac{-(-2) \pm \sqrt{(-2)^2 - (4 \cdot 1 \cdot -3)}}{2 \cdot 1} = \dfrac{2 \pm \sqrt{4 + 12}}{2} = \dfrac{2 \pm \sqrt{16}}{2} =$

$\dfrac{2 \pm 4}{2} = 1 \pm 2;$ $x = 1 + 2 = 3$ or $x = 1 - 2 = -1$

5. $x^2 + x - 20 = 0;$ $x = \dfrac{-1 \pm \sqrt{1^2 - (4 \cdot 1 \cdot -20)}}{2 \cdot 1} = \dfrac{-1 \pm \sqrt{1 + 80}}{2} = \dfrac{-1 \pm \sqrt{81}}{2} =$

$\dfrac{-1 \pm 9}{2};$ $x = \dfrac{-1 + 9}{2} = \dfrac{8}{2} = 4$ or $x = \dfrac{-1 - 9}{2} = \dfrac{-10}{2} = -5$

6. $3y^2 + 13y + 4 = 0;$ $y = \dfrac{-13 \pm \sqrt{13^2 - (4 \cdot 3 \cdot 4)}}{2 \cdot 3} = \dfrac{-13 \pm \sqrt{169 - 48}}{6} =$

$\dfrac{-13 \pm \sqrt{121}}{6} = \dfrac{-13 \pm 11}{6};$ $y = \dfrac{-13 + 11}{6} = \dfrac{-2}{6} = -\dfrac{1}{3}$ or $y = \dfrac{-13 - 11}{6} = \dfrac{-24}{6} = -4$

7. $x^2 - x - 5 = 0;$ $x = \dfrac{-(-1) \pm \sqrt{(-1)^2 - (4 \cdot 1 \cdot -5)}}{2 \cdot 1} = \dfrac{1 \pm \sqrt{1 + 20}}{2} = \dfrac{1 \pm \sqrt{21}}{2};$

$x = \dfrac{1 + \sqrt{21}}{2}$ or $x = \dfrac{1 - \sqrt{21}}{2}$

8. $2x^2 - 11x + 12 = 0;$ $x = \dfrac{-(-11) \pm \sqrt{(-11)^2 - (4 \cdot 2 \cdot 12)}}{2 \cdot 2} = \dfrac{11 \pm \sqrt{121 - 96}}{4} =$

$\dfrac{11 \pm \sqrt{25}}{4} = \dfrac{11 \pm 5}{4};$ $x = \dfrac{11 + 5}{4} = \dfrac{16}{4} = 4$ or $x = \dfrac{11 - 5}{4} = \dfrac{6}{4} = \dfrac{3}{2}$

9. $2x^2 - x - 4 = 0;$ $x = \dfrac{-(-1) \pm \sqrt{(-1)^2 - (4 \cdot 2 \cdot -4)}}{2 \cdot 2} = \dfrac{1 \pm \sqrt{1 + 32}}{4} = \dfrac{1 \pm \sqrt{33}}{4};$

$x = \dfrac{1 + \sqrt{33}}{4}$ or $x = \dfrac{1 - \sqrt{33}}{4}$

10. $x^2 - 16 = 0;$ $x = \dfrac{0 \pm \sqrt{0^2 - (4 \cdot 1 \cdot -16)}}{2 \cdot 1} = \dfrac{\pm\sqrt{64}}{2} = \dfrac{\pm 8}{2} = \pm 4;$ $x = 4$ or $x = -4$

11. $4n^2 - 3n - 1 = 0;$ $n = \dfrac{-(-3) \pm \sqrt{(-3)^2 - (4 \cdot 4 \cdot -1)}}{2 \cdot 4} = \dfrac{3 \pm \sqrt{9 + 16}}{8} = \dfrac{3 \pm \sqrt{25}}{8} =$

$\dfrac{3 \pm 5}{8};$ $x = \dfrac{3 + 5}{8} = \dfrac{8}{8} = 1$ or $x = \dfrac{3 - 5}{8} = \dfrac{-2}{8} = -\dfrac{1}{4}$

12. $x^2 - 6x - 2 = 0;$ $x = \dfrac{-(-6) \pm \sqrt{(-6)^2 - (4 \cdot 1 \cdot -2)}}{2 \cdot 1} = \dfrac{6 \pm \sqrt{36 + 8}}{2} = \dfrac{6 \pm \sqrt{44}}{2} =$

$\dfrac{6 \pm 2\sqrt{11}}{2} = 3 \pm \sqrt{11};$ $x = 3 + \sqrt{11}$ or $x = 3 - \sqrt{11}$

B   **13. a.** $x^2 - 5x - 14 = 0;$ $x = \dfrac{-(-5) \pm \sqrt{(-5)^2 - (4 \cdot 1 \cdot -14)}}{2 \cdot 1} = \dfrac{5 \pm \sqrt{25 + 56}}{2} =$

$\dfrac{5 \pm \sqrt{81}}{2} = \dfrac{5 \pm 9}{2};$ $x = \dfrac{5 + 9}{2} = \dfrac{14}{2} = 7$ or $x = \dfrac{5 - 9}{2} = \dfrac{-4}{2} = -2$

b. $x^2 - 5x - 14 = 0;$ $(x + 2)(x - 7) = 0;$ $x + 2 = 0$ or $x - 7 = 0;$ $x = -2$ or $x = 7$

14. a. $n^2 - n = 0;$ $n = \dfrac{-(-1) \pm \sqrt{(-1)^2 - (4 \cdot 1 \cdot 0)}}{2 \cdot 1} = \dfrac{1 \pm \sqrt{1}}{2} = \dfrac{1 \pm 1}{2};$ $n = \dfrac{1 + 1}{2} =$

$\dfrac{2}{2} = 1$ or $n = \dfrac{1 - 1}{2} = \dfrac{0}{2} = 0$

b. $n^2 - n = 0;$ $n(n - 1) = 0;$ $n = 0$ or $n - 1 = 0;$ $n = 0$ or $n = 1$

15. a. $y^2 - 8y + 16 = 0$; $y = \dfrac{-(-8) \pm \sqrt{(-8)^2 - (4 \cdot 1 \cdot 16)}}{2 \cdot 1} = \dfrac{8 \pm \sqrt{64 - 64}}{2} = \dfrac{8 \pm 0}{2} =$

$\dfrac{8}{2} = 4$

b. $y^2 - 8y + 16 = 0$; $(y - 4)^2 = 0$; $y - 4 = 0$; $y = 4$

16. a. $x^2 - 6x + 8 = 0$; $x = \dfrac{-(-6) \pm \sqrt{(-6)^2 - (4 \cdot 1 \cdot 8)}}{2 \cdot 1} = \dfrac{6 \pm \sqrt{36 - 32}}{2} = \dfrac{6 \pm \sqrt{4}}{2} =$

$\dfrac{6 \pm 2}{2} = 3 \pm 1$; $x = 3 + 1 = 4$ or $x = 3 - 1 = 2$

b. $x^2 - 6x + 8 = 0$; $(x - 2)(x - 4) = 0$; $x - 2 = 0$ or $x - 4 = 0$; $x = 2$ or
$x = 4$

17. a. $3x^2 - 9x + 6 = 0$; $x = \dfrac{-(-9) \pm \sqrt{(-9)^2 - (4 \cdot 3 \cdot 6)}}{2 \cdot 3} = \dfrac{9 \pm \sqrt{81 - 72}}{6} = \dfrac{9 \pm \sqrt{9}}{6} =$

$\dfrac{9 \pm 3}{6}$; $x = \dfrac{9 + 3}{6} = \dfrac{12}{6} = 2$ or $x = \dfrac{9 - 3}{6} = \dfrac{6}{6} = 1$

b. $3x^2 - 9x + 6 = 0$; $3(x^2 - 3x + 2) = 0$; $3(x - 1)(x - 2) = 0$; $x - 1 = 0$ or
$x - 2 = 0$; $x = 1$ or $x = 2$

18. a. $5x^2 - 10x - 75 = 0$; $x^2 = \dfrac{-(-10) \pm \sqrt{(-10)^2 - (4 \cdot 5 \cdot -75)}}{2 \cdot 5} =$

$\dfrac{10 \pm \sqrt{100 + 1500}}{10} = \dfrac{10 \pm \sqrt{1600}}{10} = \dfrac{10 \pm 40}{10} = 1 \pm 4$; $x = 1 + 4 = 5$ or $x =$

$1 - 4 = -3$

b. $5x^2 - 10x - 75 = 0$; $5(x^2 - 2x - 15) = 0$; $5(x + 3)(x - 5) = 0$; $x + 3 = 0$
or $x - 5 = 0$; $x = -3$ or $x = 5$

Page 480 • FOR PAGES 434–435

A

1. Let x and $x + 5 =$ the numbers. $x(x + 5) = 14$; $x^2 + 5x = 14$; $x^2 + 5x -$
$14 = 0$; $(x - 2)(x + 7) = 0$; $x - 2 = 0$ or $x + 7 = 0$; $x = 2$ or $x = -7$, so
$x + 5 = 7$ or $x + 5 = -2$. 2 and 7, or −7 and −2

2. Let x and $x - 3 =$ the numbers. $x(x - 3) = 28$; $x^2 - 3x = 28$; $x^2 - 3x -$
$28 = 0$; $(x + 4)(x - 7) = 0$; $x + 4 = 0$ or $x - 7 = 0$; $x = -4$ or $x = 7$, so
$x - 3 = -7$ or $x - 3 = 4$.   −4 and −7, or 7 and 4

3. Let x and $x + 1 =$ the consecutive numbers. $x(x + 1) = 210$; $x^2 + x = 210$;
$x^2 + x - 210 = 0$; $(x - 14)(x + 15) = 0$; $x - 14 = 0$ or $x + 15 = 0$; $x = 14$
or $x = -15$, so $x + 1 = 15$ or $x + 1 = -14$.   14 and 15, or −15 and −14

4. Let $x =$ length (in cm). Then $x - 2 =$ width. $x(x - 2) = 24$; $x^2 - 2x = 24$;
$x^2 - 2x - 24 = 0$; $(x + 4)(x - 6) = 0$; $x + 4 = 0$ or $x - 6 = 0$; $x =$
−4 (reject) or $x = 6$; $x - 2 = 4$. width: 4 cm; length: 6 cm

B

5. Let x, $x + 1$, and $x + 2 =$ the consecutive numbers. $x^2 + (x + 1)^2 + (x + 2)^2 =$
29; $x^2 + x^2 + 2x + 1 + x^2 + 4x + 4 = 29$; $3x^2 + 6x + 5 = 29$; $3x^2 + 6x -$
$24 = 0$; $x^2 + 2x - 8 = 0$; $(x - 2)(x + 4) = 0$; $x - 2 = 0$ or $x + 4 = 0$;
$x = 2$ or $x = -4$, so $x + 1 = 3$ or −3 and $x + 2 = 4$ or −2. 2, 3, 4 or −4,
−3, −2

6. Let x = length (in cm). Then $12 - x$ = width. $x(12 - x) = 32$; $12x - x^2 = 32$; $0 = x^2 - 12x + 32$; $0 = (x - 4)(x - 8)$; $x - 4 = 0$ or $x - 8 = 0$; $x = 4$ or $x = 8$, so $12 - x = 8$ or 4. 4 cm by 8 cm

Page 480 · FOR PAGES 436–439

A **1. a.**

$y = x^2 - 1$	
x	y
-2	3
-1	0
0	-1
1	0
2	3

b.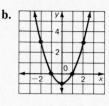

c. $(0, -1)$

2. a.

$y = x^2 - 4x$	
x	y
-1	5
0	0
1	-3
2	-4
3	-3
4	0
5	5

b.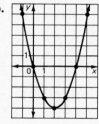

c. $(2, -4)$

3. a.

$y = -x^2$	
x	y
-2	-4
-1	-1
0	0
1	-1
2	-4

b.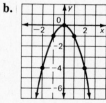

c. $(0, 0)$

Key to Extra Practice, page 480

4. a.

$y = \frac{1}{4}x^2$	
x	y
-6	9
-4	4
-2	1
0	0
2	1
4	4
6	9

b.

c. $(0, 0)$

5. a.

$y = x^2 - 2x + 1$	
x	y
-1	4
0	1
1	0
2	1
3	4

b.

c. $(1, 0)$

6. a.

$y = x^2 - 2x - 3$	
x	y
-2	5
-1	0
0	-3
1	-4
2	-3
3	0
4	5

b.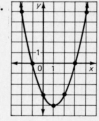

c. $(1, -4)$

7. a.

$y = x^2 + 4x + 2$	
x	y
-4	-2
-3	-1
-2	-2
-1	-1
0	2

b.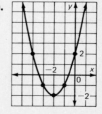

c. $(-2, -2)$

8. a.

$y = (2 - x)(2 + x)$	
x	y
-3	-5
-2	0
-1	3
0	4
1	3
2	0
3	-5

b.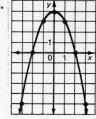

c. $(0, 4)$

9. a.

$y = x^2 + 4x - 3$	
x	y
-5	2
-4	-3
-3	-6
-2	-7
-1	-6
0	-3
1	2

b.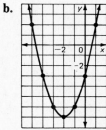

c. $(-2, -7)$

Page 481 • BASIC SKILLS

1. $(5 + 3) \cdot 8 = 8 \cdot 8 = 64$

2. $6 - (3 \cdot 1) = 6 - 3 = 3$

3. $9 + (6 - 3) = 9 + 3 = 12$

4. $6^2 = 6 \cdot 6 = 36$

5. $5^1 = 5$

6. $3^3 = 3 \cdot 3 \cdot 3 = 27$

7. $5 \cdot 0 = 0$

8. $5 + 0 = 5$

9. $\dfrac{5}{5} = 1$

10. $\dfrac{5}{1} = 5$

11. $5 \cdot 1 = 5$

12. $(5 - 5)(2 \cdot 5) = 0 \cdot 10 = 0$

13. $-1 + (-4) = -5$

14. $-1(-4) = 4$

15. $-1 - (-4) = -1 + (4) = 3$

16. $(-4)^2 = (-4)(-4) = 16$

17. $-(4)^2 = -(4 \cdot 4) = -16$

18. $-(-4)^2 = -(-4 \cdot -4) = -16$

19. $a + b = 1 + 2 = 3$

20. $-a + b = -1 + 2 = 1$

21. $-ab = -(1 \cdot 2) = -2$

22. $c \div (-a) = 3 \div (-1) = -3$

23. $2(a + c) = 2(1 + 3) = 2 \cdot 4 = 8$

24. $2a + c = 2 \cdot 1 + 3 = 2 + 3 = 5$

25. $4(a - b) = 4(1 - 2) = 4(-1) = -4$

26. $4a - b = 4 \cdot 1 - 2 = 4 - 2 = 2$

27. $-abc = -(1 \cdot 2 \cdot 3) = -6$

28. $3x + 4x = 7x$

29. $10b - b = 9b$

30. $2y - 4y = -2y$

31. $4x + 5x - 2 = 9x - 2$

32. $3a - 4a - 7a = -8a$

33. $2a + 4 - 2a = 4 + 0 = 4$

34. $3a - 5a + c - c = -2a + 0 = -2a$

35. $3 \cdot a \cdot a \cdot b \cdot b = 3a^2b^2$

36. $3(x - 1) + 2 = 3x - 3 + 2 = 3x - 1$

37. $4 \cdot \dfrac{x}{4} = \dfrac{4}{4} \cdot x = 1 \cdot x = x$

38. $3x - (-2x) = 3x + 2x = 5x$

39. $-(a - 2) = (-1 \cdot a) + (-1 \cdot -2) = -a + 2$

40. $(x + y)(-2) = (x \cdot -2) + (y \cdot -2) = -2x - 2y$

41. $4(x - 3) - x + 15 = 4x - 12 - x + 15 = 3x + 3$

42. $-(x^2 + x - 2) = (-1 \cdot x^2) + (-1 \cdot x) + (-1 \cdot -2) = -x^2 - x + 2$

Pages 481–482 • EQUATIONS AND INEQUALITIES

1. If $x = -5$, $x + 2 = -5 + 2 = -3 \neq 7$; no.

2. If $x = -1$, $x - 1 = -1 - 1 = -2 \neq -1$; no.

3. a. $1 < 3, 2 < 3, 3 = 3, 4 > 3$; 4

   b. Yes; for example, 5, 6, 7, and so on.

   c.

4. $x - 2 = 4$; $x - 2 + 2 = 4 + 2$; $x = 6$

5. $y + 6 = 5$; $y + 6 - 6 = 5 - 6$; $y = -1$

6. $x + 3 = 3$; $x + 3 - 3 = 3 - 3$; $x = 0$

7. $9 = x + 3$; $9 - 3 = x + 3 - 3$; $6 = x$

8. $3x = 18$; $x = 6$ $\qquad\qquad$ **9.** $-3x = 18$; $x = -6$

10. $\dfrac{x}{2} = 9$; $x = 18$ $\qquad\qquad$ **11.** $3x - 1 = 20$; $3x = 21$; $x = 7$

12. $5x + 25 = 0$; $5x = -25$; $x = -5$

13. $3 + 3y - y = 21$; $3 + 2y = 21$; $2y = 18$; $y = 9$

14. $7x = 25 + 2x$; $7x - 2x = 25 + 2x - 2x$; $5x = 25$; $x = 5$

15. $3n - n = -16$; $2n = -16$; $n = -8$

16. $2(x - 4) = -4$; $2x - 8 = -4$; $2x = 4$; $x = 2$

17. $5 - (3 - y) = 10$; $5 - 3 + y = 10$; $2 + y = 10$; $y = 8$

18. $4 - (2x - 1) = x - 1$; $4 - 2x + 1 = x - 1$; $5 - 2x = x - 1$; $5 = 3x - 1$;
$\quad$ $6 = 3x$; $2 = x$

19. $-(y - 4) + 5 = 9y - 1$; $-y + 4 + 5 = 9y - 1$; $-y + 9 = 9y - 1$;
$\quad$ $9 = 10y - 1$; $10 = 10y$; $1 = y$

20. $\dfrac{x}{9} = -9$; $x = -81$ $\qquad$ **21.** $\dfrac{x}{-3} = -3$; $x = 9$ $\qquad$ **22.** $>$

23. $3 - (-4) = 3 + 4 = 7$; $(-3)(-4) = 12$; $7 < 12$; $<$

24. $6 - (-2) = 6 + 2 = 8$; $(-2)(-4) = 8$; $8 = 8$; $=$

25. $\dfrac{0}{6} = 0$; $\dfrac{6}{0}$ has no value.

Page 482 ▸ PROBLEMS

1. Let n = the number. $2n + 4 = 80$; $2n = 76$; $n = 38$.   38

2. Let n and $n + 1$ = the consecutive numbers. $n + (n + 1) = 65$; $2n + 1 = 65$;
$\quad$ $2n = 64$; $n = 32$, and $n + 1 = 33$.   32 and 33

3. Let d = dimes Don has. Then $2d$ = dimes Mary has. $d + 2d = 105$; $3d = 105$;
$\quad$ $d = 35$, and $2d = 70$. Don: 35 dimes; Mary: 70 dimes

4. Let x = games lost. Then $3x$ = games won. $x + 3x = 32$; $4x = 32$; $x = 8$,
$\quad$ so $3x = 24$. 24 games won

5. Let n = the number. $3n - 5 = 25$; $3n = 30$; $n = 10$.   10

CUMULATIVE REVIEW FOR CHAPTERS 1–6

Page 483 · BASIC SKILLS

1. $15 \cdot \dfrac{0}{15x} = 15 \cdot 0 = 0$ $\qquad\qquad$ **2.** $-(-3)^2 = -(-3 \cdot -3) = -9$

3. $-4a + a - 7a = -10a$

4. $2(x - 8) + 2x = 2x - 16 + 2x = 4x - 16$

5. $10x^2 - x^2 = 9x^2$ $\qquad\qquad$ **6.** $(-a)^2 + 4a + 1 = a^2 + 4a + 1$

7. $(x - 2y) + (3x + 2y) = x + 3x - 2y + 2y = 4x$

8. $(x^2 - 2x + 1) - (x^2 + 2x - 1) = x^2 - 2x + 1 - x^2 - 2x + 1 = -4x + 2$

9. $4a \cdot a^3 = 4a^{1+3} = 4a^4$

10. $(-3xy)(4x^2y^2) = (-3 \cdot 4)(x \cdot x^2)(y \cdot y^2) = -12x^3y^3$

11. $(-3ab)^3 = (-3)^3a^3b^3 = -27a^3b^3$

12. $2x(5 - y) = (2x \cdot 5) - (2x \cdot y) = 10x - 2xy$

13. $(x + 6)(x - 1) = x^2 + 5x - 6$ 14. $(n - 4)(n + 4) = n^2 - 16$

15. $(y + 2)(y - 3) = y^2 - y - 6$

16. $(x - 3)^2 = x^2 - 2 \cdot 3x + (-3)^2 = x^2 - 6x + 9$

17. $(x + 4)^2 = x^2 + 2 \cdot 4x + 4^2 = x^2 + 8x + 16$

18. $(2x - 3)^2 = (2x)^2 - 2 \cdot 2x \cdot 3 + (-3)^2 = 4x^2 - 12x + 9$

19. $(3x - 2y)^2 = (3x)^2 - 2 \cdot 3x \cdot 2y + (-2y)^2 = 9x^2 - 12xy + 4y^2$

20. $\dfrac{6n^4}{-n^2} = \dfrac{6 \cdot n \cdot n \cdot n \cdot n}{-1 \cdot n \cdot n} = -6n^2$ 21. $\dfrac{8n^3}{2n} = \dfrac{2 \cdot 4 \cdot n \cdot n \cdot n}{2 \cdot n} = 4n^2$

22. $\dfrac{-24a^2b^2}{-12ab^2} = \dfrac{-12 \cdot 2 \cdot a \cdot a \cdot b \cdot b}{-12 \cdot a \cdot b \cdot b} = 2a$ 23. $\dfrac{-72xy^2}{8y^2} = \dfrac{-9 \cdot 8 \cdot x \cdot y \cdot y}{8 \cdot y \cdot y} = -9x$

24. $\dfrac{x^2 - 3x^3 + 5x^4}{-x^2} = \dfrac{x^2}{-x^2} - \dfrac{3x^3}{-x^2} + \dfrac{5x^4}{-x^2} = -1 + 3x - 5x^2$

25. $\dfrac{5a^2 - a^3 - a^4}{a} = \dfrac{5a^2}{a} - \dfrac{a^3}{a} - \dfrac{a^4}{a} = 5a - a^2 - a^3$

26. $\dfrac{5xy - 15x^2y^2}{-5x} = \dfrac{5xy}{-5x} - \dfrac{15x^2y^2}{-5x} = -y + 3xy^2$

27. $3xy - 6x^2y^2 = 3xy(1 - 2xy)$ 28. $x^2 + 5x + 6 = (x + 2)(x + 3)$

29. $y^2 + 7y + 6 = (y + 1)(y + 6)$ 30. $x^2 + 4x + 4 = (x + 2)^2$

31. $n^2 + 9n - 22 = (n - 2)(n + 11)$ 32. $x^2 - 9y^2 = (x + 3y)(x - 3y)$

Pages 483–484 · EQUATIONS AND FORMULAS

1. $x - 5 = 20$; $x - 5 + 5 = 20 + 5$; $x = 25$

2. $40 = 20 - n$; $40 - 20 = 20 - n - 20$; $20 = -n$; $(-1)(20) = (-1)(-n)$; $-20 = n$

3. $4a + 1 = 13$; $4a = 12$; $a = 3$ 4. $\dfrac{n}{4} = -4$; $n = -16$

5. $2(x - 1) = 14$; $2x - 2 = 14$; $2x = 16$; $x = 8$

6. $3a = 25 - 2a$; $3a + 2a = 25 - 2a + 2a$; $5a = 25$; $a = 5$

7. $12x + 8 - 10x = 16$; $2x + 8 = 16$; $2x + 8 - 8 = 16 - 8$; $2x = 8$; $x = 4$

8. $5x + 30 = 3x$; $5x + 30 - 5x = 3x - 5x$; $30 = -2x$; $-15 = x$

9. $-10 = -3x + 2$; $-12 = -3x$; $4 = x$

10. $3x - (2x + 2) = 5x + 6$; $3x - 2x - 2 = 5x + 6$; $x - 2 = 5x + 6$; $-2 = 4x + 6$; $-8 = 4x$; $-2 = x$

11. $(2a + 1) - (5a - 1) = -1$; $2a + 1 - 5a + 1 = -1$; $-3a + 2 = -1$; $-3a = -3$; $a = 1$

12. $6x = 24$; $x = 4$

13. $6x = y$; $x = \dfrac{y}{6}$

14. $kx = y$; $x = \dfrac{y}{k}$

15. $y + 7 = 10$; $y = 3$

16. $y + x = 2$; $y = 2 - x$

17. $4x - y = 0$; $4x = y$

18. $P = 8x$; $\dfrac{P}{8} = x$

19. $D = rt$; $\dfrac{D}{t} = r$

Page 484 · PROBLEMS

1. $P = 4 \cdot 4 = 16$; 16 cm

2. $A = lw = 2x \cdot 4x = 8x^2$

3. $A = \dfrac{1}{2}ba = \dfrac{1}{2} \cdot 8 \cdot 4 = 16$; 16 cm²

4. $V = 4x \cdot 3x \cdot 2x = 24x^3$

5. Let n and $n + 1 =$ the consecutive numbers. $n + (n + 1) = 25$; $2n + 1 = 25$; $2n = 24$; $n = 12$, and $n + 1 = 13$.   12 and 13

6.

	price	number	Cost
Sell	20	$x - 2$	$20(x - 2)$
Buy	12	x	$12x$
			120

$20(x - 2) - 12x = 120$; $20x - 40 - 12x = 120$; $8x - 40 = 120$; $8x = 160$; $x = 20$.   20 hooks bought

CUMULATIVE REVIEW FOR CHAPTERS 1–9

Page 485 · BASIC SKILLS

1. $\dfrac{x}{0} = \dfrac{3}{0}$; impossible

2. $\dfrac{x - 5}{x - 5} = \dfrac{3 - 5}{3 - 5} = \dfrac{-2}{-2} = 1$

3. $\dfrac{x - 3}{x - 3} = \dfrac{3 - 3}{3 - 3} = \dfrac{0}{0}$; impossible

4. $\dfrac{x}{3} = \dfrac{3}{3} = 1$

5. $\dfrac{0}{x} = \dfrac{0}{3} = 0$

6. $3n - 15n = -12n$

7. $x(3xy)(-3y) = (3 \cdot -3)(x \cdot x)(y \cdot y) = -9x^2y^2$

8. $5(2a - 3b) + 15b = 10a - 15b + 15b = 10a + 0 = 10a$

9. $(2ab^2)^2 = 2^2a^2(b^2)^2 = 4a^2b^4$

10. $(-4x^2y^2)^2 = (-4)^2(x^2)^2(y^2)^2 = 16x^4y^4$

11. $-10(-3y^2)^3 = -10(-3)^3(y^2)^3 = -10(-27)(y^6) = 270y^6$

12. $(3x - 4y) - (9x - 4y) = 3x - 4y - 9x + 4y = -6x + 0 = -6x$

13. $\dfrac{4x^2y}{-4y} = \dfrac{4 \cdot x \cdot x \cdot y}{-4 \cdot y} = -x^2$

14. $\dfrac{x}{3} - \dfrac{2x}{6} = \dfrac{x}{3} - \dfrac{x}{3} = 0$

15. $(x + 4)(x + 2) = x^2 + 6x + 8$

16. $(n - 5)(n + 5) = n^2 - 25$

17. $(2x - 1)(4x + 3) = 8x^2 + 2x - 3$

18. $(2x + 5)^2 = (2x)^2 + 2 \cdot 5 \cdot 2x + 5^2 = 4x^2 + 20x + 25$

19. $x(x - 2)(x + 3) = x(x^2 + x - 6) = x^3 + x^2 - 6x$

20. $(2a + b)(3a - 2b) = 6a^2 - ab - 2b^2$ 21. $2x^2 - 4y^2 = 2(x^2 - 2y^2)$

22. $3a^2 - 3b^2 = 3(a^2 - b^2) = 3(a + b)(a - b)$

23. $n^2 - 3n + 2 = (n - 1)(n - 2)$

24. $m^2 - 5m - 14 = (m + 2)(m - 7)$ 25. $x^2 + 8x + 16 = (x + 4)^2$

26. $x^2 - 6x - 27 = (x + 3)(x - 9)$ 27. $4y^2 - 9 = (2y + 3)(2y - 3)$

28. $n^2 - 2n - 8 = (n + 2)(n - 4)$ 29. $x^2 + x - 6 = (x - 2)(x + 3)$

Pages 485–486 • EQUATIONS AND GRAPHS

1. $4(x + 1) = 16;$ $4x + 4 = 16;$ $4x = 12;$ $x = 3$

2. $5x - 15 = 2x;$ $-15 = -3x;$ $5 = x$ 3. $x + 8 = -3x;$ $8 = -4x;$ $-2 = x$

4. $x + 3x + 2 = 18;$ $4x + 2 = 18;$ $4x = 16;$ $x = 4$

5. $6x - 1 = 3x + 11;$ $6x = 3x + 12;$ $3x = 12;$ $x = 4$

6. $-x + 9 = 4x + 4;$ $9 = 5x + 4;$ $5 = 5x;$ $1 = x$

7. $3 + y = x;$ $y = x - 3$ 8. $2x + y = 4;$ $y = 4 - 2x$

9. $5x - y = 10;$ $-y = 10 - 5x;$ $(-1)(-y) = (-1)(10 - 5x);$ $y = 5x - 10$

10. $x - y = 4;$ $-y = 4 - x;$ $(-1)(-y) = (-1)(4 - x);$ $y = x - 4$

11. $y - 2x = 8;$ $y = 2x + 8$ 12. $5x + y = 0;$ $y = -5x$

13.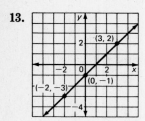

14.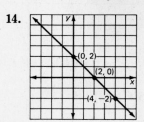

15.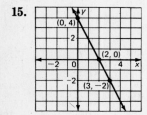

16. -1 17. $\frac{1}{2}$ 18. 2

19. $(3, 3)$

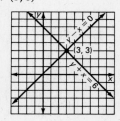

20. $(0, 2)$

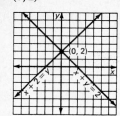

21. $(-2, 1)$

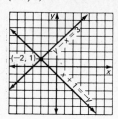

22. $y = x - 1$ and $x + y = 3$; $x + (x - 1) = 3$; $2x - 1 = 3$; $2x = 4$; $x = 2$;
$y = x - 1 = 2 - 1 = 1$. $(2, 1)$

23. $x = 2y$ and $x + y = 9$; $(2y) + y = 9$; $3y = 9$; $y = 3$; $x = 2y = 2 \cdot 3 = 6$.
$(6, 3)$

24. $x = y + 2$ and $y = 3x - 4$; $x = (3x - 4) + 2$; $x = 3x - 2$; $-2x = -2$;
$x = 1$; $y = 3x - 4 = 3 \cdot 1 - 4 = 3 - 4 = -1$. $(1, -1)$

25. $5x - y = 1$ $3 \cdot 2 + y = 15$; $6 + y = 15$; $y = 9$. $(2, 9)$
$\underline{3x + y = 15}$ Add
$8x \quad\quad = 16$; $x = 2$

26. $2x - 3y = 1 \longrightarrow 6x - 9y = \quad 3$ $2x - 3 \cdot 1 = 1$; $2x - 3 = 1$;
$\underline{3x - 2y = 4 \longrightarrow 6x - 4y = \quad 8}$ Subt. $2x = 4$; $x = 2$. $(2, 1)$
$\quad\quad\quad\quad\quad\quad\quad -5y = -5$; $y = 1$

27. $3x + \quad y = 4 \longrightarrow 6x + 2y = 8$ $1 + 2y = 3$; $2y = 2$; $y = 1$. $(1, 1)$
$\underline{x + 2y = 3 \longrightarrow \quad x + 2y = 3}$ Subt.
$\quad\quad 5x \quad\quad = 5$; $x = 1$

Page 486 · PROBLEMS

1.

	You	Jean	Together
Hours needed	2	4	n
Part done in one hour	$\frac{1}{2}$	$\frac{1}{4}$	$\frac{1}{n}$

$\frac{1}{2} + \frac{1}{4} = \frac{1}{n}$; $\frac{2}{4} + \frac{1}{4} = \frac{1}{n}$; $\frac{3}{4} = \frac{1}{n}$; $3n = 4$; $n = 1\frac{1}{3}$. $1\frac{1}{3}$ hours

2. Let x and y = the numbers. $x + y = 75$ $50 + y = 75$; $y = 25$.
$\quad\quad\quad\quad\quad\quad\quad\quad\quad\quad\quad\quad \underline{x - y = 25}$ Add   50 and 25
$\quad\quad\quad\quad\quad\quad\quad\quad\quad\quad\quad\quad 2x \quad\quad = 100$; $x = 50$

CUMULATIVE REVIEW FOR CHAPTERS 1–12

Page 487 • BASIC SKILLS

1. $(3x - 4) - (5x + 6) = 3x - 4 - 5x - 6 = -2x - 10$

2. $(n + 9) + (4n - 18) = n + 4n + 9 - 18 = 5n - 9$

3. $(a^2 + 3a + 1) - (a^2 - 2a) = a^2 + 3a + 1 - a^2 + 2a = 5a + 1$

4. $(n^2 - 3) - (n + 8) = n^2 - 3 - n - 8 = n^2 - n - 11$

5. $(5x + y) - (6x + y) = 5x + y - 6x - y = -x + 0 = -x$

6. $3(n + 2) + 4(n - 2) = 3n + 6 + 4n - 8 = 7n - 2$

7. $\dfrac{b^2 - 4}{b + 2} = \dfrac{(b + 2)(b - 2)}{b + 2} = b - 2$

8. $\dfrac{x^2 - 4}{8} \cdot \dfrac{4}{x - 2} = \dfrac{4(x + 2)(x - 2)}{8(x - 2)} = \dfrac{x + 2}{2}$

9. $\dfrac{3x^2 - 3}{2} \div \dfrac{x + 1}{16} = \dfrac{3(x^2 - 1)}{2} \cdot \dfrac{16}{x + 1} = \dfrac{16 \cdot 3(x + 1)(x - 1)}{2(x + 1)} = 24(x - 1) = 24x - 24$

10. $(3n - 1)(n + 6) = 3n^2 + 17n - 6$ 11. $(n - 4)(n + 4) = n^2 - 16$

12. $(n + 3)(n + 3) = n^2 + 6n + 9$

13. $(5n - 2)(2n + 5) = 10n^2 + 21n - 10$

14. $(4x - 1)(x - 4) = 4x^2 - 17x + 4$

15. $(2x + 7)(7 - x) = 14x - 2x^2 + 49 - 7x = -2x^2 + 7x + 49$

16. $5n^2 + 25n + 100 = 5(n^2 + 5n + 20)$

17. $3n^2 + 9n + 6 = 3(n^2 + 3n + 2) = 3(n + 1)(n + 2)$

18. $2n^2 - 8 = 2(n^2 - 4) = 2(n + 2)(n - 2)$

19. $x^2 - 7x + 6 = (x - 1)(x - 6)$

20. $x^2 - 5x - 6 = (x + 1)(x - 6)$ 21. $x^2 - 5x - 14 = (x + 2)(x - 7)$

22. $\sqrt{\dfrac{2}{16}} = \dfrac{\sqrt{2}}{\sqrt{16}} = \dfrac{\sqrt{2}}{4}$ 23. $\sqrt{\dfrac{4y^2}{9}} = \dfrac{\sqrt{4y^2}}{\sqrt{9}} = \dfrac{2y}{3}$

24. $\sqrt{\dfrac{1}{200}} = \dfrac{\sqrt{1}}{\sqrt{200}} = \dfrac{1}{10\sqrt{2}} = \dfrac{1 \cdot \sqrt{2}}{10\sqrt{2} \cdot \sqrt{2}} = \dfrac{\sqrt{2}}{20}$

25. $\dfrac{1}{4}\sqrt{\dfrac{16}{x^2}} = \dfrac{1}{4} \cdot \dfrac{\sqrt{16}}{\sqrt{x^2}} = \dfrac{1}{4} \cdot \dfrac{4}{x} = \dfrac{1}{x}$ 26. $\sqrt{3} \cdot \sqrt{12} = \sqrt{3 \cdot 12} = \sqrt{36} = 6$

27. $3\sqrt{2} \cdot \sqrt{2x^2} = 3\sqrt{2 \cdot 2x^2} = 3\sqrt{4x^2} = 3 \cdot 2x = 6x$

28. $\sqrt{\dfrac{1}{4}} \cdot \sqrt{\dfrac{1}{3}} = \sqrt{\dfrac{1}{4} \cdot \dfrac{1}{3}} = \sqrt{\dfrac{1}{12}} = \dfrac{\sqrt{1}}{\sqrt{12}} = \dfrac{1 \cdot \sqrt{12}}{\sqrt{12} \cdot \sqrt{12}} = \dfrac{\sqrt{12}}{12} = \dfrac{2\sqrt{3}}{12} = \dfrac{\sqrt{3}}{6}$

29. $\dfrac{\sqrt{2}}{\sqrt{5}} = \dfrac{\sqrt{2} \cdot \sqrt{5}}{\sqrt{5} \cdot \sqrt{5}} = \dfrac{\sqrt{10}}{5}$ 30. $\sqrt{8} - \sqrt{2} = 2\sqrt{2} - \sqrt{2} = \sqrt{2}$

31. $3\sqrt{3} - 2\sqrt{12} = 3\sqrt{3} - 2 \cdot 2\sqrt{3} = 3\sqrt{3} - 4\sqrt{3} = -\sqrt{3}$

32. $2\sqrt{27} - \sqrt{3} = 2 \cdot 3\sqrt{3} - \sqrt{3} = 6\sqrt{3} - \sqrt{3} = 5\sqrt{3}$

33. $4\sqrt{2} - \sqrt{18} = 4\sqrt{2} - 3\sqrt{2} = \sqrt{2}$

Pages 487–488 • EQUATIONS AND INEQUALITIES

1. $3x - 1 = 4x - 9$; $3x + 8 = 4x$; $8 = x$

2. $2(2x - 1) = 6$; $4x - 2 = 6$; $4x = 8$; $x = 2$

3. $2n + 3 - 3n + 5 = 0$; $-n + 8 = 0$; $8 = n$

4.

5. $\dfrac{2}{x} = \dfrac{x}{8}$; $x^2 = 16$; $x = \pm 4$; $x = 4$ or $x = -4$

6. $\dfrac{n}{3} - \dfrac{n}{5} = 16$; $\dfrac{5n}{15} - \dfrac{3n}{15} = 16$; $\dfrac{2n}{15} = \dfrac{16}{1}$; $2n = 240$; $n = 120$

7. $\dfrac{x}{x + 2} = \dfrac{4}{3}$; $3x = 4(x + 2)$; $3x = 4x + 8$; $-x = 8$; $x = -8$

8. $a^2 - 3 = 13$; $a^2 = 16$; $a = \pm 4$; $a = 4$ or $a = -4$

9. $y^2 = 50$; $y = \pm\sqrt{50} = \pm 5\sqrt{2}$; $y = 5\sqrt{2}$ or $y = -5\sqrt{2}$

10. $0.3x + 0.9x = 3.6$; $10(0.3x + 0.9x) = 10(3.6)$; $3x + 9x = 36$; $12x = 36$; $x = 3$

11. $0.4x = 4.8$; $10(0.4x) = 10(4.8)$; $4x = 48$; $x = 12$

12. $(n + 4)(n - 6) = 0$; $n + 4 = 0$ or $n - 6 = 0$; $n = -4$ or $n = 6$

13. $0.04(100 - x) = 12 - 0.2x$; $100[0.04(100 - x)] = 100(12 - 0.2x)$; $4(100 - x) = 1200 - 20x$; $400 - 4x = 1200 - 20x$; $400 + 16x = 1200$; $16x = 800$; $x = 50$

14. $(x + 2)(x - 2) = 0$; $x + 2 = 0$ or $x - 2 = 0$; $x = -2$ or $x = 2$

15. $x(x - 4) = 0$; $x = 0$ or $x - 4 = 0$; $x = 0$ or $x = 4$

16. $(3x - 6)(2x + 14) = 0$; $3x - 6 = 0$ or $2x + 14 = 0$; $3x = 6$ or $2x = -14$; $x = 2$ or $x = -7$

17. **a.** $3y = 24$; $y = 8$ **b.** $ay = 24$; $y = \dfrac{24}{a}$ **c.** $ay = c$; $y = \dfrac{c}{a}$

18. $\begin{aligned} x + y &= 4 \\ x - y &= 2 \quad \text{Add} \\ \hline 2x &= 6; \quad x = 3 \end{aligned}$ $3 + y = 4$; $y = 1$. $(3, 1)$

19. $\begin{aligned} x + 3y &= 4 \longrightarrow 2x + 6y = 8 \\ 2x - 5y &= -3 \longrightarrow \underline{2x - 5y = -3} \quad \text{Subt.} \\ & 11y = 11; \quad y = 1 \end{aligned}$ $x + 3 \cdot 1 = 4$; $x + 3 = 4$; $x = 1$. $(1, 1)$

20. $\begin{aligned} 4x - 3y &= 1 \longrightarrow 16x - 12y = 4 \\ 3x + 4y &= 7 \longrightarrow \underline{9x + 12y = 21} \quad \text{Add} \\ & 25x = 25; \quad x = 1 \end{aligned}$ $3 \cdot 1 + 4y = 7$; $3 + 4y = 7$; $4y = 4$; $y = 1$. $(1, 1)$

21. $n^2 - 7n + 6 = 0$; $(n - 1)(n - 6) = 0$; $n - 1 = 0$ or $n - 6 = 0$; $n = 1$ or $n = 6$

22. $y^2 - 2y - 3 = 0$; $(y + 1)(y - 3) = 0$; $y + 1 = 0$ or $y - 3 = 0$; $y = -1$ or $y = 3$

23. $x^2 - 16 = 0$; $(x + 4)(x - 4) = 0$; $x + 4 = 0$ or $x - 4 = 0$; $x = -4$ or $x = 4$

24. $2n^2 + 3n + 1 = 0$; $n = \dfrac{-3 \pm \sqrt{3^2 - (4 \cdot 2 \cdot 1)}}{2 \cdot 2} = \dfrac{-3 \pm \sqrt{9 - 8}}{4} = \dfrac{-3 \pm \sqrt{1}}{4} =$

$\dfrac{-3 \pm 1}{4}$; $n = \dfrac{-3 + 1}{4} = \dfrac{-2}{4} = -\dfrac{1}{2}$ or $n = \dfrac{-3 - 1}{4} = \dfrac{-4}{4} = -1$

25. $3x^2 - 2x - 3 = 0$; $x = \dfrac{-(-2) \pm \sqrt{(-2)^2 - (4 \cdot 3 \cdot -3)}}{2 \cdot 3} = \dfrac{2 \pm \sqrt{4 + 36}}{6} = \dfrac{2 \pm \sqrt{40}}{6} =$

$\dfrac{2 \pm 2\sqrt{10}}{6} = \dfrac{1 \pm \sqrt{10}}{3}$; $x = \dfrac{1 + \sqrt{10}}{3}$ or $x = \dfrac{1 - \sqrt{10}}{3}$

26. $2n^2 - 3n = 5$; $2n^2 - 3n - 5 = 0$; $n = \dfrac{-(-3) \pm \sqrt{(-3)^2 - (4 \cdot 2 \cdot -5)}}{2 \cdot 2} =$

$\dfrac{3 \pm \sqrt{9 + 40}}{4} = \dfrac{3 \pm \sqrt{49}}{4} = \dfrac{3 \pm 7}{4}$; $n = \dfrac{3 + 7}{4} = \dfrac{10}{4} = \dfrac{5}{2}$ or $n = \dfrac{3 - 7}{4} = \dfrac{-4}{4} = -1$

Page 488 · PROBLEMS

1.

	p	n	C
65¢ kind	65	x	$65x$
40¢ kind	40	$100 - x$	$40(100 - x)$
Mixture	50	100	5000

$65x + 40(100 - x) = 5000$; $65x + 4000 - 40x = 5000$; $25x = 1000$; $x = 40$, and $100 - x = 60$.   40 kg of 65¢ kind, 60 kg of 40¢ kind

2. Let $x =$ the length (in m). $x^2 = 30^2 + 40^2$; $x^2 = 900 + 1600$; $x^2 = 2500$; $x = \sqrt{2500} = 50$.   50 m

CUMULATIVE REVIEW OF WORD PROBLEMS

Pages 489–491

1. $(x + 7)4 = 68$; $4x = 40$; $x = 10$

2. Let $x =$ the number of games won.   Then $x - 5 =$ the number lost. $x + x - 5 = 27$; $2x - 5 = 27$; $x = 16$

3. Let $x =$ Gregory's age now.   Then $x - 6 =$ Lara's age.   In 2 years their ages will be $x + 2$ and $x - 6 + 2$ or $x - 4$. $x + 2 + x - 4 = 30$; $2x = 32$; $x = 16$. Gregory is 16.

4.

	rate	time	Distance
Armando	r	$2\frac{1}{2}$	$2\frac{1}{2}r$
Pat	$r + 2$	$2\frac{1}{2}$	$2\frac{1}{2}(r + 2)$
			80

$2\frac{1}{2}r + 2\frac{1}{2}(r + 2) = 80$

$2\frac{1}{2}r + 2\frac{1}{2}r + 5 = 80$

$5r = 75$

$r = 15$

Armando travels $2\frac{1}{2}(15)$ or 37.5 km;   Pat travels $2\frac{1}{2}(15 + 2)$ or 42.5 km

5.

	price	number	Cost
Sell	20	$n - 2$	$20(n - 2)$
Buy	12	n	$12n$
			152

$20(n - 2) - 12n = 152$

$20n - 40 - 12n = 152$

$8n - 40 \qquad = 150$

$n \qquad\qquad = 24$

24 erasers bought

6. Let x = amount Leon has. Then $2x$ = amount Carol has, and $2x + 3$ = amount Harry has. $x + 2x + 2x + 3 = 108$; $5x + 3 = 108$; $5x = 105$; $x = 21$. Leon has \$21. Carol has \$42. Harry has \$45.

7. $5n = 2n + 36$; $3n = 36$; $n = 12$

8.

	p	r	t	I
Amount at 6%	x	0.06	1	$0.06x$
Amount at 5%	$6000 - x$	0.05	1	$0.05(6000 - x)$
				335

$0.06x + 300 - 0.05x = 335$

$0.01x + 300 \qquad = 335$

$0.01x \qquad\qquad = 35$

$x \qquad\qquad\quad = 3500$

\$3500 at 6%

9.

Distance	80 km	112 km
Gasoline	5 L	x

$\dfrac{80}{5} = \dfrac{112}{x}$

$80x = 560$

$x = 7$

7 L

10. Let $4x$ = Carl's money and $5x$ = Rita's money. $4x + 5x = 279$; $9x = 279$; $x = 31$. \$155

11.

	rate	time	Distance
To Stadium	10	t	$10t$
From Stadium	60	$t - 1$	$60(t - 1)$

$10t = 60t - 60$; $60 = 50t$

$t = 1.2$, so Distance $= 10 \times 1.2$

Distance $= 12$ km

12.

	pay per hour	number of hours	Weekly pay
Kelly	x	8	$8x$
Juanita	$x + 1$	8	$8(x + 1)$
			112

$$8x + 8x + 8 = 112$$
$$16x + 8 = 112$$
$$16x = 104$$
$$x = 6.5$$

Kelly: \$6.50;   Juanita: \$7.50

13.

	Simon	Roger	Together
Hours needed	3	2	n
Part done in one hour	$\dfrac{1}{3}$	$\dfrac{1}{2}$	$\dfrac{1}{n}$

$$\frac{1}{3} + \frac{1}{2} = \frac{1}{n}$$
$$\frac{2}{6} + \frac{3}{6} = \frac{1}{n}$$
$$\frac{5}{6} = \frac{1}{n}$$
$$5n = 6$$
$$n = 1\frac{1}{5}$$

$1\frac{1}{5}$ hours

14. Let $3x$ = cubic meters of sand, $4x$ = cubic meters of gravel, and $2x$ = cubic meters of concrete. $3x + 4x + 2x = 45$; $9x = 45$; $x = 5$. 15 m^3

15. $120\% \times 55 = x$; $1.20 \times 55 = x$; $66 = x$. 66

16. $n \times 75 = 30$; $n = \dfrac{30}{75}$; $n = 0.40$. 40%

17. $30\% \times n = 72$; $0.30n = 72$; $n = 240$

18. Let x = rate of boat in still water.   Let y = rate of boat in flowing water.

	rate	time	Distance
Downstream	$x + y$	2	32
Upstream	$x - y$	4	32

$$2(x + y) = 32$$
$$4(x - y) = 32$$
$$2x + 2y = 32$$
$$4x - 4y = 32$$
$$4x + 4y = 64$$
$$\underline{4x - 4y = 32}$$
$$8x = 96$$
$$x = 12$$
$$y = 4$$

4 km/h

19. Let x and y = the numbers. $x + y = 73$ and $x - y = 11$; $x = -y + 73$ and $x - y = 11$; $(-y + 73) - y = 11$; $-2y = -62$; $y = 31$. $x - y = 11$; $x - 31 = 11$; $x = 42$.   31, 42

20. Let x and y = the numbers. $x - y = 32$ and $2y - 12 = x$; $(2y - 12) - y = 32$; $y = 44$. $x - 4 = 32$; $x - 44 = 32$; $x = 76$. 44, 76

21. Let x = the width.   Then $2x + 5$ = the length. $2(x + 2x + 5) = 46$; $6x + 10 = 46$; $6x = 36$; $x = 6$. $2(6) + 5 = 17$.   width = 6 cm, length = 17 cm

22. Let x, $x + 1$, and $x + 2$ = the integers. $x + x + 1 + x + 2 = 363$; $3x + 3 = 363$; $3x = 360$; $x = 120$; $x + 1 = 121$ and $x + 2 = 122$. **120, 121, 122**

23. Let s = the cost of a shirt and p = the cost of a pair of pants.

$2s + 3p = 96 \longrightarrow 4s + 6p = 192$ $4s + 2(26) = 88$; $4s = 36$; $s = 9$.
$4s + 2p = 88 \longrightarrow \underline{4s + 2p = \ \ 88}$ Subt.
$\qquad\qquad\qquad\qquad\qquad 4p = 104$; $p = 26$

shirt: \$9;   pants: \$26

24. Let x = Marcia's age. $x + 4$ = Emile's age and $\dfrac{x}{2}$ = Ken's age.

$x + x + 4 + \dfrac{x}{2} = 64$; $\dfrac{5}{2}x = 60$; $x = 24$.   **Marcia is 24.**

25. Let x = width, then $x + 5$ = length. $x(x + 5) = 150$; $x^2 + 5x - 150 = 0$;
$(x - 10)(x + 15) = 0$; $x = 10$.   width: 10 cm;   length: 15 cm

26. Let x = positive number. $x^2 - 20 = 8x$; $x^2 - 8x - 20 = 0$;
$(x - 10)(x + 2) = 0$; $x - 10 = 0$; $x = 10$.   **10**

27.

	p	n	C
Brazil nuts	4.50	x	$4.50x$
cashews	6.00	$30 - x$	$6(30 - x)$
mixture	5.00	30	150

$4.5x + 180 - 6x = 150$
$\qquad\quad 30 \qquad\quad = 1.5x$
$\qquad\quad 20 \qquad\quad = x$

Brazil nuts: 20 kg
cashews: 10 kg

28.

	price	number	Cost
Sell	3.00	$x - 20$	$3(x - 20)$
	2.40	20	48
Buy	2.00	x	$2x$
			18

$3x - 60 + 48 - 2x = 18$
$\qquad\qquad\qquad\quad x = 30$

30 notebooks

29.

	Hillary	Sylvia	Together
Hours needed	5	n	3
Part done in one hour	$\dfrac{1}{5}$	$\dfrac{1}{n}$	$\dfrac{1}{3}$

$\dfrac{1}{5} + \dfrac{1}{n} = \dfrac{1}{3}$; $\dfrac{n}{5n} + \dfrac{5}{5n} = \dfrac{1}{3}$;

$\dfrac{n + 5}{5n} = \dfrac{1}{3}$; $3(n + 5) = 5n$;

$3n + 15 = 5n$; $15 = 2n$;

$7\dfrac{1}{2} = n$. $7\dfrac{1}{2}$ hours

30. Let g = cost for a dozen game balls and p = cost for a dozen practice balls.
$12g + 15p = 1440$ and $8g + 20p = 1440$;

$24g + 30p = \ \ 2880$
$\underline{24g + 60p = \ \ 4320}$ Subt.
$\quad -30p = -1440$; $p = 48$; $12g + 15(48) = 1440$; $12g + 720 = 1440$;
$\qquad\qquad 12g = 720$; $g = 60$.   \$60

31. $20\% \times 48 = n$; $0.20 \times 48 = 9.60$; $48 - 9.60 = 38.40$; $5\% \times 38.40 = n$;
$0.05 \times 38.40 = 1.92$; $38.40 + 1.92 = 40.32$. $40.32

32.

Quantity	2700 L	1800 L
Time	36 min	x

$\dfrac{2700}{36} = \dfrac{1800}{x}$; $2700x = 64{,}800$

$x = 24$.   24 min

33. $2l + 2w = 40$; $l + w = 20$; $l = 20 - w$. $A = lw$, so $64 = (20 - w)w$;
$64 = 20w - w^2$; $w^2 - 20w + 64 = 0$; $(w - 4)(w - 16) = 0$; $w = 4$ or $w = 16$;
so $l = 16$ or $l = 4$.   length: 16 cm;   width: 4 cm

34.

	Amount of solution	% of salt	Amount of salt
Original solution	12	25% or 0.25	3
Water added	x	0%	0
New solution	$12 + x$	20% or 0.20	$0.20(12 + x)$

$3 + 0 = 0.20(12 + x)$

$3 \quad\;\; = 2.4 + 0.20x$

$0.60 \;\; = 0.20x$

$3 \quad\;\; = x$

3 liters water

35. Let $x =$ one even integer. Then $x + 2 =$ the other even integer.
$x^2 + (x + 2)^2 = 340$; $x^2 + x^2 + 4x + 4 = 340$; $2x^2 + 4x - 336 = 0$;
$(2x - 24)(x + 14) = 0$; $2x - 24 = 0$; $2x = 24$; $x = 12$. $12 + 2 = 14$.   12, 14

Page 492 · EXERCISES

1. $x - 5 = 9;\ x - 5 + 5 = 9 + 5;\ x = 14$

2. $x - 5 > 9;\ x - 5 + 5 > 9 + 5;\ x > 14$

3. $x - 5 = 1;\ x - 5 + 5 = 1 + 5;\ x = 6$

4. $x - 5 < 1;\ x - 5 + 5 < 1 + 5;\ x < 6$

5. $x - 6 = 2;\ x - 6 + 6 = 2 + 6;\ x = 8$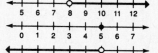

6. $x - 6 > 2;\ x - 6 + 6 > 2 + 6;\ x > 8$

7. $x - 3 = 2;\ x - 3 + 3 = 2 + 3;\ x = 5$

8. $x - 3 < 2;\ x - 3 + 3 < 2 + 3;\ x < 5$

9. $x + 3 = 5;\ x + 3 - 3 = 5 - 3;\ x = 2$

10. $x + 3 > 5;\ x + 3 - 3 > 5 - 3;\ x > 2$

11. $x + 2 = 4;\ x + 2 - 2 = 4 - 2;\ x = 2$

12. $x + 2 < 4;\ x + 2 - 2 < 4 - 2;\ x < 2$

13. $x - 5 \le 1;\ x - 5 < 1$ or $x - 5 = 1;\ x < 6$ or $x = 6$

14. $x - 6 \ge 2;\ x - 6 > 2$ or $x - 6 = 2;\ x > 8$ or $x = 8$

15. $x - 3 \le 2;\ x - 3 < 2$ or $x - 3 = 2;\ x < 5$ or $x = 5$

16. $x + 1 \le 4;\ x + 1 < 4$ or $x + 1 = 4;\ x < 3$ or $x = 3$

17. $x + 3 \ge 3;\ x + 3 > 3$ or $x + 3 = 3;\ x > 0$ or $x = 0$

18. $x - 5 \le -7;\ x - 5 < -7$ or $x - 5 = 7;\ x < -2$ or $x = -2$

Page 494 · EXERCISES

1. $-3y = 12;\ \dfrac{-3y}{-3} = \dfrac{12}{-3};\ y = -4$

2. $-3y < 12;\ \dfrac{-3y}{-3} > \dfrac{12}{-3};\ y > -4$

3. $-2b = 10;\ \dfrac{-2b}{-2} = \dfrac{10}{-2};\ b = -5$

4. $-2b > 10;\ \dfrac{-2b}{-2} < \dfrac{10}{-2};\ b < -5$

5. $-2x > -6;\ \dfrac{-2x}{-2} < \dfrac{-6}{-2};\ x < 3$

6. $-2a - a = 3;\ -3a = 3;\ \dfrac{-3a}{-3} = \dfrac{3}{-3};\ a = -1$

7. $\dfrac{x}{2} > 5;\ 2 \cdot \dfrac{x}{2} > 2 \cdot 5;\ x > 10$

8. $\dfrac{-x}{2} > 5$; $2 \cdot \dfrac{-x}{2} > 2 \cdot 5$; $-x > 10$; $-1 \cdot -x < -1 \cdot 10$; $x < -10$

9. $\dfrac{-x}{6} < 1$; $6 \cdot \dfrac{-x}{6} < 6 \cdot 1$; $-x < 6$; $-1 \cdot -x > -1 \cdot 6$; $x > -6$

10. $-n > 4$; $-1 \cdot -n < -1 \cdot 4$; $n < -4$

11. $\dfrac{-2n}{3} > 1$; $3 \cdot \dfrac{-2n}{3} > 3 \cdot 1$; $-2n > 3$; $\dfrac{-2n}{-2} < \dfrac{3}{-2}$; $n < -\dfrac{3}{2}$

12. $4 = -4x$; $\dfrac{4}{-4} = \dfrac{-4x}{-4}$; $-1 = x$

13. $2x + 7 \geq 11$; $2x + 7 > 11$ or $2x + 7 = 11$; $2x > 4$ or $2x = 4$; $x > 2$ or $x = 2$

14. $-2x + 7 \geq 11$; $-2x + 7 > 11$ or $-2x + 7 = 11$; $-2x > 4$ or $-2x = 4$; $x < -2$ or $x = -2$

15. $-3x - 1 \leq 5$; $-3x - 1 < 5$ or $-3x - 1 = 5$; $-3x < 6$ or $-3x = 6$; $x > -2$ or $x = -2$

APPENDIX B • Modeling Expressions

Note: Tile diagrams are given for selected exercises. Check students' models or drawings for all exercises. Students' arrangements of tiles in some exercises may vary.

Page 496 • EXERCISES

1. $3x + 5$

2. $2x^2 + x + 3$

3. $2(x^2 + 3)$

4. $2(x + 2) = 2x + 4$

5. $3x + 1 \rightarrow$ □□□ □

6. $2x^2 + 5 \rightarrow$ □□ □□□□

7. $3x^2 + 4x + 2 \rightarrow$ □□□ □□□□ □□

8. $4x + x + 2x \rightarrow$ □□□□ □ □□ → (□□□□□□□) → $7x$

9. $9x + 5 + 9 \rightarrow$ □□□□□□□□□ □□□□ □□□□□ →

□□□□□□□□□ (□□□□□□) → $9x + 14$

10. $x^2 + 3x + 2x + 4 = x^2 + 5x + 4$

11. $2x^2 + 5x^2 + x + 1 \rightarrow$ □□ □□□□□ □ □ →

(□□□□□□□) □ □ → $7x^2 + x + 1$

12. $4x^2 + 3x^2 + 6x + 2x = (4x^2 + 3x^2) + (6x + 2x) = 7x^2 + 8x$

13. $5(x + 4) \rightarrow$ 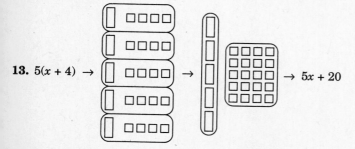 $\rightarrow 5x + 20$

14. $2(x + 1) = 2x + 2$

15. $4(2x + 1) = 8x + 4$

16. $3(x^2 + 5) = 3x^2 + 15$

Page 497 • EXERCISES

1. $x^2 + 2x + 3; x + 6$

2. $x^2 + 2x + 1; 2x^2 + 3x + 3$

3. $x^2 + 3x + 9$

4. $3x^2 + 5x + 4$

5. $(2x + 5) + (3x + 4) \rightarrow$ 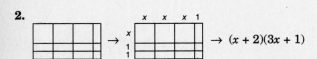 $\rightarrow 5x + 9$

6. $(x + 2) + (x^2 + 3x + 2) \rightarrow$ $\rightarrow x^2 + 4x + 4$

7. $(x^2 + 2x + 5) + (x^2 + 5x + 1) = 2x^2 + 7x + 6$

8. $(x^2 + 7) + (x^2 + 7x + 4) = 2x^2 + 7x + 11$

9. $(3x^2 + x + 7) + (2x^2 + 4x + 1) = 5x^2 + 5x + 8$

10. $(2x^2 + 4x) + (8x^2 + 12) = 10x^2 + 4x + 12$

Page 499 • EXERCISES

1. $\rightarrow$ $\rightarrow (2x + 1)(x + 1)$

2. $\rightarrow$ $\rightarrow (x + 2)(3x + 1)$

3. $\rightarrow$ $\rightarrow$ two 1 tiles

4. $\rightarrow$ $\rightarrow$ two x tiles and six 1 tiles

5. $x(3x) \rightarrow$ $\rightarrow 3x^2$

6. $(2x)(4x) = 8x^2$

7. $x(x + 4) \rightarrow$ $\rightarrow x^2 + 4x$

8. $2x(x + 3) \rightarrow$ 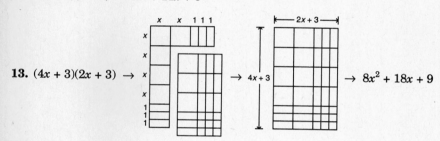 $\rightarrow 2x^2 + 6x$

9. $(x + 1)(x + 2) \rightarrow$ $\rightarrow x^2 + 3x + 2$

10. $(x + 3)(x + 4) = x^2 + 7x + 12$

11. $(x + 2)(2x + 1) \rightarrow$ $\rightarrow 2x^2 + 5x + 2$

12. $(3x + 2)(x + 3) = 3x^2 + 11x + 6$

13. $(4x + 3)(2x + 3) \rightarrow$ $\rightarrow 8x^2 + 18x + 9$

14. $(5x + 1)(2x + 7) = 10x^2 + 37x + 7$

15. $(6x + 5)(4x + 3) = 24x^2 + 38x + 15$

16. $(3x + 2)(3x + 2) = 9x^2 + 12x + 4$

Page 500 • EXERCISES

1. $x^2 + 5x + 4 \rightarrow$

2. $x^2 + 3x \rightarrow$ $\rightarrow x(x + 3)$

3. $x^2 + 5x + 6 \rightarrow$

$\rightarrow (x + 2)(x + 3)$

4. $x^2 + 7x + 12 = (x + 4)(x + 3)$ **5.** $x^2 + 12x + 20 = (x + 10)(x + 2)$

6. $x^2 + 2x + 1 = (x + 1)(x + 1)$, or $(x + 1)^2$

7. $x^2 + 10x + 25 = (x + 5)(x + 5)$, or $(x + 5)^2$

8. $x^2 + 3x + 4 \rightarrow$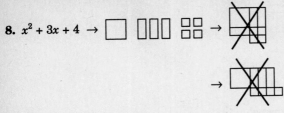

The tiles cannot be used to form a rectangle so the polynomial $x^2 + 3x + 4$ cannot be factored.